The Lanczos and Conjugate Gradient Algorithms

SOFTWARE • ENVIRONMENTS • TOOLS

The series includes handbooks and software guides as well as monographs on practical implementation of computational methods, environments, and tools. The focus is on making recent developments available in a practical format to researchers and other users of these methods and tools.

Software, Environments, and Tools

Gérard Meurant, *The Lanczos and Conjugate Gradient Algorithms: From Theory to Finite Precision Computations*

Bo Einarsson, editor, *Accuracy and Reliability in Scientific Computing*

Michael W. Berry and Murray Browne, *Understanding Search Engines: Mathematical Modeling and Text Retrieval, Second Edition*

Craig C. Douglas, Gundolf Haase, and Ulrich Langer, *A Tutorial on Elliptic PDE Solvers and Their Parallelization*

Louis Komzsik, *The Lanczos Method: Evolution and Application*

Bard Ermentrout, *Simulating, Analyzing, and Animating Dynamical Systems: A Guide to XPPAUT for Researchers and Students*

V. A. Barker, L. S. Blackford, J. Dongarra, J. Du Croz, S. Hammarling, M. Marinova, J. Waśniewski, and P. Yalamov, *LAPACK95 Users' Guide*

Stefan Goedecker and Adolfy Hoisie, *Performance Optimization of Numerically Intensive Codes*

Zhaojun Bai, James Demmel, Jack Dongarra, Axel Ruhe, and Henk van der Vorst, *Templates for the Solution of Algebraic Eigenvalue Problems: A Practical Guide*

Lloyd N. Trefethen, *Spectral Methods in MATLAB*

E. Anderson, Z. Bai, C. Bischof, S. Blackford, J. Demmel, J. Dongarra, J. Du Croz, A. Greenbaum, S. Hammarling, A. McKenney, and D. Sorensen, *LAPACK Users' Guide, Third Edition*

Michael W. Berry and Murray Browne, *Understanding Search Engines: Mathematical Modeling and Text Retrieval*

Jack J. Dongarra, Iain S. Duff, Danny C. Sorensen, and Henk A. van der Vorst, *Numerical Linear Algebra for High-Performance Computers*

R. B. Lehoucq, D. C. Sorensen, and C. Yang, *ARPACK Users' Guide: Solution of Large-Scale Eigenvalue Problems with Implicitly Restarted Arnoldi Methods*

Randolph E. Bank, *PLTMG: A Software Package for Solving Elliptic Partial Differential Equations, Users' Guide 8.0*

L. S. Blackford, J. Choi, A. Cleary, E. D'Azevedo, J. Demmel, I. Dhillon, J. Dongarra, S. Hammarling, G. Henry, A. Petitet, K. Stanley, D. Walker, and R. C. Whaley, *ScaLAPACK Users' Guide*

Greg Astfalk, editor, *Applications on Advanced Architecture Computers*

Françoise Chaitin-Chatelin and Valérie Frayssé, *Lectures on Finite Precision Computations*

Roger W. Hockney, *The Science of Computer Benchmarking*

Richard Barrett, Michael Berry, Tony F. Chan, James Demmel, June Donato, Jack Dongarra, Victor Eijkhout, Roldan Pozo, Charles Romine, and Henk van der Vorst, *Templates for the Solution of Linear Systems: Building Blocks for Iterative Methods*

E. Anderson, Z. Bai, C. Bischof, J. Demmel, J. Dongarra, J. Du Croz, A. Greenbaum, S. Hammarling, A. McKenney, S. Ostrouchov, and D. Sorensen, *LAPACK Users' Guide, Second Edition*

Jack J. Dongarra, Iain S. Duff, Danny C. Sorensen, and Henk van der Vorst, *Solving Linear Systems on Vector and Shared Memory Computers*

J. J. Dongarra, J. R. Bunch, C. B. Moler, and G. W. Stewart, *Linpack Users' Guide*

The Lanczos and Conjugate Gradient Algorithms

From Theory to Finite Precision Computations

Gérard Meurant
CEA/DIF
Bruyères le Châtel, France

siam
Society for Industrial and Applied Mathematics
Philadelphia

10 9 8 7 6 5 4 3 2 1

Library of Congress Cataloging-in-Publication Data

Meurant, Gérard A.

The Lanczos and conjugate gradient algorithms : from theory to finite precision computations / Gérard Meurant.

p. cm. — (Software, environments, and tools)

Includes bibliographical references and index.

ISBN-13: 978-0-898716-16-0

ISBN-10: 0-89871-616-0

1. Conjugate gradient methods. 2. Algorithms—Methodology. I. Title.

QA218.M48 2006

518'.1—dc22

2006044391

ISBN-13: 978-0-898716-16-0

ISBN-10: 0-89871-616-0

This book is dedicated to the memories
of my mother Jacqueline (1926–2001)
and my goddaughter Melina (1978–2005)

Many thanks to Franz Schubert,
Erroll Garner, and Ella Fitzgerald,
whose music helped me during the writing.

Je distingue deux moyens de cultiver les sciences :
l'un d'augmenter la masse des connaissances par des découvertes,
et c'est ainsi qu'on mérite le nom d'inventeur,
l'autre de rapprocher les découvertes et de les ordonner entre elles,
afin que plus d'hommes soient éclairés, et que chacun participe,
selon sa portée, à la lumière de son siècle.
Denis Diderot, Discours préliminaire à l'Encyclopédie (1751)

Wenn die Könige baun, haben die Kärrner zu thun.
Johann Wolfgang von Goethe and Friedrich Schiller, Xenien (1796)

Contents

Preface **xi**

1 The Lanczos algorithm in exact arithmetic **1**

1.1 Introduction to the Lanczos algorithm . . . 2
1.2 Lanczos polynomials . . . 9
1.3 Interlacing properties and approximations of eigenvalues . . . 16
1.4 The components of the eigenvectors of T_k . . . 20
1.5 Study of the pivot functions . . . 22
1.6 Bounds for the approximation of eigenvalues . . . 25
1.7 A priori bounds . . . 41
1.8 Computation of the approximate eigenvalues . . . 43
1.9 Harmonic Ritz values . . . 43

2 The CG algorithm in exact arithmetic **45**

2.1 Derivation of the CG algorithm from the Lanczos algorithm . . . 46
2.2 Relations between residuals and descent directions . . . 53
2.3 The norm of the residual . . . 55
2.4 The A-norm of the error . . . 58
2.5 The l_2 norm of the error . . . 68
2.6 Other forms of the CG algorithm . . . 74
2.7 Bounds for the norms of the error . . . 77

3 A historical perspective on the Lanczos algorithm in finite precision **81**

3.1 The tools of the trade . . . 83
3.2 Numerical example . . . 89
3.3 The work of Chris Paige . . . 93
3.4 Illustration of the work of Chris Paige . . . 102
3.5 The work of Joe Grcar . . . 105
3.6 Illustration of the work of Joe Grcar . . . 109
3.7 The work of Horst Simon . . . 110
3.8 Illustration of the work of Horst Simon . . . 118
3.9 The work of Anne Greenbaum and Zdenek Strakoš . . . 121
3.10 The work of J. Cullum and R. Willoughby . . . 130
3.11 The work of V. Druskin and L. Knizhnerman . . . 130
3.12 Recent related results . . . 136

4 The Lanczos algorithm in finite precision 139
4.1 Numerical examples . . . 139
4.2 Solution of three-term recurrences . . . 149
4.3 The Lanczos vectors in finite precision . . . 152
4.4 Another solution to three-term recurrences . . . 166
4.5 Bounds for the perturbation terms . . . 174
4.6 Some more examples . . . 176
4.7 Conclusions of this chapter . . . 184

5 The CG algorithm in finite precision 187
5.1 Numerical examples . . . 187
5.2 Relations between CG and Lanczos in finite precision . . . 191
5.3 Study of the CG three-term recurrences . . . 198
5.4 Study of the CG two-term recurrences . . . 210
5.5 Relations between p^k and r^k . . . 214
5.6 Local orthogonality in finite precision . . . 215
5.7 CG convergence in finite precision . . . 223
5.8 Numerical examples of convergence in variable precision . . . 230

6 The maximum attainable accuracy 239
6.1 Difference of residuals . . . 239
6.2 Numerical experiments for $Ax = 0$ . . . 241
6.3 Estimate of the maximum attainable accuracy . . . 249
6.4 Some ways to improve the maximum attainable accuracy . . . 252

7 Estimates of norms of the error in finite precision 257
7.1 Computation of estimates of the norms of the error in exact arithmetic 257
7.2 The A-norm of the error in finite precision . . . 263
7.3 The l_2 norm of the error in finite precision . . . 265
7.4 Numerical examples . . . 268
7.5 Comparison of two-term and three-term CG . . . 275

8 The preconditioned CG algorithm 281
8.1 PCG in exact arithmetic . . . 281
8.2 Formulas for norms of the error . . . 282
8.3 PCG in finite precision . . . 284
8.4 Numerical examples of convergence . . . 287
8.5 Numerical examples of estimation of norms . . . 292

9 Miscellaneous 297
9.1 Choice of the starting vector . . . 297
9.2 Variants of CG and multiple right-hand sides . . . 298
9.2.1 Constrained CG . . . 298
9.2.2 Block CG . . . 299
9.2.3 Init, Augmented, and Deflated CG . . . 300
9.2.4 Global CG . . . 302

9.3 Residual smoothing . . . 302
9.4 Inner-outer iterations and relaxation strategies . . . 304
9.5 Numerical experiments with starting vectors . . . 305
9.6 Shifted matrices . . . 308
9.7 CG on indefinite systems . . . 310
9.8 Examples with PCG . . . 313

Appendix **323**
A.1 A short biography of Cornelius Lanczos . . . 323
A.2 A short biography of M.R. Hestenes and E. Stiefel . . . 324
A.3 Examples in "exact" arithmetic . . . 325

Bibliography **349**

Index **363**

Preface

The Lanczos algorithm is one of the most frequently used numerical methods for computing a few eigenvalues (and eventually eigenvectors) of a large sparse symmetric matrix A. If the extreme eigenvalues of A are well separated, the Lanczos algorithm usually delivers good approximations to these eigenvalues in a few iterations. This method is interesting because it needs only the matrix A whose eigenvalues are sought in the form of matrix-vector products. Hence, it can even be used in some applications for which the matrix cannot be stored in the main memory of the computer as long as one is able to produce the result of the matrix A times a given vector. Another interesting property is that when one needs just the eigenvalues, the Lanczos algorithm requires only a very small storage of a few vectors besides eventually storing the matrix. At iteration k the algorithm constructs an orthogonal basis of a Krylov subspace of dimension k, which we are going to define below, whose basis vectors are columns of an orthogonal matrix V_k and a tridiagonal matrix T_k whose eigenvalues are approximations of the eigenvalues of A. Moreover, in exact arithmetic $AV_m = V_m T_m$ for an integer m such that $m \leq n$, n being the dimension of the problem which means that the eigenvalues of T_m are distinct eigenvalues of A. But, it should be noticed that in many cases, some eigenvalues of T_k are good approximations of some eigenvalues of A for $k \ll m$.

All these properties are quite nice. However, it has been known since the introduction of the method in 1950 by Cornelius Lanczos [108] that when used in finite precision arithmetic this algorithm does not fulfill its theoretical properties. In particular, the computed basis vectors which are mathematically orthogonal lose their orthogonality in practical computations. Moreover, numerically some additional copies of the eigenvalues which have already converged reappear within the computed approximate eigenvalues when we continue the iterations. This leads to a delay in the computation of the eigenvalues which have not yet converged. In fact, in some cases, to obtain all the distinct eigenvalues of A, one has to do many more iterations than the order n of the matrix. It is because of these problems in finite precision arithmetic that the algorithm was dismissed or ignored for some time until some people, mainly Chris Paige in the seventies, showed that despite its departure from theory, the algorithm can be used efficiently for computing eigenvalues and eigenvectors, especially of large and sparse matrices. Therefore, it is important to understand why the problems we mentioned before are happening and to have a view as clear as possible about the behavior of the Lanczos algorithm in finite precision arithmetic. To try to reach this goal, we shall consider the problem from different points of view. We shall describe the matrix point of view that has been used by Paige to derive elegant matrix relations describing the propagation of rounding errors in the algorithm. But, we shall also consider the projections of the Lanczos vectors on the eigenvectors of A and derive solutions for their recurrences. Even though this may seem more difficult and less fruitful than the matrix approach, we shall obtain some additional insights on the behavior of the algorithm in finite precision arithmetic.

The Lanczos algorithm is closely linked to the conjugate gradient (CG) method for solving linear systems $Ax = b$. This numerical method was introduced by Magnus Hestenes and Eduard Stiefel at about the same time Lanczos published his own algorithm. In 2006 CG is the state-of-the-art algorithm for solving large sparse linear systems with a positive definite symmetric matrix A. Today CG is generally used in connection with a preconditioner to speed up convergence. As we shall see there is a very close relationship between CG convergence and approximation of the eigenvalues of A in the Lanczos algorithm. In finite precision arithmetic, like the Lanczos algorithm, CG does not fulfill its mathematical properties. The residual vectors which must be orthogonal lose their orthogonality. Moreover, when compared to what would be obtained in exact arithmetic, in most cases convergence is delayed in finite precision computations. Hence, here also it is of interest to understand the origin of these problems which are, of course, closely linked to the ones encountered with the Lanczos algorithm.

This book presents some known and a few new results about the Lanczos and CG algorithms both in exact and finite precision arithmetics. Therefore, we are following the ideas expressed in the quote from Diderot. When great men like Lanczos, Hestenes, and Stiefel invent methods like the Lanczos and CG algorithms, we, ordinary people, have much work to do, as said in the quote from Goethe and Schiller. Our aim is to describe and explain the "generic" behavior of these algorithms, that is, what happens in most cases. Therefore, we shall not pay too much attention to some subtleties that may arise in some contrived examples. We are particularly interested in the eigenvalues of the Lanczos tridiagonal matrices T_k and how well they approximate the eigenvalues of the matrix A. This is because our main goal is to study the convergence rate of the CG algorithm, and we shall see that the behavior of the norms of the error and the residual depends on how well the eigenvalues of A are approximated in the Lanczos algorithm. Therefore, when studying the Lanczos algorithm we shall put the emphasis on the eigenvalue approximations and not too much on eigenvectors.

The Lanczos and CG algorithms are among the most fascinating numerical algorithms because even though their finite precision arithmetic behavior is far from theory and there can be a very large growth of the rounding errors, they can nevertheless be used very efficiently to compute eigenvalues and for solving linear systems with large sparse matrices.

In Chapters 1 and 2 we consider both algorithms in exact arithmetic and we describe their mathematical properties. The first chapter is devoted to the Lanczos algorithm. Given a vector v and a symmetric matrix A, the algorithm constructs an orthonormal basis of the Krylov subspace which is spanned by the vectors

$$\{v, Av, A^2v, \dots, A^{k-1}v\}.$$

We show how to derive the Lanczos algorithm from orthogonalizing the natural basis of the Krylov subspace by a Gram–Schmidt process. Then we explain in detail how the algorithm is related to the QR factorization of the Krylov matrix whose columns are the Krylov vectors A^iv. We show the close relationship of the Lanczos algorithm with orthogonal polynomials. We study the relation between the approximations of the eigenvalues (the so-called Ritz values) at iterations k and $k+1$. An important ingredient in these relations are the components of the eigenvectors of the Lanczos tridiagonal matrix T_k containing the coefficients defined by the algorithm. Therefore, we give analytical expressions for the components of these eigenvectors. They involve the diagonal entries of the Cholesky decomposition of $T_k - \lambda I$ that we study as a function of λ. Then we derive bounds for the distances between the Ritz values and the eigenvalues of A by first recalling classical mathematical results. We establish new bounds for these distances. We also give some insights on which eigenvalues of A are first approximated by the Lanczos algorithm.

Chapter 2 derives the CG algorithm from the Lanczos algorithm. This derivation is based on the LU decomposition of the Lanczos matrix T_k. We exhibit the relations between the Lanczos coefficients and the CG coefficients as well as the orthogonality properties of the residuals r^k, the conjugacy of the descent directions p^k, and the relations between r^k and p^k. Then we study the norms of the CG residuals which often exhibit a very oscillatory behavior. A very natural measure of the error in CG is the A-norm of the error because this norm is minimized at each iteration. Moreover, in certain problems arising from partial differential equations it corresponds to the energy norm. We exhibit the relations between the A-norm of the error and Gauss quadrature. This leads to being able to estimate the norm of the error during the iterations. We derive expressions for the A-norm of the error involving the eigenvalues of A and the Ritz values showing precisely that CG convergence depends on how the eigenvalues of A are approximated by Ritz values. We also establish expressions for the l_2 norm of the error which are, unfortunately, more complicated than those for the A-norm. We also consider the three-term recurrence form of CG and the relations between the coefficients of the two forms of the algorithm. Some CG mathematical optimality properties are explained and the classical bounds of the A-norm of the error involving the condition number of A are derived.

Chapter 3 gives a historical perspective on the Lanczos and CG algorithms in finite precision arithmetic. We first recall the classical model of finite precision arithmetic (corresponding to IEEE arithmetic) and derive results about rounding errors for the operations that are involved in the Lanczos and CG algorithms. These results will be used in the next chapters. We summarize and illustrate by examples some of the most important results that were obtained over the years by different people for the Lanczos and CG algorithms in finite precision arithmetic. The main contributions to these fields were given by C. Paige and A. Greenbaum. We also briefly review the works of J. Grcar, H. Simon, Z. Strakŏs, J. Cullum and R. Willoughby, V. Druskin, and L. Knizhnerman. The book and papers of B.N. Parlett have also been very influential over the last 30 years.

Chapter 4 is devoted to some new results on the Lanczos algorithm in finite precision arithmetic. We first show the results of some numerical experiments which reveal some interesting structures involving the perturbations of the Lanczos basis vectors caused by roundoff errors. Then we study the growth of those perturbations in the algorithm three-term recurrences by looking at the components of the Lanczos vectors on the eigenvectors of A, and we give a description of the "generic" behavior of the Lanczos algorithm in finite precision arithmetic. Specifically, we explain, from a different viewpoint than usual, the origin of the loss of orthogonality and why the Lanczos algorithm computes multiple copies of the eigenvalues in finite precision arithmetic. As it is known since Paige's work, the growth of the perturbations is linked to convergence of Ritz values. The results of this chapter are obtained by exploiting the solutions of nonhomogeneous three-term recurrences in terms of polynomials and associated polynomials. We also derive new solutions to this problem involving elements of inverses of tridiagonal matrices $T_k - \lambda I$. This helps to prove that, as long as there is no convergence of Ritz values, the perturbations stay bounded.

Chapter 5 studies CG in finite precision arithmetic. We consider the relationship of Lanczos and CG algorithms in finite precision. This allows us to obtain equivalent perturbed three-term recurrences for CG. By applying our results for the Lanczos algorithm, we derive expressions for the components of the residual vectors on the eigenvectors of A and we obtain results for the local orthogonality properties. We also give expressions directly obtained from the two-term recurrences. We study the relations between the residuals r^k and the descent directions p^k in finite precision. Finally, we consider CG convergence in finite precision arithmetic. We show that the A-norm of the error is strictly decreasing, provided the condition number of A is not too large, and we establish results on the convergence of $\|r^k\|$. Finally, we give numerical examples of CG convergence when varying the precision of the computations and we compare these results to what is obtained with full reorthogonalization of the residual vectors at every iteration.

It is well known that if we perform a large enough number of iterations, at some point the norms of the residual $b - Ax^k$ and of the error stagnate. This is called the maximum attainable accuracy. This means that a finite precision CG computation does not provide the approximate solution with an arbitrary small error. Chapter 6 deals with the problem of estimating the maximum attainable accuracy. The level of stagnation depends on the matrix and the initial vector norms. We also describe some ways to slightly improve the maximum attainable accuracy.

Chapter 7 shows that the expressions for the A-norm of the error we derived in Chapter 2 are still valid in finite precision arithmetic up to small perturbation terms. We also obtain results for the l_2 norm of the error. From this we can derive estimates of the error norms. Using these bounds or estimates of the norms of the error and the knowledge of the maximum attainable accuracy that can both be computed when running the algorithm, we have a complete control of CG convergence that can be incorporated into a reliable implementation of CG. We also give numerical comparisons between the two-term and three-term variants of CG.

Chapter 8 is concerned with the modifications that have to be made in the results of the previous chapters when CG is used with a preconditioner. These preconditioning matrices are designed to speed up CG convergence by obtaining a better condition number and/or a better distribution of the eigenvalues. The results from the previous chapters help in understanding why some preconditioners give better results than others mainly because they give favorable distributions of eigenvalues.

Chapter 9 deals with various topics that would not fit naturally in previous chapters. The first one is the choice of the starting vector for CG. Then we describe some variants of CG, some of them being used when solving several linear systems in sequence with the same matrix A but with different right-hand sides. Sometimes, when using a preconditioner the solutions of the linear systems we have to solve at each iteration are provided with an iterative method. This leads to inner-outer iterations for which we describe the work of Golub and Ye, who give results on when to stop the inner iterations. We also consider the work of Frayssé et al., where the precision with which the matrix-vector product is computed can be relaxed during the iterations. We give some examples of the use of CG on indefinite linear systems. Finally, we describe some numerical experiments with preconditioned CG.

Although the main audience of this book are researchers in numerical linear algebra and more generally in numerical analysis, it could be of interest to all people, engineers and scientists, using the Lanczos algorithm to compute eigenvalues and the CG algorithm to solve linear systems. Moreover, it can be used in advanced courses about iterative methods or as an example of the study of a well-known numerical method in finite precision arithmetic.

Generally we denote matrices by capital letters (like A), their elements by lowercase letters (like $a_{i,j}$), and vectors by lowercase letters. Elements of sequences of vectors are denoted by an upper index like v^k and their components by v_i^k. Generally, scalars are denoted by lowercase Greek letters. In any case, scalars are always denoted by lowercase letters, except some constants C_j in upper bounds.

All the computations were done with MATLAB version 6 software on a PC, some of them using the Symbolic Toolbox, which uses a Maple kernel for extended precision computations. Throughout the book we shall use an example that was defined by Strakoš [184] and extensively studied in a paper by Greenbaum and Strakoš [81]. The matrix of dimension n is diagonal with eigenvalues

$$\lambda_i = \lambda_1 + \left(\frac{i-1}{n-1}\right)(\lambda_n - \lambda_1)\rho^{n-i}, \quad i = 1, \dots, n.$$

The parameters λ_1 and λ_n are, respectively, the smallest and largest eigenvalues. The parameter ρ controls the distribution of the eigenvalues. We shall often use $\lambda_1 = 0.1$, $\lambda_n = 100$, and a value $\rho = 0.9$, which gives well-separated large eigenvalues. We shall denote this example as Strakos$n$. This matrix is interesting since, for some choices of parameters (like the ones above), orthogonality is lost quite rapidly.

We shall also use some symmetric matrices from the Matrix Market [115]. Even though the Lanczos and CG algorithms are used today with large matrices, these small examples are appropriate to demonstrate their properties and to illustrate their behavior in finite precision arithmetic.

Acknowledgments. The author thanks Chris Paige for very interesting comments and suggestions. Many thanks also to Zdeněk Strakoš for helping to clarify some points. Thanks to J. Grcar, A. Greenbaum, H. Simon, L. Knizhnerman, and J.-P.M. Zemke for their reading and comments.

The author is also grateful to Gene Golub, who encouraged him to write this book.

It has been a pleasure working with the SIAM publication staff, in particular Linda Thiel, Elizabeth Greenspan, Sara Murphy, and David Riegelhaupt.

Chapter 1

The Lanczos algorithm in exact arithmetic

This numerical method was introduced by Cornelius Lanczos (see the biographical notes in the appendix) in 1950 [108]. We should mention in passing that this paper is still worth reading more than half a century later. The algorithm was intended to compute the coefficients of the characteristic polynomial of a matrix as a mean to compute the eigenvalues of a symmetric matrix as it was common in those days. The Lanczos algorithm is a projection method on a Krylov subspace [106]. The Krylov subspace $\mathcal{K}_k(v, A)$ of order k based on the vector v and the matrix A of order n is $span\{v, Av, \dots, A^{k-1}v\}$. Lanczos first introduced a method using these basis vectors. He noticed that, even though this was an improvement over previous methods, this can lead to very severe coefficient growth problems when the condition number of the matrix is large (by 1950 standards). This led him to propose a method he called "minimized iterations," which is what we now call the Lanczos algorithm (although he used a different normalization). This method computed an orthogonal basis of the Krylov subspace. Lanczos in a footnote attributed the successive orthogonalization of a set of vectors to O. Száz (1910), a Hungarian mathematician. However, this process that we now call the Gram–Schmidt algorithm was introduced for sets of functions by J.P. Gram (1883) and E. Schmidt (1907), although there is a mostly unnoticed result of Laplace on this subject (1820) and this was also essentially used by Cauchy in 1836. According to Stewart [178] it was used by G. Kowalevski for finite dimensional spaces in 1909. Actually, in his 1950 paper Lanczos also introduced the nonsymmetric Lanczos method for nonsymmetric matrices using two sets of biorthogonal vectors defined with the help of A^T. He noticed the connection of his method with orthogonal polynomials and the problems of rounding errors, mentioning that the problem can be solved by full reorthogonalization of the computed basis vectors.

Since computing the characteristic polynomial is not a reliable way of computing the eigenvalues, the Lanczos algorithm was afterwards considered as an algorithm to reduce a matrix to tridiagonal form. Unfortunately (for Lanczos), soon after the Lanczos method was discovered, W. Givens and A.S. Householder proposed more efficient and cheaper methods to compute triangular or tridiagonal forms of a matrix using orthogonal transformations; see, for instance, [96], [68]. Therefore, the Lanczos algorithm was dismissed as a method for computing all the eigenvalues of a symmetric matrix. Nevertheless, Lanczos was aware that the method can be used to compute approximations of a few of the eigenvalues of the matrix and to compute solutions of linear equations (see [109]) which are the ways in which we are using it today. The Lanczos algorithm was put back to life through the efforts of C. Paige and B.N. Parlett in the seventies and the eighties. In this chapter we are going to describe the mathematical properties of the Lanczos algorithm in exact arithmetic.

1.1 Introduction to the Lanczos algorithm

In this section we first show how the Lanczos algorithm is derived by orthogonalizing the natural basis of the Krylov subspace. Then we shall exhibit the relationship of the Lanczos algorithm with the QR factorization of the Krylov matrix where Q is an orthogonal matrix and R is an upper triangular matrix.

The matrix A whose eigenvalues (and eventually eigenvectors) are sought is supposed to be real symmetric of order n. We can introduce the Lanczos algorithm by looking at the Krylov basis. Let v be a given vector and

$$K_k = (v \quad Av \quad \cdots \quad A^{k-1}v)$$

be the Krylov matrix of dimension $n \times k$ where the columns are obtained by successive multiplication with the matrix A starting from v. There is a maximal dimension $k = m \le n$ for which the rank of K_k is k. This is the order of the minimal polynomial of v with respect to A. For any v this degree is always less than or equal to the degree of the minimal polynomial of A. The basis given by the columns of K_k is numerically badly behaved since, when i increases, the vectors $A^i v$ tend to be aligned in the direction of the eigenvector associated with the eigenvalue of A of maximum modulus. When the vectors are normalized, this is known as the power method (see [68], [179]), which is used to compute the largest modulus eigenvalue of a matrix and the corresponding eigenvector. Hence, sooner or later numerically the computed vectors of the Krylov basis are going to lose their linear independence. Therefore, it is useful to look for a better behaved basis for the Krylov subspace.

The goal is to construct an orthonormal basis of the Krylov subspace. Although historically things did not proceed in this way, let us consider what is now called the Arnoldi process [3]. This is a variant of the Gram–Schmidt orthogonalization process applied to the Krylov basis without the assumption of A being symmetric. For constructing basis vectors v^{k+1}, instead of orthogonalizing $A^k v$ against the previous vectors, we can orthogonalize Av^k. Starting from $v^1 = v$ (normalized if necessary), the algorithm for computing the $k+1$st vector of the basis using the previous ones is

$$h_{i,k} = (Av^k, v^i), \quad i = 1, \dots, k,$$

$$\bar{v}^k = Av^k - \sum_{i=1}^{k} h_{i,k} v^i,$$

$$h_{k+1,k} = \|\bar{v}^k\|, \quad \text{if } h_{k+1,k} = 0, \text{ then stop},$$

$$v^{k+1} = \frac{\bar{v}^k}{h_{k+1,k}}.$$

It is easy to verify that the vectors v^j span the Krylov subspace and that they are orthonormal. Collecting the basis vectors v^j up to iteration k in a matrix V_k, the relations defining the vector v^{k+1} can be written in matrix form as

$$AV_k = V_k H_k + h_{k+1,k} v^{k+1} (e^k)^T,$$

where H_k is an upper Hessenberg matrix with elements $h_{i,j}$, which means that $h_{i,j} = 0,\ j = 1, \dots, i-2,\ i > 2$; that is, the upper triangular part is nonzero and there is one nonzero subdiagonal next to the main diagonal. The vector e^k is the kth column of the $k \times k$ identity matrix. Throughout this book e^j will denote the jth column of identity matrices of different orders. From the orthogonality of the basis vectors, we have $V_k^T v^{k+1} = 0$ and

$$V_k^T A V_k = H_k.$$

If we now suppose that the matrix A is symmetric, H_k is also symmetric owing to the last relation. It is obvious that a symmetric Hessenberg matrix is tridiagonal. Therefore, we have $h_{i,j} = 0,\ j = i+2, \dots, k$, and we denote H_k by T_k. This means that $\bar{v}^k$ and hence v^{k+1} can be computed by only using the two previous vectors v^k and v^{k-1}. This essentially describes the Lanczos algorithm. A more formal definition will be given later.

Let us now study the relationship of this algorithm to the QR factorization (see [68]) of the Krylov matrix. This is another (more complicated) way to introduce the Lanczos algorithm but which explains in more detail how things are working. If we denote by E the $k \times k$ shift matrix

$$E = \begin{pmatrix} 0 & 0 & \cdots & & 0 \\ 1 & 0 & \cdots & & 0 \\ 0 & 1 & \cdots & & 0 \\ 0 & 0 & \ddots & & 0 \\ 0 & 0 & \cdots & 1 & 0 \end{pmatrix},$$

we have the relation

$$AK_k = K_k E - (0 \quad A^k v).$$

Such a decomposition has been called a Krylov decomposition by Stewart [179]. We are going to see how this can be applied to the Lanczos algorithm. If $\text{rank}(K_k) < k$, then we have found an invariant subspace for A. Therefore, let us suppose that K_k has full rank k. Let $K_k = V_k R_k$ be a QR factorization (see [68]) of the Krylov matrix where V_k is $n \times k$ and orthonormal ($V_k^T V_k = I$, the identity matrix) and R_k is $k \times k$ nonsingular and upper triangular (say, with positive elements on the diagonal). We can see that V_k is obviously an orthonormal basis of the Krylov space and it is perfectly conditioned. It is interesting to see if we can cheaply compute V_k. By plugging the QR factorization of K_k into the Krylov decomposition and multiplying from the right with R_k^{-1}, we obtain

$$AV_k = V_k R_k E R_k^{-1} + (0 \quad A^k v) R_k^{-1}.$$

Multiplying from the left by V_k^T it follows that

$$V_k^T A V_k = R_k C_k R_k^{-1},$$

where $C_k = E + (\,0 \quad R_k^{-1} V_k^T A^k v\,)$. The matrix C_k is called a companion matrix. It has the following structure (the x's denoting nonzero components):

$$C_k = \begin{pmatrix} 0 & 0 & \cdots & 0 & x \\ 1 & 0 & \cdots & 0 & x \\ 0 & 1 & \cdots & 0 & x \\ 0 & 0 & \ddots & 0 & x \\ 0 & 0 & \cdots & 1 & x \end{pmatrix}.$$

It is not too difficult to see that $R_k C_k R_k^{-1}$ is an upper Hessenberg matrix in the general case, but since here $V_k^T A V_k$ is symmetric, it is a tridiagonal matrix that we denote by T_k. It means that T_k is the projection of A on the Krylov subspace. Let $w^k = R_k^{-1} V_k^T A^k v$ be the last column of C_k; then the relation for AV_k can be written as

$$AV_k = V_k(T_k - (\,0 \quad R_k w^k\,)\, R_k^{-1}) + (\,0 \quad A^k v\,)\, R_k^{-1}.$$

Hence

$$AV_k = V_k T_k + (\,0 \quad A^k v - V_k R_k w^k\,)\, R_k^{-1}$$

and

$$AV_k = V_k T_k + (\,0 \quad (I - V_k V_k^T) A^k v\,)\, R_k^{-1}.$$

The matrix $I - V_k V_k^T$ is a projector on the orthogonal complement of V_k. If $A^k v \in \mathcal{K}_k(v, A)$, then $AV_k = V_k T_k$ and we have found an invariant subspace of A. In this case the eigenvalues of T_k are eigenvalues of A.

The QR factorization of V_k can be easily updated in the following way when we add a new vector to the Krylov basis; see [178]. Let

$$K_{k+1} = (\,K_k \quad A^k v\,) = (\,V_k R_k \quad A^k v\,) = V_{k+1} R_{k+1}.$$

Let $(\,z^k \quad \xi\,)^T$ be the last column of R_{k+1}, where z^k is a vector of length k and ξ is a real number,

$$R_{k+1} = \begin{pmatrix} R_k & z^k \\ 0 & \xi \end{pmatrix}.$$

Moreover let $V_{k+1} = (\,V_k \quad v^{k+1}\,)$; then writing the last column of $V_{k+1} R_{k+1}$, we have

$$V_k z^k + \xi v^{k+1} = A^k v.$$

Multiplying on the left by V_k^T,

$$z^k + \xi V_k^T v^{k+1} = V_k^T A^k v,$$

but the second term on the left-hand side is zero. This gives $z^k = V_k^T A^k v$. Then,

$$\xi v^{k+1} = (I - V_k V_k^T) A^k v,$$

and ξ is chosen such that $\|v^{k+1}\| = 1$. Therefore,

$$\xi = \|(I - V_k V_k^T) A^k v\|.$$

Notice that this is a constructive proof of the existence of the QR factorization (by induction). Hence, we can drop the index k on R_k since it can be extended at each step by one row and one column. The matrix R is also the Cholesky factor of the moment matrix $K_k^T K_k$. That is,

$$R^T R = \begin{pmatrix} 1 & (v, Av) & \cdots & (v, A^{k-1}v) \\ (v, Av) & (v, A^2 v) & \cdots & (v, A^k v) \\ \vdots & \vdots & & \vdots \\ (v, A^{k-1}v) & \cdots & & (v, A^{2k-2}v) \end{pmatrix}.$$

The elements $r_{i,j}$ of R can also be related to those of T_k. If we denote the matrix T_k by

$$T_k = \begin{pmatrix} \alpha_1 & \eta_2 & & & \\ \eta_2 & \alpha_2 & \eta_3 & & \\ & \ddots & \ddots & \ddots & \\ & & \eta_{k-1} & \alpha_{k-1} & \eta_k \\ & & & \eta_k & \alpha_k \end{pmatrix},$$

by direct multiplication in $T_k = RC_k R^{-1}$ we obtain $r_{1,1} = 1$ and

$$\eta_i = \frac{r_{i,i}}{r_{i-1,i-1}}.$$

This shows that

$$\eta_2 \cdots \eta_k = r_{k,k}.$$

We also find

$$\alpha_1 = r_{1,2}, \quad \alpha_i = \frac{r_{i,i+1}}{r_{i,i}} - \frac{r_{i-1,i}}{r_{i-1,i-1}}, \quad i = 2, \dots, k-1, \quad \alpha_k = w_k^k - \frac{r_{k-1,k}}{r_{k-1,k-1}}.$$

We verify easily that $w_k^k = (v^k, A^k v)/r_{k,k}$ and $r_{k,k+1} = (v^k, A^k v)$. So we remark that if we want to compute the α's and the η's from the QR factorization all we need are the diagonal elements of R as well as the elements of the first superdiagonal and $(v^k, A^k v)$.

By definition, $\xi = r_{k+1,k+1}$ and it follows that $\xi = \eta_{k+1} r_{k,k}$. Therefore,

$$\eta_{k+1} r_{k,k} v^{k+1} = (I - V_k V_k^T) A^k v$$

and since

$$AV_k = V_k T_k + (\, 0 \quad (I - V_k V_k^T) A^k v \,)\, R^{-1},$$

we have the following relation:

$$AV_k = V_k T_k + \eta_{k+1} (\, 0 \quad v^{k+1} \,).$$

Hence we can compute V_k by using this relation without computing K_k or R, provided we can compute T_k. As we have seen before, this is the most important matrix relation of the Lanczos algorithm. Returning for a moment to the Krylov matrix, we remark (see [150]) that

$$AK_k = K_k C_k + (\, 0 \quad -K_k w^k + A^k v \,).$$

Denoting by f^k the residual $f^k = -K_k w^k + A^k v$, we have $f^k = (I - V_k V_k^T)A^k v = \eta_{k+1} r_{k,k} v^{k+1}$. We consider the solution of $K_k c^k = A^k v$ in the least squares sense for which the vector c^k is given by the solution of the normal equations (see [68])

$$K_k^T K_k c^k = K_k^T A^k v,$$

which can be written as

$$R^T R c^k = R^T V_k^T A^k v.$$

Since R is square and nonsingular, this gives

$$R c^k = V_k^T A^k v,$$

which is $V_k^T K_k c^k = V_k^T A^k v$ and $c^k = w^k$. Therefore f^k is the residual and w^k is the solution of the least squares problem

$$\min_c \| K_k c - A^k v \|.$$

Clearly the residual can be written as $f^k = \psi_k(A)v$, where ψ_k is a monic polynomial of degree k. It is easy to see that

$$\psi_k(\lambda) = \lambda^k - [1 \cdots \lambda^{k-1}]c^k = \det(\lambda I - C_k) = (-1)^k \det(C_k - \lambda I).$$

The elements of the vector c^k are the coefficients of the characteristic polynomial. The polynomial $\psi_k(A)$ applied to v minimizes the norm of the least squares problem over all monic polynomials of degree k. The zeros of ψ_k are the eigenvalues of C_k and therefore those of T_k. Moreover,

$$A^k v = (I - V_k V_k^T)A^k v + V_k V_k^T A^k v = \eta_{k+1} r_{k,k} v^{k+1} + V_k R w^k.$$

The first term is in $\mathcal{K}_k^\perp$ and the second one in $\mathcal{K}_k$. Since $A^k v - V_k R w^k = (I - V_k V_k^T)A^k v = f^k$ and $\eta_{k+1} r_{k,k} = \eta_2 \cdots \eta_{k+1}$ we have

$$v^{k+1} = \frac{\psi_k(A)v}{\eta_2 \cdots \eta_{k+1}}$$

and $\| f^k \| = \eta_2 \cdots \eta_{k+1}$.

What we have done so far is nothing other than constructing an orthonormal basis of the Krylov subspace by the Gram–Schmidt process. The matrix relation for the matrix V_k whose columns are the basis vectors can be written more compactly as

$$A V_k = V_k T_k + \eta_{k+1} v^{k+1} (e^k)^T,$$

where e^k is the kth column of the $k \times k$ identity matrix. The previous relation can also be written as

$$A V_k = V_{k+1} \tilde{T}_k,$$

with a matrix $\tilde{T}_k$ of dimension $(k+1) \times k$

$$\tilde{T}_k = \begin{pmatrix} T_k \\ \eta_{k+1}(e^k)^T \end{pmatrix}.$$

We also have $AV_n = V_n T_n$ if no v^j is zero before step n since $v^{n+1} = 0$, v^{n+1} being a vector orthogonal to a set of n orthogonal vectors in a space of dimension n. Otherwise there exists an $m < n$ for which $AV_m = V_m T_m$ and we have found an invariant subspace of A, the eigenvalues of T_m being eigenvalues of A.

The previous matrix relations imply that the basis vectors v^k (which we call the Lanczos vectors) can be computed by a three-term recurrence. This gives the elegant Lanczos algorithm which is written, starting from a vector $\tilde{v}^1 = v$ for $k = 1, 2, \dots,$ as

$$\eta_k = \|\tilde{v}^k\|,$$
$$v^k = \frac{\tilde{v}^k}{\eta_k},$$
$$\alpha_k = (v^k, Av^k) = (v^k)^T A v^k,$$
$$\tilde{v}^{k+1} = Av^k - \alpha_k v^k - \eta_k v^{k-1}.$$

The Lanczos coefficients of the matrix T_k are computed directly without resorting to the upper triangular matrix R to obtain the orthonormality of the basis vectors and therefore

$$\eta_k = (Av^k, v^{k-1}) = \|Av^{k-1} - \alpha_{k-1}v^{k-1} - \eta_{k-1}v^{k-2}\|.$$

The second equality can be proved by remarking that

$$\|Av^{k-1} - \alpha_{k-1}v^{k-1} - \eta_{k-1}v^{k-2}\|^2 = (Av^{k-1}, Av^{k-1}) - \alpha_{k-1}^2 - \eta_{k-1}^2.$$

We have $\eta_k Av^k = A^2 v^{k-1} - \alpha_{k-1}Av^{k-1} - \eta_{k-1}Av^{k-2}$. Multiplying by v^{k-1} we obtain

$$\eta_k(Av^k, v^{k-1}) = (Av^{k-1}, Av^{k-1}) - \alpha_{k-1}^2 - \eta_{k-1}(Av^{k-2}, v^{k-1}).$$

If $\eta_k = (Av^k, v^{k-1})$, we see that $\eta_k^2 = \|Av^{k-1} - \alpha_{k-1}v^{k-1} - \eta_{k-1}v^{k-2}\|^2$. The scalar η_k can always be taken to be positive by changing the signs.

We note that since $\|v^k\| = 1$, α_k is a so-called Rayleigh quotient. This implies that

$$\lambda_{min}(A) \leq \alpha_k \leq \lambda_{max}(A),$$

where λ_{min} and λ_{max} are the smallest and largest eigenvalues of A (which are real since A is a symmetric matrix). We shall see later on that we can find better bounds on α_k. Throughout this book we denote the eigenvalues of A by

$$\lambda_{min} = \lambda_1 \leq \lambda_2 \leq \cdots \leq \lambda_n = \lambda_{max}.$$

Another version of the Lanczos algorithm has been advocated by Paige [132] to enforce the local orthogonality (with the previous vector) in finite precision computations. It replaces the third and fourth steps by

$$\alpha_k = (v^k)^T (Av^k - \eta_k v^{k-1}),$$
$$\tilde{v}^{k+1} = (Av^k - \eta_k v^{k-1}) - \alpha_k v^k.$$

Notice that this variant can be implemented by using only two vectors of storage instead of three for the basic formulation. It corresponds to using the modified Gram–Schmidt process; see [68]. In exact arithmetic both versions are mathematically equivalent since v^k is orthogonal to v^{k-1}, but Paige's formulation better preserves local orthogonality in finite precision arithmetic. Therefore, we see that we have several available versions of the Lanczos algorithm differing by the ways they compute the coefficients. We shall also see that the Lanczos coefficients can be computed when running the conjugate gradient (CG) algorithm. The variant most used today is Paige's version. Therefore, the recommended Lanczos algorithm is the following.

Algorithm 1.1.

$v^1 = v/\|v\|$
$\alpha_1 = (v^1)^T A v^1$
$\tilde{v}^2 = Av^1 - \alpha_1 v^1$
for $k = 2, 3, \dots$
$\eta_k = \|\tilde{v}^k\|$
$v^k = \tilde{v}^k/\eta_k$
$u^k = Av^k - \eta_k v^{k-1}$
$\alpha_k = (v^k)^T u^k$
$\tilde{v}^{k+1} = u^k - \alpha_k v^k$
end

The algorithm computes the coefficients of the matrix T_k with $(T_k)_{k,k} = \alpha_k$, $(T_k)_{k,k-1} = \eta_k$, and the Lanczos vectors v^k. The approximations of the eigenvalues of A are obtained by computing the eigenvalues of T_k.

The Lanczos algorithm must terminate at some step m, $m \le n$ with $v^{m+1} = 0$. The value of m is generally n depending on the initial vector if A has only simple eigenvalues. It is less than m otherwise since unreduced tridiagonal matrices (with $\eta_j \neq 0$) have simple eigenvalues. If A has multiple eigenvalues, it cannot be similar to a matrix with only simple eigenvalues; therefore one of the η_j has to be zero before step n, which means that the corresponding v^j is zero and we have found an invariant subspace for A. The Lanczos algorithm cannot detect the fact that there are multiple eigenvalues. In exact arithmetic they are found once. Lanczos was already aware of that fact.

Since we can suppose that $\eta_i \neq 0$, T_k has real and simple eigenvalues which we denote by $\theta_j^{(k)}$. They are known as the Ritz values that we order as

$$\theta_1^{(k)} < \theta_2^{(k)} < \cdots < \theta_k^{(k)}.$$

We shall see that the Ritz values are approximations of the eigenvalues of A. We recall that the eigenvalues of T_k are the same as those of C_k, that is, the roots of the monic polynomial ψ_k. We denote the corresponding normalized eigenvectors of T_k as $z_{(k)}^j$ or simply z^j with components z_i^j when no confusion is possible and denote the matrix of the eigenvectors as Z_k. The vectors $x_{(k)}^j = V_k z_{(k)}^j$ are known as the Ritz vectors. They are the approximations of the eigenvectors of A given by the algorithm. The residual associated with an eigenpair $(\theta_j^{(k)}, x_{(k)}^j)$ of T_k is

$$r_{(k)}^j = Ax_{(k)}^j - \theta_j^{(k)} x_{(k)}^j = (AV_k - V_k T_k) z_{(k)}^j = \eta_{k+1} (z_{(k)}^j)_k v^{k+1},$$

where $(z^j_{(k)})_k$ is the last component of $z^j_{(k)}$. It will be denoted sometimes as z^j_k when it is clear from the context that we are referring to the eigenvectors of T_k. Therefore,

$$\|r^j_{(k)}\| = \eta_{k+1}|(z^j_{(k)})_k|.$$

This means that $r^j_{(k)}/\|r^j_{(k)}\|$ is independent of j and all the residual vectors are in the direction of v^{k+1}. Moreover, we have a small residual, when the product of η_{k+1} and $|(z^j_{(k)})_k|$ is small. This residual vector $r^j_{(k)}$ is proportional to the one for the least squares problem with the Krylov matrix. We have

$$r^j_{(k)} = (z^j_{(k)})_k \frac{\psi_k(A)v}{\eta_2 \cdots \eta_k}.$$

We also remark that

$$\sum_{j=1}^{k+1} \alpha_j = \text{trace}(T_{k+1}) = \sum_{j=1}^{k+1} \theta_j^{(k+1)} = \text{trace}(T_k) + \alpha_{k+1} = \sum_{j=1}^{k} \theta_j^{(k)} + \alpha_{k+1}.$$

This relation was used to suggest that one can directly compare the eigenvalues of T_{k+1} and those of T_k by inserting α_{k+1} in the latter set. From this we also have

$$\alpha_k = \sum_{j=1}^{k} \theta_j^{(k)} - \sum_{j=1}^{k-1} \theta_j^{(k-1)}.$$

By grouping the Ritz values two by two and using the Cauchy interlacing property that will be stated later on, we can prove that

$$\theta_1^{(k)} \le \alpha_k \le \theta_k^{(k)}.$$

Unfortunately, since we shall see that $\theta_1^{(k)}$ is a decreasing sequence of k and $\theta_k^{(k)}$ an increasing one, the interval containing α_k gets larger and larger when k increases. Finally, notice that when A is positive definite, T_k is also positive definite for all k and the Ritz values $\theta_j^{(k)}$ and the coefficients α_k are also positive.

In the next sections we are going to study some properties of the Ritz values which are the eigenvalues of the tridiagonal Lanczos matrix T_k generated during the algorithm and the corresponding eigenvectors. This is important since some properties of the Lanczos algorithm are closely linked to components of the eigenvectors of T_k both in exact and finite precision arithmetics. In fact, we are interested in shifted matrices $T_k - \lambda I$ and their Cholesky decomposition (when they exist).

1.2 Lanczos polynomials

In this section we characterize the Lanczos vectors v^k as polynomials in the matrix A applied to the initial vector $v^1 = v$. First, we notice that the determinant of T_k verifies a three-term recurrence.

Lemma 1.1.

$$\det(T_{k+1}) = \alpha_{k+1}\det(T_k) - \eta_{k+1}^2 \det(T_{k-1})$$

with initial conditions

$$\det(T_1) = \alpha_1, \quad \det(T_2) = \alpha_1\alpha_2 - \eta_2^2.$$

Proof. This is obtained by expanding the determinant along the last row or column of T_{k+1}. □

We denote by $\chi_{1,k}(\lambda)$ the determinant of $T_k - \lambda I$ (which is the same as the monic polynomial $\psi_k(\lambda)$ we have seen before) and for further reference $\chi_{j,k}(\lambda)$ the determinant of $T_{j,k} - \lambda I$, where the tridiagonal matrix $T_{j,k}$ of order $k - j + 1$ is

$$T_{j,k} = \begin{pmatrix} \alpha_j & \eta_{j+1} & & & \\ \eta_{j+1} & \alpha_{j+1} & \eta_{j+2} & & \\ & \ddots & \ddots & \ddots & \\ & & \eta_{k-1} & \alpha_{k-1} & \eta_k \\ & & & \eta_k & \alpha_k \end{pmatrix}.$$

This is obtained from the matrix T_k by deleting the first $j - 1$ rows and columns. Notice that we also have

$$\det(T_{j,k}) = \alpha_j \det(T_{j+1,k}) - \eta_{j+1}^2 \det(T_{j+2,k}).$$

The relation for the determinants of T_k in Lemma 1.1 is linked to the Cholesky decomposition of T_k when it exists. Let us write $T_k = L_k \Delta_k^{-1} L_k^T$, where L_k is a lower bidiagonal matrix

$$L_k = \begin{pmatrix} \delta_1 & & & & \\ \eta_2 & \delta_2 & & & \\ & \ddots & \ddots & & \\ & & \eta_{k-1} & \delta_{k-1} & \\ & & & \eta_k & \delta_k \end{pmatrix},$$

and Δ_k is a diagonal matrix with entries $\delta_1, \ldots, \delta_k$ on the diagonal. Similarly we denote as $\delta_1(\lambda), \ldots, \delta_k(\lambda)$ the diagonal entries of the Cholesky decomposition of $T_k - \lambda I$ when it exists. The zeros of $\det(T_k - \lambda I)$ are the eigenvalues of T_k and we have the following result.

Proposition 1.2.

$$\det(T_k - \lambda I) = \delta_1(\lambda) \cdots \delta_k(\lambda),$$

where

$$\delta_1(\lambda) = \alpha_1 - \lambda, \quad \delta_l(\lambda) = \alpha_l - \lambda - \frac{\eta_l^2}{\delta_{l-1}(\lambda)}, \quad l = 2, \ldots, k.$$

If there exists an $m \le k$ such that $\delta_m(\lambda) = 0$, λ is an eigenvalue of T_m.

Proof. The first relation is obvious, the determinant of a triangular matrix being the product of its diagonal elements. The relations for $\delta_i(\lambda)$ are obtained by identification. If there exists an m such that $\delta_m(\lambda) = 0$ (and, therefore, the Cholesky decomposition does not exist), then $\det(T_m - \lambda I) = 0$ and λ is an eigenvalue of T_m. We shall see later on that this cannot happen before the Lanczos algorithm stops with a null vector and that the zeros of $\det(T_k - \lambda I)$ are those of $\delta_k(\lambda)$. $\square$

We easily see that δ_j, $j = 1, \ldots, k$, is a rational function of λ. By induction we can prove that the numerator of $\delta_k(\lambda)$ is a polynomial of degree k and the denominator a polynomial of degree $k-1$. Actually, we can verify that the rational function $\delta_k(\lambda)$ is

$$\delta_k(\lambda) = \frac{\det(T_k - \lambda I)}{\det(T_{k-1} - \lambda I)}.$$

The function δ_k has been used by Parlett in [144], where it is called the "last pivot" function, referring to the fact that it is obtained from a recurrence starting with δ_1 and ending with δ_k. We shall also consider a UL decomposition of T_k, where U is upper triangular and L lower triangular (actually, they are upper and lower bidiagonal matrices) [117], that we write as $T_k = \bar{L}_k^T D_k^{-1} \bar{L}_k$, D_k being diagonal and $\bar{L}_k$ lower bidiagonal. The diagonal elements of the decomposition, that is, those of D_k denoted by $d_j^{(k)}$, are given (when they exist) by

$$d_k^{(k)}(\lambda) = \alpha_k - \lambda, \quad d_l^{(k)}(\lambda) = \alpha_l - \lambda - \frac{\eta_{l+1}^2}{d_{l+1}^{(k)}(\lambda)}, \quad l = k-1, \ldots, 1.$$

We add an upper index (k) because contrary to the LU decomposition, all the diagonal elements of the UL decomposition change when we go from step k to step $k+1$ adding a row and a column to T_k. Similarly to what we had before,

$$\det(T_k - \lambda I) = d_1^{(k)}(\lambda) \cdots d_k^{(k)}(\lambda).$$

We shall see that the zeros of $d_1^{(k)}$ are the eigenvalues of T_k. Moreover, $d_j^{(k)}$ is a rational function and

$$d_1^{(k)}(\lambda) = \frac{\det(T_k - \lambda I)}{\det(T_{2,k} - \lambda I)}.$$

By analogy to [144], $d_1^{(k)}$ can be called the "first pivot" function referring to the final index 1. To summarize, the last (resp., first) pivot function arises from the Cholesky decomposition starting from the top (resp., the bottom) of $T_k - \lambda I$. The derivatives of the first and last pivot functions also play an important role in the Lanczos algorithm and we shall study them later on.

It is also interesting to characterize the Lanczos basis vectors v^k as polynomials in the matrix A acting on the initial vector v^1. From what we have seen before or from Lemma 8 of [73] we have the following result.

Proposition 1.3.

$$v^k = p_{1,k}(A)v^1, \quad p_{1,k}(\lambda) = (-1)^{k-1} \frac{\chi_{1,k-1}(\lambda)}{\eta_2 \cdots \eta_k}.$$

The polynomials $p_{1,k}$ of degree $k-1$ are denoted as the normalized Lanczos polynomials.

Proof. Compare the three-term recurrence for v^k and for the determinant $\chi_{1,k-1}(\lambda)$. $\square$

As we said, this last result is a consequence of what we have seen before for the monic polynomial ψ_{k-1} which is involved in the solution of the least squares problem,

$$p_{1,k} = (-1)^{k-1} \frac{\psi_{k-1}}{\eta_k r_{k-1,k-1}} = (-1)^{k-1} \frac{\psi_{k-1}}{\eta_2 \cdots \eta_k}.$$

Obviously, the polynomials $p_{1,k}$ (or p_k to simplify the notation; the first index will be needed when we study the Lanczos algorithm in finite precision arithmetic) satisfy a scalar three-term recurrence,

$$\eta_{k+1} p_{k+1}(\lambda) = (\lambda - \alpha_k) p_k(\lambda) - \eta_k p_{k-1}(\lambda), \quad k = 1, 2, \ldots,$$

with initial conditions, $p_0 \equiv 0$, $p_1 \equiv 1$. Considering the inner product of two Lanczos vectors, we have

$$\begin{aligned}(v^i, v^j) &= (p_i(A)v^1, p_j(A)v^1), \\ &= \sum_{l=1}^{n} p_i(\lambda_l) p_j(\lambda_l) [\bar{v}_l^1]^2.\end{aligned}$$

This is written as a Riemann–Stieltjes integral for a measure α which is piecewise constant (here we suppose for the sake of simplicity that the eigenvalues of A are distinct):

$$\alpha(\lambda) = \begin{cases} 0 & \text{if } \lambda < \lambda_1, \\ \sum_{j=1}^{i} [\bar{v}_j^1]^2 & \text{if } \lambda_i \leq \lambda < \lambda_{i+1}, \\ \sum_{j=1}^{n} [\bar{v}_j^1]^2 & \text{if } \lambda_n \leq \lambda, \end{cases}$$

where $\bar{v}_j^1$ is the jth component of $Q^T v^1$, that is, (q^j, v^1), where Q is the matrix whose columns are the eigenvectors q^j of A. Therefore,

$$(v^i, v^j) = (p^i, p^j) = \int_{\lambda_1}^{\lambda_n} p^i(\lambda) p^j(\lambda) \, d\alpha(\lambda) = \delta_{i,j}.$$

The Lanczos polynomials are orthonormal with respect to the measure α; see [188] for properties of orthogonal polynomials. The measure has a finite number of points of increase at the eigenvalues of A. This implies that p_k is orthogonal to any polynomial of degree less than $k - 1$.

These polynomials are closely related to Gauss quadrature; see [64]. We shall come back to this important point in the next chapters. If we write $P_k(\lambda) = (p_1(\lambda) \cdots p_k(\lambda))^T$ with $p_1 \equiv 1$ we have

$$\lambda P_k(\lambda) = T_k P_k(\lambda) + \eta_{k+1} p_{k+1}(\lambda) e^k.$$

This shows a fact that we have already proved—that the roots of the polynomial p_{k+1} are the eigenvalues $\theta_j^{(k)}$ of T_k; see [69], [199]. This also tells us that the eigenvectors of T_k are

$$(p_1(\theta_j^{(k)}) \quad \cdots \quad p_k(\theta_j^{(k)}))^T,$$

although they are unnormalized. An interesting property that we shall also use several times is the Christoffel–Darboux relation.

Theorem 1.4.

$$(\lambda-\mu)\sum_{j=1}^{k} p_j(\lambda)p_j(\mu) = \eta_{k+1}[p_k(\mu)p_{k+1}(\lambda) - p_k(\lambda)p_{k+1}(\mu)],$$

$$\sum_{j=1}^{k} p_j(\lambda)^2 = \eta_{k+1}[p_k(\lambda)p'_{k+1}(\lambda) - p_{k+1}(\lambda)p'_k(\lambda)].$$

Proof. See, for instance, [199]. The first relation can be proved by using the vector relation for the polynomials for λ and μ. The second relation is proved by dividing the first one by $\lambda-\mu$ and letting $\mu \to \lambda$. □

The relations of Theorem 1.4 show that taking $\mu = \theta_i^{(k)}$ and since the Ritz values $\theta_i^{(k)}$ are the roots of p_{k+1},

$$(\lambda-\theta_i^{(k)})\sum_{j=1}^{k} p_j(\lambda)p_j(\theta_i^{(k)}) = \eta_{k+1}p_k(\theta_i^{(k)})p_{k+1}(\lambda) \quad \forall i,$$

$$\sum_{j=1}^{k} p_j(\theta_i^{(k)})^2 = \eta_{k+1}p_k(\theta_i^{(k)})p'_{k+1}(\theta_i^{(k)}) \quad \forall i.$$

Notice that $p'_{k+1}(\theta_i^{(k)}) \neq 0$ since there are no multiple roots, all the eigenvalues of T_k being simple. We also have

$$(\theta_i^{(k-1)}-\lambda)\sum_{j=1}^{k-1} p_j(\theta_i^{(k-1)})p_j(\lambda) = \eta_{k+1}p_k(\lambda)p_{k+1}(\theta_i^{(k-1)}) \quad \forall i.$$

The first relation gives an expression for the distances from eigenvalues to Ritz values $\lambda_l - \theta_i^{(k)}$ for all l and i and the second one shows that

$$\eta_{k+1} = \frac{\sum_{j=1}^{k} p_j(\theta_i^{(k)})^2}{p_k(\theta_i^{(k)})p'_{k+1}(\theta_i^{(k)})} \quad \forall i.$$

Using the three-term recurrence with $\lambda = \theta_i^{(k)}$, we obtain

$$\alpha_k = \theta_i^{(k)} - \eta_k\frac{p_{k-1}(\theta_i^{(k)})}{p_k(\theta_i^{(k)})} \quad \forall i.$$

The last relation is a nonlinear equation for $\theta_i^{(k)}$. When we simplify it using the definition of p_k we find the relation defining $\delta_k(\lambda)$. The relation for η_{k+1} shows that $p_k(\theta_i^{(k)})$ and $p'_{k+1}(\theta_i^{(k)})$ have the same sign since $\eta_{k+1} > 0$.

An example of Lanczos polynomial is given in Figure 1.1 for the matrix Strakos10. The stars are $\theta_i^{(k-1)}$ the eigenvalues of T_{k-1}, which are the roots of p_k of degree $k-1$. What is important for the Lanczos vectors are the values of the Lanczos polynomial at the eigenvalues of A. This is shown in Figure 1.2. We see that the Lanczos polynomial

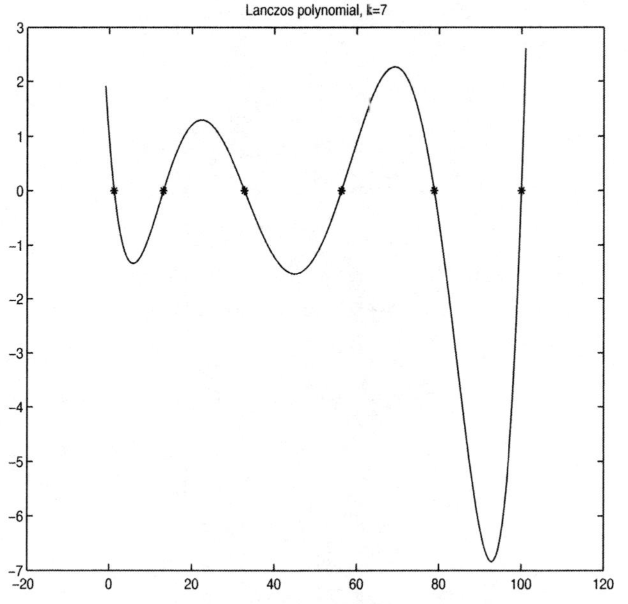

Figure 1.1. *Lanczos polynomial p_7 for the Strakos10 matrix*

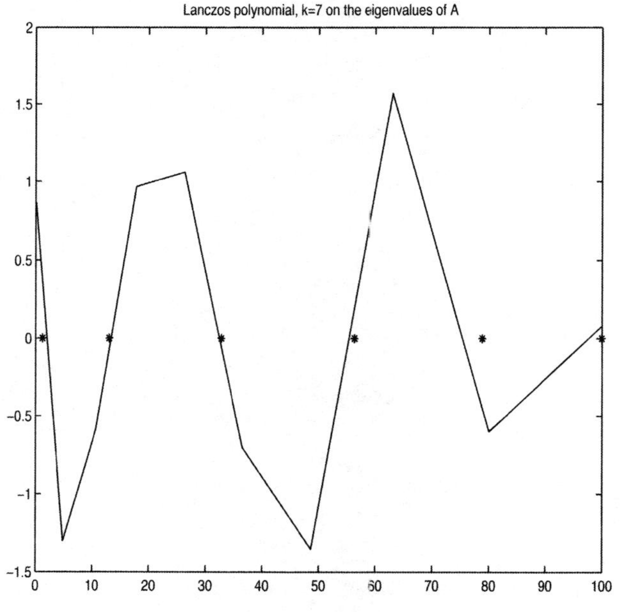

Figure 1.2. *Values of the Lanczos polynomial $p_7(\lambda_i)$ for the Strakos10 matrix at the eigenvalues λ_i of A*

oscillates and has large values between the roots. However, the values on the spectra of A are much smaller, especially when k is large, because the Ritz values are then close to the eigenvalues of A.

Other interesting polynomials for the Lanczos algorithm are the so-called Ritz polynomials which are related to the Ritz vectors. They are defined by

$$x^j_{(k)} = q^{(k)}_j(A)v^1.$$

From Parlett [141] we have that the Ritz values $\theta^{(k)}_i$ without $\theta^{(k)}_j$ are the zeros of $q^{(k)}_j$. The Lanczos polynomial p_{k+1} is a scalar multiple of $\chi_{1,k}$ and therefore has zeros $\theta^{(k)}_i$. Therefore,

$$p_{k+1}(\lambda) = c^{(k)}_j(\lambda - \theta^{(k)}_j)q^{(k)}_j(\lambda),$$

and from [141] we have $|c^{(k)}_j| = 1/\|r^j_{(k)}\|$. The Ritz polynomial can also be written as

$$q^{(k)}_j(\lambda) = \frac{z^j_k}{\eta_2 \cdots \eta_k} \prod_{\substack{i=1 \\ i \neq j}}^{k} (\lambda - \theta^{(k)}_i).$$

So far we know the Lanczos polynomials only through their zeros which are the eigenvalues of the matrices T_k. We now state a result of Cybenko [29] which gives an expression for the Lanczos polynomials involving only the eigenvalues of A; see also [71].

Theorem 1.5. *Let $V(\omega_1, \dots, \omega_k)$ be the Vandermonde determinant based on $\omega_1, \dots, \omega_k$ and Δ_k be the collection of k-tuples $I_k = (i_1, \dots, i_k)$ with $1 \le i_1 \le \cdots \le i_k \le k$. Then the Lanczos polynomial p_{k+1} of degree k, which is a multiple of the characteristic polynomial of T_k, can be written as*

$$\sigma \sum_{I_k \in \Delta_k} \gamma_{i_1} \cdots \gamma_{i_k} V^2(\lambda_{i_1}, \dots, \lambda_{i_k}) \prod_{j=1}^{k} (\lambda - \lambda_{i_j}),$$

where σ is a constant and $\gamma_i = (q^i, v)^2$, q^i being the ith eigenvector of A.

The Vandermonde determinant is given by

$$V(\omega_1, \dots, \omega_l) = \begin{vmatrix} 1 & 1 & \cdots & 1 \\ \omega_1 & \omega_2 & \cdots & \omega_l \\ \vdots & \vdots & & \vdots \\ \omega_1^{l-1} & \omega_2^{l-1} & \cdots & \omega_l^{l-1} \end{vmatrix} = \prod_{\substack{i,j \\ i>j}} (\omega_i - \omega_j).$$

If we look at the value of this polynomial for an eigenvalue λ_i of A which is of interest for the Lanczos vectors, we see that in the sum there remain only the tuples not containing i. The value of the polynomial at λ_i clearly depends on the products of the distances to the other eigenvalues.

1.3 Interlacing properties and approximations of eigenvalues

In this section we consider what happens to the Ritz values when we go from step k to step $k+1$ in the Lanczos algorithm. We look for an eigenvalue λ and an eigenvector $x = (y \quad \zeta)^T$ of T_{k+1}, where y is a vector of dimension k and ζ is a real number. This gives the two equations

$$T_k y + \eta_{k+1} \zeta e^k = \lambda y,$$
$$\eta_{k+1} y_k + \alpha_{k+1} \zeta = \lambda \zeta,$$

where y_k is the last component of y. By eliminating the vector y from these equations we have

$$(\alpha_{k+1} - \eta_{k+1}^2 ((e^k)^T (T_k - \lambda I)^{-1} e^k)) \zeta = \lambda \zeta.$$

We can divide by ζ if it is nonzero. Otherwise, λ is an eigenvalue of T_k. By using the spectral decomposition of $T_k - \lambda I$ we obtain the following result.

Theorem 1.6. *The eigenvalues of T_{k+1} are solutions of the so-called "secular equation" for λ*

$$\alpha_{k+1} - \eta_{k+1}^2 \sum_{j=1}^{k} \frac{\xi_j^2}{\theta_j - \lambda} = \lambda,$$

where $\xi_j = z_k^j = (Z_k^T e^k)_j$, the θ_j's are the eigenvalues of T_k (we drop the upper index (k) for simplicity), and Z_k is the orthogonal matrix in the spectral decomposition $T_k = Z_k \Theta_k Z_k^T$, which means the ξ_j's are the last components of the eigenvectors of T_k.

The secular equation is in fact nothing other than $\delta_{k+1}(\lambda) = 0$ since it is easy to prove (provided $\delta_k(\lambda) \neq 0$) that

$$(e^k)^T (T_k - \lambda I)^{-1} e^k = \frac{1}{\delta_k(\lambda)}.$$

Therefore, for obtaining the eigenvalues of T_{k+1} from those of T_k we have to solve

$$f(\lambda) = \lambda - \alpha_{k+1} + \eta_{k+1}^2 \sum_{j=1}^{k} \frac{\xi_j^2}{\theta_j - \lambda} = 0.$$

The secular function f has poles at the eigenvalues of T_k for $\lambda = \theta_j = \theta_j^{(k)}$, $j = 1 \dots, k$. We also easily see that it is a strictly increasing function between two consecutive poles; see Figure 1.3 for a small example. There is only one zero in each interval between poles. The problem of obtaining the zeros of f can also be considered as finding the intersections of the rational function

$$g(\lambda) = \eta_{k+1}^2 \sum_{j=1}^{k} \frac{\xi_j^2}{\theta_j - \lambda} = \frac{\eta_{k+1}^2}{\delta_k(\lambda)}$$

and the straight line of equation $\alpha_{k+1} - \lambda$ with slope -1; see Figure 1.4. The shape of the function g in each interval depends on the values of η_{k+1} and the last components of the eigenvectors of T_k. By looking at Figure 1.4 we can understand why the extreme eigenvalues

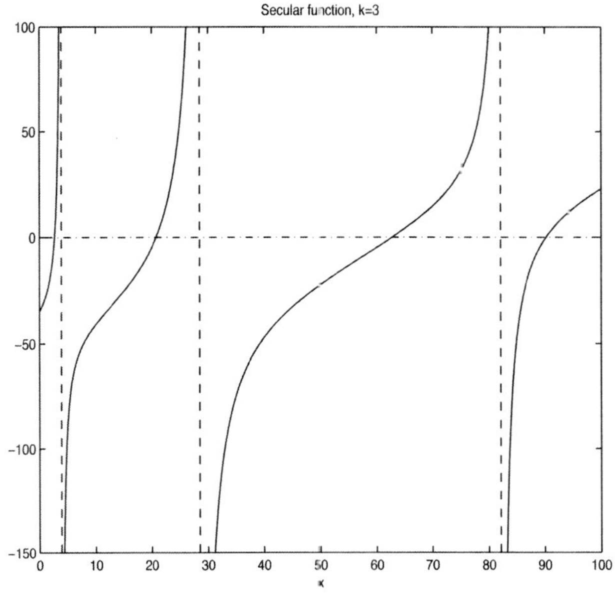

Figure 1.3. *An example of secular function f for the Strakos10 matrix*

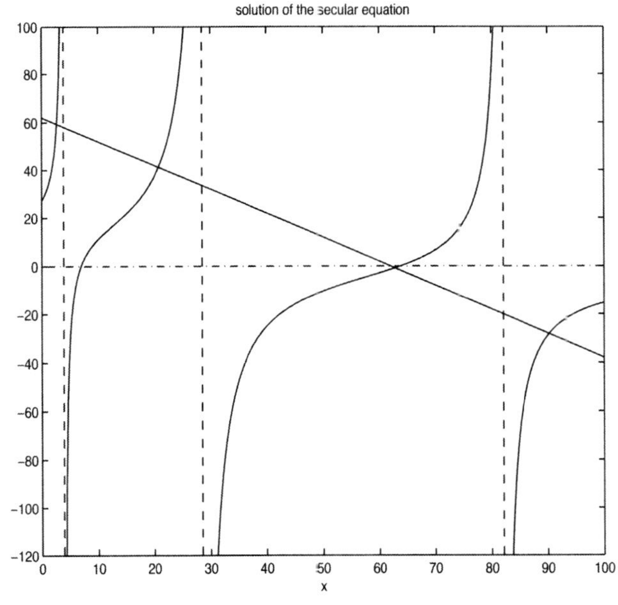

Figure 1.4. *Intersections of $g(\lambda)$ and $\alpha_{k+1} - \lambda$, Strakos10 matrix*

of A are generally the first to be well approximated by Ritz values, especially when they are well separated. On this topic using other methods, see Parlett and Reid [145], and Kuijlaars [107]. Figure 1.4 also illustrates the well-known Cauchy interlacing property.

Lemma 1.7. *The eigenvalues of T_{k+1} strictly interlace the eigenvalues of T_k,*

$$\theta_1^{(k+1)} < \theta_1^{(k)} < \theta_2^{(k+1)} < \theta_2^{(k)} < \cdots < \theta_k^{(k)} < \theta_{k+1}^{(k+1)}.$$

Corollary 1.8.

$$\lambda_j < \theta_j^{(k)}, \quad \theta_{k+1-j}^{(k)} < \lambda_{n+1-j}, \quad 1 \le j \le k.$$

Lemma 1.7 is a consequence of the famous Cauchy theorem, which was proved at the beginning of the 19th century. To understand how the Lanczos algorithm is working it is useful to have more refined interlacing properties that arise for tridiagonal matrices. In [95] Hill and Parlett proved results that were extended by Bar-On [6]. These results use the leading and trailing submatrices of T_k. The set of Ritz values $\theta_j^{(k)}$ is extended by $\theta_0^{(k)} = -\infty$ and $\theta_{k+1}^{(k)} = +\infty$. Bar-On proved the following result for unreduced (that is, with $\eta_j \neq 0,\ j = 2, \ldots$) tridiagonal matrices.

Theorem 1.9. *Let $\vartheta^{(j)}$ denote the extended set of eigenvalues of T_j, $j < k-1$, and let $\vartheta^{(j+2,k)}$ be the extended set of eigenvalues of $T_{j+2,k}$. Let*

$$\vartheta = \{\vartheta_0 = -\infty < \vartheta_1 \le \cdots \le \vartheta_{k-1} < -\infty = \vartheta_k\}$$

be the union of $\vartheta^{(j)}$ and $\vartheta^{(j+2,k)}$. Then there is exactly one eigenvalue of T_k in each open interval $(\vartheta_{i-1}, \vartheta_i)$, $i = 1, \ldots, k$.

Notice that if $\vartheta_{i-1} = \vartheta_i$, then this value is an eigenvalue of T_k. A similar result is also presented in [10]. The result of Hill and Parlett is a consequence of Theorem 1.9.

Corollary 1.10. *Let $j < k$; in each interval $(\theta_{i-1}^{(j)}, \theta_i^{(j)})$, $i = 1, \ldots, j+1$, there is at least one eigenvalue of T_k.*

For instance, in $(\theta_2^{(3)}, \theta_3^{(3)})$ there is one eigenvalue of T_5. By using a permutation of the rows and columns, one can also show the following result.

Corollary 1.11. *Let $j < k$ and m be such that $j + m \le k$; in each interval $(\theta_{i-1}^{(j+m,k)}, \theta_i^{(j+m,k)})$, $i = 1, \ldots, k - j - m + 2$, there is at least one eigenvalue of $T_{j,k}$.*

The paper [6] contains more refined results that can be useful to locate Ritz values.

Theorem 1.12. *Suppose that we have r sequences such as ϑ defined in Theorem 1.9 that are such that there is exactly one eigenvalue of T_k in each interval. Let ν denote their union,*

$$\nu = \{\nu_0 = -\infty < \nu_1 \le \cdots \le \nu_{r(k-1)} < \infty = \nu_{r(k-1)+1}\}.$$

Then in each interval (ν_{ri}, ν_{ri+1}), $i = 0, \ldots, k-1$, there is exactly one eigenvalue of T_k.

So, for instance, using the last theorem, we can construct sequences by choosing pairs of matrices $[T_1, T_{3,k}], [T_2, T_{4,k}], \ldots, [T_{k-2}, T_{k,k}]$. In each interval of the union of their eigenvalues there is exactly one eigenvalue of T_k.

When we study the Lanczos algorithm in finite precision arithmetic, we shall see that the computed matrices T_k may have clusters of very close eigenvalues, even though they are mathematically distinct. In [204] Ye proved that two close eigenvalues of two complementary diagonal blocks of a tridiagonal matrix may give rise to a pair of close eigenvalues even though the coupling entry η_{j+1} is not small. Let us write

$$T_k = \begin{pmatrix} T_j & E \\ E^T & T_{j+1,k} \end{pmatrix},$$

the matrix E having only the bottom left entry which is nonzero and equal to η_{j+1}. We have the following result.

Theorem 1.13. *Let θ_1 be an eigenvalue of T_j with an eigenvector z^j and θ_2 be an eigenvalue of $T_{j+1,k}$ with an eigenvector y^j. Then there are two eigenvalues $\theta_1^{(k)}$ and $\theta_2^{(k)}$ of T_k such that*

$$|\theta_1^{(k)} - \theta_2^{(k)}| \leq |\theta_1 - \theta_2| + \eta_{j+1}\sqrt{2((z_j^j)^2 + (y_1^j)^2)}.$$

So, if θ_1 is close to θ_2, the same is true for $\theta_1^{(k)}$ and $\theta_2^{(k)}$ if z_j^j and y_1^j are small. Ye also proved that having close eigenvalues from complementary diagonal blocks is the only cause of appearance of close eigenvalues for a tridiagonal matrix.

Theorem 1.14. *Let $\theta_1^{(k)}$ and $\theta_2^{(k)}$ be two eigenvalues of T_k. Then there exist j and an eigenvalue θ_1 of T_j and an eigenvalue θ_2 of $T_{j+1,k}$ such that*

$$|\theta_1 - \mu| \leq |\theta_1^{(k)} - \theta_2^{(k)}|/\sqrt{2},$$

$$|\theta_2 - \mu| \leq |\theta_1^{(k)} - \theta_2^{(k)}|/\sqrt{2},$$

where $\mu = (\theta_1^{(k)} + \theta_2^{(k)})/2$.

This theorem shows that if we have two close eigenvalues of T_k, there are eigenvalues which are close to the midvalue for T_j and $T_{j+1,k}$, $j < k$. Therefore, if we have two eigenvalues of T_k which are both close to an eigenvalue of A, there was already an eigenvalue of T_j, $j < k$, and an eigenvalue of $T_{j+1,k}$ which were close to that eigenvalue of A.

Lemma 1.7 shows that the minimum eigenvalues of T_k for $k = 1, 2, \ldots$ are a decreasing sequence which ends with the minimum eigenvalue of T_m which is generally the minimum eigenvalue of A, so things improve at each iteration. Similarly, the maximum eigenvalues of T_k are an increasing sequence towards the maximum eigenvalue of A. This improvement is, so to speak, "automatic" and comes from simple geometrical considerations using $g(\lambda)$, but it can be very small until the last step, as we shall see later on; see the examples in the appendix. For a given i, $\theta_i^{(k)}$ is a decreasing sequence when k increases. Lemma 1.7 also shows that the eigenvalues of T_k are the zeros of the rational function $\delta_k(\lambda)$, since the $\theta_i^{(k)}$ cannot be roots of the $\delta_j(\lambda)$ for $j < k$ because of the interlacing property. The location of the eigenvalues of T_{k+1} depends on the values of α_{k+1}, η_{k+1} and the last components of the eigenvectors of T_k.

1.4 The components of the eigenvectors of T_k

We have seen previously that some components of the eigenvectors of T_k play an important role in the Lanczos algorithm. In this section we give expressions for these components. The result can be expressed in different ways and can be found in many papers; see [135], [141], [144], [143], [160], [45]. However, it seems that one of its earliest appearance is in a paper by Thompson [189]; see also [190]. Actually, the result of Thompson is more general than just for eigenvectors of tridiagonal matrices.

Theorem 1.15. *Let H be a symmetric matrix of order n with distinct eigenvalues λ_i such that*

$$\lambda_1 < \lambda_2 < \cdots < \lambda_n,$$

and let $\xi_{i,j}$, $j = 1, \ldots, n-1$, be the eigenvalues of the matrix obtained from H by deleting row i and column i. The Cauchy interlace theorem gives

$$\lambda_1 \le \xi_{i,1} \le \lambda_2 \le \xi_{i,2} \le \cdots \le \xi_{i,n-1} \le \lambda_n.$$

Let U be the matrix of the eigenvectors of H; then

$$|u_{i,j}|^2 = \frac{\lambda_j - \xi_{i,1}}{\lambda_j - \lambda_1} \cdots \frac{\lambda_j - \xi_{i,j-1}}{\lambda_j - \lambda_{j-1}} \frac{\xi_{i,j} - \lambda_j}{\lambda_{j+1} - \lambda_j} \cdots \frac{\xi_{i,n-1} - \lambda_j}{\lambda_n - \lambda_j}.$$

Considering the eigenvectors of the tridiagonal matrix T_k, this leads to the following corollary.

Corollary 1.16. *The last components of the eigenvectors $z^i = z^i_{(k)}$ of T_k satisfy*

$$(z^i_k)^2 = \left| \frac{\chi_{1,k-1}(\theta_i^{(k)})}{\chi'_{1,k}(\theta_i^{(k)})} \right|,$$

that is,

$$(z^i_k)^2 = \frac{\theta_i^{(k)} - \theta_1^{(k-1)}}{\theta_i^{(k)} - \theta_1^{(k)}} \cdots \frac{\theta_i^{(k)} - \theta_{i-1}^{(k-1)}}{\theta_i^{(k)} - \theta_{i-1}^{(k)}} \frac{\theta_i^{(k-1)} - \theta_i^{(k)}}{\theta_{i+1}^{(k)} - \theta_i^{(k)}} \cdots \frac{\theta_{k-1}^{(k-1)} - \theta_i^{(k)}}{\theta_k^{(k)} - \theta_i^{(k)}}.$$

The first components of the eigenvectors are

$$(z^i_1)^2 = \left| \frac{\chi_{2,k}(\theta_i^{(k)})}{\chi'_{1,k}(\theta_i^{(k)})} \right|,$$

that is,

$$(z^i_1)^2 = \frac{\theta_i^{(k)} - \xi_1}{\theta_i^{(k)} - \theta_1^{(k)}} \cdots \frac{\theta_i^{(k)} - \xi_{i-1}}{\theta_i^{(k)} - \theta_{i-1}^{(k)}} \frac{\xi_i - \theta_i^{(k)}}{\theta_{i+1}^{(k)} - \theta_i^{(k)}} \cdots \frac{\xi_{k-1} - \theta_i^{(k)}}{\theta_k^{(k)} - \theta_i^{(k)}},$$

where the $\xi_j = \theta_j^{(2,k)}$ are the eigenvalues of $T_{2,k}$. Moreover,

$$|z^i_1 z^i_k| = \left| \frac{\eta_2 \cdots \eta_k}{\chi'_{1,k}(\theta_i^{(k)})} \right|,$$

and more generally for $1 < j < k$,

$$(z_j^i)^2 = \left| \frac{\chi_{1,j-1}(\theta_i^{(k)})\chi_{j+1,k}(\theta_i^{(k)})}{\chi'_{1,k}(\theta_i^{(k)})} \right|.$$

The last equality of the corollary comes from a result proved by Paige [132] by computing determinants of the minors of the tridiagonal matrix saying that the element (r, t), $r \le t$ of the adjugate (the transpose of the matrix of cofactors) is

$$\chi_{1,r-1}\eta_{r+1}\cdots\eta_t\chi_{t+1,k}.$$

From the previous corollary we see that if the ith eigenvalue converges (that is, there is one eigenvalue $\theta_i^{(k-1)}$ of T_{k-1} and one eigenvalue $\theta_j^{(k)}$ of T_k whose difference is small), then the last element z_k^i of the corresponding eigenvector must be small. Because of the interlacing property the index j is restricted to the neighbors of i. We shall see more on this point later. The elements of the eigenvectors can also be characterized using the first and last pivot functions. To our knowledge, these results appear to be new.

Theorem 1.17. *The first components of the eigenvectors of* T_k *are given by*

$$(z_1^j)^2 = \left| \frac{1}{[d_1^{(k)}]'(\theta_j^{(k)})} \right|.$$

For the last components we have

$$(z_k^j)^2 = \left| \frac{1}{\delta'_k(\theta_j^{(k)})} \right|.$$

Proof. We have

$$(z_1^j)^2 = \left| \frac{\chi_{2,k}(\theta_j^{(k)})}{\chi'_{1,k}(\theta_j^{(k)})} \right|.$$

But $\chi_{2,k}(\theta_j^{(k)}) = d_2^{(k)}(\theta_j^{(k)})\cdots d_k^{(k)}(\theta_j^{(k)})$ and

$$\chi'_k(\lambda) = [d_1^{(k)}(\lambda)\cdots d_k^{(k)}(\lambda)]' = [d_1^{(k)}]'(\lambda)d_2^{(k)}(\lambda)\cdots d_k^{(k)}(\lambda) + d_1^{(k)}(\lambda)[d_2^{(k)}(\lambda)\cdots d_k^{(k)}(\lambda)]'.$$

Hence,

$$\chi'_k(\theta_j^{(k)}) = [d_1^{(k)}]'(\theta_j^{(k)})d_2^{(k)}(\theta_j^{(k)})\cdots d_k^{(k)}(\theta_j^{(k)}).$$

After simplification, we obtain the result. The proof is more or less the same for the last components since

$$(z_k^j)^2 = \left| \frac{\chi_{k-1}(\theta_j^{(k)})}{\chi'_{1,k}(\theta_j^{(k)})} \right|,$$

and $\chi_{k-1}(\lambda) = \delta_1(\lambda)\cdots\delta_{k-1}(\lambda)$. □

The absolute values are necessary because we shall see later that the derivatives are negative (or we should put a minus sign). The other elements of the eigenvectors can be handled in the same way by using a twisted factorization (see [117]), starting from both the top and the bottom of the matrix.

Theorem 1.18. *The components of the eigenvectors of T_k are given by*

$$(z_i^j)^2 = \left| \frac{1}{[\omega_i^{(k)}]'(\theta_j^{(k)})} \right|, \quad 1 < i < k,$$

where

$$\omega_i^{(k)}(\lambda) = \alpha_i - \lambda - \frac{\eta_i^2}{\delta_{i-1}(\lambda)} - \frac{\eta_{i+1}^2}{d_{i+1}^{(k)}(\lambda)}.$$

Proof. We have

$$(z_i^j)^2 = \left| \frac{\chi_{i-1}(\theta_j^{(k)})\chi_{i+1,k}(\theta_j^{(k)})}{\chi_k'(\theta_j^{(k)})} \right|.$$

We already know that

$$\chi_{i-1}(\theta_j^{(k)}) = \delta_1(\theta_j^{(k)}) \cdots \delta_{i-1}(\theta_j^{(k)}),$$

$$\chi_{i+1,k}(\theta_j^{(k)}) = d_{i+1}^{(k)}(\theta_j^{(k)}) \cdots d_k^{(k)}(\theta_j^{(k)}).$$

For the derivative of the characteristic polynomial, we use a twisted factorization of T_k (see [117]); that is, T_k is the product of two matrices, the first one being lower triangular up to row i and then upper triangular from row i to the end and the second matrix having a structure which is the transpose of the previous one. Every twisted factorization can be obtained by using (parts of) the LU and UL factorizations introduced before. We have

$$\chi_k(\lambda) = \delta_1(\lambda) \cdots \delta_{i-1}(\lambda)\omega_i(\lambda)d_{i+1}^{(k)}(\lambda) \cdots d_k^{(k)}(\lambda).$$

We differentiate this relation and remark that $\omega_i(\theta_j^{(k)}) = 0$ since by the interlacing properties none of the other factors can be zero. Therefore,

$$\chi_k'(\theta_j^{(k)}) = \delta_1(\theta_j^{(k)}) \cdots \delta_{i-1}(\theta_j^{(k)})\omega_i'(\theta_j^{(k)})d_{i+1}^{(k)}(\theta_j^{(k)}) \cdots d_k^{(k)}(\theta_j^{(k)}),$$

which proves the result. □

In conclusion, we can obtain all the elements of an eigenvector of T_k corresponding to $\theta_j^{(k)}$ by knowing the derivatives of the first and last pivot functions at $\theta_j^{(k)}$.

1.5 Study of the pivot functions

We have already seen that the first pivot function is involved when we go from step k to step $k+1$ of the Lanczos algorithm. We shall see later that it appears also in other quantities linked to the Lanczos algorithm behavior. The last pivot function and its derivative are involved in expressions for the A-norm of the error in CG. In this section we shall study both functions and their derivatives.

The last pivot function δ_k, $k > 1$, has zeros at $\theta_i^{(k)}$, $i = 1, \ldots, k$, and poles at $\theta_i^{(k-1)}$, $i = 1, \ldots, k-1$. Its derivative is recursively defined by

$$\delta_k'(\lambda) = -1 + \eta_k^2 \frac{\delta_{k-1}'(\lambda)}{\delta_{k-1}^2(\lambda)}, \quad \delta_1'(\lambda) \equiv -1.$$

It may seem strange to look at the derivative of a function which has poles, but we can look at the limits of the derivative when the variable λ tends to the poles by positive and negative values. By induction, we see that $\delta_k'(\lambda) < 0$ between the roots of δ_{k-1} which are $\theta_j^{(k-1)}$. Moreover when $\lambda \to \theta_j^{(k-1)}$, $\delta_k' \to -\infty$. What we are really interested in, as we have seen, is $1/|\delta_k'(\lambda)|$. This is a smooth function whose zeros are the $\theta_j^{(k-1)}$ because the absolute value of the inverse of the derivative tends to zero when the variable tends to the pole either from the left or from the right. Hence, this function is continuous. It goes to 1 when $\lambda \to \pm\infty$. Between two zeros it is increasing and then decreasing and it is bounded by 1 since the absolute value of $\delta_k'(\lambda)$ is strictly larger than 1. This is illustrated in Figures 1.5 and 1.6 for the Strakoš matrix of dimension 10. The stars on the axis are the $\theta_i^{(k)}$ and the dashed vertical lines are the $\theta_i^{(k-1)}$. The interesting behavior is that if $\theta_j^{(k)}$ is close to $\theta_j^{(k-1)}$, then $1/|\delta_k'(\theta_j^{(k)})|$ is small, as seen in Figure 1.6 for the first and last stars. Therefore, when an eigenvalue "converges," then $z_k^j \to 0$. Moreover, $\alpha_{k+1} - \lambda$ is the asymptote of $\delta_k(\lambda)$ when $\lambda \to \pm\infty$.

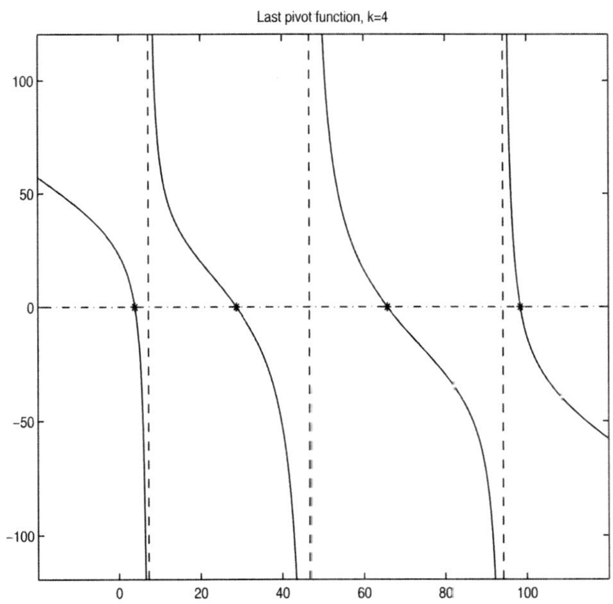

Figure 1.5. *Last pivot function δ_4, Strakos10 matrix*

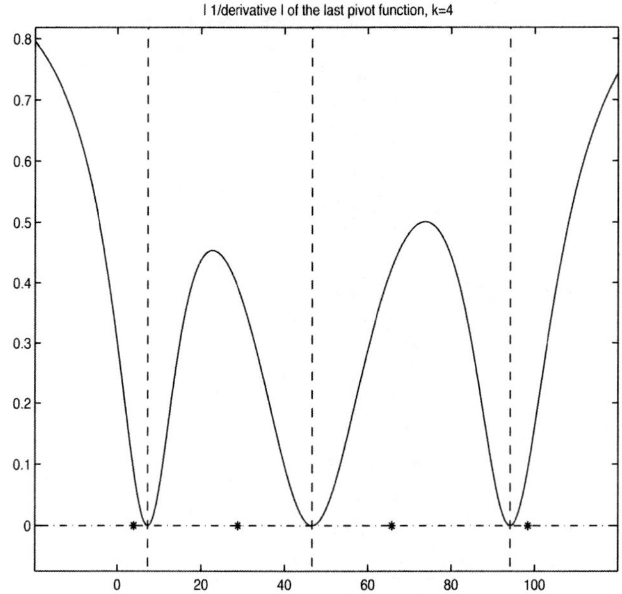

Figure 1.6. $1/|$ *derivative of last pivot function* δ_4 $|$, *Strakos*10 *matrix*

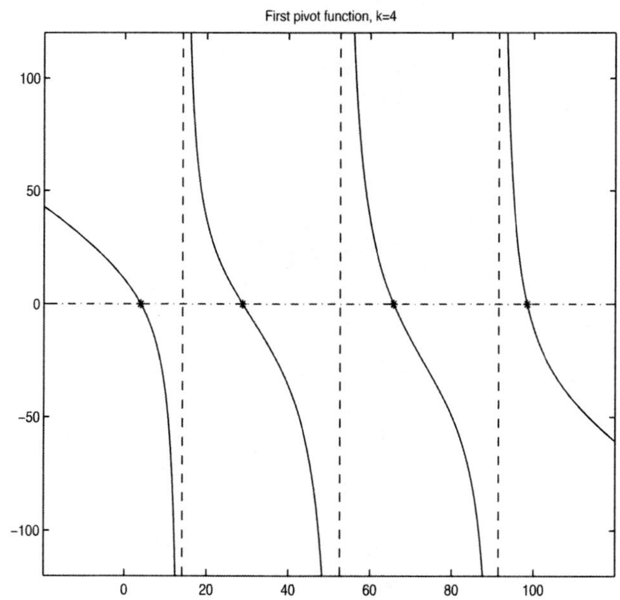

Figure 1.7. *First pivot function* $d_1^{(4)}$

As we said it is also interesting for CG convergence to study the function $1/|[d_1^{(k)}]'(\lambda)|$. Things are almost the same as what we have done previously for δ_k. The zeros of $d_1^{(k)}$ are the eigenvalues of T_k and the poles are the eigenvalues of $T_{2,k}$. Between the poles $d_1^{(k)}$ is a strictly decreasing function. The derivative is always negative and tends to $-\infty$ when λ tends to the poles. When taking the reciprocal and the absolute value, we obtain a function which is continuously differentiable and smaller than 1. At the poles of $d_1^{(k)}$ the reciprocal of the derivative is zero. This is illustrated in Figures 1.7 and 1.8 for the Strakoš matrix of dimension 10. The stars are the $\theta_i^{(k)}$ and the vertical lines are the eigenvalues of $T_{2,k}$. For CG convergence we shall see that the function which has to be considered is really $1/(\lambda|[d_1^{(k)}]'(\lambda)|)$, which has only one pole at 0.

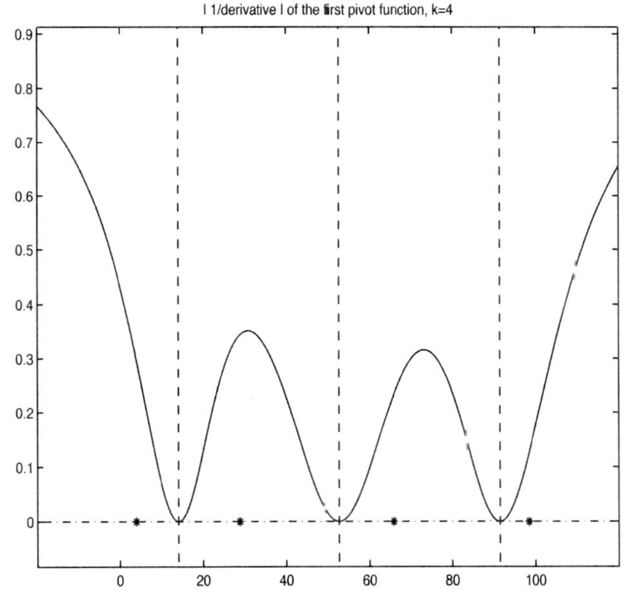

Figure 1.8. $1/|$ *derivative of first pivot function* $d_1^{(4)}$ $|$

1.6 Bounds for the approximation of eigenvalues

In this section we shall give some results about eigenvalue "convergence" in exact arithmetic. Strictly speaking this is not really convergence because one more Ritz value appears at each Lanczos iteration and, in exact arithmetic, we obtain all the eigenvalues of A at the last iteration m, $m \le n$. But, for instance, we can monitor the moves of the smallest and largest Ritz values and therefore speak about convergence until we reach the last step. In fact, most of the results give bounds on the distances of Ritz values to eigenvalues of A. There are two types of results: some of them use quantities which are computable during the algorithm and others can be called "a priori" bounds since they just rely on eigenvalues of A and the initial vector. Most of these latter results are based on general theory about Rayleigh–Ritz approximations. Good expositions of the theory are given by Stewart [179] and Parlett [141].

We have seen that for the Lanczos algorithm, at least for the smallest and largest eigenvalues, there is an improvement at each iteration. However, we are going to see that the initial vector is really important and there is no hope to obtain a theorem saying that the eigenvalues (at least the smallest and/or the largest ones) converge nicely and fast to the eigenvalues of A whatever the initial vector. In fact, there is a result by Scott [160] showing that we can choose the initial vector of the Lanczos recurrence to delay the convergence as much as we wish up to the last step. In choosing this initial vector there is no "convergence" during $n-1$ steps and all the eigenvalues are obtained at the last step. To see this, we need to show that for constructing this vector, up to a spectral transformation, we can choose A to be diagonal.

Proposition 1.19. *Let $A = Q\Lambda Q^T$ be the spectral decomposition of A. Then the Lanczos algorithm run with A starting from v^1 generates the same matrix T_k as when run with Λ starting from $Q^T v^1$.*

Proof. The Lanczos recurrence gives

$$Q\Lambda Q^T V_k = V_k T_k + \eta_{k+1} v^{k+1} (e^k)^T.$$

Multiplying on the left by Q^T and noticing that $Q^T Q = I$,

$$\Lambda Q^T V_k = Q^T V_k T_k + \eta_{k+1} Q^T v^{k+1} (e^k)^T.$$

Let $\bar{V}_k = Q^T V_k$. Then we obtain that

$$\Lambda \bar{V}_k = \bar{V}_k T_k + \eta_{k+1} \bar{v}^{k+1} (e^k)^T,$$

$$\alpha_k = (Av^k, v^k) = (\Lambda Q^T v^k, Q^T v^k) = (\Lambda \bar{v}^k, \bar{v}^k),$$

and

$$\eta_k = (\Lambda \bar{v}^k, \bar{v}^{k-1}). \quad \square$$

We remark that $\bar{v}^k = Q^T v^k$ is the vector of the projections of v^k on the eigenvectors of A, the ith entry being (q^i, v^k).

Proposition 1.20.

$$\alpha_k = (\Lambda \bar{v}^k, \bar{v}^k) = \sum_{i=1}^{n} \lambda_i (\bar{v}_i^k)^2$$

and since $\sum_{i=1}^{n} (\bar{v}_i^k)^2 = 1$, α_k is a convex combination of the eigenvalues of A for all k. We also have that

$$\eta_k = \sum_{i=1}^{n} \lambda_i \bar{v}_i^k \bar{v}_i^{k-1},$$

with $\sum_{i=1}^{n} \bar{v}_i^k \bar{v}_i^{k-1} = 0$ *because of the local orthogonality of the Lanczos vectors. This can also be written using the Lanczos polynomials as*

$$\alpha_k = \sum_{i=1}^{n} \lambda_i p_k^2(\lambda_i)(\bar{v}_i^1)^2 = \int_{\lambda_1}^{\lambda_n} \lambda p_k^2(\lambda)\, d\alpha(\lambda),$$

$$\eta_k = \sum_{i=1}^{n} \lambda_i p_k(\lambda_i) p_{k-1}(\lambda_i)(\bar{v}_i^1)^2.$$

The components of the vectors $\bar{v}^k$ are given by a scalar three-term recurrence

$$\eta_{k+1}\bar{v}_i^{k+1} = \lambda_i \bar{v}_i^k - \alpha_k \bar{v}_i^k - \eta_k \bar{v}_i^{k-1}.$$

The components of $\bar{v}_i^{k+1}$ are coupled (nonlinearly) only through the coefficients α_k and η_k. At the end of the algorithm, supposing that the eigenvalues of A are distinct, we have

$$\Lambda \bar{V}_n = \bar{V}_n T_n.$$

Let $Z = Z_n$ be the matrix of normalized eigenvectors of T_n which has the same eigenvalues as A; then $T_n = Z \Lambda Z^T$. By comparison we have

$$\bar{V}_n = Z^T,$$

where $\bar{V}_n = Q^T V_n = (\, \bar{v}^1 \quad \cdots \quad \bar{v}^n \,)$. This shows that the components of the initial vector $\bar{v}^1$ are (eventually up to a normalization constant) the first components of the eigenvectors of T_n. This leads to a result of Scott [160] showing that if we know the eigenvalues of A, we can choose the initial vector in such a way that we could obtain any prescribed set of eigenvalues for T_{n-1}, provided they satisfy the strict interlacing property with the eigenvalues of A (or Λ in our diagonal case), see also [161].

Theorem 1.21. *Let* $\{\mu_i\}$ *a set of numbers satisfying*

$$\lambda_1 < \mu_1 < \lambda_2 < \mu_2 < \cdots < \lambda_{n-1} < \mu_{n-1} < \lambda_n.$$

Then we can choose $\bar{v}^1$ *to obtain* $\theta_j^{(n-1)} = \mu_j$ *as eigenvalues of* T_{n-1}.

Proof. Let $c_n = \eta_2 \cdots \eta_n$. We have seen that

$$(z_1^i)^2 = \frac{c_n^2}{\chi_{1,n-1}(\lambda_i)\chi'_{1,n}(\lambda_i)}.$$

This can be written as

$$(z_1^i)^2 = \frac{c_n^2}{|\prod_{\substack{j=1 \\ j\neq i}}^{n}(\lambda_i - \lambda_j)\prod_{j=1}^{n}(\lambda_i - \mu_j)|}.$$

Therefore $(z_1^i)^2$ can be computed when the μ_j's are given. This directly gives the components of the first Lanczos vector in the diagonal case, the constant c_n which is unknown at this point, being determined by requiring that the norm of the vector $\bar{v}^1$ must be equal to 1. $\square$

Knowing the eigenvalues λ_i, we can impose any set of eigenvalues $\theta_i^{(n-1)}$ by choosing appropriately the starting vector. An example is given in the appendix. In the general case for nondiagonal A, the starting vector is $Q\bar{v}^1$. The case of multiple eigenvalues is discussed in [160]. This last theorem is closely related to the work of de Boor and Golub [14] about reconstruction of tridiagonal matrices from spectral data; see also Gragg and Harrod [72].

We are now going to review some results on "convergence" of the Ritz values. We shall look for relations between the Ritz values and the eigenvalues of A. For a different approach involving polynomials, see van der Sluis and van der Vorst [197], who did a careful study of convergence when A has some close eigenvalues. For other results, see Ipsen [97]. The first result relates the eigenvalues and the Ritz values.

Theorem 1.22. *Let $A = Q\Lambda Q^T$ and $T_k = Z_k\Theta_k Z_k^T$ be the spectral decompositions of A and T_k and $W_k = Q^T V_k Z_k = \bar{V}_k Z_k$ of dimension $n \times k$. Then,*

$$\Theta_k = W_k^T \Lambda W_k.$$

Therefore,

$$\theta_i^{(k)} = \sum_{j=1}^{n} \lambda_j (w_j^i)^2, \quad \sum_{j=1}^{n} (w_j^i)^2 = 1,$$

where w^i is the ith column of W_k.

Proof. Since $\bar{V}_k^T \Lambda \bar{V}_k = T_k$, the result is straightforward by using the spectral decompositions of A and T_k. □

We remark that $W_n = \bar{V}_n Z_n = Z_n^T Z_n = I$. The eigenvalues $\theta_i^{(k)}$ of T_k are a convex combination of the eigenvalues of A and the elements of W_k govern the convergence of the Ritz values. But, notice that the entries of w^i depend on $\theta_j^{(k)}$, $j \neq i$. These vectors w^i are the projections of the Ritz vectors on the eigenvectors of A. The matrix W_k changes completely every iteration because Z_k changes. Returning to the residual equation for an approximate eigenpair we have

$$r_{(k)}^j = Q\Lambda Q^T V_k z^j - \theta_j^{(k)} V_k z^j.$$

Then,

$$Q^T r_{(k)}^j = (\Lambda - \theta_j^{(k)} I) w^j.$$

Taking norms we have the following theorem.

Theorem 1.23.

$$(\lambda_i - \theta_j^{(k)})(w_i^j)^2 \leq \sum_{l=1}^{n} (\lambda_l - \theta_j^{(k)})^2 (w_l^j)^2 = \|r_{(k)}^j\|^2 = \eta_{k+1}^2 (z_k^j)^2 \quad \forall i, j,$$

and $\exists\, i$ such that

$$|\lambda_i - \theta_j^{(k)}| \leq \eta_{k+1} |z_k^j|.$$

The second relation is a well-known result; see Parlett [141]. It is obtained by taking the minimum of $|\lambda_l - \theta_j^{(k)}|$ in the sum. We notice that the right-hand side of the inequality $\eta_{k+1}|z_k^j|$ is computable during the Lanczos iterations and can be used to monitor Ritz value convergence. We can write the relation of Theorem 1.22 in another way to exhibit the differences of Ritz values and eigenvalues.

Theorem 1.24.

$$(\lambda_i - \theta_j^{(k)})w_i^j = \eta_{k+1} z_k^j \bar{v}_i^{k+1} \quad \forall i, j.$$

Proof. We have

$$Q\Lambda Q^T V_k = V_k Z_k \Theta_k Z_k^T + \eta_{k+1} v^{k+1} (e^k)^T,$$

which we can write as

$$\Lambda W_k = W_k \Theta_k + \eta_{k+1} \bar{v}^{k+1} (e^k)^T Z_k.$$

Writing the ith element of the jth column we obtain the result. □

Noticing that $\|w^i\| = 1$ and bounding $|w_i^j|$, we obtain the following corollary, which gives a lower bound for the distances between Ritz values and eigenvalues.

Corollary 1.25.

$$|\lambda_i - \theta_j^{(k)}| \geq \eta_{k+1} |z_k^j \bar{v}_i^{k+1}| \quad \forall i, j.$$

Unfortunately, this result is not computationally useful because, as we shall see, the term $\bar{v}_i^{k+1}$ contains the factor $\lambda_i - \theta_j^{(k)}$ and moreover the right-hand side is not available during the algorithm. We can write the result of Theorem 1.24 as follows.

Theorem 1.26.

$$\lambda_i - \theta_j^{(k)} = \eta_{k+1} \frac{z_k^j \bar{v}_i^{k+1}}{\sum_{l=1}^k z_l^j \bar{v}_i^l}.$$

Proof. We have

$$w_i^j = (W_k)_{i,j} = (\bar{V}_k Z_k)_{i,j} = \sum_{l=1}^{k} \bar{v}_i^l z_l^j. \quad \square$$

It is interesting to notice that this last result is nothing other than the Christoffel–Darboux relation for the polynomials p_k (see Theorem 1.4) written for the points λ_i and $\theta_j^{(k)}$. It shows that, provided the denominator is not too small, the "convergence" of the Ritz values depend on $z_k^j \bar{v}_i^{k+1}$, both factors eventually converging to 0, because $\bar{v}_i^{k+1} = p_{k+1}(\lambda_i)\bar{v}_i^1$ and z_k^j is proportional to $p_k(\theta_j^{(k)})$. We also see that if a value z_k^j is small because the Ritz value $\theta_j^{(k)}$ is converging to an eigenvalue of A, say λ_m, the term $w_i^j = \sum_{l=1}^k z_l^j \bar{v}_i^l$ must be small for the other eigenvalues λ_i, $i \neq m$, which are not close to $\theta_j^{(k)}$ to compensate for the small numerator (because of z_k^j) and w_m^j must be $O(1)$. So, the vectors of the ith components of the (projections of the) Lanczos vectors from iterations 1 to k

$$(\bar{v}_i^1 \quad \cdots \quad \bar{v}_i^k)^T$$

for $i \neq m$ must be almost orthogonal to the jth eigenvector of T_k. There is a strong relationship between the projections of the Lanczos vectors and the eigenvectors of T_k. Supposing that the eigenvalues of A are distinct, at the last iteration the matrix W_n is the identity matrix.

At this point, it is of interest to look for the solution of the scalar recurrence equation

$$\eta_{k+1}\bar{v}_i^{k+1} = \lambda_i \bar{v}_i^k - \alpha_k \bar{v}_i^k - \eta_k \bar{v}_i^{k-1}.$$

General solutions of second order recurrence relations of this type are known (see, for instance, [114]), but this is not so useful here since they are complicated expressions that depend on the coefficients of the recurrence. We shall return to this point when we study the algorithm in finite precision arithmetic. However, in exact arithmetic the solution of the recurrence is known a posteriori since we have seen that $\bar{V}_n = Z_n^T$. Another way to write this is

$$\bar{v}_i^{k+1} = p_{k+1}(\lambda_i)\bar{v}_i^1.$$

Hence, $\bar{v}_i^{k+1}$ contains the factor $\lambda_i - \theta_j^{(k)}$ since $\theta_j^{(k)}$ is a root of p_{k+1}. This shows that if $\theta_j^{(k)}$ converges to λ_i, the corresponding component $\bar{v}_i^k$ goes to zero. This is also implied by the following result, which involves the last pivot function.

Proposition 1.27.

$$\frac{\bar{v}_i^{k+1}}{\bar{v}_i^k} = -\frac{\delta_k(\lambda_i)}{\eta_{k+1}}.$$

Proof. Let $y_{k+1} = \frac{\bar{v}_i^{k+1}}{\bar{v}_i^k}$. Then using the three-term recurrence of the Lanczos vectors,

$$\eta_{k+1} y_{k+1} = \lambda_i - \alpha_k - \frac{\eta_k}{y_k}.$$

We have

$$y_2 = \frac{\lambda_i - \alpha_1}{\eta_2} = -\frac{\delta_1(\lambda_i)}{\eta_2}.$$

By induction, we obtain

$$\eta_{k+1} y_{k+1} = \lambda_i - \alpha_k + \frac{\eta_k^2}{\delta_{k-1}(\lambda_i)},$$

which proves the result by using the equation defining $\delta_k(\lambda_i)$. □

Another way to look at the increase or decrease of the components of $\bar{v}^k$ is to write

$$\frac{\bar{v}_i^{k+1}}{\bar{v}_i^k} = \frac{p_{1,k+1}(\lambda_i)}{p_{1,k}(\lambda_i)} = \frac{\prod_{j=1}^k (\lambda_i - \theta_j^{(k)})}{\eta_{k+1}\prod_{j=1}^{k-1}(\lambda_i - \theta_j^{(k-1)})}.$$

So, the fact that the components of $\bar{v}^k$ are growing or decreasing depends on the distances of the Ritz values to the eigenvalues of A. But these distances are constrained by the interlacing property of the Ritz values. This is most noticeable for the largest eigenvalue λ_n. Let

$$\varpi_j^{(k)} = \frac{\lambda_n - \theta_j^{(k)}}{\lambda_n - \theta_j^{(k-1)}} > 1, \quad j = 1, \dots, k-1,$$

and

$$\varpi^{(k)} = \prod_{j=1}^{k-1} \varpi_j^{(k)} > 1.$$

But

$$\frac{\bar{v}_n^{k+1}}{\bar{v}_n^k} = \varpi^{(k)} \frac{\lambda_n - \theta_k^{(k)}}{\eta_{k+1}}.$$

As long as $\theta_k^{(k)}$ is far enough from λ_n (that is, $\lambda_n - \theta_k^{(k)} > \eta_{k+1}$) the absolute value of the nth component of the projection of the Lanczos vector is growing. Of course, since its absolute value has to be smaller than 1 and $\lambda_n - \theta_k^{(k)}$ is decreasing as a function of k, at some point the absolute value of the last component will start to decrease since η_{k+1} is bounded. Remember that

$$|z_k^k \bar{v}_n^{k+1}| \le \frac{\lambda_n - \theta_k^{(k)}}{\eta_{k+1}} \le |z_k^k|.$$

This result can be generalized to the other eigenvalues λ_i, $i \ne n$.

Theorem 1.28. *Let i be given and j be such that*

$$\theta_{j-1}^{(k-1)} < \lambda_i < \theta_j^{(k-1)}.$$

If

$$|\lambda_i - \theta_j^{(k)}| > \eta_{k+1},$$

the ith component $|\bar{v}_i^k|$ is increasing as a function of k.

Proof. We have

$$\frac{|\bar{v}_i^{k+1}|}{|\bar{v}_i^k|} = \frac{|\lambda_i - \theta_j^{(k)}|}{\eta_{k+1}} \frac{|\lambda_i - \theta_1^{(k)}|}{|\lambda_i - \theta_1^{(k-1)}|} \cdots \frac{|\lambda_i - \theta_{j-1}^{(k)}|}{|\lambda_i - \theta_{j-1}^{(k-1)}|} \frac{|\lambda_i - \theta_{j+1}^{(k)}|}{|\lambda_i - \theta_j^{(k-1)}|} \cdots \frac{|\lambda_i - \theta_k^{(k)}|}{|\lambda_i - \theta_{k-1}^{(k-1)}|}.$$

By the interlacing property all the ratio of distances of Ritz values to λ_i in the right-hand side are larger than one. Therefore, if

$$|\lambda_i - \theta_j^{(k)} > \eta_{k+1},$$

the ith component is growing as a function of k. If the distance is small enough, the component may start decreasing. □

A way to obtain an upper bound for the distance of Ritz values to eigenvalues of A is to bound $|\bar{v}_i^{k+1}|$ by 1 since $\bar{v}^{k+1}$ has norm 1.

Theorem 1.29. *If* $|w_{i,j}| \neq 0$,

$$|\lambda_i - \theta_j^{(k)}| \leq \frac{|z_k^j|}{|w_{i,j}|}\eta_{k+1} \quad \forall i, j.$$

This may seem weaker than the result of Theorem 1.23 since $|w_{i,j}| \leq 1$, but here we have a comparison of $\theta_j^{(k)}$ with all the eigenvalues of A when Theorem 1.23 says that there exists an eigenvalue of A such that the given bound is satisfied. We can also use the Ritz polynomials.

Theorem 1.30.

$$\theta_i^{(k)} = \sum_{l=1}^{n} \lambda_l [q_i^{(k)}(\lambda_l)]^2 (\bar{v}_l^1)^2 = \int_{\lambda_1}^{\lambda_n} \lambda [q_i^{(k)}(\lambda)]^2 \, d\alpha(\lambda),$$

$$\sum_{l=1}^{n} \lambda_l q_i^{(k)}(\lambda_l) q_j^{(k)}(\lambda_l)(\bar{v}_l^1)^2 = \int_{\lambda_1}^{\lambda_n} \lambda q_i^{(k)}(\lambda) q_j^{(k)}(\lambda) \, d\alpha(\lambda) = 0, \quad i \neq j.$$

Proof. From the definition of W_k, $w^i = Q^T V_k z^i = Q^T x_{(k)}^i = Q^T q_i^{(k)}(A) v^1$, $q_i^{(k)}$ being the Ritz polynomial. From this, it is easy to see that $w^i = q_i^{(k)}(\Lambda)\bar{v}^1$. Writing the entries of $W_k^T \Lambda W_k$ we obtain the characterization of the Ritz values. □

From the previous theorem, it is of interest to have an expression for $[q_j^{(k)}(\lambda_l)]^2$. This is given by

$$[q_j^{(k)}(\lambda)]^2 = \left| \frac{1}{(\eta_2 \cdots \eta_k)^2} \frac{\prod_{i=1}^{k-1}(\theta_j^{(k)} - \theta_i^{(k-1)})}{\prod_{\substack{i=1 \\ i \neq j}}^{k}(\theta_j^{(k)} - \theta_i^{(k)})} \prod_{\substack{i=1 \\ i \neq j}}^{k} (\lambda - \theta_i^{(k)})^2 \right|.$$

We remark that it would be interesting to have an expression for the Ritz polynomial involving only the eigenvalues of A. From Theorem 1.30 we have the following results for the distances between Ritz values and eigenvalues of A.

Proposition 1.31.

$$\theta_i^{(k)} - \lambda_j = \sum_{\substack{l=1 \\ l \neq j}}^{n} (\lambda_l - \lambda_j) q_i^{(k)}(\lambda_l)^2 (\bar{v}_l^1)^2 \quad \forall j.$$

We note that

$$\theta_1^{(k)} - \lambda_1 = \sum_{l=2}^{n} (\lambda_l - \lambda_1) q_1^{(k)}(\lambda_l)^2 (\bar{v}_l^1)^2$$

and

$$\lambda_n - \theta_k^{(k)} = \sum_{l=1}^{n-1} (\lambda_n - \lambda_l) q_k^{(k)}(\lambda_l)^2 (\bar{v}_l^1)^2.$$

In these last two sums, the differences of the eigenvalues of A are all positive.

Corollary 1.32.

$$|\theta_j^{(k)} - \lambda_l| \leq \max(\lambda_n - \lambda_l, \lambda_l - \lambda_1) \max_{i \neq l} [q_j^{(k)}(\lambda_i)]^2 \quad \forall l.$$

In particular

$$\theta_1^{(k)} - \lambda_1 \leq (\lambda_n - \lambda_1) \max_{i \neq 1} [q_1^{(k)}(\lambda_i)]^2$$

and

$$\lambda_n - \theta_k^{(k)} \leq (\lambda_n - \lambda_1) \max_{i \neq n} [q_k^{(k)}(\lambda_i)]^2.$$

However, these upper bounds are usually quite pessimistic because of the (large) bound we used for the difference of the eigenvalues of A. Moreover, since the eigenvalues of A are involved, the bounds are not computable during the algorithm. Paige [135] proved important results about the difference of Ritz values at different iterations.

Theorem 1.33. *Let $l < k$. Then,*

$$\min_j |\theta_j^{(k)} - \theta_i^{(l)}| \leq \eta_{l+1} |(z_{(l)}^i)_l|.$$

Proof. Let $z^i = z_{(l)}^i$ be an eigenvector of T_l corresponding to the eigenvalue $\theta_i^{(l)}$. Then we apply T_k to the vector w of dimension k consisting of z^i completed with zeros. Hence,

$$T_k w = \begin{pmatrix} T_l z^i \\ \eta_{l+1} z_l^i \\ 0 \end{pmatrix} = \begin{pmatrix} \theta_i^{(l)} z^i \\ \eta_{l+1} z_l^i \\ 0 \end{pmatrix}.$$

Therefore,

$$(T_k - \theta_i^{(l)}) w = \begin{pmatrix} 0 \\ \eta_{l+1} z_l^i \\ 0 \end{pmatrix}.$$

We apply a well-known result proved in [200, p. 171] (easily proved by using the spectral decomposition of T_k) saying that there is an eigenvalue $\theta_j^{(k)}$ of T_k such that its distance to $\theta_i^{(l)}$ is bounded by the norm of the residual. This gives the result. □

Theorem 1.33, which is sometimes called the persistence theorem, has important consequences. It implies that for any $k > l$ there is an eigenvalue $\theta_j^{(k)}$ of T_k within $\delta = \eta_{l+1} |(z_{(l)}^i)_l|$. It is said that $\theta_i^{(l)}$ is stabilized within δ. In other words, when an eigenvalue of A is approximated by a Ritz value at step l, it is approximated by a Ritz value with at least the same accuracy at all subsequent steps. Another result of Paige [135] is the following.

Theorem 1.34. *Using the same notations as in Theorem* 1.33, *we have*

$$(\theta_j^{(k)} - \theta_i^{(l)}) (z_{(k)}^j)^T \begin{pmatrix} z_{(l)}^i \\ 0 \end{pmatrix} = \eta_{l+1} (z_{(l)}^i)_l (z_{(k)}^j)_{l+1}.$$

Proof. We multiply from the left the equation for $T_k w$ by $(z_{(k)}^j)^T$. □

This theorem is interesting to compare Ritz values on successive steps of the Lanczos algorithm; that is, we take $l = k - 1$. Because of the interlacing property it is enough to consider $i = j$ or $i = j - 1$. Using the index l in a different meaning as in the last theorem we have

$$(\theta_j^{(k)} - \theta_i^{(k-1)}) \sum_{l=1}^{k-1} (z_{(k)}^j)_l (z_{(k-1)}^i)_l = \eta_k (z_{(k)}^j)_k (z_{(k-1)}^i)_{k-1}.$$

In particular this leads to

$$\eta_k |(z_{(k)}^j)_k (z_{(k-1)}^i)_{k-1}| \le |\theta_j^{(k)} - \theta_i^{(k-1)}|$$

for $i = j$ or $i = j - 1$. This shows that if $|\theta_j^{(k)} - \theta_i^{(k-1)}|$ is small, then the product of the last elements of the eigenvectors is small, assuming that η_k is not too small. We can also obtain bounds for the difference of the approximate eigenvalues by using the secular equation whose solutions give the eigenvalues of T_{k+1} in terms of eigenvalues of T_k.

Proposition 1.35.

$$\theta_i^{(k)} - \theta_i^{(k+1)} \le \frac{\eta_{k+1}^2 (z_k^i)^2}{\alpha_{k+1} - \theta_i^{(k)} - \eta_{k+1}^2 \sum_{j \ne i} \frac{(z_k^j)^2}{\theta_j^{(k)} - \theta_i^{(k)}}}.$$

Proof. $\theta_i^{(k+1)}$ is a solution of the secular equation

$$f(\lambda) = \lambda - \alpha_{k+1} + \eta_{k+1}^2 \sum_j \frac{(z_k^j)^2}{\theta_j^{(k)} - \lambda} = 0.$$

Taking out the term for $j = i$ in the sum leads to the result because

$$\theta_i^{(k)} - \theta_i^{(k+1)} = \frac{\eta_{k+1}^2 (z_k^i)^2}{\alpha_{k+1} - \theta_i^{(k+1)} - \eta_{k+1}^2 \sum_{j \ne i} \frac{(z_k^j)^2}{\theta_j^{(k)} - \theta_i^{(k+1)}}},$$

and $\theta_i^{(k+1)} < \theta_i^{(k)}$. □

Other bounds can be derived using the secular function. To explain how they can be obtained it is enough to look at the case $k = 2$. For the sake of simplicity we denote $\theta_i = \theta_i^{(2)}$ and $z_i = z_{(2)}^i$, $i = 1, 2$, so the upper index 2 in z_i^2 is a square for the following derivation. Then the secular equation is

$$\alpha_3 - \lambda = \eta_3^2 \left[\frac{z_1^2}{\theta_1 - \lambda} + \frac{z_2^2}{\theta_2 - \lambda} \right].$$

First, we are interested in a solution $\lambda < \theta_1$, corresponding to $\theta_1^{(3)}$. The two fractions are positive and

$$\alpha_3 - \lambda > \eta_3^2 \frac{z_1^2}{\theta_1 - \lambda}.$$

This shows that the solution λ is less than the smallest root of the quadratic equation $\lambda^2 - (\alpha_3 + \theta_1)\lambda + \alpha_3\theta_1 - \eta_3^2 z_1^2 = 0$. Therefore, the solution $\theta_1^{(3)}$ is such that

$$\lambda_1 \leq \theta_1^{(3)} \leq \frac{1}{2}\left(\alpha_3 + \theta_1 - \sqrt{(\alpha_3 - \theta_1)^2 + 4\eta_3^2 z_1^2}\right) \leq \theta_1.$$

As a consequence we have

$$\theta_1 - \theta_1^{(3)} \geq \frac{1}{2}\left(\theta_1 - \alpha_3 + \sqrt{(\alpha_3 - \theta_1)^2 + 4\eta_3^2 z_1^2}\right) \geq 0$$

and

$$0 \leq \theta_1^{(3)} - \lambda_1 \leq \theta_1 - \lambda_1 + \frac{1}{2}\left(\alpha_3 - \theta_1 - \sqrt{(\alpha_3 - \theta_1)^2 + 4\eta_3^2 z_1^2}\right).$$

The term within parentheses on the right-hand side is negative, giving a bound of the decrease of the distance to λ_1. We can use the same kind of argument at the other end of the spectrum. All the terms in the right-hand side of the secular equation are negative and we write it as

$$\lambda - \alpha_3 = \eta_3^2 \left[\frac{z_1^2}{\lambda - \theta_1} + \frac{z_2^2}{\lambda - \theta_2}\right].$$

Then,

$$\lambda - \alpha_3 \geq \eta_3^2 \frac{z_2^2}{\lambda - \theta_2},$$

and we obtain

$$\lambda_n \geq \theta_3^{(3)} \geq \frac{1}{2}\left(\alpha_3 + \theta_2 + \sqrt{(\alpha_3 - \theta_2)^2 + 4\eta_3^2 z_2^2}\right) \geq \theta_2.$$

Moreover,

$$0 \leq \lambda_n - \theta_3^{(3)} \leq \lambda_n - \theta_2 - \frac{1}{2}\left(\alpha_3 - \theta_2 + \sqrt{(\alpha_3 - \theta_2)^2 + 4\eta_3^2 z_2^2}\right).$$

Let us now consider the case where we look for $\theta_1 < \lambda < \theta_2$ for which things are a little bit more difficult. In the right-hand side of the secular equation one term is positive and the other negative. Therefore, we first write

$$\alpha_3 - \lambda \leq \eta_3^2 \frac{z_2^2}{\theta_2 - \lambda}.$$

The root $\theta_2^{(3)}$ is located in between the roots of the quadratic equation corresponding to this inequality, but the largest root is larger than θ_2 and

$$\frac{1}{2}\left(\alpha_3 + \theta_2 - \sqrt{(\alpha_3 - \theta_2)^2 + 4\eta_3^2 z_2^2}\right) \leq \theta_2^{(3)}.$$

Now we change the signs in the secular equation and write

$$\lambda - \alpha_3 \leq \eta_3^2 \frac{z_1^2}{\lambda - \theta_1}.$$

Once again $\theta_2^{(3)}$ is located between the roots, but the smallest root is smaller than θ_1 and

$$\theta_2^{(3)} \leq \frac{1}{2}\left(\alpha_3 + \theta_1 + \sqrt{(\alpha_3 - \theta_1)^2 + 4\eta_3^2 z_1^2}\right).$$

Finally,

$$\max\left[\theta_1, \frac{1}{2}\left(\alpha_3 + \theta_2 - \sqrt{(\alpha_3 - \theta_2)^2 + 4\eta_3^2 z_2^2}\right)\right] \leq \theta_2^{(3)},$$

$$\theta_2^{(3)} \leq \min\left[\theta_2, \frac{1}{2}\left(\alpha_3 + \theta_1 + \sqrt{(\alpha_3 - \theta_1)^2 + 4\eta_3^2 z_1^2}\right)\right].$$

The lower and upper bounds depend on the position of α_3 relative to θ_1 and θ_2. It is not so obvious to know to which eigenvalue of A we have to compare $\theta_2^{(3)}$. If we denote the lower and upper bounds by

$$\underline{\theta} \leq \theta_2^{(3)} \leq \overline{\theta},$$

we have for all $j = 2, \dots, n-1$

$$\underline{\theta} - \lambda_j \leq \theta_2^{(3)} - \lambda_j \leq \overline{\theta} - \lambda_j,$$

and we can rearrange the terms on the left and the right. These results can be generalized to the case $k > 2$.

Theorem 1.36. *For the smallest eigenvalue we have*

$$\lambda_1 \leq \theta_1^{(k+1)} \leq \frac{1}{2}\left(\alpha_{k+1} + \theta_1^{(k)} - \sqrt{(\alpha_{k+1} - \theta_1^{(k)})^2 + 4\eta_{k+1}^2 (z_k^1)^2}\right) \leq \theta_1^{(k)}.$$

As a consequence we have

$$0 \leq \theta_1^{(k+1)} - \lambda_1 \leq \theta_1^{(k)} - \lambda_1 + \frac{1}{2}\left(\alpha_{k+1} - \theta_1^{(k)} - \sqrt{(\alpha_{k+1} - \theta_1^{(k)})^2 + 4\eta_{k+1}^2 (z_k^1)^2}\right).$$

For the largest eigenvalue we have

$$\lambda_n \geq \theta_{k+1}^{(k+1)} \geq \frac{1}{2}\left(\alpha_{k+1} + \theta_k^{(k)} + \sqrt{(\alpha_{k+1} - \theta_k^{(k)})^2 + 4\eta_{k+1}^2 (z_k^k)^2}\right) \geq \theta_k^{(k)}.$$

Moreover,

$$0 \leq \lambda_n - \theta_{k+1}^{(k+1)} \leq \lambda_n - \theta_k^{(k)} - \frac{1}{2}\left(\alpha_{k+1} - \theta_k^{(k)} + \sqrt{(\alpha_{k+1} - \theta_k^{(k)})^2 + 4\eta_{k+1}^2 (z_k^k)^2}\right).$$

For the other eigenvalues $i \neq 1$ or $k+1$, suppose we are considering the interval $\theta_i^{(k)} < \lambda < \theta_{i+1}^{(k)}$. We write the secular equation as

$$\alpha_{k+1} - \lambda = \eta_{k+1}^2 \left(\sum_{j=1}^{i} \frac{(z_k^j)^2}{\theta_j^{(k)} - \lambda} + \sum_{j=i+1}^{k} \frac{(z_k^k)^2}{\theta_j^{(k)} - \lambda}\right).$$

The first sum within the parentheses is negative and the second one is positive. Therefore,

$$\alpha_{k+1} - \lambda \le \eta_{k+1}^2 \sum_{j=i+1}^{k} \frac{(z_k^j)^2}{\theta_j^{(k)} - \lambda}.$$

But $\theta_j^{(k)} - \lambda \ge \theta_{i+1}^{(k)} - \lambda,\ j > i$. Hence

$$\alpha_{k+1} - \lambda \le \frac{\eta_{k+1}^2}{\theta_{i+1}^{(k)} - \lambda} \sum_{j=i+1}^{k} (z_k^j)^2.$$

So, we can do exactly the same as before, replacing $(z_1)^2$ by the sum of squares. On the other hand, we have

$$\lambda - \alpha_{k+1} \le \frac{\eta_{k+1}^2}{\lambda - \theta_i^{(k)}} \sum_{j=1}^{i} (z_k^j)^2.$$

Theorem 1.37. *Let i such that $1 \le i \le k+1$ and $\underline{z} = \sum_{j=1}^{i} (z_k^j)^2$, $\overline{z} = \sum_{j=i+1}^{k} (z_k^j)^2$. Then*

$$\max\left[\theta_i^{(k)}, \frac{1}{2}\left(\alpha_{k+1} + \theta_{i+1}^{(k)} - \sqrt{(\alpha_{k+1} - \theta_{i+1}^{(k)})^2 + 4\eta_{k+1}^2 \overline{z}}\right)\right] \le \theta_i^{(k+1)},$$

$$\theta_i^{(k+1)} \le \min\left[\theta_{i+1}^{(k)}, \frac{1}{2}\left(\alpha_{k+1} + \theta_i^{(k)} + \sqrt{(\alpha_{k+1} - \theta_i^{(k)})^2 + 4\eta_{k+1}^2 \underline{z}}\right)\right].$$

We have seen that there exists an integer j such that

$$|\theta_j^{(k+1)} - \theta_i^{(k)}| \le \eta_{k+1} |z_k^i|.$$

Quite often these eigenvalue intervals are overlapping and this result does not tell too much about the location of the Ritz values at iteration $k+1$. However, we can use it for the first and last eigenvalues. Then we obtain the following result.

Proposition 1.38.

$$\theta_1^{(k)} - \eta_{k+1}|z_k^1| \le \theta_1^{(k+1)} \le \frac{1}{2}\left(\alpha_{k+1} + \theta_1^{(k)} - \sqrt{(\alpha_{k+1} - \theta_1^{(k)})^2 + 4\eta_{k+1}^2 (z_k^1)^2}\right),$$

$$\frac{1}{2}\left(\alpha_{k+1} + \theta_k^{(k)} + \sqrt{(\alpha_{k+1} - \theta_k^{(k)})^2 + 4\eta_{k+1}^2 (z_k^k)^2}\right) \le \theta_{k+1}^{(k+1)} \le \theta_k^{(k)} + \eta_{k+1}|z_k^k|.$$

These results can eventually be improved by using refined bounds; see Parlett and Nour Omid [144]. One has to consider a gap where there is no approximate eigenvalue. Let gap_1 be such that

$$\theta_1^{(k)} < \theta_1^{(k)} + gap_1 < \theta_2^{(k)}.$$

Then,

$$\theta_1^{(k+1)} \geq \theta_1^{(k)} - \frac{\eta_{k+1}^2 (z_k^2)^2}{gap_1}.$$

For defining gap_1 we can use our lower bound for the second interval. If the lower bound for $\theta_2^{(k+1)}$ is equal to $\theta_2^{(k)}$, then $gap_1 = 0$ and we cannot use this. But otherwise, we obtain the next result.

Proposition 1.39. *Let* $gap_1 = \frac{1}{2}(\alpha_{k+1} + \theta_2^{(k)} - \sqrt{(\alpha_{k+1} - \theta_2^{(k)})^2 + 4\eta_{k+1}^2 \bar{z}}) - \theta_1^{(k)} > 0.$ *Then*

$$\theta_1^{(k+1)} \geq \theta_1^{(k)} - \frac{\eta_{k+1}^2 (z_k^1)^2}{gap_1}.$$

Something similar can be done at the other end of the spectrum by using the gap in the next-to-last interval. Other bounds can be obtained by using a decomposition like

$$\frac{1}{\delta_k(\lambda)} = \sum_{j=1}^{k} \frac{\varpi_j}{\theta_j^{(k)} - \lambda}.$$

It is easy to see that $\varpi_j = -1/\delta_k'(\theta_j^{(k)}) \geq 0$. By taking the first term out of the sum and setting $\lambda = \lambda_1$ we obtain the next result.

Proposition 1.40.

$$\theta_1^{(k)} - \lambda_1 \geq \varpi_1 \delta_k(\lambda_1) = \frac{\delta_k(\lambda_1)}{|\delta_k'(\theta_1^{(k)})|}.$$

Unfortunately, these bounds on $\theta_1^{(k)} - \lambda$ involve λ_1, which is generally unknown. Therefore, they are only of theoretical interest. When k increases, $\delta_k(\lambda)$ tends to behave at both ends of the spectrum like $\alpha_{k+1} - \lambda$, making the numerator of the lower bound $\alpha_{k+1} - \lambda_1$. The derivative in the denominator is getting larger and larger, so the lower bound is going to zero. At the other end of the spectrum we have

$$\lambda_n - \theta_k^{(k)} \geq \varpi_k |\delta_k(\lambda_n)| = \left| \frac{\delta_k(\lambda_n)}{\delta_k'(\theta_k^{(k)})} \right|.$$

It is also interesting to note that the possible values of α_{k+1} are constrained by the secular equation $\alpha_{k+1} - \lambda = g(\lambda)$ since the intersection of $y = \alpha_{k+1} - \lambda$ with the leftmost branch of $g(\lambda)$, which is $\theta_1^{(k)}$, cannot be smaller than λ_1; otherwise the interlacing property will be violated. At the other end of the spectrum, the intersection with the rightmost branch cannot be larger than λ_n. This gives the following bounds.

Proposition 1.41. *Let* g *be the function involved in the secular equation. Then,*

$$g(\lambda_1) + \lambda_1 \leq \alpha_{k+1} \leq g(\lambda_n) + \lambda_n.$$

We remark that $g(\lambda_1) > 0$ and $g(\lambda_n) < 0$; hence this improves the obvious bounds $\lambda_1 \leq \alpha_{k+1} \leq \lambda_n$. Most of these bounds and estimates are summarized in Figure 1.9 for a Strakoš matrix of dimension 10. The initial vector has all its components equal. The vertical dashed lines mark the location of the poles of g whose graph is the solid curve. The solid line is $\alpha_{k+1} - \lambda$ and the circles are the intersections with g, that is, the eigenvalues of T_{k+1}. The stars are our lower and upper bounds. The segments ending with crosses are the intervals containing at least one eigenvalue of T_{k+1} using $\eta_{k+1}|z_k^i|$. The small segments above are obtained using the refined bounds for the first and last eigenvalues. The diamonds are the lower and upper bounds for α_{k+1}. The crosses on the x-axis are the eigenvalues of A.

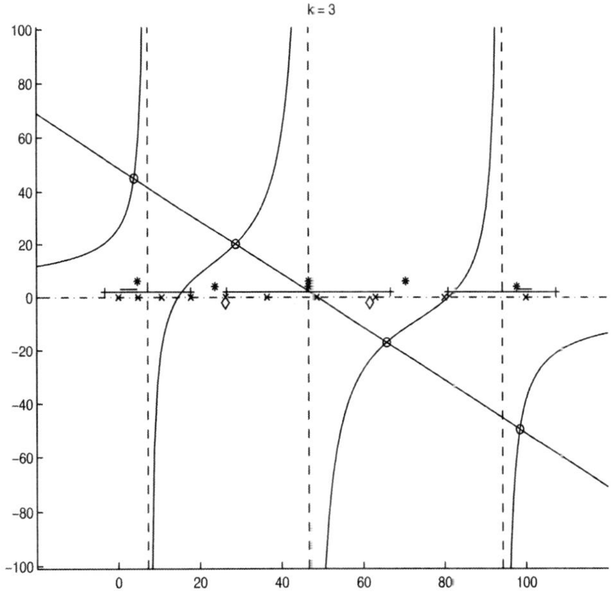

Figure 1.9. *Example of bounds and estimates*

Figure 1.10 shows for the same example the $\log_{10}$ of distances $\theta_1^{(k)} - \lambda_1$ and $\lambda_n - \theta_k^{(k)}$ as the middle solid curves as a function of the number of Lanczos iterations. Notice the scales are not the same for the two figures. The dashed curves are the upper bounds obtained with the secular function in Theorem 1.36. The upper solid curves are the upper bounds of Theorem 1.23. The dot-dashed curves are the lower bounds of Corollary 1.25, which are, in fact, not computable if we do not know the eigenvectors of A. We see that the bounds using the secular function are quite good. However, since we do not know the distances to the exact eigenvalues at the previous iteration, what we would really be able to compute during the iterations are only the increments to the distances at the previous iteration, whose $\log_{10}$ are shown in Figure 1.11, where the dashed curve corresponds to the largest eigenvalue.

Finally, we note that all these figures display the results of finite precision computations. However, in this small example of dimension 10 the influence of rounding errors is negligible.

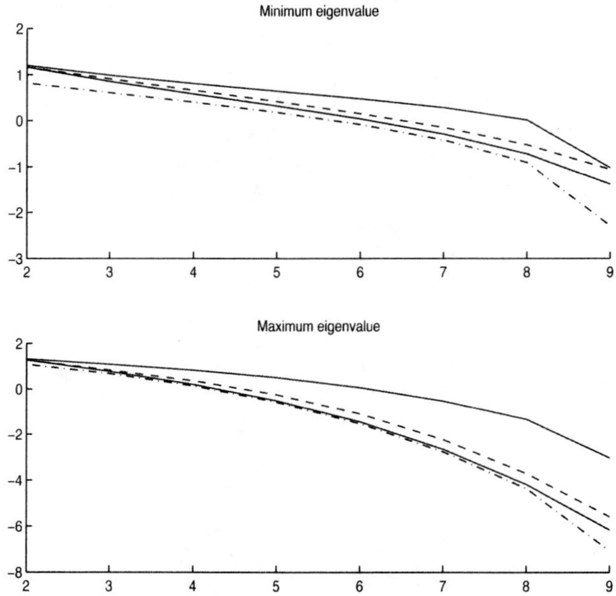

Figure 1.10. $\log_{10}$ *of bounds for the distances to the min and max eigenvalues of A*

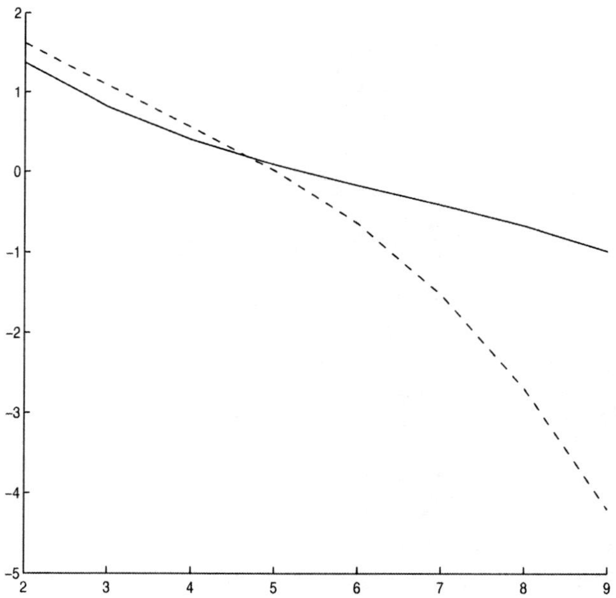

Figure 1.11. $\log_{10}$ *of increments to the distances to the min (solid) and max (dashed) eigenvalues of A*

1.7 A priori bounds

This section reviews the classical bounds on the behavior of the Ritz values which were given by Kaniel [102], corrected by Paige [132], and extended by Saad [153]; see also [173], [175]–[177]. The nice thing with these bounds is that they just depend on the distribution of the eigenvalues of A and the initial vector For these bounds one uses the Chebyshev polynomials, which we define for an interval $[a, b]$ by

$$C_k(\lambda; a, b) = C_k\left(\frac{2\lambda - (a+b)}{b-a}\right).$$

This transformation maps the given interval to $[-1, 1]$, where the polynomials are classically defined. For $|x| \le 1$ the definition is $C_k(x) = \cos(k \arccos x)$. They are orthogonal polynomials satisfying a three-term recurrence

$$C_0(x) = 1, \quad C_1(x) = x, \quad C_{k+1}(x) = 2xC_k(x) - C_{k-1}(x).$$

On $[a, b]$ their absolute values are less than or equal to 1, which is the maximum value.

Theorem 1.42.

$$\theta_1^{(k)} - \lambda_1 \le (\lambda_n - \lambda_1)\left[\frac{\tan \angle(q^1, v^1)}{C_{k-1}(\lambda_1; \lambda_2, \lambda_n)}\right]^2,$$

$$\theta_1^{(k)} - \lambda_1 \le \frac{(\lambda_n - \lambda_1)(\lambda_n - \lambda_2)^2}{(\lambda_2 - \lambda_1)^2 C_{k-2}^2(\lambda_1; \lambda_3, \lambda_n)} \frac{\sum_{j\ge 3}(\bar{v}_j)^2}{(\bar{v}_1)^2}.$$

Proof. From [197] we have

$$\theta_1^{(k)} - \lambda_1 \le \frac{\sum_{j=2}^n (\lambda_j - \lambda_1)\varphi^2(\lambda_j)(\bar{v}_j)^2}{\sum_{j=1}^n \varphi^2(\lambda_j)(\bar{v}_j)^2},$$

where φ is a polynomial of degree less than $k-1$. We first take $\varphi = C_{k-1}$. On the numerator we bound $\lambda_j - \lambda_1$ by $\lambda_n - \lambda_1$ and $\varphi^2(\lambda_j)$ by 1. The denominator can be written as $(\bar{v}_1)^2 C_{k-1}^2$ plus something positive. Therefore,

$$\theta_1^{(k)} - \lambda_1 \le \frac{(\lambda_n - \lambda_1)\sum_{j=2}^n (\bar{v}_j)^2}{C_{k-1}^2(\lambda_1; \lambda_2, \lambda_n)(\bar{v}_1)^2}.$$

Now, we have $\cos \angle(q^1, v^1) = (q^1, v^1) = \bar{v}_1$. Hence,

$$\sin^2 \angle(q^1, v^1) = 1 - (\bar{v}_1)^2 = \sum_{j=2}^n (\bar{v}_j)^2.$$

Therefore, the ratio of the terms involving $\bar{v}$ is $\tan^2 \angle(q^1, v^1)$.

For the other inequality (see [197]) we use $\varphi = (\lambda - \lambda_2)C_{k-2}(\lambda; \lambda_3, \lambda_n)$ and similar bounds. □

Notice that we can continue in the same way, taking out more eigenvalues by using $\varphi = (\lambda - \lambda_2)(\lambda - \lambda_3)C_{k-3}(\lambda; \lambda_4, \lambda_n)$, etc. A more general result has been proved by Saad [153].

Theorem 1.43.

$$\theta_i^{(k)} - \lambda_i \leq (\lambda_n - \lambda_1)\left[\kappa_i^{(k)} \frac{\tan \angle(q^i, v^1)}{C_{k-1}(\lambda_i; \lambda_{i+1}, \lambda_n)}\right]^2,$$

with

$$\kappa_i^{(k)} = \prod_{j<i} \frac{\theta_j^{(k)} - \lambda_n}{\theta_j^{(k)} - \lambda_i}.$$

These results can also be proved by using the min-max characterization of the eigenvalues of A, that is, the Courant–Fischer theorems; see [155], [141]. Figure 1.12 gives an example of the a priori bounds for the Strakos10 example. The solid line is the $\log_{10}$ of the true distance to the minimum eigenvalue as a function of the iteration number. The dashed line shows the bound involving only λ_1 and the stars show the bound with λ_1 and λ_2 which is worst in this case. The bounds are not sharp; the only thing they show in this example is that the distance of the smallest Ritz value to the minimum eigenvalue must finally decrease.

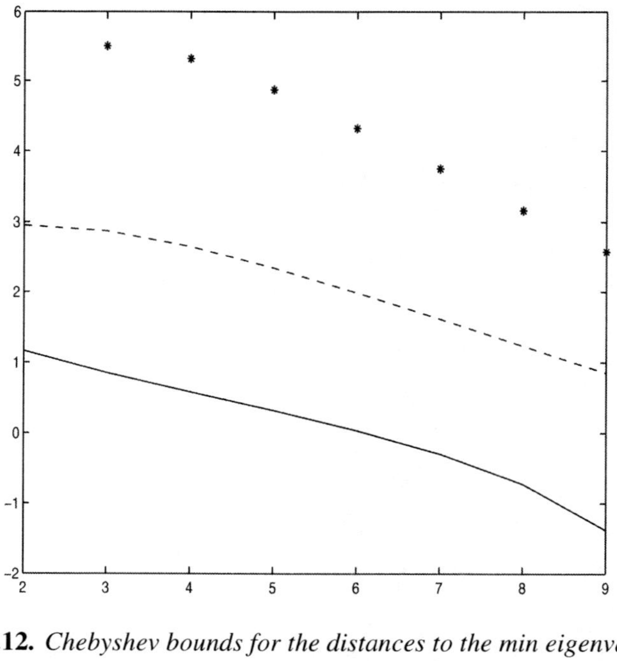

Figure 1.12. *Chebyshev bounds for the distances to the min eigenvalue of A*

1.8 Computation of the approximate eigenvalues

At iteration $k+1$ of the Lanczos algorithm we obtain a tridiagonal matrix T_{k+1}. If we want to have the Ritz values $\theta_j^{(k+1)}$, we have to compute the eigenvalues (and eventually eigenvectors) of a tridiagonal matrix. This is a well-known problem; see [68]. One way to do this is to use the QR algorithm.

However, it can be tempting to use the fact that the Ritz values are solution of the secular equation

$$f(\lambda) = \lambda - \alpha_{k+1} + \eta_{k+1}^2 \sum_{j=1}^{k} \frac{\xi_j^2}{\theta_j - \lambda} = 0.$$

The secular function f has poles at the eigenvalues of T_k for $\lambda = \theta_j = \theta_j^{(k)}$, $j = 1, \ldots, k$. We also know that it is a strictly increasing function between two consecutive poles. We would like to be able to compute the zeros of this function. One way to do this is to use Newton's method. But, we have to be careful about how to implement Newton's method for this problem, especially when two poles are close. These problems are, in fact, the same as the ones we face when implementing a very well-known algorithm for computing the eigenvalues of a tridiagonal matrix, the Cuppen's method [28], which is also known as the divide and conquer algorithm [36]. In this method the tridiagonal matrix is partitioned into two pieces and the eigenvalues are obtained from those of the two pieces by solving a secular equation. This is used recursively until the pieces are small enough for the eigenvalues to be computed easily. Clever methods have been devised to efficiently solve the secular equation. One of the most interesting papers about this problem is from Li [113]. There are many details which have to be carefully considered to obtain a reliable algorithm like, for instance, the choice of the initial guess of the iterations. For details, see [113]. See also [83], [152], [191].

Another possibility to compute the Ritz values is to use efficient algorithms recently introduced by Parlett and Dhillon based on variants of the qd algorithm and relatively robust representations of tridiagonals [33], [34], [35], [146], [9].

1.9 Harmonic Ritz values

Using the Krylov subspace built by the Lanczos algorithm, there are some other possibilities for approximating the eigenvalues. Harmonic Ritz values are defined as inverses of the Ritz values of A^{-1} with respect to $A\mathcal{K}(v^1, A)$. A characterization is given in [194]. They are characterized as the local extrema of

$$\frac{(A^{-1}Ax)^T Ax}{(Ax)^T Ax} = \frac{x^T Ax}{x^T A^2 x}.$$

Let $T_{k+1,k}$ be the $k+1 \times k$ tridiagonal matrix obtained by the $k+1$st iteration; the harmonic Ritz values θ_H are given by

$$T_{k+1,k}^T T_{k+1,k} z = \theta_H T_{k,k} z.$$

They satisfy the equation

$$\det(T_{k,k}(T_{k,k} - \theta_H I) + \eta_{k+1}^2 e^k (e^k)^T) = 0.$$

Generally, harmonic Ritz values are obtained with a shift σ to compute interior eigenvalues of A. Another characterization is the following. Let

$$\frac{1}{\delta_k} = (e^k)^T (T_{k,k} - \sigma I)^{-1} e^k.$$

Then, the eigenvalues of the matrix

$$\begin{pmatrix} T_{k,k} - \sigma I & \eta_{k+1} \\ \eta_{k+1} & \frac{\eta_{k+1}^2}{\delta_k} \end{pmatrix}$$

are 0 and the harmonic Ritz values for $A - \sigma I$. This shows that the harmonic Ritz values and the shifted Ritz values interlace. For more results on harmonic Ritz values see [172]. They are also linked to the study of Krylov methods for solving indefinite linear systems.

Chapter 2

The CG algorithm in exact arithmetic

Besides computing approximations of eigenvalues and eigenvectors, the Lanczos algorithm can be used to solve symmetric linear systems $Ax = b$, as it was shown by Lanczos in 1952; see [109]. The orthogonal basis vectors of the Krylov subspace given by the Lanczos algorithm are related to the residuals in the CG algorithm (see Hestenes and Stiefel [93]) for solving $Ax = b$ when A is symmetric and positive definite. In this chapter we shall first derive CG from the Lanczos algorithm. Then we exhibit the relationship between the residual vectors and the descent vectors of CG. We study the norm of the residual giving conditions showing when the norm of the residual vector can oscillate as it is sometimes observed in practical problems. We show the close relationship between the A-norm of the error and Gauss quadrature and use this to obtain expressions of the A-norm of the error which show that CG convergence depends on the way the Ritz values approximate the eigenvalues of A. We also give expressions for the l_2 norm of the error. We consider the three-term form of CG and recall CG optimality properties as well as the classical upper bounds of the A-norm of the error involving the condition number of A.

If we have an initial vector x^0 and an initial residual $r^0 = b - Ax^0$, the approximate solution of $Ax = b$ with the Lanczos algorithm is sought as $x^k = x^0 + V_k y^k$, where y^k is to be determined. We ask for the residual $r^k = b - Ax^k$ to be orthogonal to V_k, which is equivalent to solving a projected equation. This will give $r^{n+1} = 0$ in exact arithmetic. Since $r^k = r^0 - AV_k y^k$, we have $V_k^T r^0 - T_k y^k = 0$ and this implies that the vector y^k is obtained by solving

$$T_k y^k = \|r^0\| e^1.$$

Then the corresponding CG residual r^k is

$$r^k = r^0 - (V_k T_k + \eta_{k+1} v^{k+1} (e^k)^T) y^k = -\eta_{k+1} y_k^k v^{k+1},$$

and $\|r^k\| = \eta_{k+1} \|r^0\| \, |(T_k^{-1})_{1,k}|$.

2.1 Derivation of the CG algorithm from the Lanczos algorithm

The CG algorithm was developed independently by M. Hestenes in the U.S. and E. Stiefel in Switzerland in the early fifties. Later, they met during the "Symposium on Simultaneous Linear Equations and the Determination of Eigenvalues," organized in Los Angeles in 1951 by the Institute of Numerical Analysis of the U.S. National Bureau of Standards. Then they realized that their algorithms were the same and wrote a famous joint paper [93], which was published in 1952. In [93, p. 409] one can read, "The method of Conjugate Gradients was developed independently by E. Stiefel of the Institute of Applied Mathematics at Zurich and by M.R. Hestenes with the cooperation of J.B. Rosser, G. Forsythe and L. Paige of the Institute for Numerical Analysis, National Bureau of Standards. The present account was prepared jointly by M.R. Hestenes and E. Stiefel during the latter's stay at the National Bureau of Standards. The first papers on this method were given by E. Stiefel [1952] and by M.R. Hestenes [1951]. Reports on this method were given by E. Stiefel and J.B. Rosser at a symposium on August 23–25, 1951. Recently, C. Lanczos [1952] developed a closely related routine based on his earlier paper on eigenvalue problems [1950]. Examples and numerical tests of the method have been by R. Hayes, U. Hoschstrasser and M. Stein." For other details, see the short biography of Hestenes and Stiefel at the end of this book. See also Hestenes' papers and book [52], [88], [89], [90], [91], and [92].

Although the CG algorithm was first derived in a completely different way using conjugacy and minimization of functionals, it turns out that it is equivalent to the Lanczos algorithm when the matrix A is symmetric positive definite, as we shall see below. For earlier papers on that topic, see Householder [96] and Paige and Saunders [137]. If we already know the result, the equivalence between CG and Lanczos algorithms is easy to prove. In this section, for pedagogical reasons, we are going to pretend that we are not aware of the result and that we are just looking at simpler relations for computing the solution given by the Lanczos algorithm. Although the derivation is more involved it can be useful for the reader to see most of the details of this process.

If A is positive definite, so is the tridiagonal matrix T_k produced by the Lanczos algorithm, and we can use the Cholesky (-like) decomposition $T_k = \bar{L}_k \Delta_k^{-1} \bar{L}_k^T$, where Δ_k is a diagonal matrix with diagonal elements $\delta_i (= \delta_i(0))$ and $\bar{L}_k$ is lower bidiagonal with the same elements as those of T_k on the subdiagonal and the δ_i's on the diagonal. Even though it seems natural to use a symmetric factorization of the symmetric matrix T_k, it turns out that it is easier to consider a nonsymmetric factorization which leads to simpler formulas. We write

$$T_k = \bar{L}_k \Delta_k^{-1} \Omega_k^{-1} \Omega_k \bar{L}_k^T,$$

where Ω_k is a diagonal matrix with diagonal elements ω_i that we shall choose later on. We denote this factorization as

$$T_k = L_k \Omega_k^{-1} U_k, \quad L_k = \bar{L}_k \Delta_k^{-1}, \quad U_k = \Omega_k \bar{L}_k^T,$$

where L_k (resp., U_k) is lower (resp., upper) triangular and Ω_k is diagonal. Replacing T_k by its factorization, we have the Lanczos matrix relation

$$AV_k = V_k L_k \Omega_k^{-1} U_k + G_k,$$

where G_k is zero except for the last column. We multiply this relation to the right by U_k^{-1} to obtain

$$AV_kU_k^{-1} = V_kL_k\Omega_k^{-1} + G_kU_k^{-1}.$$

We introduce a new matrix $P_k = V_kU_k^{-1}$ and we denote the columns of P_k as $p^0, p^1, \ldots, p^{k-1}$. The vectors p^k must not be confused with the polynomials p_k involved in the Lanczos algorithm. Therefore,

$$AP_k = V_kL_k\Omega_k^{-1} + G_kU_k^{-1}.$$

Let us now consider the iterates x^k,

$$x^k = x^0 + \|r^0\|\; V_kT_k^{-1}e^1.$$

We introduce the factorization of T_k in this relation and we obtain

$$x^k = x^0 + \|r^0\|\; V_kU_k^{-1}\Omega_kL_k^{-1}e^1 = x^0 + \|r^0\|\; P_k\Omega_kL_k^{-1}e^1.$$

We see that we are interested in updating the first column $L_k^{-1}e^1$ of the inverse of the lower triangular matrix L_k. It can be expressed with the inverse of L_{k-1} as

$$L_k^{-1}e^1 = \begin{pmatrix} L_{k-1}^{-1}e^1 \\ t_k \end{pmatrix},$$

where t_k is the last element of the first column t of the inverse. Introducing this last result in the expression of x^k and noticing that $x^{k-1} = x^0 + \|r^0\|\; P_{k-1}\Omega_{k-1}L_{k-1}^{-1}e^1$, we have

$$x^k = x^{k-1} + \|r^0\|\; \omega_kt_kp^{k-1}$$

Provided we can easily compute the vectors p^j, we see that the new iterate x^k is directly obtained from the previous one x^{k-1} and p^{k-1}. Let us now compute the element t_k. We have to solve the triangular system

$$\begin{pmatrix} 1 & & & \\ \frac{\eta_2}{\delta_1} & 1 & & \\ & \ddots & \ddots & \\ & & \frac{\eta_k}{\delta_{k-1}} & 1 \end{pmatrix} t = e^1.$$

This gives the last element of the vector t,

$$t_k = (-1)^{k-1}\frac{\eta_k\cdots\eta_2}{\delta_{k-1}\cdots\delta_1}.$$

Let us denote the coefficient of p^{k-1} in the equation of x^k by γ_{k-1}. Then

$$x^k = x^{k-1} + \gamma_{k-1}p^{k-1}, \qquad \gamma_{k-1} = \|r^0\|\omega_kt_k.$$

The residual $r^k = b - Ax^k$ is given by the recurrence

$$r^k = r^{k-1} - \gamma_{k-1}Ap^{k-1}.$$

We return to the relationship between V_k and AP_k,

$$AP_k = V_k L_k \Omega_k^{-1} + G_k U_k^{-1}.$$

Writing the last column of this matrix identity, we have

$$Ap^{k-1} - \frac{1}{\omega_k} v^k = \frac{\eta_{k+1}}{\delta_k \omega_k} v^{k+1}.$$

We have seen before that the Lanczos vector v^{k+1} is a scalar multiple of the residual r^k. Since the Lanczos vectors are of norm 1, we have

$$v^{k+1} = (-1)^k \frac{r^k}{\|r^k\|},$$

the sign being chosen to have $v^1 = r^0/\|r^0\|$. We identify the coefficients in the equation for r^k and v^{k+1} and we obtain

$$\frac{\eta_{k+1}}{\delta_k} = \frac{\|r^k\|}{\|r^{k-1}\|}, \qquad \gamma_{k-1} = (-1)^{k-1} \omega_k \|r^{k-1}\|.$$

Remember that the definition of γ_{k-1} was

$$\gamma_{k-1} = (-1)^{k-1} \|r^0\| \omega_k \frac{\eta_2 \cdots \eta_k}{\delta_1 \cdots \delta_{k-1}}.$$

But,

$$\frac{\eta_2 \cdots \eta_k}{\delta_1 \cdots \delta_{k-1}} = \frac{\|r^{k-1}\|}{\|r^0\|}.$$

Hence, we choose

$$\omega_k = (-1)^{k-1} \frac{1}{\|\delta_k r^{k-1}\|},$$

because it shall simplify the equation for p^k. This choice gives $\gamma_{k-1} = 1/\delta_k$. Now, let us look at the definition of P_k to see how we can find a relation to compute p^k. By definition, we have $P_{k+1} U_{k+1} = V_{k+1}$. Writing the last column of this matrix identity, we have

$$v^{k+1} = \omega_{k+1} \delta_{k+1} p^k + \eta_{k+1} \omega_k p^{k-1}.$$

Then,

$$p^k = (-1)^k \frac{1}{\|r^k\| \omega_{k+1} \delta_{k+1}} r^k - \frac{\eta_{k+1} \omega_k}{\delta_{k+1} \omega_{k+1}} p^{k-1}.$$

Since $\omega_{k+1}\delta_{k+1} = (-1)^k/\|r^k\|$ the coefficient of r^k is 1. For the other term,

$$\frac{\eta_{k+1} \omega_k}{\delta_{k+1} \omega_{k+1}} = \frac{\eta_{k+1}}{\eta_k} \frac{\delta_k \omega_k}{\delta_{k+1} \omega_{k+1}} = -\frac{\|r^k\|^2}{\|r^{k-1}\|^2}.$$

If we denote $\beta_k = \|r^k\|^2/\|r^{k-1}\|^2$, we have

$$p^k = r^k + \beta_k p^{k-1}.$$

This shows that the vector p^k can also be computed in a very simple way. It remains to compute the value of $\gamma_k = 1/\delta_{k+1}$. We can compute it using the Cholesky factorization of T_{k+1}. However, there is a simpler way to obtain γ_k. We have

$$P_k^T A P_k = U_k^{-T} V_k^T A V_k U_k^{-1} = U_k^{-T} T_k U_k^{-1} = U_k^{-T} L_k \Omega_k^{-1}.$$

Therefore,

$$P_k^T A P_k = \Omega_k^{-1} \bar{L}_k^{-1} \bar{L}_k \Delta_k^{-1} \Omega_k^{-1}.$$

This shows that $P_k^T A P_k$ is a diagonal matrix and

$$(Ap^k, p^k) = \frac{1}{\delta_{k+1}\omega_{k+1}^2}.$$

We obtain

$$\gamma_k = \frac{\|r^k\|^2}{(Ap^k, p^k)}.$$

Finally, there is a relation between α_k and the other coefficients. This is a consequence of the definition of δ_k, which can also be written as

$$\alpha_k = \delta_k + \beta_{k-1}\delta_{k-1}.$$

Summarizing all these results we have the CG algorithm.

ALGORITHM 2.1.
Let x^0 be given and $r^0 = b - Ax^0$
for $k = 0, 1, \dots$ until convergence
$\beta_k = (r^k, r^k)/(r^{k-1}, r^{k-1}), \ \beta_0 = 0$
$p^k = r^k + \beta_k p^{k-1}$
$\gamma_k = (r^k, r^k)/(Ap^k, p^k)$
$x^{k+1} = x^k + \gamma_k p^k$
$r^{k+1} = r^k - \gamma_k A p^k$
end

Concerning the equivalence of CG and Lanczos algorithms in exact arithmetic the previous derivation shows that we have the following result.

Theorem 2.1. *If x^0 and v with $\|v\| = 1$ are such that $r^0 = b - Ax^0 = \|r^0\| v$ the Lanczos algorithm started from v generates the same iterates as the CG algorithm started from x^0 when solving the linear system $Ax = b$ with A symmetric and positive definite and we have the following relations between the coefficients:*

$$\alpha_k = \frac{1}{\gamma_{k-1}} + \frac{\beta_{k-1}}{\gamma_{k-2}}, \quad \beta_0 = 0, \quad \gamma_{-1} = 1,$$

$$\eta_{k+1} = \frac{\sqrt{\beta_k}}{\gamma_{k-1}},$$

and the Lanczos vectors are

$$v^{k+1} = (-1)^k \frac{r^k}{\|r^k\|}.$$

Moreover, γ_k is related to the diagonal elements of the Cholesky decomposition of the Lanczos matrix T_k

$$\gamma_k = \frac{1}{\delta_{k+1}}.$$

During the derivation of CG we have proved that the vectors p^k are mutually conjugate and that the residuals r^k are orthogonal. The matrix relation $V_k^T P_k \geq U_k^{-1}$ shows that $(v^i, p^j) = 0$ when $i > j$. Therefore $(r^{i-1}, p^j) = 0$.

As we said at the beginning of this chapter the CG algorithm can also be seen as a minimization algorithm. Since A is symmetric, the solution x of $Ax = b$ also minimizes the functional

$$\varphi(x) = \frac{1}{2}(Ax, x) - (b, x).$$

The gradient of this functional is $\delta\varphi(x) = Ax - b$, which is the opposite of the residual. Suppose we have a sequence of vectors $\{p^0, p^1, \ldots\}$ and an approximation x^k; if we want to minimize φ in the direction of p^k, we look for γ minimizing $\varphi(x^k + \gamma p^k)$. The solution is given by

$$\gamma = \frac{(p^k, r^{k-1})}{(p^k, Ap^k)}.$$

This is just a local minimization. The vector p^k has to be chosen carefully to achieve a global minimization. If the vectors p^k are chosen to be mutually A-orthogonal, then the local minimization is a global minimization; see, for instance, [68]. This is what is done by the CG algorithm.

We remark that if $\eta_{k+1} = 0$, then $\beta_k = 0$, and this leads to $v^{k+1} = 0$. If A is positive definite, the CG and Lanczos coefficients are all positive. There cannot be any breakdown of the algorithm because if $r^k = 0$ or $p^k = 0$, then the algorithm has found the solution. Moreover, we remark that to be able to go from the Lanczos algorithm to the CG formulas it is not necessary for A to be positive definite. All we need is the existence of the Cholesky-like factorization $T_k = \bar{L}_k \Delta_k^{-1} \bar{L}_k^T$, which can be obtained if no δ_i is zero, but this can eventually happen since (Ap^k, p^k) can be zero without p^k being zero when A is not positive definite. This is unlikely in finite precision arithmetic, but this term can be small, leading to some numerical difficulties. A way to get around these possible troubles is to use either a block factorization or a QR factorization of T_k. The latter leads to the SYMMLQ algorithm of Paige and Saunders [137]; see also [50].

By comparing the Lanczos residuals $r_{(k)}^j$ for the eigenvalue problem and the CG residuals r^k when solving a linear system (with compatible initial vectors) we obtain

$$r_{(k)}^j = -\frac{(z_{(k)}^j)_k}{y_k^k} r^k \quad \forall j.$$

This is an interesting relation since we have seen that when an eigenpair of T_k converges, the last element of the eigenvector becomes small. For "almost" converged eigenpairs the

Lanczos residual can be much smaller than the CG residual at the same iteration. The fact that some eigenvalues have already converged does not always imply a small CG residual, nor a small error, as we shall see. The solution y^k of the linear system $T_k y^k = \|r^0\| e^1$ gives

$$y_k^k = \|r^0\| (T_k^{-1} e^1, e^k).$$

It can be proved that when l increases, $(T_k^{-1} e^1, e^l)$ has alternate signs and its absolute value increases with k for a given l. To prove this, the UL decomposition of T_k is useful in computing the first column of the inverse of T_k. The diagonal elements of this decomposition that for the sake of simplicity we denote by d_j instead of $d_j^{(k)}$ are given by the first pivot function at 0,

$$d_k = \alpha_k, \quad d_l = \alpha_l - \frac{\eta_{l-1}^2}{d_{l-1}}, \quad l = k-1, \ldots, 1.$$

Proposition 2.2. *Let T_k be a positive definite tridiagonal matrix. Then*

$$|(T_k^{-1} e^1)_l| < |(T_{k+1}^{-1} e^1)_l|, \quad l = 1, \ldots, k,$$

$$|(T_{k+1}^{-1} e^1)_{k+1}| = \frac{\eta_{k+1}}{\delta_{k+1}} |(T_k^{-1} e^1)_k|.$$

Proof. Let $x = T_k^{-1} e^1$ (resp., $\bar{x} = T_{k+1}^{-1} e^1$) be the first column of T_k^{-1} (resp., T_{k+1}^{-1}). Then it is easy to see [117] that

$$x_1 = \frac{1}{d_1}, \quad x_l = -\frac{\eta_l}{d_l} x_{l-1} = (-1)^{l-1} \frac{\eta_l \cdots \eta_2}{d_l \cdots d_1}.$$

This shows that $x_1 = (T_k^{-1} e^1, e^1) > 0$ and the signs of the elements x_l alternate. Moreover, let $\bar{d}_l$ be the diagonal elements of the UL decomposition of T_{k+1} (that is, $d_l^{(k+1)}(0)$); then we have

$$d_k - \bar{d}_k = \frac{\eta_{k+1}^2}{\alpha_{k+1}} > 0.$$

We remark that both recurrences for d_k and $\bar{d}_k$ are the same; only the initial values differ and at the next step

$$d_{k-1} - \bar{d}_{k-1} = \eta_k^2 \left(\frac{d_k - \bar{d}_k}{d_k \bar{d}_k} \right) > 0.$$

Recursively, we have

$$d_{j-1} - \bar{d}_{j-1} = \eta_j^2 \left(\frac{d_j - \bar{d}_j}{d_j \bar{d}_j} \right) > 0, \quad j = k-1, \ldots, 2.$$

Therefore if all the $\eta_i \neq 0$, $\bar{d}_l < d_l$, $l = 1, \ldots, k$. Therefore, by induction

$$|x_l| = \frac{\eta_l}{d_l} |x_{l-1}| \leq \frac{\eta_l}{\bar{d}_l} |x_{l-1}| \leq \frac{\eta_l}{\bar{d}_l} |\bar{x}_{l-1}| = |\bar{x}_l|.$$

This shows that

$$|(T_k^{-1} e^1)_l| < |(T_{k+1}^{-1} e^1)_l|, \quad l = 1, \ldots, k.$$

However, this does not tell what $(T_{k+1}^{-1}e^1)_{k+1}$ is going to be. Many results are known about inverses of tridiagonal matrices; see [117]. More results on the elements of the inverse of the tridiagonal matrix T_k can be obtained by using the Sherman–Morrison formula; see [68]. We have

$$T_{k+1} = \begin{pmatrix} T_k & \eta_{k+1}e^k \\ \eta_{k+1}(e^k)^T & \alpha_{k+1} \end{pmatrix}.$$

The upper left block of T_{k+1}^{-1} is the inverse of a Schur complement,

$$\left(T_k - \frac{\eta_{k+1}^2}{\alpha_{k+1}} e^k (e^k)^T \right)^{-1}.$$

Then we apply the Sherman–Morrison formula

$$\left(T_k - \frac{\eta_{k+1}^2}{\alpha_{k+1}} e^k (e^k)^T \right)^{-1} = T_k^{-1} + \frac{(T_k^{-1}e^k)((e^k)^T T_k^{-1})}{\frac{\alpha_{k+1}}{\eta_{k+1}^2} - (e^k)^T T_k^{-1} e^k}.$$

The element $(k+1, 1)$ of T_{k+1}^{-1} is

$$(T_{k+1}^{-1})_{k+1,1} = -\frac{\eta_{k+1}}{\alpha_{k+1}} \left(e^1, \left(T_k - \frac{\eta_{k+1}^2}{\alpha_{k+1}} e^k (e^k)^T \right)^{-1} e^k \right).$$

After a few manipulations we have

$$(T_{k+1}^{-1})_{k+1,1} = -(e^1, T_k^{-1} e^k) \left[\frac{\eta_{k+1}}{\alpha_{k+1} - \eta_{k+1}^2 (e^k, T_k^{-1} e^k)} \right].$$

Now, consider the $(k+1, k+1)$ element of T_{k+1}^{-1}. It is the inverse of the Schur complement, that is,

$$\frac{1}{\alpha_{k+1} - \eta_{k+1}^2 (e^k, T_k^{-1} e^k)}.$$

Therefore,

$$(T_{k+1}^{-1})_{k+1,1} = -\eta_{k+1} (T_k^{-1})_{k,1} (T_{k+1}^{-1})_{k+1,k+1}.$$

But $(T_{k+1}^{-1})_{k+1,k+1} = 1/\delta_{k+1}$ and

$$|\bar{x}_{k+1}| = \frac{\eta_{k+1}}{\delta_{k+1}} |x_k| \geq \frac{\eta_{k+1}}{\alpha_{k+1}} |x_k|. \quad \square$$

When we increase k the corresponding elements of the first column of T_k^{-1} keep the same sign and their absolute values increase towards the values of those for T_n. If $\delta_{k+1} \leq \eta_{k+1}$, the value of the last element of the solution of the tridiagonal system increases when going from k to $k+1$. Using the Cholesky decomposition of T_k we have

$$y_k^k = (-1)^{k-1} \|r^0\| \frac{\eta_2 \cdots \eta_k}{\delta_1 \cdots \delta_k} = (-1)^{k-1} \frac{\|r^{k-1}\|}{\delta_k},$$

and as we have seen in the last proposition,

$$|y_{k+1}^{k+1}| = \frac{\eta_{k+1}}{\varepsilon_{k+1}} |y_k^k|.$$

We have $r^k = -\eta_{k+1} y_k^k v^{k+1}$ and also

$$\|r^k\| = \|r^0\| \frac{\eta_2 \cdots \eta_{k+1}}{\delta_1 \cdots \delta_k} = \|r^0\| \sqrt{\beta_1 \cdots \beta_k}.$$

This means that, when running the Lanczos algorithm, we can compute the norm of the CG residual without having to compute the iterates x^k. However, when A is positive definite, CG is the method of choice over the Lanczos algorithm because we do not need to keep (or store) all the basis vectors to compute the approximate solution. But, the Lanczos algorithm can be used for indefinite symmetric matrices (see [140]), even though it may be preferable to use the SYMMLQ algorithm [137].

2.2 Relations between residuals and descent directions

In this section we investigate the relations between the vectors p^k and r^k. We have $p^0 = r^0$ and

$$p^k = r^k + \beta_k p^{k-1}.$$

By induction we obtain

$$p^k = r^k + \beta_k r^{k-1} + \beta_k \beta_{k-1} r^{k-2} + \cdots + \beta_k \cdots \beta_1 r^0.$$

Hence

$$p^k = r^k + \frac{\|r^k\|^2}{\|r^{k-1}\|^2} r^{k-1} + \cdots + \frac{\|r^k\|^2}{\|r^0\|^2} r^0,$$

and using the relation between the residuals and the Lanczos basis vectors

$$p^k = r^k + \|r^k\|^2 \left[\frac{v^1}{\|r^0\|} - \frac{v^2}{\|r^1\|} + \cdots + (-1)^{k-1} \frac{v^k}{\|r^{k-1}\|} \right].$$

This shows that we have the following result.

Proposition 2.3. *In matrix form we have*

$$p^k = r^k + \|r^k\|^2 V_k D_k^r w,$$

where $w = (1 \quad -1 \quad 1 \quad \cdots)^T$ *and* D_k^r *is a diagonal matrix*

$$D_k^r = \begin{pmatrix} \frac{1}{\|r^0\|} & & & \\ & \frac{1}{\|r^1\|} & & \\ & & \ddots & \\ & & & \frac{1}{\|r^{k-1}\|} \end{pmatrix}.$$

We also have

$$\frac{p^k}{\|r^k\|^2} = (-1)^k \frac{v^{k+1}}{\|r^k\|} + \frac{v^1}{\|r^0\|} - \frac{v^2}{\|r^1\|} + \cdots + (-1)^{k-1} \frac{v^k}{\|r^{k-1}\|}.$$

Hence,

$$p^k = \|r^k\|^2 V_{k+1} D^r_{k+1} \tilde{w},$$

with $\tilde{w}^T = (w^T \; (-1)^k)$. This leads to

$$\|p^k\| \le \sqrt{k+1}\|r^k\|^2 \, \|D^r_{k+1}\| \le \sqrt{k+1}\|r^k\| \max_{i=0,\ldots,k} \frac{\|r^k\|}{\|r^i\|}.$$

Notice that because of the local orthogonality properties,we have

$$(p^k, r^k) = (r^k, r^k).$$

By the Cauchy–Schwarz inequality

$$\|r^k\|^2 \le \|r^k\| \, \|p^k\|$$

and

$$\|r^k\| \le \|p^k\|.$$

We shall show later on that we have $(Ap^k, p^k) \le (Ar^k, r^k)$. Then we have the following; see Bollen [11].

Proposition 2.4.

$$\|p^k\| \le \kappa(A)^{\frac{1}{2}} \|r^k\|,$$

where $\kappa(A) = \lambda_n/\lambda_1$ is the condition number of A.

Proof. Since A is positive definite, we can write

$$\begin{aligned}\frac{\|p^k\|}{\|r^k\|} &= \frac{\|A^{-\frac{1}{2}} A^{\frac{1}{2}} p^k\|}{\|A^{\frac{1}{2}} r^k\|} \frac{\|A^{\frac{1}{2}} r^k\|}{\|r^k\|} \\ &\le \|A^{-\frac{1}{2}}\| \frac{(Ap^k, p^k)}{(Ar^k, r^k)} \frac{\|A^{\frac{1}{2}} r^k\|}{\|r^k\|} \\ &\le \sqrt{\frac{\lambda_n}{\lambda_1}}. \quad \square\end{aligned}$$

Therefore, we have

$$\|r^k\| \le \|p^k\| \le \kappa(A)^{\frac{1}{2}} \|r^k\|.$$

2.3 The norm of the residual

We suppose that A is positive definite, so all the Lanczos and CG coefficients are positive. In this section we look at how the norm of the residuals behave. CG residuals sometimes exhibit a very erratic oscillatory behavior. The example in Figure 2.1 is using the matrix Bcsstk01 of dimension 48 from the Matrix Market [115] with 400 nonzero entries, $\|A\|_2 = 3.0152\ 10^9$, and a condition number $\kappa(A) = 8.82\ 10^5$. There are other examples where oscillations can even be worse than this. Notice that this computation was done in finite precision arithmetic. We shall see later that for this example some oscillations may be caused by finite precision. Nevertheless, it is of interest to know when the ratio $\pi_k = \delta_k/\eta_{k+1}$ is smaller or larger than 1. If $\pi_k > 1$, the norm of the residual decreases; otherwise it increases. As a starting point we have the following result.

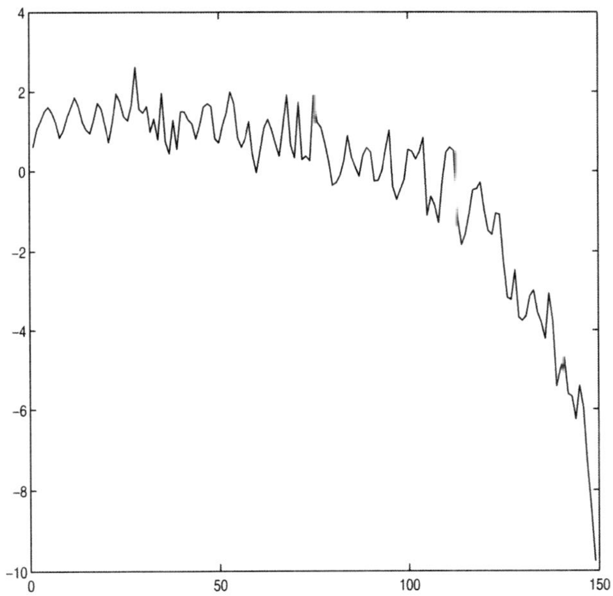

Figure 2.1. $\log_{10}$ *of the norm of the residual for Bcsstk01*

Proposition 2.5. *If* $\pi_{k-1} \geq 1$ *and* $\alpha_k \geq \eta_k + \eta_{k+1}$, *then* $\pi_k > 1$. *If* $\pi_{k-1} \leq 1$ *and* $\alpha_k \leq \eta_k + \eta_{k+1}$, *then* $\pi_k < 1$.

Proof. By the definition of δ_k's, which are the diagonal elements of the Cholesky decompositions of the Lanczos tridiagonal matrices, we have

$$\frac{\delta_k}{\eta_{k+1}} = \frac{\alpha_k}{\eta_{k+1}} - \frac{\eta_k}{\eta_{k+1}} \frac{\eta_k}{\delta_{k-1}},$$

which is

$$\pi_k = \frac{\alpha_k}{\eta_{k+1}} - \frac{\eta_k}{\eta_{k+1}} \frac{1}{\pi_{k-1}}.$$

The initial condition is $\pi_1 = \alpha_1/\eta_2$. We shall see later on that we can obtain the solution of this recurrence. But, if the (local) diagonal dominance hypothesis is satisfied, then

$$\frac{\alpha_k}{\eta_{k+1}} \geq 1 + \frac{\eta_k}{\eta_{k+1}},$$

and

$$\pi_k > 1 + \frac{\eta_k}{\eta_{k+1}} \left(1 - \frac{1}{\pi_{k-1}}\right) > 1.$$

The proof for the other assertion is the same. □

The condition $\alpha_k \geq \eta_k + \eta_{k+1}$ is having diagonal dominance for row (or column) k of the Lanczos tridiagonal matrix. Since $\pi_k > 1$ corresponds to the nice situation where $\|r^k\| < \|r^{k-1}\|$, that is, a decrease of the l_2 norm of the residual, the larger the π_k the better it is. We can also prove that $\pi_{k-1} < 1$ and $\pi_k > 1$ (going from a bad to a good situation) give diagonal dominance and that $\pi_{k-1} > 1$ and $\pi_k < 1$ (going from good to bad) imply that we do not have diagonal dominance. To summarize, if we are in a good (resp., bad) situation and we have (resp., do not have) diagonal dominance, we stay in that situation. So, to have oscillations of the norms of the residuals we must alternate diagonal dominance and nondiagonal dominance in T_k, that is, at least having a nonmonotone behavior of the Lanczos coefficients.

We can also express the conditions to have $\pi_k > 1$ in different ways. The recurrence for π_k is

$$\pi_k = \frac{\alpha_k}{\eta_{k+1}} - \frac{\eta_k}{\eta_{k+1}} \frac{1}{\pi_{k-1}}.$$

It can be solved by setting $\pi_k = \omega_{k+1}/\omega_k$, which gives

$$\eta_{k+1}\omega_{k+1} = \alpha_k \omega_k - \eta_k \omega_{k-1}.$$

By comparing to the recurrence for the Lanczos polynomials, this shows that

$$\omega_k = (-1)^{k+1} p_k(0)$$

and

$$\frac{\|r^{k-1}\|}{\|r^k\|} = \frac{\delta_k}{\eta_{k+1}} = \pi_k = -\frac{p_{k+1}(0)}{p_k(0)} = \left|\frac{p_{k+1}(0)}{p_k(0)}\right|.$$

The condition $\pi_k > 1$ can be written using the Lanczos polynomials at 0,

$$\pi_k > 1 \Leftrightarrow (-1)^k p_{k+1}(0) > (-1)^{k-1} p_k(0).$$

Since $p_k(0) > 0$ if k is odd and $p_k(0) < 0$ if k is even, this translates into the following condition.

Proposition 2.6.

$$\pi_k > 1 \Leftrightarrow |p_{k+1}(0)| > |p_k(0)|.$$

Therefore, the norm of the CG residual oscillates if the value of the Lanczos polynomial at 0 oscillates. The value of the Lanczos polynomial at 0 is given by Cybenko's theorem,

Theorem 1.5. It is completely determined by the eigenvalues of A and the projections of the initial residual on the eigenvectors of A.

Exploiting the previous results we can also show that the residual is given by a polynomial acting on the initial residual. This was already obvious from the fact that the residual is a scalar multiple of the Lanczos vectors. But, now we can use our results on the norm of the residual. By remembering that

$$\frac{\|r^k\|}{\|r^0\|} = \frac{1}{\pi_1 \cdots \pi_k},$$

and putting in the solution for π_j, we obtain

$$\|r^k\| = \frac{\|r^0\|}{|p_{k+1}(0)|}.$$

Since $v^{k+1} = p_{k+1}(A)v^1$ and

$$v^{k+1} = (-1)^k \frac{r^k}{\|r^k\|},$$

we have

$$r^k = (-1)^k \frac{1}{|p_{k+1}(0)|} p_{k+1}(A)r^0.$$

This implies that the polynomial giving r^k has a value of 1 at 0. We shall see later some relations between the residual and error norms that will shed some more light on the behavior of the residual. Finally, we have a bound for the (possible) increase of the residual.

Proposition 2.7.

$$\frac{\|r^{k+1}\|}{\|r^k\|} \leq 1 + \kappa(A),$$

where $\kappa(A) = \lambda_n/\lambda_1$ is the condition number of A.

Proof. Since $r^{k+1} = r^k - \gamma_k A p^k$ we have

$$\|r^{k+1}\| \leq \|r^k\| + \frac{\|r^k\|^2}{(p^k, Ap^k)} \|A\| \, \|p^k\|.$$

Therefore,

$$\frac{\|r^{k+1}\|}{\|r^k\|} \leq 1 + \frac{\|A\|}{\frac{(p^k, Ap^k)}{\|r^k\| \, \|p^k\|}} \leq 1 + \frac{\|A\|}{\frac{(p^k, Ap^k)}{\|p^k\|^2}} \leq 1 + \frac{\|A\|}{\lambda_1} \qquad \square$$

We also have some relations for the components $\bar{r}_i^k$ of the projections of the residual vector on the eigenvectors of A.

Proposition 2.8.

$$\frac{\bar{r}_i^k}{\bar{r}_i^{k-1}} = \frac{\delta_k(\lambda_i)}{\delta_k(0)}.$$

Proof. From Proposition 1.27, we have for the Lanczos vectors,

$$\frac{\bar{v}_i^{k+1}}{\bar{v}_i^k} = -\frac{\delta_k(\lambda_i)}{\eta_{k+1}}.$$

Then,

$$\frac{\bar{r}_i^k}{\bar{r}_i^{k-1}} = \frac{\|r^k\|}{\|r^{k-1}\|}\frac{\delta_k(\lambda_i)}{\eta_{k+1}} = \frac{\delta_k(\lambda_i)}{\delta_k(0)} = \frac{p_{k+1}(\lambda_i)}{p_{k+1}(0)}\frac{p_k(0)}{p_k(\lambda_i)}. \quad \square$$

The growth or decrease of the ith component of $\bar{r}^k$ depends on the last pivot function at λ_i being larger or smaller than the value of the function at 0. This leads to the following bounds for the norms of the residual:

$$\frac{\min_i \delta_k(\lambda_i)^2}{\delta_k(0)^2}\|r^{k-1}\|^2 \le \|r^k\|^2 \le \frac{\max_i \delta_k(\lambda_i)^2}{\delta_k(0)^2}\|r^{k-1}\|^2.$$

2.4 The A-norm of the error

The A-norm of the error defined as $\|\epsilon^k\|_A = (A(x - x^k), x - x^k)^{1/2}$ is the most important measure of the error for CG, because, as we shall see, CG minimizes the A-norm of the error at each iteration. In this section we show that computing the A-norm of the error is closely related to Gauss quadrature for quadratic forms. This was studied extensively by Golub during the last 30 years and was summarized by Golub and Meurant in [64], from which most of the following is taken; see also [65]. The matrix A being real symmetric positive definite, the problem considered in [64] was to find upper and lower bounds (or approximations) for the entries of a function of a matrix. This problem leads us to consider the bilinear form

$$u^T f(A) v,$$

where u and v are given vectors and f is some smooth function on a given interval of the real line. As an example, if $f(x) = \frac{1}{x}$ and $u^T = (e^i)^T = (0, \ldots, 0, 1, 0, \ldots, 0)$, the nonzero element being in the ith position and $v = e^j$, we can obtain bounds on the elements of the inverse A^{-1}; see [64]. This is related to the problem of computing the A-norm of the error by noticing that the error ϵ^k is related to the residual r^k by the equation

$$A\epsilon^k = r^k.$$

Therefore,

$$\|\epsilon^k\|_A^2 = (A\epsilon^k, \epsilon^k) = (A^{-1}r^k, r^k) = (r^k)^T A^{-1} r^k.$$

So, here the function of interest is also $f(x) = \frac{1}{x}$, but we are concerned with the case of a quadratic form where $u = v = r^k$. Since it is symmetric, we write A as

$$A = Q\Lambda Q^T,$$

where Q is the orthonormal matrix whose columns are the normalized eigenvectors of A and Λ is a diagonal matrix whose diagonal elements are the eigenvalues λ_i. By definition, we have

$$f(A) = Qf(\Lambda)Q^T.$$

Therefore,

$$
\begin{aligned}
u^T f(A) v &= u^T Q f(\Lambda) Q^T v \\
&= y^T f(\Lambda) s, \\
&= \sum_{i=1}^{n} f(\lambda_i) y_i s_i .
\end{aligned}
$$

The last sum can be considered as a Riemann–Stieltjes integral

$$
I[f] = u^T f(A) v = \int_{\lambda_1}^{\lambda_n} f(\lambda)\, d\alpha(\lambda),
$$

where the measure α is piecewise constant and (supposing A has distinct eigenvalues) defined by

$$
\alpha(\lambda) = \begin{cases} 0 & \text{if } \lambda < \lambda_1, \\ \sum_{j=1}^{i} y_j s_j & \text{if } \lambda_i \le \lambda < \lambda_{i+1}, \\ \sum_{j=1}^{n} y_j s_j & \text{if } \lambda_n \le \lambda. \end{cases}
$$

In the case of interest when $u = v$, we note that α is a nondecreasing positive function. We are looking for methods to obtain upper and lower bounds m_L and m_U for $I[f]$,

$$
m_L \le I[f] \le m_U .
$$

A way to obtain bounds for the Stieltjes integrals is to use Gauss, Gauss–Radau, and Gauss–Lobatto quadrature formulas. This point of view was mainly developed by Golub; see [30], [62], [63]. For the Stieltjes integral, the general formula we shall use is

$$
\int_{\lambda_1}^{\lambda_n} f(\lambda)\, d\alpha(\lambda) = \sum_{j=1}^{N} w_j f(t_j) + \sum_{k=1}^{M} v_k f(z_k) + R[f],
$$

where the weights $[w_j]_{j=1}^N$, $[v_k]_{k=1}^M$ and the nodes $[t_j]_{j=1}^N$ are unknowns and the nodes $[z_k]_{k=1}^M$ are prescribed; see [32], [53], [54], [55], [69]. When $u = v$, which we consider in the following, it is known (see, for instance, [183]) that

$$
R[f] = \frac{f^{(2N+M)}(\eta)}{(2N+M)!} \int_{\lambda_1}^{\lambda_n} \prod_{k=1}^{M} (\lambda - z_k) \left[\prod_{j=1}^{N} (\lambda - t_j) \right]^2 d\alpha(\lambda), \qquad \lambda_1 < \eta < \lambda_n .
$$

If $M = 0$, this leads to the Gauss rule with no prescribed nodes. If $M = 1$ and we fix a node at one of the end points, $z_1 = \lambda_1$ or $z_1 = \lambda_n$, we have the Gauss–Radau formula. If $M = 2$ and $z_1 = \lambda_1$, $z_2 = \lambda_n$, this is the Gauss–Lobatto formula. Here, for simplicity we shall consider only the Gauss rule.

Let us recall briefly how the nodes and weights are obtained in the Gauss rule. For the measure α, it is possible to define a sequence of polynomials $p_1(\lambda)$, $p_2(\lambda)$, ... that are orthonormal with respect to α. It turns out (see [64]) that these polynomials are the Lanczos polynomials obtained by running the Lanczos algorithm starting from u normalized to 1. Looking at the Lanczos vectors v^k we have

$$(v^k, v^l) = \int_{\lambda_1}^{\lambda_n} p_k(\lambda) p_l(\lambda_l)\, d\alpha(\lambda).$$

We have already seen that the roots of p_k (the Ritz values) are distinct and real and lie in the interval $[\lambda_1, \lambda_n]$. In matrix form, the relation for the Lanczos polynomials can be written as

$$\lambda P(\lambda) = T_N P(\lambda) + \eta_{N+1} p_{N+1}(\lambda) e^N,$$

where

$$P(\lambda)^T = [p_1(\lambda)\ p_2(\lambda) \cdots\ p_N(\lambda)],$$

T_N being the tridiagonal matrix of the Lanczos coefficients. The eigenvalues of T_N (the Ritz values $\theta_i^{(N)}$) are the nodes of the Gauss quadrature rule (i.e., $M = 0$). The weights are the squares of the first elements of the normalized eigenvectors of T_N; see [69]. For the Gauss quadrature rule (renaming the weights and nodes w_j^G and t_j^G) we have

$$\int_{\lambda_1}^{\lambda_n} f(\lambda)\, d\alpha(\lambda) = \sum_{j=1}^{N} w_j^G f(t_j^G) + R_G[f],$$

and the next theorem follows.

Theorem 2.9. *Suppose $u = v$ and f is such that $f^{(2j)}(\xi) > 0$ for all j, for all ξ, $\lambda_1 < \xi < \lambda_n$, and let*

$$L_G[f] = \sum_{j=1}^{N} w_j^G f(t_j^G).$$

Then for all N, $\exists \eta \in [\lambda_1, \lambda_n]$ such that

$$L_G[f] \le I[f],$$

$$I[f] - L_G[f] = \frac{f^{(2N)}(\eta)}{(2N)!} \int_{\lambda_1}^{\lambda_n} \left[\prod_{j=1}^{N} (\lambda - t_j^G) \right]^2 d\alpha(\lambda).$$

Proof. See [183]. □

We remark that for obtaining bounds we need not always compute the eigenvalues and eigenvectors of the tridiagonal matrix. Let Z_N be the matrix of the eigenvectors of T_N whose columns we denote by z^i and Θ_N be the diagonal matrix of the eigenvalues $t_i^G = \theta_i^{(N)}$ (the Ritz values) which are the nodes of the Gauss quadrature rule. The weights are

$$w_i^G = (z_1^i)^2 = ((e^1)^T z^i)^2.$$

Theorem 2.10.

$$\sum_{l=1}^{N} w_l^G f(t_l^G) = (e^1)^T f(T_N) e^1.$$

Proof.

$$\begin{aligned}\sum_{l=1}^{N} w_l^G f(t_l^G) &= \sum_{l=1}^{N} (e^1)^T z^l f(t_l^G)(z^l)^T e^1 \\ &= (e^1)^T \left(\sum_{l=1}^{N} z^l f(t_l^G)(z^l)^T \right) e^1 \\ &= (e^1)^T Z_N f(\Theta_N) Z_N^T e^1 \\ &= (e^1)^T f(T_N) e^1. \quad \square\end{aligned}$$

In some cases where $f(T_N)$ is easily computable (for instance, if $f(\lambda) = 1/\lambda$), we do not need to compute the eigenvalues and eigenvectors of T_N to obtain bounds.

The previous developments show that when we have an approximation x^k and the corresponding residual r^k (wherever they come from) we can obtain bounds of the A-norm of the error by running some iterations of the Lanczos algorithm. Of course, this does not make too much sense when x^k is obtained by the Lanczos algorithm itself or CG. Therefore, we have to use something else to compute bounds or approximations of the norms of the error. In [119] the following results are proved concerning the A-norm of the error $\epsilon^k = x - x^k$ in CG.

Theorem 2.11.

$$\|\epsilon^k\|_A^2 = \|r^0\|^2[(T_n^{-1}e^1, e^1) - (T_k^{-1}e^1, e^1)]$$

and

$$\|\epsilon^k\|_A^2 = \|r^0\|^2 \left[\sum_{j=1}^{n} \frac{[(z_{(n)}^j)_1]^2}{\lambda_j} - \sum_{j=1}^{k} \frac{[(z_{(k)}^j)_1]^2}{\theta_j^{(k)}} \right],$$

where $z_{(k)}^j$ is the jth eigenvector of T_k.

Proof. The first relation has been well known for quite a long time; see the papers of Golub and his coauthors [30], [31]. It is also mentioned in a slightly different form in a paper by Paige, Parlett, and van der Vorst [138]. By using the definition of the A-norm and the relation $A\epsilon^k = r^k = r^0 - AV_k y^k$ we have

$$\|\epsilon^k\|_A^2 = (A\epsilon^k, \epsilon^k) = (A^{-1}r^0, r^0) - 2(r^0, V_k y^k) + (AV_k y^k, V_k y^k).$$

The first term of the right-hand side is easy to evaluate since $AV_n = V_n T_n$. The square matrix V_n of order n is orthogonal; hence this gives $A^{-1}V_n = V_n T_n^{-1}$. Now,

$$r^0 = \|r^0\| v^1 = \|r^0\| V_n e^1.$$

Therefore,

$$A^{-1}r^0 = \|r^0\| A^{-1} V_n e^1 = \|r^0\| V_n T_n^{-1} e^1$$

and

$$(A^{-1}r^0, r^0) = \|r^0\|^2(V_n T_n^{-1} e^1, V_n e^1) = \|r^0\|^2(T_n^{-1} e^1, e^1).$$

For the second term we have to compute $(r^0, V_k y^k)$. But since $r^0 = \|r^0\| v^1 = \|r^0\| V_k e^1$, this term is equal to $\|r^0\|^2(e^1, T_k^{-1} e^1)$ by using the orthogonality of the Lanczos vectors. The third term is $(A V_k y^k, V_k y^k)$. Using $V_k^T A V_k = T_k$ we have

$$(A V_k y^k, V_k y^k) = (V_k^T A V_k y^k, y^k) = (T_k y^k, y^k).$$

Hence $(A V_k y^k, V_k y^k) = \|r^0\|^2(T_k^{-1} e^1, e^1)$. This proves the formula in the theorem. The second relation is obtained by using the spectral decomposition of T_n and T_k. □

This formula is closely related to Gauss quadrature since, as we have seen before, the inner product $(T_n^{-1} e^1, e^1)$ (or $(A^{-1}r^0, r^0)$) can be written as a Riemann–Stieltjes integral. In fact this is exactly Gauss quadrature since $(T_k^{-1} e^1, e^1)$ is nothing other than the Gauss quadrature approximation to this integral; see [64]. It is interesting to consider this point of view because it allows us to compute the remainder of the quadrature rule whose sign is known and also lower and upper bounds (if we have estimates of λ_1) for the A-norm of the error. This formula has been used in [65], [117] to compute bounds for the A-norm of the error during the CG iterations by introducing an integer delay d writing

$$\|\epsilon^{k-d}\|_A^2 \simeq \|r^0\|^2[(T_k^{-1} e^1, e^1) - (T_{k-d}^{-1} e^1, e^1)]$$

and using additive relations between T_k^{-1} and T_{k-d}^{-1} to compute the difference. We shall go back to this point later. By using the derivative of the first pivot function we have the following result.

Theorem 2.12.

$$\|\epsilon^k\|_A^2 = \|r^0\|^2 \left[\sum_{j=1}^{n} \frac{1}{\lambda_j |[d_1^{(n)}]'(\lambda_j)|} - \sum_{j=1}^{k} \frac{1}{\theta_j^{(k)} |[d_1^{(k)}]'(\theta_j^{(k)})|} \right].$$

This shows exactly how the norm of the error depends on the eigenvalues of A and their approximation the Ritz values. For CG convergence the function which has to be considered is really $1/(\lambda |[d_1^{(k)}]'(\lambda)|)$, which has only one pole at 0. This is shown in Figure 2.2 for the Strakoš matrix of dimension 10, where this function is the solid curve on which the stars give the value of the function for $\theta_j^{(k)}$, which are the stars on the axis. The stars without a curve are the values of the function for step n at the eigenvalues of A, which are the crosses on the axis. These are the targets that the stars on the solid curve have to reach. We see that it is almost done for the largest eigenvalue. But this is far from being done for the smallest eigenvalue which gives the largest part of the error. In fact, the stars for the first eigenvalue are out of the picture since λ_1 is close to 0, which is a pole of the function. This explains why the CG error is usually dominated by the terms for the small eigenvalues when they are close to zero.

We now give results for the error vector and other expressions for its norm. We can express the error vector in the following way. Since $x^k = x^0 + V_k y^k$ and $r^k = r^0 - A V_k y^k$, we have

$$\epsilon^k = A^{-1} r^k = A^{-1} r^0 - V_k y^k.$$

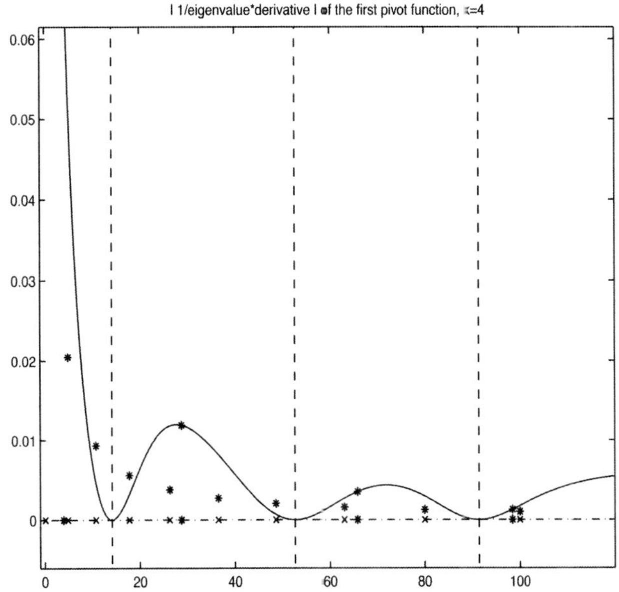

Figure 2.2. $1/\lambda|$ *derivative of first pivot function* $d_1^{(k)}|$

Using $A^{-1} = Q\Lambda^{-1}Q^T$ and $r^0 = \|r^0\| V_k e^1$, we have

$$\epsilon^k = \|r^0\|(A^{-1}V_k e^1 - V_k T_k^{-1} e^1)$$

and the projections of the error on the eigenvectors of A are

$$Q^T \epsilon^k = \|r^0\|(\Lambda^{-1}\bar{V}_k - \bar{V}_k T_k^{-1})e^1.$$

Looking at the ith component of $Q^T \epsilon_k$

$$(Q^T \epsilon^k)_i = \|r^0\| \sum_{j=1}^{k} \xi_j^{(i)} \bar{v}_i^j,$$

with

$$\xi_1^{(i)} = \frac{1}{\lambda_i} - (T_k^{-1}e^1, e^1), \quad \xi_j^{(i)} = -(T_k^{-1}e^1, e^j), \quad j = 2, \ldots, k.$$

We can write $r^0 = \|r^0\| V_n e^1$ and $A^{-1}V_n = V_n T_n^{-1}$. Therefore

$$\epsilon^k = \|r^0\| \left(V_n T_n^{-1} e^1 - \frac{1}{\|r^0\|} V_k y^k \right).$$

Theorem 2.13.

$$\epsilon^k = \|r^0\|(V_n T_n^{-1} e^1 - V_k T_k^{-1} e^1).$$

Proof. The result is obtained using the definition of $y^k = \|r^0\| T_k^{-1} e^1$. $\square$

This gives the decomposition of the error at iteration k over the vectors v^j of the orthonormal basis. This relation can be decomposed into two parts with $i = 1, \dots, k$ and $i = k+1, \dots, n$ since $V_n = (\, V_k \quad \tilde{V}_{n-k}\,)$,

$$\epsilon^k = \|r^0\|^2 (V_k[(T_n^{-1} e^1)_{1:k} - T_k^{-1} e^1] + \tilde{V}_{n-k}(T_n^{-1} e^1)_{k+1:n}),$$

where $y_{i:j}$ denotes the vector made of the components of y from i to j. Using the spectral decompositions of T_n^{-1} and T_k^{-1},

$$\epsilon^k = \|r^0\|(V_n Z_n \Lambda^{-1} Z_n^T e^1 - V_k Z_k \Theta_k^{-1} Z_k^T e^1),$$

and using the Ritz vectors which are the columns of $X_k = V_k Z_k$,

$$\epsilon_i^k = \|r^0\| \left(\sum_{j=1}^{n} \frac{(x_{(n)}^j)_i (z_{(n)}^j)_i}{\lambda_j} - \sum_{j=1}^{k} \frac{(x_{(k)}^j)_i (z_{(k)}^j)_i}{\theta_j^{(k)}} \right).$$

Following what we have done before, we can obtain another expression for the A-norm of the error by relating it to the CG coefficients and the norm of the residual.

Theorem 2.14.

$$\|\epsilon^k\|_A^2 = \|r^k\|^2 [\gamma_k + \beta_{k+1}\gamma_{k+1} + \beta_{k+1}\beta_{k+2}\gamma_{k+2} + \cdots + \beta_{k+1}\cdots\beta_{n-1}\gamma_{n-1}],$$

$$\|\epsilon^k\|_A^2 = \sum_{j=k}^{n-1} \gamma_j \|r^j\|^2.$$

Proof. By induction using the notations of Proposition 2.2 we have

$$d_1 - \bar{d}_1 = \frac{\eta_2^2 \cdots \eta_k^2}{(d_2 \cdots d_k)(\bar{d}_2 \cdots \bar{d}_k)} \frac{\eta_{k+1}^2}{\alpha_{k+1}} = \frac{(\eta_2^2 \cdots \eta_{k+1}^2) d_1 \bar{d}_1}{\det(T_k)\det(T_{k+1})}.$$

Therefore,

$$\bar{x}_1 - x_1 = \frac{\eta_2^2 \cdots \eta_{k+1}^2}{\det(T_k)\det(T_{k+1})}.$$

By using this formula for all the indices between k and n, if we denote by $x_1^{(j)}$ the first element of the first column of the inverse of T_j and $\chi_j = \det(T_j)$, we obtain by summing up these relations,

$$x_1^{(n)} - x_1^{(k)} = \eta_2^2 \cdots \eta_{k+1}^2 \left[\frac{1}{\chi_k \chi_{k+1}} + \frac{\eta_{k+2}^2}{\chi_{k+1}\chi_{k+2}} + \cdots + \frac{\eta_{k+2}^2 \cdots \eta_n^2}{\chi_{n-1}\chi_n} \right].$$

But

$$\chi_k = \frac{1}{\gamma_0 \cdots \gamma_{k-1}}, \qquad \chi_k \chi_{k+1} = \frac{\chi_k^2}{\gamma_k},$$

and

$$\chi_{k+j}\chi_{k+j+1} = \frac{1}{\gamma_0 \cdots \gamma_{k+j-1}} \frac{1}{\gamma_0 \cdots \gamma_{k+j}} = \frac{\chi_k^2}{(\gamma_k \cdots \gamma_{k+j-1})^2 \gamma_{k+j}}.$$

Therefore,

$$\frac{\eta_{k+2}^2 \cdots \eta_{k+j+1}^2}{\chi_{k+j}\chi_{k+j+1}} = \frac{1}{\chi_k^2} \eta_{k+2}^2 \cdots \eta_{k+j+1}^2 (\gamma_k \cdots \gamma_{k+j-1})^2 \gamma_{k+j} = \frac{1}{\chi_k^2} \beta_{k+1} \cdots \beta_{k+j} \gamma_{k+j}.$$

We also have

$$\frac{\eta_2^2 \cdots \eta_{k+1}^2}{\chi_k^2} = \beta_1 \cdots \beta_k = \frac{\|r^k\|^2}{\|r^0\|^2}.$$

Hence

$$x_1^{(n)} - x_1^{(k)} = \frac{\|r^k\|^2}{\|r^0\|^2} [\gamma_k + \beta_{k+1}\gamma_{k+1} + \beta_{k+1}\beta_{k+2}\gamma_{k+2} + \cdots + \beta_{k+1} \cdots \beta_{n-1}\gamma_{n-1}].$$

This shows that

$$\|\epsilon^k\|_A^2 = \|r^k\|^2 [\gamma_k + \beta_{k+1}\gamma_{k+1} + \beta_{k+1}\beta_{k+2}\gamma_{k+2} + \cdots + \beta_{k+1} \cdots \beta_{n-1}\gamma_{n-1}].$$

The other relation in the theorem is proved from the first one by using the definition of β_k. □

The first relation relates the A-norm of the error to the l_2 norm of the residual. It tells us when the norm of the residual is close to the norm of the error. However, at step k of the CG algorithm the right-hand side is not computable since it involves terms from iterations k to n. This result tells us that $\|\epsilon^k\|_A < \|r^k\|$ when

$$\gamma_k + \beta_{k+1}\gamma_{k+1} + \beta_{k+1}\beta_{k+2}\gamma_{k+2} + \cdots + \beta_{k+1} \cdots \beta_{n-1}\gamma_{n-1} < 1$$

or

$$\sum_{j=k}^{n-1} \gamma_j \|r^j\|^2 < \|r^k\|^2.$$

Since the sum is larger than $\gamma_k \|r^k\|^2$, a necessary condition for this inequality to hold is $\gamma_k < 1$.

The second relation of Theorem 2.14 was implicit in some results of Golub and Meurant [65]. It is related to a result proved in Hestenes and Stiefel [93, Theorem 6.1, p. 416], which was stated as

$$\|\epsilon^k\|_A^2 - \|\epsilon^l\|_A^2 = \sum_{j=k}^{l-1} \gamma_j \|r^j\|^2, \quad l > k.$$

An elementary proof was given in a nice paper by Strakoš and Tichý [186],

$$\begin{aligned} \|\epsilon^k\|_A^2 - \|\epsilon^{k+1}\|_A^2 &= \|x - x^{k+1} + x^{k+1} - x^k\|_A^2 - \|\epsilon^{k+1}\|_A^2 \\ &= \|x^{k+1} - x^k\|_A^2 + 2(x - x^{k+1})^T A(x^{k-1} - x^k) \\ &= \gamma_k^2 (p_k, Ap^k) + 2(r^{k+1}, x^{k+1} - x^k) = \gamma_k \|r^k\|^2. \end{aligned}$$

The Hestenes and Stiefel result was proved completely differently from ours and it gives only the difference of norms at different iterations, but since $\epsilon^n = 0$ in exact arithmetic, it is equivalent to our result. Unfortunately, this result has been mostly unnoticed during the years since the discovery of CG, although it can also be used to derive approximations of the A-norm.

Since we have $\|\epsilon^k\|_A^2 - \|\epsilon^{k+1}\|_A^2 = \gamma_k \|r^k\|^2$ we see that if the difference of the A-norms of the errors is small (resp., large), then $\gamma_k \|r^k\|^2$ is small (resp., large). Notice that we have lower and upper bounds on γ_k. Hence the difference of the norms behaves more or less like the norm of the residual. It is likely that oscillations in $\|r^k\|$ correspond to changes between almost stagnation and more rapid decrease of the A-norm of the error, although this depends also on the values of γ_k. This is illustrated in Figure 2.3 for the matrix Bcsstk01. The rapid decrease of the error norm corresponds more or less to peaks of the residual norm. Other ways to express the A-norm of the error are the following.

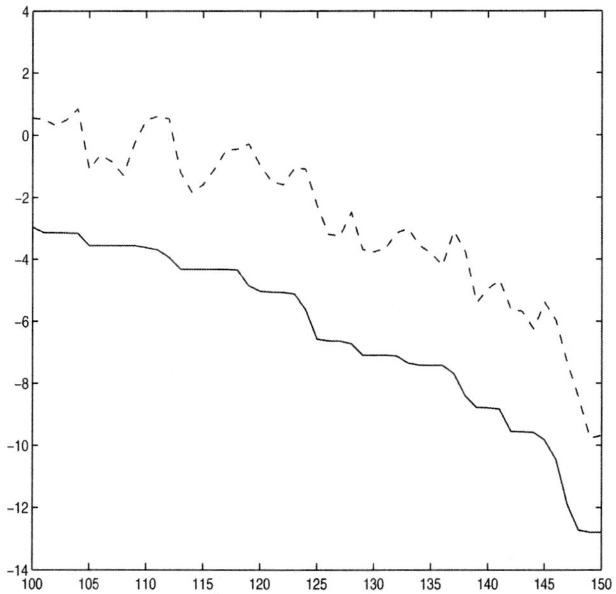

Figure 2.3. $\log_{10}$ *of the norm of the residual (dashed) and the A-norm of the error (solid) for Bcsstk01*

Theorem 2.15.

$$\|\epsilon^k\|_A^2 = -\eta_{k+1}\|r^0\|^2 (e^k, T_k^{-1} e^1)(T_n^{-1} e^1, f^{k+1}),$$

where f^{k+1} is a vector whose components are all zero except for the $k+1$st one, which is 1. Moreover, for $k < n$

$$\|\epsilon^k\|_A^2 = \|r^k\|^2 |(T_n^{-1})_{k+1,k+1}| = \|r^k\|^2 \sum_{j=1}^{n} \frac{[(z_{(n)}^j)_{k+1}]^2}{\lambda_j}.$$

Proof. We remark that

$$\|\epsilon^k\|_A^2 = (A^{-1}r^k, r^k) = (A^{-1}r^0, r^k) - (V_k y^k, r^k) = (A^{-1}r^0, r^k).$$

Since $AV_n = V_n T_n$, we have $V_n T_n^{-1} = A^{-1}V_n$ and $A^{-1}r^0 = \|r^0\| V_n T_n^{-1} e^1$. Now since

$$r^k = b - Ax^k = r^0 - AV_k y^k,$$

using the Lanczos relation, we have

$$r^k = r^0 - V_k T_k y^k - \eta_{k+1} v^{k+1} (e^k)^T y^k.$$

The first two terms of the right-hand side cancel by the definition of y^k and we obtain that

$$r^k = -\eta_{k+1} \|r^0\| (e^k, T_k^{-1} e^1) v^{k+1}.$$

This again shows that r^k is a scalar multiple of v^{k+1}. Then,

$$\|\epsilon^k\|_A^2 = -\eta_{k+1} \|r^0\|^2 (e^k, T_k^{-1} e^1)(V_n T_n^{-1} e^1, v^{k+1}).$$

But $V_n^T v^{k+1} = f^{k+1}$. This proves the first result. We remark that this proof also shows that

$$\frac{\|r^k\|}{\|r^0\|} = (-1)^{k+1} \eta_{k+1} (e^k, T_k^{-1} e^1),$$

as we already know, and

$$\|\epsilon^k\|_A^2 = (-1)^k \|r^0\| \, \|r^k\| (T_n^{-1} e^1)_{k+1}.$$

From the previous results we can also write

$$\|\epsilon^k\|_A^2 = \eta_{k+1}^2 \|r^0\|^2 (e^k, T_k^{-1} e^1)^2 (v^{k+1}, A^{-1} v^{k+1}).$$

But,

$$(v^{k+1}, A^{-1} v^{k+1}) = (V_n f^{k+1}, V_n T_n^{-1} f^{k+1}) = (f^{k+1}, T_n^{-1} f^{k+1}),$$

which shows the result of the theorem. This also shows that

$$\frac{1}{p_{k+1}(0)} = (-1)^{k+1} \eta_{k+1} (e^k, T_k^{-1} e^1)$$

and

$$\eta_{k+1} (T_k^{-1})_{1,k} (T_n^{-1})_{k+1,k+1} = -(T_n^{-1})_{1,k+1}. \quad \square$$

2.5 The l_2 norm of the error

In this section we shall obtain expressions for the l_2 norm of the error. Since this measure of the error is not as closely linked to the algorithm as the A-norm of the error, the expressions we shall obtain are more complicated. However, some of them can be used to obtain approximations of the l_2 norm of the error if it is needed. Hestenes and Stiefel [93] proved the following result relating the l_2 norm and the A-norm of the error.

Theorem 2.16.

$$\|\epsilon^k\|^2 - \|\epsilon^{k+1}\|^2 = \frac{\|\epsilon^k\|_A^2 + \|\epsilon^{k+1}\|_A^2}{\mu(p^k)},$$

with

$$\mu(p^k) = \frac{(p^k, Ap^k)}{\|p^k\|^2}.$$

Proof. The following proof is different from [93]. Since

$$p^k = r^k + \frac{\|r^k\|^2}{\|r^{k-1}\|^2} r^{k-1} + \cdots + \frac{\|r^k\|^2}{\|r^0\|^2} r^0,$$

and

$$\epsilon^{k+1} = \epsilon^k - \gamma_k p^k,$$

we have

$$\epsilon^{k+1} = \epsilon^k - \gamma_k \left(r^k + \frac{\|r^k\|^2}{\|r^{k-1}\|^2} r^{k-1} + \cdots + \frac{\|r^k\|^2}{\|r^0\|^2} r^0 \right).$$

The l_2 norm is expressed as

$$\|\epsilon^{k+1}\|^2 = \|\epsilon^k\|^2 - 2\gamma_k(\epsilon^k, p^k) + \gamma_k^2(p^k, p^k).$$

It is not too difficult to see that

$$(\epsilon^k, r^j) = (A^{-1}r^k, r^j) = (A^{-1}r^k, r^k) = \|\epsilon^k\|_A^2, \quad j \le k.$$

Therefore,

$$(\epsilon^k, p^k) = \|\epsilon^k\|_A^2 \|r^k\|^2 \left(\frac{1}{\|r^k\|^2} + \frac{1}{\|r^{k-1}\|^2} + \cdots + \frac{1}{\|r^0\|^2} \right).$$

But,

$$(p^k, p^k) = \|r^k\|^4 \left(\frac{1}{\|r^k\|^2} + \frac{1}{\|r^{k-1}\|^2} + \cdots + \frac{1}{\|r^0\|^2} \right).$$

Hence,

$$\|\epsilon^{k+1}\|^2 = \|\epsilon^k\|^2 + \gamma_k \frac{\|p^k\|^2}{\|r^k\|^2} (\gamma_k \|r^k\|^2 - 2\|\epsilon^k\|_A^2).$$

Since $\gamma_k \|r^k\|^2 = \|\epsilon^k\|_A^2 - \|\epsilon^{k+1}\|_A^2$,

$$\|\epsilon^{k+1}\|^2 = \|\epsilon^k\|^2 - \frac{\|p^k\|^2}{(Ap^k, p^k)} (\|\epsilon^{k+1}\|_A^2 + \|\epsilon^k\|_A^2). \quad \square$$

This result proves that the l_2 norm of the error is monotonically decreasing when k increases.

Corollary 2.17.

$$\|\epsilon^k\|^2 = \|\epsilon^0\|^2 - \sum_{j=0}^{k-1} \gamma_k \frac{\|p^j\|^2}{\|r^j\|^2}(\|\epsilon^{j+1}\|_A^2 + \|\epsilon^j\|_A^2).$$

We can also obtain expressions for the l_2 norm of the error using the same techniques as before for the A-norm, even though it is more complicated.

Theorem 2.18.

$$\|\epsilon^k\|^2 = \|r^0\|^2[(e^1, T_n^{-2}e^1) - (e^1, T_k^{-2}e^1)] + (-1)^k 2\eta_{k+1}\frac{\|r^0\|}{\|r^k\|}(e^k, T_k^{-2}e^1)\|\epsilon^k\|_A^2.$$

Proof. We have

$$\|\epsilon^k\|^2 = (A^{-1}r^0, A^{-1}r^0) - 2(A^{-1}r^0, V_k y^k) + (V_k y^k, V_k y^k).$$

The first and last terms of the right-hand side are easy to handle. They are, respectively, $\|r^0\|^2(e^1, T_n^{-2}e^1)$ and $\|r^0\|^2(e^1, T_k^{-2}e^1)$. The troubles come from the term $(A^{-1}r^0, V_k y^k)$. The matrix A is symmetric; therefore

$$(A^{-1}r^0, V_k y^k) = (r^0, A^{-1}V_k y^k).$$

Then we use

$$V_k T_k^{-1} = A^{-1}V_k + \eta_{k+1}A^{-1}v^{k+1}(e^k)^T T_k^{-1}$$

and

$$(r^0, A^{-1}V_k y^k) = (r^0, V_k T_k^{-1} y^k) - \eta_{k+1}(e^k)^T T_k^{-1} y_k (r^0, A^{-1}v^{k+1}),$$

but $V_k r^0 = \|r^0\|e^1$ and $T_k^{-1}y^k = \|r^0\|T_k^{-2}e^1$. Then we have to consider $(r^0, A^{-1}v^{k+1})$. We have seen that

$$(r^0, A^{-1}v^{k+1}) = \frac{(-1)^k}{\|r^k\|}(r^0, A^{-1}r^k).$$

But

$$(r^0, A^{-1}r^k) = (r^0, \epsilon^k) = (r^0, \epsilon^0) - (r^0, V_k y^k).$$

Therefore,

$$(r^0, A^{-1}r^k) = (r^0, A^{-1}r^0) - \|r^0\|^2(e^1, T_k^{-1}e^1) = \|\epsilon^k\|_A^2. \quad \square$$

The formula for the l_2 norm can be written in an alternate way.

Corollary 2.19.

$$\|\epsilon^k\|^2 = \|r^0\|^2[(e^1, T_n^{-2}e^1) - (e^1, T_k^{-2}e^1)] - 2\frac{(e^k, T_k^{-2}e^1)}{(e^k, T_k^{-1}e^1)}\|\epsilon^k\|_A^2.$$

Proof. We have seen before that

$$r^k = -\eta_{k+1}\|r^0\|(e^k, T_k^{-1}e^1)v^{k+1}.$$

Therefore

$$\frac{(-1)^{k+1}\|r^k\|}{\eta_{k+1}} = \|r^0\|(e^k, T_k^{-1}e^1). \quad \square$$

We shall see later that we can use the QR factorization of T_k to approximate $\|\epsilon^k\|^2$. However, it is interesting to look for expressions for $\|r^0\|^2[(e^1, T_{k+1}^{-2}e^1) - (e^1, T_k^{-2}e^1)]$.

Proposition 2.20.

$$\|r^0\|^2[(e^1, T_{k+1}^{-2}e^1) - (e^1, T_k^{-2}e^1)] = 2\gamma_k\|r^k\|^2\sum_{j=0}^{k-1}\gamma_j\frac{\|p^j\|^2}{\|r^j\|^2} + \gamma_k^2\|p^k\|^2.$$

Proof. Let D_k be a diagonal matrix with diagonal elements plus or minus the inverses of the norms of the residual vectors and $\bar{T}_k$ the tridiagonal matrix such that $T_k = D_k^{-1}\bar{T}_k D_k$. Let s^k be the solution of $T_k s^k = e^1$. We have

$$\bar{T}_k D_k s^k = D_k e^1 = \frac{1}{\|r^0\|}e^1$$

and

$$y^k = \|r^0\|T_k^{-1}e^1, \quad s^k = \frac{1}{\|r^0\|}y^k.$$

Therefore,

$$\begin{aligned}
s_1^k &= \gamma_0 + \frac{\|r^1\|^2}{\|r^0\|^2}\gamma_1 + \cdots + \frac{\|r^{k-1}\|^2}{\|r^0\|^2}\gamma_{k-1},\\
s_2^k &= -\frac{\|r^1\|}{\|r^0\|}\left[\gamma_1 + \frac{\|r^2\|^2}{\|r^1\|^2}\gamma_2 + \cdots + \frac{\|r^{k-1}\|^2}{\|r^1\|^2}\gamma_{k-1}\right],\\
\vdots &= \vdots\\
s_{k-1}^k &= (-1)^{k-2}\frac{\|r^{k-2}\|}{\|r^0\|}\left[\gamma_{k-2} + \frac{\|r^{k-1}\|^2}{\|r^{k-2}\|^2}\right],\\
s_k^k &= (-1)^{k-1}\frac{\|r^{k-1}\|}{\|r^0\|}\gamma_{k-1}.
\end{aligned}$$

Let e be the vector of all ones. We have

$$s^{k+1} = \begin{pmatrix} s^k + \gamma_k\frac{\|r^k\|^2}{\|r^0\|}D_k e \\ (-1)^k\frac{\|r^k\|}{\|r^0\|}\gamma_k \end{pmatrix}.$$

Then,

$$\|r^0\|^2[(e^1, T_{k+1}^{-2}e^1)-(e^1, T_k^{-2}e^1)] = 2\|r^0\|\gamma_k\|r^k\|^2(D_k e, s^k)+\|r^k\|^4\gamma_k^2(D_k e, D_k e)+\gamma_k^2\|r^k\|^2.$$

Thus, we have reduced this computation to quantities only involving $T_k^{-1}e^1$. Moreover, we have

$$2\|r^0\|\gamma_k\|r^k\|^2(D_k e, s^k) = 2\gamma_k\|r^k\|^2\left[\gamma_0 + \gamma_1\|r^1\|^2\left(\frac{1}{\|r^0\|^2} + \frac{1}{\|r^1\|^2}\right)\right.$$
$$+\gamma_2\|r^2\|^2\left(\frac{1}{\|r^0\|^2} + \frac{1}{\|r^1\|^2} + \frac{1}{\|r^2\|^2}\right)$$
$$\left. + \cdots + \gamma_{k-1}\|r^{k-1}\|^2\left(\frac{1}{\|r^0\|^2} + \cdots + \frac{1}{\|r^{k-1}\|^2}\right)\right].$$

Using the expression for $\|p^k\|^2$ this is

$$2\gamma_k\|r^k\|^2\sum_{j=0}^{k-1}\gamma_j\frac{\|p^j\|^2}{\|r^j\|^2}.$$

Moreover,

$$\gamma_k^2\|r^k\|^4(D_k e, D_k e) + \gamma_k^2\|r^k\|^2 = \gamma_k^2\|p^k\|^2 \quad \square$$

We can also compute the term involving $\|\epsilon^k\|_A^2$ in the l_2 norm of the error.

Proposition 2.21.

$$\frac{(T_k^{-2})_{1,k}}{(T_k^{-1})_{1,k}} = \sum_{j=0}^{k-1}\gamma_j\frac{\|p^j\|^2}{\|r^j\|^2}.$$

Proof. We need the $(1,k)$ entry of T_k^{-2}. This can be written as $(T_k^{-1}e^k, T_k^{-1}e^1)$. We have to compute the last column t^k of T_k^{-1}. We can do this by using the LU decomposition of $\bar{T}_k$. We obtain

$$t_1^k = (-1)^{k-1}\frac{\|r^{k-1}\|}{\|r^0\|}\gamma_{k-1},$$
$$t_2^k = (-1)^{k-2}\frac{\|r^{k-1}\|}{\|r^1\|}\gamma_{k-1},$$
$$\vdots = \vdots$$
$$t_{k-1}^k = -\frac{\|r^{k-1}\|}{\|r^{k-2}\|}\gamma_{k-1},$$
$$t_k^k = \gamma_{k-1}.$$

We multiply by $(s^k)^T$ and divide by

$$(T_k^{-1})_{1,k} = (-1)^{k-1}\frac{\|r^{k-1}\|}{\|r^0\|}\gamma_{k-1}$$

to obtain

$$\frac{(T_k^{-2})_{1,k}}{(T_k^{-1})_{1,k}} = \gamma_0 + \gamma_1 \|r^1\|^2 \left(\frac{1}{\|r^0\|^2} + \frac{1}{\|r^1\|^2} \right) + \cdots$$
$$+ \gamma_{k-1} \|r^{k-1}\|^2 \left(\frac{1}{\|r^0\|^2} + \cdots + \frac{1}{\|r^{k-1}\|^2} \right).$$

This can also be written as

$$\gamma_0 + \gamma_1 \frac{\|p^1\|^2}{\|r^1\|^2} + \cdots + \gamma_{k-1} \frac{\|p^{k-1}\|^2}{\|r^{k-1}\|^2}. \quad \square$$

The last two results can be used to prove the following result that we have already seen in the proof of Theorem 2.16.

Proposition 2.22.

$$\|\epsilon^k\|^2 - \|\epsilon^{k+1}\|^2 = 2\gamma_k \frac{\|p^k\|^2}{\|r^k\|^2} \|\epsilon^k\|_A^2 - \gamma_k^2 \|p^k\|^2.$$

Proof. We have

$$\|\epsilon^k\|^2 - \|\epsilon^{k+1}\|^2 = \|r^0\|^2 ((T_{k+1}^{-1})_{1,1} - (T_k^{-1})_{1,1}) - 2 \frac{(T_k^{-2})_{1,k}}{(T_k^{-1})_{1,k}} \|\epsilon^k\|_A^2 + 2 \frac{(T_{k+1}^{-2})_{1,k+1}}{(T_{k+1}^{-1})_{1,k+1}} \|\epsilon^{k+1}\|_A^2.$$

By using the previous results the first set of parentheses on the right-hand side is equal to

$$2\gamma_k \|r^k\|^2 \sum_{j=0}^{k-1} \gamma_j \frac{\|p^j\|^2}{\|r^j\|^2} + \gamma_k^2 \|p^k\|^2.$$

The terms involving the A-norm of the error are

$$2\|\epsilon^{k+1}\|_A^2 \sum_{j=0}^{k} \gamma_j \frac{\|p^j\|^2}{\|r^j\|^2} - 2\|\epsilon^k\|_A^2 \sum_{j=0}^{k-1} \gamma_j \frac{\|p^j\|^2}{\|r^j\|^2}.$$

By using $\|\epsilon^{k+1}\|_A^2 = \|\epsilon^k\|_A^2 - \gamma_k \|r^k\|^2$, this is

$$2\|\epsilon^k\|_A^2 \gamma_k \frac{\|p^k\|^2}{\|r^k\|^2} - 2\gamma_k \|r^k\|^2 \sum_{j=0}^{k} \gamma_j \frac{\|p^j\|^2}{\|r^j\|^2}.$$

Putting all this together, we obtain the result. $\square$

The last result proves that the formulation involving the inverses of tridiagonal matrices is (of course) equivalent to what was proved by Hestenes and Stiefel. Finally, we give expressions of the residual and error norms involving the Ritz values and the eigenvalues of A. The expressions of the residual and the error using the Lanczos polynomials are

$$r^k = (-1)^k \frac{1}{|p_{k+1}(0)|} p_{k+1}(A) r^0,$$

$$\epsilon^k = (-1)^k \frac{1}{|p_{k+1}(0)|} p_{k+1}(A) A^{-1} r^0.$$

Taking norms and using the spectral decomposition of A, this leads to

$$\|r^k\|^2 = \frac{1}{[p_{k+1}(0)]^2} \sum_{j=1}^{n} p_{k+1}(\lambda_j)^2 (\bar{r}_j^0)^2,$$

$$\|\epsilon^k\|^2 = \frac{1}{[p_{k+1}(0)]^2} \sum_{j=1}^{n} \frac{p_{k+1}(\lambda_j)^2}{\lambda_j^2} (\bar{r}_j^0)^2,$$

$$\|\epsilon^k\|_A^2 = \frac{1}{[p_{k+1}(0)]^2} \sum_{j=1}^{n} \frac{p_{k+1}(\lambda_j)^2}{\lambda_j} (\bar{r}_j^0)^2,$$

where $\bar{r}^0 = Q^T r^0$ is the vector of the projections of the initial residual on the eigenvectors of A. We see that the three norms differ only by the weights of $p_{k+1}(\lambda_j)^2$. The Lanczos polynomial has the Ritz values as roots. Therefore, we have the following result.

Theorem 2.23.

$$\|r^k\|^2 = \sum_{j=1}^{n} \prod_{i=1}^{k} \left(1 - \frac{\lambda_j}{\theta_i^{(k)}}\right)^2 (\bar{r}_j^0)^2,$$

$$\|\epsilon^k\|^2 = \sum_{j=1}^{n} \prod_{i=1}^{k} \left(\frac{1}{\lambda_j} - \frac{1}{\theta_i^{(k)}}\right)^2 (\bar{r}_j^0)^2,$$

$$\|\epsilon^k\|_A^2 = \sum_{j=1}^{n} \prod_{i=1}^{k} \left(\frac{1}{\sqrt{\lambda_j}} - \frac{\sqrt{\lambda_j}}{\theta_i^{(k)}}\right)^2 (\bar{r}_j^0)^2.$$

Proof. We remark that

$$\frac{p_{k+1}(\lambda_j)^2}{p_{k+1}(0)^2} = \prod_{i=1}^{k} \left(1 - \frac{\lambda_j}{\theta_i^{(k)}}\right)^2.$$

This proves the results. □

2.6 Other forms of the CG algorithm

In this section we derive alternate forms of CG and prove optimality properties. Since CG can be derived from the Lanczos algorithm, it does not come as a surprise that there is a three-term recurrence form of CG. It can be obtained from the two-term form by eliminating vectors p^k or derived directly. This is what we are going to do in the following. We postulate that there exists a relation

$$x^{k+1} = \nu_{k+1}(\mu_k r^k + x^k - x^{k-1}) + x^{k-1},$$

where ν_{k+1} and μ_k are real parameters to be determined. This gives us a relation for the residuals $r^k = b - Ax^k$,

$$r^{k+1} = r^{k-1} - \nu_{k+1}(\mu_k Ar^k - r^k + r^{k-1}).$$

The parameters are computed as in the two-term form by requiring that r^{k+1} is orthogonal to r^k and r^{k-1}.

Proposition 2.24. *If μ_k is chosen as*

$$\mu_k = \frac{(r^k, r^k)}{(r^k, Ar^k)},$$

then $(r^k, r^{k+1}) = 0$. If ν_{k+1} is chosen as

$$\nu_{k+1} = \frac{1}{1 + \mu_k \frac{(r^{k-1}, Ar^k)}{(r^{k-1}, r^{k-1})}},$$

then $(r^{k-1}, r^{k+1}) = 0$.

Proof. See [26] or [120]. □

It is easily seen (for instance in [120]) that there is an alternate expression for ν_{k+1},

$$\nu_{k+1} = \frac{1}{1 - \frac{\mu_k (r^k, r^k)}{\nu_k \mu_{k-1} (r^{k-1}, r^{k-1})}}.$$

This last formula is computationally more efficient since for computing the two coefficients μ_k and ν_{k+1} we have to compute only two scalars products instead of three. The iterations are started by taking $\nu_1 = 1$; then

$$x^1 = \mu_0 r^0 + x^0,$$

and we need to define only x^0 and $r^0 = b - Ax^0$. The first step is only a steepest descent iteration. Now, we must show that we have global orthogonality; that is, the new vector r^{k+1} is orthogonal not only to the last two but to all the previous vectors: $(r^{k+1}, r^j) = 0$, $0 \le j < k - 1$. This is done by induction; multiplying the equation defining r^{k+1} by r^j, $0 \le j < k - 1$, we have

$$(r^j, r^{k+1}) = (r^j, r^{k-1}) - \nu_{k+1}[\mu_k (r^j, Ar^k) - (r^j, r^k) + (r^j, r^{k-1})].$$

But, since $j < k-1$,

$$(r^j, r^{k+1}) = \nu_{k+1}\mu_k(r^j, Ar^k).$$

Now, we write the definition of r^{j+1} to get

$$r^{j+1} = r^{j-1} - \nu_{j+1}(\mu_j Ar^j - r^j + r^{j-1}).$$

Multiplying by r^k and taking into account that $j+1 < k$,

$$\nu_{j+1}\mu_j(r^k, Ar^j) = 0.$$

Then because A is symmetric we obtain $(Ar^k, r^j) = 0$. This shows that $(r^j, r^{k+1}) = 0$ for all j such that $j < k-1$. Therefore, as in the Lanczos algorithm and because A is symmetric, the local orthogonality with r^k and r^{k-1} implies the global orthogonality with all r^j, $j = k-2, \ldots, 0$. This particular form of the method has been popularized by Concus, Golub, and O'Leary [26]. We summarize the three-term variant of CG in the following algorithm.

ALGORITHM 2.2.

Let x^{-1} be arbitrary and x^0 be given, $r^0 = b - Ax^0$
for $k = 0, 1, \ldots$ until convergence do
$\mu_k = (r^k, r^k)/(r^k, Ar^k)$
$\nu_1 = 1, \quad \nu_{k+1} = 1/\left[1 - (\mu_k(r^k, r^k)/\{\nu_k\mu_{k-1}(r^{k-1}, r^{k-1})\})\right], k > 0$
$x^{k+1} = \nu_{k+1}(\mu_k r^k + x^k - x^{k-1}) + x^{k-1}$
$r^{k+1} = r^{k-1} - \nu_{k+1}(\mu_k Ar^k - r^k + r^{k-1})$
end

The three-term form of CG is more expensive than the two-term form. Here we also have two inner products and a matrix-vector product, but there are $10n$ other floating point additions and multiplications compared to $6n$ for the two-term version. The three-term recurrence is also reputed to be less stable (see Gutknecht and Strakoš [86]), but we shall come back to this point later when we study the algorithms in finite precision arithmetic. However, it has some interest for parallel computation. In the two-term form of CG there must be a synchronization point of all the processors after almost all the steps of the algorithm. The only things that can be done in parallel are the recursive computations of x^{k+1} and r^{k+1}. In the three-term variant the two inner products can be computed concurrently, and then all the other computations are parallel. Other variants suited for parallel computation were studied in [116].

Of course there are some relations between the coefficients of the two-term and three-term CG recurrences. To obtain them, we can eliminate p^k in the two-term recurrence and it follows that

$$\nu_{k+1} = 1 + \frac{\gamma_k\beta_k}{\gamma_{k-1}},$$

$$\mu_k = \frac{\gamma_k}{1 + \frac{\gamma_k\beta_k}{\gamma_{k-1}}}.$$

There are also relations between the three-term recurrence CG coefficients and those of the Lanczos algorithm. We write the three-term recurrence for the residuals as

$$r^{k+1} = -\nu_{k+1}\mu_k Ar^k + \nu_{k+1}r^k + (1 - \nu_{k+1})r^{k-1},$$

and we use the relation between the residuals and the Lanczos basis vectors $v^{k+1} = (-1)^k r^k/\|r^k\|$. This leads to

$$\mu_k = \frac{1}{\alpha_{k+1}}, \qquad \nu_{k+1} = 1 + \frac{\eta_{k+1}}{\eta_{k+2}} \frac{\|r^{k+1}\|}{\|r^{k-1}\|}.$$

These relations show that

$$\nu_k \geq 1, \quad \mu_k \leq \gamma_k.$$

This shows that

$$\gamma_k > \frac{1}{\lambda_n},$$

since

$$\frac{1}{\mu_k} = \frac{(Ar^k, r^k)}{(r^k, r^k)} \leq \lambda_n.$$

We are now going to see that CG has some optimality properties. First, we show that CG is a polynomial method as the Lanczos algorithm. Using the relation of the residuals and basis vectors, we have

$$r^k = (-1)^k \frac{\|r^k\|}{\|r^0\|} p_{k+1}(A) r^0.$$

From the three-term CG recurrence it is easy to show the following.

Lemma 2.25. r^{k+1} *is a polynomial in* A,

$$r^{k+1} = [I - As_k(A)]r^0,$$

where s_k *is a kth degree polynomial satisfying*

$$s_k(\lambda) = \mu_k \nu_{k+1} + \nu_{k+1}(1 - \mu_k \lambda) s_{k-1}(\lambda) - (\nu_{k+1} - 1) s_{k-2}(\lambda),$$

$$s_0(\lambda) = \mu_0, \quad s_1(\lambda) = \nu_2(\mu_0 + \mu_1 - \mu_0 \mu_1 \lambda).$$

Proof. The proof is straightforward by induction on k. □

Therefore we have a relation between the Lanczos and the CG polynomials

$$1 - \lambda s_{k-1}(\lambda) = (-1)^k \frac{1}{|p_{k+1}(0)|} p_{k+1}(\lambda).$$

Remember that p_{k+1} is a polynomial of degree k.

Proposition 2.26. *Let* s_k *be the polynomial defined in Lemma* 2.25. *The iterates of CG satisfy*

$$x^{k+1} = x^0 + s_k(A) r^0.$$

Proof. We have

$$x^{k+1} = x^{k-1} + \nu_{k+1}(\mu_k z^k + x^k - x^{k-1}).$$

By induction and with the help of Lemma 2.25 this is written as

$$x^{k+1} = x^0 + s_{k-2}(A)r^0 + \nu_{k+1}(\mu_k[I - As_{k-1}(A)]r^0 + s_{k-1}(A)r^0 - s_{k-2}(A)r^0),$$

$$x^{k+1} = x^0 + s_k(A)r^0,$$

because of the recurrence relation satisfied by s_k. □

Concerning the A-norm of the error we have an optimality result.

Theorem 2.27. *Consider all the iterative methods that can be written as*

$$\overline{x}^{k+1} = \overline{x}^0 + q_k(A)\overline{r}^0, \quad \overline{x}^0 = x^0, \quad \overline{r}^0 = b - A\overline{x}^0,$$

where q_k is a kth degree polynomial. Of all these methods, CG is the one which minimizes $\|\epsilon^k\|_A$ at each iteration.

Proof. See [120] for a proof. □

There exist other forms of the CG algorithm. For instance, the coefficients of the two-term form can be computed with other (mathematically equivalent) formulas. This was considered some time ago by Reid [149], who performed some numerical comparisons. The conclusion was that the formulas we gave before for the two-term version are the best ones. The other formulas are

$$\gamma_k = \frac{(r^k, p^k)}{(p^k, Ap^k)}, \quad \beta_k = -\frac{(r^k, Ap^{k-1})}{(p^{k-1}, Ap^{k-1})}.$$

Another variant of CG was given by Rutishauser [151]. The formulas are

$$\Delta r^k = (-Ar^k + \Delta r^{k-1}\zeta_{k-1})\tau_k^{-1},$$

$$\Delta x^k = (r^k + \Delta x^{k-1}\zeta_{k-1})\tau_k^{-1},$$

$$r^{k+1} = r^k + \Delta r^k,$$

$$x^{k+1} = x^k + \Delta x^k,$$

starting from $r^0 = b - Ax^0$, $\Delta r^{-1} = 0$, $\Delta x^{-1} = 0$, and $\zeta_{-1} = 0$. The coefficients are given by

$$\tau_k = \frac{(r^k, Ar^k)}{(r^k, r^k)} - \eta_{k-1}, \quad \zeta_k = \tau_k \frac{(r^{k+1}, r^{k+1})}{(r^k, r^k)}.$$

2.7 Bounds for the norms of the error

In this section we recall some of the classical bounds for the A-norm of the error only involving the condition number of the matrix A. The optimality property of CG in Theorem 2.27 leads to the classical bounds for the A-norm of the error which are the analogue for CG of the Kaniel–Paige–Saad bounds for the Lanczos algorithm; see [120] or [195].

Theorem 2.28.

$$\|\epsilon^k\|_A^2 \le \max_{1\le i\le n} (t_k(\lambda_i))^2 \|\epsilon^0\|_A^2$$

for all polynomials t_k of degree k such that $t_k(0) = 1$.

Proof. In Theorem 2.27, we show that the CG polynomial s_k minimizes $\|\epsilon^k\|_A$. Replacing the polynomial s_k by any other kth degree polynomial, we shall get an upper bound. This can be written as

$$\|\epsilon^k\|_A^2 \le \sum_{i=1}^{n} (t_k(\lambda_i))^2 (\bar{\epsilon}_i^0)^2,$$

where $\bar{\epsilon}^j = \Lambda^{1/2} Q^T \epsilon^j$. This holds for all polynomials t_k of degree k, such that $t_k(0) = 1$, equality holding only if $t_k(\lambda) = 1 - \lambda s_{k-1}(\lambda)$. Therefore,

$$\|\epsilon^k\|_A^2 \le \max_{1\le i\le n} (t_k(\lambda_i))^2 \sum_{i=1}^{n} (\bar{\epsilon}_i^0)^2.$$

But, $(\bar{\epsilon}^0, \bar{\epsilon}^0) = \|\epsilon^0\|_A^2$, which proves the result. □

We are free to choose the polynomial in these bounds, provided it satisfies the constraint at 0.

Proposition 2.29. *If A has only p distinct eigenvalues, then $\|\epsilon^p\|_A = 0$.*

Proof. We choose

$$t_k(\lambda) = \prod_{i=1}^{k} \left(1 - \frac{\lambda}{\lambda_i}\right).$$

Hence, $t_p(\lambda_i) = 0$ for all $i,\ 1 \le i \le n,$ so $\|\epsilon^p\|_A = 0$, by taking into account the distinct eigenvalues of A. □

This is also proved by the formula for the A-norm of the error involving the eigenvalues of A and T_k. The next result is the most well known bound on the A-norm of the error. It uses the condition number of A.

Theorem 2.30.

$$\|\epsilon^k\|_A \le 2\left(\frac{\sqrt{\kappa}-1}{\sqrt{\kappa}+1}\right)^k \|\epsilon^0\|_A,$$

where $\kappa = \frac{\lambda_n}{\lambda_1}$ is the condition number of A.

Proof. $\max_{1\le i\le n}(t_k(\lambda_i))^2$ is bounded by $\max_{\lambda_1\le\lambda\le\lambda_n}(t_k(\lambda))^2$. For t_k we choose the kth degree polynomial such that $t_k(0) = 1$, which minimizes the maximum. The solution to this problem is given by the shifted Chebyshev polynomials:

$$t_k(\lambda) = \frac{C_k(\frac{\lambda_1+\lambda_n-2\lambda}{\lambda_n-\lambda_1})}{C_k(\frac{\lambda_1+\lambda_n}{\lambda_n-\lambda_1})}.$$

By the properties of the Chebyshev polynomials,

$$\max_{\lambda_1 \le \lambda \le \lambda_n} |t_k(\lambda)| \le 2\left(\frac{\sqrt{\kappa}-1}{\sqrt{\kappa}+1}\right)^k.$$

This proves the result. □

In exact arithmetic, $\|\epsilon^k\|_A$ is bounded above by a decreasing sequence that converges to 0; see also Powell [147]. In most cases this bound is overly pessimistic. Studies of convergence have been done assuming some particular distribution of eigenvalues mainly well separated small or large eigenvalues or both; see, for instance, [4].

Bonnet and Meurant [13] proved that if we assume only exact local orthogonality, we can prove that CG still converges and as fast as the steepest descent method for which the descent direction is chosen as $p^k = r^k$. For the sake of simplicity let us suppose that we have

$$\beta_k = \frac{(r^k, r^k)}{(r^{k-1}, r^{k-1})},$$
$$(p^k, r^{k+1}) = 0,$$
$$(p^k, r^k) = (r^k, r^k),$$
$$(r^{k+1}, r^k) = 0.$$

Lemma 2.31. *Supposing only the previous relations, we have*

$$(Ap^k, p^k) = (Ar^k, r^k) - \beta_k^2(Ap^{k-1}, p^{k-1})$$

and so $(Ap^k, p^k) \le (Ar^k, r^k)$.

Proof. We have

$$Ap^k = Ar^k + \beta_k Ap^{k-1}.$$

Therefore,

$$(Ap^k, p^k) = (Ar^k, r^k) + 2\beta_k(Ar^k, p^{k-1}) + \beta_k^2(Ap^{k-1}, p^{k-1}),$$

but

$$(Ar^k, p^{k-1}) = (r^k, Ap^{k-1}) = (p^k - \beta_k p^{k-1}, Ap^{k-1}).$$

This shows that $(Ar^k, p^{k-1}) = -\beta_k(Ap^{k-1}, p^{k-1})$. Therefore, the results holds. □

Theorem 2.32. *Using only local orthogonality, CG converges and we have*

$$\|\epsilon^k\|_A \le \left(\frac{\kappa(A)-1}{\kappa(A)+1}\right)^k \|\epsilon^0\|_A.$$

Proof. We have

$$A^{-1}r^{k+1} = A^{-1}r^k - \gamma_k p^k.$$

Therefore

$$\begin{aligned}\|\epsilon^{k+1}\|_A^2 &= (A^{-1}r^{k+1}, r^{k+1}) \\ &= (A^{-1}r^k, r^{k+1}) - \gamma_k(p^k, r^{k+1}) \\ &= (A^{-1}r^k, r^{k+1}) \\ &= (A^{-1}r^k, r^k - \gamma_k A p^k) \\ &= (A^{-1}r^k, r^k) - \gamma_k(r^k, p^k).\end{aligned}$$

This shows that

$$\|\epsilon^{k+1}\|_A^2 = \|\epsilon^k\|_A^2 - \gamma_k\|r^k\|^2,$$

a relation already proved by Hestenes and Stiefel [93] that we have already seen. Since it involves only local orthogonality, it is likely that it will be approximately satisfied in finite precision arithmetic. Incidentally, this shows that $\|\epsilon^k\|_A^2$ is strictly decreasing unless $\|r^k\| = 0$. Hence,

$$\|\epsilon^{k+1}\|_A^2 = \|\epsilon^k\|_A^2 - \frac{\|r^k\|^4}{(Ap^k, p^k)} \le \|\epsilon^k\|_A^2 - \frac{\|r^k\|^4}{(Ar^k, r^k)}.$$

The proof is ended by using the Kantorovich inequality (for a proof of this inequality, see, for instance, [192]). □

Note that when using only local orthogonality the upper bound involves $\kappa(A)$ and not $\sqrt{\kappa(A)}$. This result shows that if we preserve local orthogonality, the convergence rate could be at worst that of steepest descent. Of course, this result is only of theoretical interest since the convergence rate of the steepest descent method can be quite slow. We shall see that even with a loss of orthogonality, CG usually converges much faster than steepest descent.

More refined bounds for the A-norm of the error than the one using Chebyshev polynomials were obtained by Greenbaum in her Ph.D. thesis [75] and were published in [74]. This used the kth degree minimax polynomial on the eigenvalues of A. This is a polynomial with value one at the origin, which minimizes the maximum deviation from zero on the set $\{\lambda_1, \dots, \lambda_n\}$. It takes its maximum absolute value on a set of $k+1$ points $\{\lambda_{\pi_1}, \dots, \lambda_{\pi_{k+1}}\}$, where π is a permutation. The polynomial is written as

$$t_k(\lambda) = \frac{\sum_{j=1}^{k+1}(-1)^{j-1}\prod_{\substack{i=1\\ i\neq j}}^{k+1}\frac{\lambda_{\pi_i}-\lambda}{\lambda_{\pi_i}-\lambda_{\pi_j}}}{\sum_{j=1}^{k+1}(-1)^{j-1}\prod_{\substack{i=1\\ i\neq j}}^{k+1}\frac{\lambda_{\pi_i}}{\lambda_{\pi_i}-\lambda_{\pi_j}}}.$$

There are weights for which t_k is the weighted least squares approximation to zero on $\{\lambda_{\pi_1}, \dots, \lambda_{\pi_{k+1}}\}$. This polynomial gives

$$\|\epsilon^k\|_A \le \frac{\|\epsilon^0\|_A}{\left|\sum_{j=1}^{k+1}(-1)^{j-1}\prod_{\substack{i=1\\ i\neq j}}^{k+1}\frac{\lambda_{\pi_i}}{\lambda_{\pi_i}-\lambda_{\pi_j}}\right|}.$$

This bound is optimal in the sense that for each k there is an initial error for which equality holds.

Chapter 3

A historical perspective on the Lanczos algorithm in finite precision

It has been known since Lanczos [108] that the properties of the algorithm in finite precision arithmetic are far from the theoretical ones. In particular, as a consequence of rounding errors, the Lanczos vectors do not stay orthogonal as they should. This also means that $V_k^T A V_k$ is no longer a tridiagonal matrix. However, the algorithm still computes a tridiagonal matrix T_k which is not the projection of A on the Krylov subspace. The matrix T_n is not similar to A and even after n iterations the algorithm may not deliver all the eigenvalues of A. This problem is related to what happens with the Gram–Schmidt orthogonalization algorithm. Remember we have seen that the Lanczos algorithm is mathematically equivalent to the Gram–Schmidt process on the Krylov matrix. For recent results about orthogonality in the Gram–Schmidt process, see [57], [58], [59], [60], [61]. Another annoying problem (which is also a consequence of the rounding errors) is the appearance of multiple copies of (approximations of) some eigenvalues of A in the set of converged Ritz values. These multiple copies form clusters of close eigenvalues of the computed Lanczos matrices T_k. It is sometimes difficult to decide if they are good approximations of eventually close eigenvalues of A or just an artifact caused by finite precision arithmetic.

Despite these problems that are described (but not explained) in most textbooks, when we want to compute only a few of the extreme or isolated eigenvalues of A, it is likely that the Lanczos algorithm (in finite precision arithmetic) will deliver them in a few iterations in many cases. There are, of course, thousands of examples of this nice behavior in the literature and this is the goal for which the algorithm is most useful. Today, there still are two schools about what to do with the Lanczos algorithm in finite precision arithmetic if one wants to compute all (or a large number of) the eigenvalues of A. The first possibility is to accept doing more (and sometimes many more) iterations than n, the dimension of the problem. This was particularly advocated by Cullum and Willoughby [27]. The other possibility is to use some forms of reorthogonalization to maintain orthogonality between the Lanczos vectors, at least in a weak way. Lanczos [108] advocated the use of full reorthogonalization for dealing with rounding errors. Cheaper methods were proposed by Grcar [73], Parlett and Scott [142], and Simon [162]. Today, the trend is to use restarted methods. This has been introduced for nonsymmetric matrices using the Arnoldi algorithm by Sorensen and Lehoucq and is well described in Lehoucq's Ph.D. thesis [111]; see also the user's guide of the ARPACK software [112]. For application to symmetric matrices see Calvetti, Reichel, and Sorensen [19] and also [5]. For another approach, see Simon and Wu [165].

The first (and still most) significant results for explaining the behavior of the Lanczos algorithm in finite precision arithmetic were obtained by Chris Paige in his Ph.D. thesis in 1971 [132] and strengthened and extended in his subsequent papers in journals [133], [134], [135]. He proved the important result that loss of orthogonality has a close relationship with convergence of Ritz values. In fact, he derived a matrix equation describing the propagation of the rounding errors in the algorithm. One of the consequence of this equation is to give a relation for the inner product of the new Lanczos vector and the Ritz vectors showing that the new vector stays (approximately) orthogonal to all unconverged Ritz vectors. This relation also shows that orthogonality is lost with a Ritz vector if there is convergence of the corresponding Ritz value to an eigenvalue of A. It can be said that Paige was the first to show that, despite the rounding error problems, the Lanczos algorithm can be used to compute accurate approximations of some eigenvalues of A. His papers renewed the interest in the Lanczos algorithm and are beautiful pieces of mathematical work that are still worth reading.

The Ph.D. thesis of Grcar in 1981 [73] attempted a more classical forward analysis of the Lanczos algorithm. In this work Grcar attributed the behavior of the Lanczos algorithm (particularly the loss of orthogonality) to a growth, through the use of the three-term recurrence, of the local rounding errors which are of the order of the unit roundoff.

An interesting work on the finite precision Lanczos algorithm is the Ph.D. thesis of Simon [162], who was at that time a student of Beresford Parlett at U.C. Berkeley. He derived a recurrence for the level of orthogonality (to be defined later; it measures how orthogonal are the Lanczos vectors) and used the fact that to obtain a behavior of the Lanczos algorithm close to the exact one, it is enough to maintain semiorthogonality that is at a level of the square root of the unit roundoff. This led to an efficient algorithm called partial reorthogonalization.

Many results on the Lanczos algorithm in finite precision are summarized in the nice book by Parlett [141]. See also Scott's Ph.D. thesis [159]. Results on the finite precision Lanczos algorithm can also be found in the papers and the book of Cullum and Willoughby [27].

On the foundations provided by Paige, a model and an explanation of the behavior of Lanczos and CG algorithms in finite precision arithmetic were given by Greenbaum [75], [76], [77], [78]. Interesting examples were studied by Greenbaum and Strakoš in [81], [184], [185]. These papers have inspired many other works. More recently, results in the same direction were published by Druskin and Knizhnerman [37], [38], [39], [40], [41]. See also Druskin, Greenbaum, and Knizhnerman [42], and Knizhnerman [103], [104], [105]. For some recent results, see Zemke [205], [206], [207] and Wülling [202], [203].

The questions we would like to (partially) address here and beyond are the following:

- What theoretical properties of the Lanczos algorithm remain (approximately) true in finite precision arithmetic?
- What is the cause of the loss of orthogonality of the Lanczos vectors?
- What happens to the equivalence of the Lanczos and CG algorithms in finite precision?
- What is the consequence of finite precision arithmetic on CG convergence?

We already have some answers to these questions from the papers we have cited before. We shall see a summary of the most significant results in this chapter. It may also be considered as a tribute to the people who had contributed the most to the understanding of the Lanczos algorithm. After establishing the hypothesis we are going to use here and in the next chapters about floating point arithmetic, we shall review the main results in the works of Paige and Greenbaum and to a lesser extent of Grcar, Simon, and a few others; see also [123]. Throughout, we shall try to illustrate their results on the small example of a Strakoš matrix.

3.1 The tools of the trade

In this section we shall review the properties of the basic operations involved in the Lanczos and CG algorithms in finite precision arithmetic. Most of the following formulas are obtained from results given in the nice book by Higham [94]. For interesting information about IEEE floating point arithmetic, see the small but enlightening book by Overton [127]. In what follows we shall sometimes denote by a tilde (˜) the computed quantities or sometimes by $fl(x)$ the result of the computation of x. In the results that will be quoted or described in this chapter there are constants appearing, like 1.01 or 3.02, etc. It must be clear that the precise values of all these constants are not really important. The main point is that they are small and independent of the dimension of the problem and/or the number of iterations. Looking at rounding errors is a delicate topic since we are faced with two dangers: if we want to be mathematically correct, we have to retain all the terms involving rounding errors and then the proofs of results can become lengthy and cumbersome; the other danger is to be too lax and inaccurate. We shall try to avoid these two pitfalls.

We use the standard model corresponding to IEEE arithmetic. For any of the four basic operations $(+, -, *, /)$ denoted by op and two floating point numbers x and y, we have

$$fl(x \; op \; y) = (x \; op \; y)(1 + \delta), \quad |\delta| \le u,$$

u being throughout this book the unit roundoff which is $(1/2)\beta^{1-t}$, where β is the base and t is the number of digits in the mantissa of floating point numbers. This bound is obtained when rounding to the nearest floating point number is used, but this is generally the case. Otherwise, u is twice this value. In IEEE double precision ($\beta = 2, t = 53$) and

$$u = 1.110223024625157 \; 10^{-16}.$$

It is half of the machine epsilon $\epsilon_M = \beta^{1-t}$, which is the distance from 1 to the next larger floating point number.

The operations we are interested in for the Lanczos and CG algorithms are inner products, ratios of inner products, matrix-vector multiplications, and additions of scalar multiples of vectors. For inner products the backward error analysis, writing the computed result as a perturbation of the data, is (see [94])

$$fl(x^T y) = (x + \Delta x)^T y = x^T (y + \Delta y),$$

where x and y are floating point vectors of length n and

$$|\Delta x| \le \bar{\gamma}_n |x|, \quad |\Delta y| \le \bar{\gamma}_n |y|, \quad \bar{\gamma}_n = \frac{nu}{1 - nu}, \quad nu < 1.$$

This relates the result of $fl(x^T y)$ to perturbations of the data x or y, $fl(x^T y)$ being the exact result of an inner product with perturbed data. Therefore we can write

$$fl(x^T y) = x^T y + \Delta x^T y = x^T y + Cnu,$$

where C is a real number depending on x, y, and n satisfying

$$|C| \le \frac{1}{1 - nu} |x^T|\,|y|.$$

This can also be written as

$$fl(x^T y) = x^T y + C' nu + O(u^2),$$

with $|C'| \le |x|^T\,|y|$. Note that $|x|^T|y|$ can be much different from $x^T y$ if the signs of the elements of x and y are not all positive or negative. This result gives a bound on the difference between the true value and the computed value of the inner product. The forward error analysis comparing the exact and floating point results is

$$|x^T y - fl(x^T y)| \le \bar{\gamma}_n |x^T|\,|y|.$$

We are also interested in computing (squares of) norms for which we take $x = y$ in the inner product and we have for $x \ne 0$,

$$fl(x^T x) = x^T x + C_1 nu\, x^T x = (1 + C_1 nu) x^T x,$$

with $|C_1| \le 1/(1 - nu)$. This is obtained because $|x^T|\,|x| = x^T x$ and for $x \ne 0$ we can write

$$C = \frac{C}{x^T x} x^T x,$$

so $C_1 = C / x^T x$, with

$$|C| \le \frac{1}{1 - nu} x^T x.$$

We can also write this result as

$$fl(x^T x) = x^T x + C_2 nu\, x^T x + O(u^2), \quad |C_2| \le 1.$$

The next step is to compute ratios of (nonzero) inner products. We have

$$fl\left(\frac{x^T y}{w^T z}\right) = \frac{fl(x^T y)}{fl(w^T z)}(1 + \delta), \quad |\delta| \le u.$$

The right-hand side is equal to

$$\frac{x^T y + C_1 nu}{w^T z + C_2 nu}(1 + \delta),$$

with

$$|C_1| \le \frac{1}{1 - nu} |x^T|\,|y|, \quad |C_2| \le \frac{1}{1 - nu} |w^T|\,|z|.$$

But, if C_2nu/w^Tz is small enough, we have

$$\frac{1}{w^Tz + C_2nu} = \frac{1}{w^Tz}\left(1 - \frac{C_2nu}{w^Tz}\right) + O(u^2).$$

Putting all this together we obtain

$$fl\left(\frac{x^Ty}{w^Tz}\right) = \frac{x^Ty}{w^Tz} + nu\left(\frac{C_1}{w^Tz} - \frac{C_2x^Ty}{(w^Tz)^2}\right) + \delta\frac{x^Ty}{w^Tz} + O(u^2).$$

This can be written as

$$fl\left(\frac{x^Ty}{w^Tz}\right) = \frac{x^Ty}{w^Tz} + C_3u + O(u^2),$$

with

$$|C_3| \le \frac{|x|^T\,|y|}{|w^Tz|} + \frac{n}{1-nu}\left(\frac{(|x|^T\,|y|}{|w^Tz|} + \frac{|w|^T\,|z|\,|x^Ty|}{(w^Tz)^2}\right).$$

Notice that if $|w^Tz|$ is small, $|C_3|$ can be large. Applying this result to the ratio of squares of norms, we have

$$fl\left(\frac{\|x\|^2}{\|y\|^2}\right) = \frac{\|x\|^2}{\|y\|^2}(1 + Cu) + O(u^2),$$

$$|C| \le 1 + \frac{2n}{1-nu}.$$

This can be written as

$$fl\left(\frac{\|x\|^2}{\|y\|^2}\right) = \frac{\|x\|^2}{\|y\|^2}[1 + C'(1+2n)u] + O(u^2), \quad |C'| \le 1.$$

For a matrix-vector multiply $y = Ax$, where A is a square sparse matrix with at most m nonzero entries per row, we can use the previous results about inner products to obtain (see [94])

$$fl(y) = (A + \Delta A)x, \quad |\Delta A| \le \bar{\gamma}_m|A|.$$

Componentwise,

$$|y - fl(y)| \le \bar{\gamma}_m|A|\,|x|$$

or

$$|y - fl(y)| \le mu|A|\,|x| + O(u^2).$$

If we look more closely at the components of the product, we have

$$fl(y_i) = fl(a_{i,:}x) = a_{i,:}x + C_iu,$$

where $a_{i,:}$ is the row vector of the ith row of A and

$$|C_i| \le \frac{m_i}{1 - m_iu}|a_{i,:}|\,|x|.$$

So, we can write

$$fl(y) = Ax + Cu,$$

where C is now a vector, $C = (C_1 \quad \cdots \quad C_n)^T$, and, of course, componentwise we have the forward result given before,

$$|C| \le \frac{m}{1 - mu} |A|\,|x|.$$

This bound can be pessimistic for some components C_i if the number of nonzero entries per row vary greatly. For the l_2 norm we have

$$\|y - fl(y)\| \le \|\Delta A\|\,\|x\|.$$

We can bound $\|\Delta A\|$ either by $\|A\|$ or by $\|\,|A|\,\|$. This leads to

$$\|y - fl(y)\| \le n^{\frac{1}{2}} \bar{\gamma}_m \|A\|$$

or

$$\|y - fl(y)\| \le \bar{\gamma}_m \|\,|A|\,\|.$$

We notice that there are cases where $\|\,|A|\,\| = \|A\|$, for instance, if A is a diagonally dominant matrix. We are also interested in inner products involving a matrix-vector product like $(Ay, y) = y^T Ay$. We have

$$fl((Ay, y)) = (Ay + C_A u, y) + C_1 nu,$$

$$|C_1| \le \frac{1}{1 - nu} (|y|, |Ay + C_A u|), \quad |C_A| \le \frac{m}{1 - mu} |A|\,|y|.$$

This is written as

$$fl((Ay, y)) = (Ay, y) + C_2 u + O(u^2), \quad |C_2| \le (m + n)(|A|\,|y|, |y|).$$

Supposing that A is positive definite, that is, $(Ay, y) > 0$ for $y \neq 0$, in the CG algorithm this will be used in a ratio to compute one of the coefficients

$$fl\left(\frac{(x, x)}{(Ay, y)}\right) = \frac{(x, x) + C_1 nu + O(u^2)}{(Ay, y) + C_2 u + O(u^2)} (1 + \delta).$$

After some manipulations this can be written as

$$fl\left(\frac{(x, x)}{(Ay, y)}\right) = \frac{(x, x)}{(Ay, y)} (1 + C_3 u) + O(u^2), \quad |C_3| \le \frac{|C_2|}{(Ay, y)} + n + 1,$$

$$|C_2| \le (m + n)(|A|\,|y|, |y|).$$

Let us now turn to linear combinations of vectors. Let α be a scalar and x be a vector. Then,

$$fl(\alpha x) = \alpha x(1 + \delta), \quad |\delta| \le u.$$

Hence,

$$\|fl(\alpha x) - \alpha x\| \le u|\alpha|\,\|x\|, \quad |fl(\alpha x) - \alpha x| \le u|\alpha|\,|x|,$$

and the backward analysis is

$$fl(\alpha x) = \alpha(x + \Delta x) = (\alpha + \Delta\alpha)x, \quad |\Delta x| \le u|x|, \quad |\Delta\alpha| \le u|\alpha|.$$

We now consider $\alpha x + \beta y$, where α and β are scalars and x and y are vectors. We have

$$fl(\alpha x + \beta y) = (fl(\alpha x) + fl(\beta y))(1 + \delta).$$

Then,

$$fl(\alpha x) = \alpha x(1 + \delta_1), \quad fl(\beta x) = \beta x(1 + \delta_2), \quad |\delta_i| \le u,$$
$$fl(\alpha x + \beta y) = \alpha x + \beta y + \alpha x(\delta_1 + \delta) + \beta y(\delta_2 + \delta) + O(u^2).$$

Therefore,

$$fl(\alpha x + \beta y) = \alpha x + \beta y + Cu + O(u^2),$$
$$|C| \le 2(|\alpha x| + |\beta y|).$$

This gives

$$|fl(\alpha x + \beta y) - (\alpha x + \beta y)| \le 2u(|\alpha|\,|x| + |\beta|\,|y|) + O(u^2)$$

and

$$\|fl(\alpha x + \beta y) - (\alpha x + \beta y)\| \le 2u(|\alpha|\,\|x\| + |\beta|\,\|y\|) + O(u^2).$$

Let us now consider operations that are involved in the Lanczos and CG algorithms. For CG, we have to compute expressions like $x + \beta y$. In this case we have

$$fl(x + \beta y) = x + \beta y + Cu + O(u^2), \quad |C| \le |x| + 2|\beta|\,|y|.$$

If we write, C_A being a vector,

$$fl(Ax) = Ax + uC_A + O(u^2),$$

we have

$$fl(\alpha Ax + \beta y) = \alpha Ax + \beta y + Cu + O(u^2),$$

where the vector C is such that

$$|C| \le (m + 2)|\alpha|\,|A|\,|x| + 2|\beta|\,|y|.$$

For the CG algorithm, we have to compute expressions like $y - \alpha Ax$. In the previous computation, if $\beta = 1$, we have

$$|C| \le (m + 2)|\alpha|\,|A|\,|x| + |y|.$$

The computation of a vector like $\alpha Ax + \beta y + \gamma z$ depends on the order that is chosen, either $(\alpha Ax + \beta y) + \gamma z$ or $(\alpha Ax + \gamma z) + \beta y$. We have

$$fl(\alpha Ax + \beta y + \gamma z) = \alpha Ax + \beta y + \gamma z + Cu + O(u^2)$$

with

$$|C| \leq (m+3)|\alpha|\,|A|\,|x| + 3|\beta|\,|y| + |\gamma|\,|z|$$

in the first case or

$$|C| \leq (m+3)|\alpha|\,|A|\,|x| + 3|\gamma|\,|z| + |\beta|\,|y|$$

in the second case. In the Lanczos algorithm we would like to evaluate the entries of the vector f in expressions like

$$\eta z = Ax - \beta x - \gamma y + f,$$

where all quantities besides f are the computed ones. Thus, ηz is the exact product of two floating point numbers. Let

$$w = Ax - \beta x - \gamma y.$$

In the Lanczos algorithm, one first computes $\tilde{w} = fl(w)$. Then η is computed as the floating point norm of $\tilde{w}$ and z is the floating point result of the division of $\tilde{w}$ by η. Therefore,

$$z = \frac{fl(w)}{\eta}(1+\delta), \quad |\delta| \leq u.$$

Hence,

$$\eta z = fl(w)(1+\delta).$$

Since $fl(w) = w + C_w u$, we have

$$f = C_w u + \delta w + O(u^2),$$

with

$$|C_w| \leq (m+2)|A|\,|x| + 3|\beta|\,|x| + |\gamma|\,|y|,$$

where we have a 2 instead of a 3 because the factor of Ax in w is 1. This can be written as $f = Cu + O(u^2)$, where C is a vector such that

$$|C| \leq (m+3)|A|\,|x| + 4|\beta|\,|x| + 2|\gamma|\,|y|.$$

If we want to bound the l_2 norm of f we have

$$\|f\| \leq u[(m+3)\|\,|A|\,\|\,\|x\| + 4|\beta|\,\|x\| + 2|\gamma|\,\|y\|] + O(u^2).$$

In the Lanczos algorithm we shall see that we have

$$|\beta| \leq \|A\| + C_\beta u, \quad |\gamma| \leq \|A\| + C_\gamma u,$$

$$\|x\| \leq 1 + C_x u, \quad \|y\| \leq 1 + C_y u.$$

Therefore,

$$\|f\| \leq u[(m+3)\|\,|A|\,\| + 6\|A\|] + O(u^2).$$

There are some different results in the literature about $\|f\|$. Several papers of Chris Paige contain different results. However, it is fair to say that he analyzed several different versions of the Lanczos algorithm. There are also results in Grcar's Ph.D. thesis [73]. His result is

$$u[m\| \, |A| \, \| - 13\|A\|].$$

It is difficult to trace the origin of these differences. This illustrates the dangers that we have stressed at the beginning of this section. Even though this is not very satisfactory, it does not matter too much for the analysis of the method if we have a 7 or a 14 (Paige) or a 13 (Grcar) or a 6 (here) in front of $\|A\|$ since the bounds that we shall establish subsequently are most often large overestimates of the actual quantities. What is important is that we have a small constant times the norm of A.

In everything we are going to do about finite precision computations we shall always suppose that we know the exact Ritz values and eigenvectors of the computed tridiagonal Lanczos matrix $\tilde{T}_k$. The reason is that this has no influence on the computations of the Lanczos and CG algorithms. Moreover, there exist numerical methods to compute these eigenvalues and eigenvectors to working precision; see [141]. Let us now summarize the computations we need for the Lanczos and CG algorithms.

Theorem 3.1. *For the Lanczos algorithm, we have*

$$fl((Av, v)) = (Av, v) + Cu + O(u^2), \quad |C| \leq (m+n)(|A| \, |v|, |v|).$$

$$\eta z = Ax - \beta x - \gamma y + f, \quad f = Cu + O(u^2),$$

with

$$|C| \leq (m+3)|A| \, |x| + 4|\beta| \, |x| + 2|\gamma| \, |y|.$$

In the Lanczos algorithm, because of the bounds we shall establish, we have

$$\|f\| \leq u[(m+3)\| \, |A| \, \| + 6\|A\|] + O(u^2).$$

Theorem 3.2. *For the CG algorithm we have*

$$fl\left(\frac{\|x\|^2}{\|y\|^2}\right) = \frac{\|x\|^2}{\|y\|^2}[1 + C(1+2n)u] + O(u^2), \quad |C| \leq 1,$$

$$fl\left(\frac{\|x\|^2}{(Ay, y)}\right) = \frac{\|x\|^2}{(Ay, y)}(1 + Cu) + O(u^2), \quad |C| \leq \frac{(m+n)(|A| \, |y|, |y)}{(Ay, y)} + n + 1,$$

$$fl(x + \beta y) = x + \beta y + Cu + O(u^2), \quad |C| \leq |x| + 2|\beta| \, |y|,$$

$$fl(\alpha Ax + y) = \alpha Ax + y + Cu + O(u^2), \quad |C| \leq (m+2)|\alpha| \, |A| \, |x| + |y|.$$

3.2 Numerical example

To illustrate the results of Paige and others we choose a diagonal Strakoš matrix. Remember that the matrix is diagonal with eigenvalues

$$\lambda_i = \lambda_1 + \left(\frac{i-1}{n-1}\right)(\lambda_n - \lambda_1)\rho^{n-i}, \quad i = 1, \ldots, n.$$

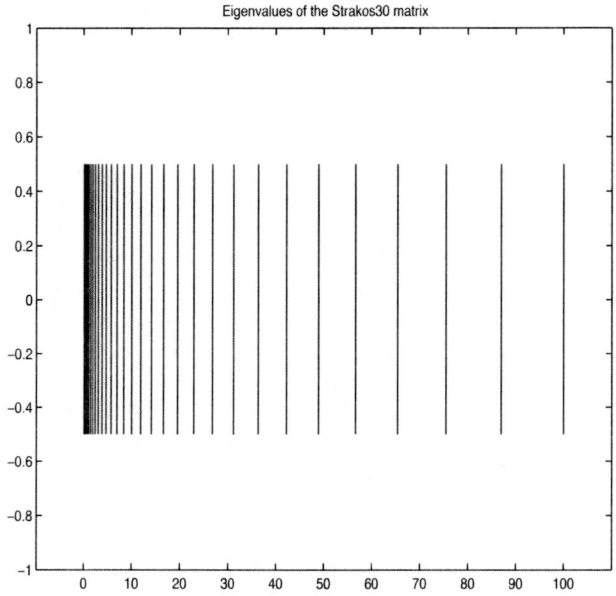

Figure 3.1. *Eigenvalues of the Strakos*30 *matrix*

Here we choose $n = 30$, $\lambda_1 = 0.1$, $\lambda_n = 100$, $\rho = 0.9$. Notice that this is not the worst example in this family of matrices regarding the loss of orthogonality and convergence; see [81]. The eigenvalues of A which are all distinct are shown in Figure 3.1. The largest eigenvalues are well separated and the smallest eigenvalues are quite close to each other. The loss of orthogonality of the basis vectors in the Lanczos algorithm is exemplified in Figure 3.2, which shows the matrix $\log_{10}(|\tilde{V}_{30}^T \tilde{V}_{30}|)$ as a function of the row and column indices. Ideally, this should be a representation of an identity matrix up to working precision, but here, some nondiagonal elements are $O(1)$. We remark that local orthogonality is more or less preserved. The loss of orthogonality is also shown in Figure 3.3, where the curve is $\log_{10}(\|I - \tilde{V}_k^T \tilde{V}_k\|)$ as a function of k. We see that $\|I - \tilde{V}_k^T \tilde{V}_k\|$ grows from the beginning starting at the roundoff level and going up to $O(1)$.

Figure 3.4 shows the base 10 logarithm of the distance of the minimum eigenvalue of T_k to the minimum eigenvalue of A. Figure 3.5 shows the same for the distance of the maximum Ritz value to the maximum eigenvalue. The solid curves are computed with double reorthogonalization where the residual vectors are orthogonalized twice against all the previous residuals at every iteration and the dashed curve is the Lanczos algorithm result. Both curves show the maximum with 10^{-20} since some distances were exactly 0. We see that the finite precision algorithm converges to the largest eigenvalue at the same time as the "exact" one and there is almost no difference between the standard algorithm and the algorithm with reorthogonalization. Even in finite precision we have an approximation of the largest eigenvalue (which is the first to converge) up to roundoff at iteration 18. In finite precision the minimum eigenvalue has not yet converged at iteration 30. The exact and finite precision curves start to differ significantly at iteration 23.

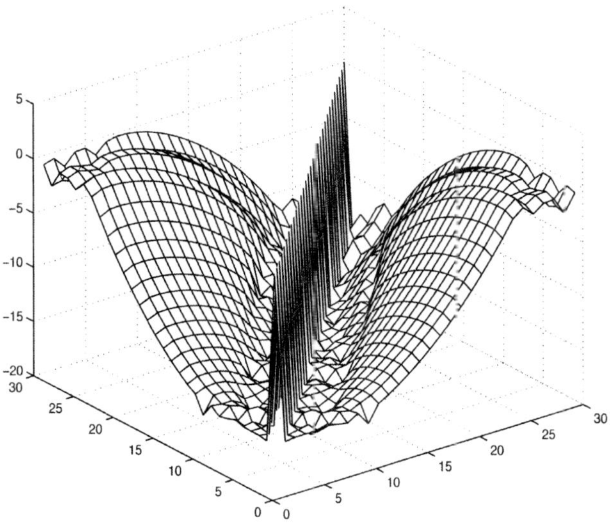

Figure 3.2. $\log_{10}(|\tilde{V}_{30}^T \tilde{V}_{30}|)$ *for the Strakos*30 *matrix*

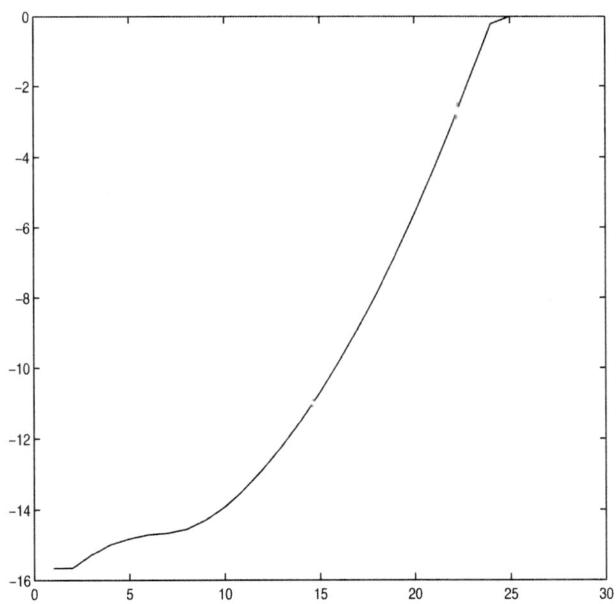

Figure 3.3. $\log_{10}(\|I - \tilde{V}_k^T \tilde{V}_k\|)$ *for the Strakos*30 *matrix as a function of k*

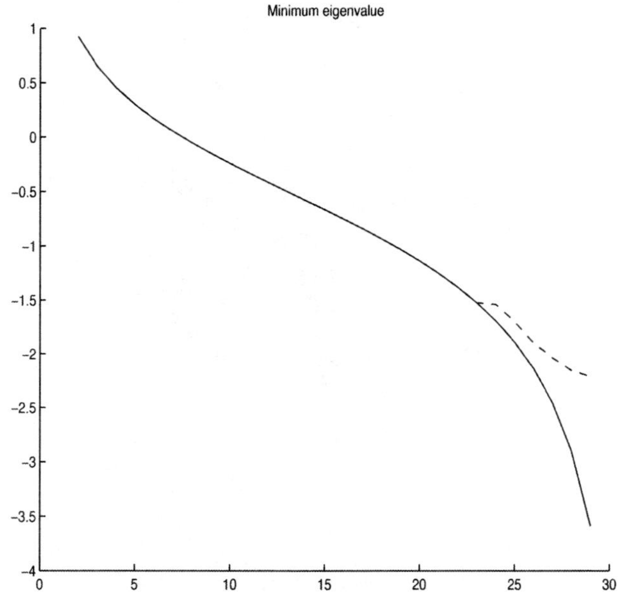

Figure 3.4. $\log_{10}$ *of the distance to the minimum eigenvalue of A, with orthogonalization (solid) and finite precision (dashed)*

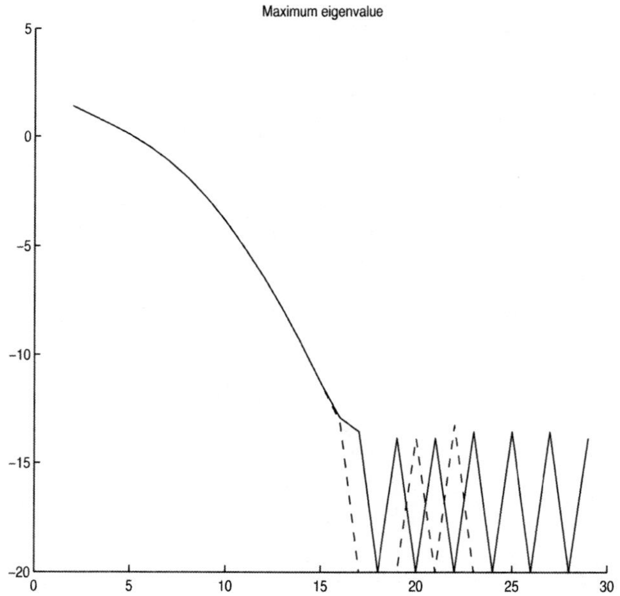

Figure 3.5. $\log_{10}$ *of the distance to the maximum eigenvalue of A, with orthogonalization (solid) and finite precision (dashed)*

3.3 The work of Chris Paige

The fundamental work of Chris Paige on the Lanczos algorithm and more generally on Krylov methods started at the end of the sixties with some technical reports [128], [129], [130], whose results led to his Ph.D. thesis [132] in 1971. Even though Lanczos [108] was aware of the problems of his algorithm in finite precision arithmetic, Paige's thesis is the first work in which it is clearly stated and proved that even though the Lanczos algorithm in finite precision does not fulfill its theoretical mathematical properties, it is nevertheless of use for computing some approximations of eigenvalues of large sparse matrices. Almost at the same time an equivalent statement was made for CG by Reid [149]. A few years later, Paige published some papers, [131], [133], [134], [135] summarizing and improving the results contained in his thesis. Let us now describe some of these results. In Paige's own words, "the purpose of this thesis will not be to introduce new methods but to concentrate mainly on the Lanczos method for real symmetric matrices with the intention of showing that it has been unjustly discarded and has a lot to offer in practice" [132].

Before studying this method Paige gave some results for floating point errors in basic operations. He used repeatedly the following result:

$$1-(p+q)u \le \frac{\prod_{i=1}^{p}(1+\epsilon_i)}{\prod_{i=1}^{q}(1+\epsilon_i')} \le 1+1.01(p+q)u,$$

where $|\epsilon_i|,\ |\epsilon_i'| \le u$. Most of the time, second order terms involving u^2 are discarded. Paige used a handy notation to bound the errors. Let $|\epsilon_i| \le u$; then there exists a value α such that

$$\prod_{i=1}^{p}(1+\epsilon_i)=\alpha^p, \quad |\alpha-1| \le u.$$

$D(\alpha)$ denotes a diagonal matrix with elements not necessarily equal but satisfying the above bounds. The rules for manipulating such quantities are

$$\alpha^p\alpha^q=\alpha^{p+q},$$
$$x=\alpha(y+z) \Rightarrow x=\alpha y+\alpha z,$$
$$x=\left(\frac{\alpha^p}{\alpha^q}\right) y \text{ or } x=\alpha^p\alpha^q y \Rightarrow x=[1+(p+q)\epsilon]y,$$

where $|\epsilon y| \le 1.01u|y|$. Using these notations, for the inner product we have

$$fl(x^Ty)=x^TD(\alpha^n)y=x^Ty+n\epsilon|x^T|\,|y|,$$

and for the computation of the l_2 norm

$$fl(x^Tx)=\alpha^n x^Tx.$$

For the matrix-vector product Paige proposed to use

$$fl(Ax)=(A+\delta A)x, \quad |\delta A| \le m\epsilon|A|,$$

where m is the maximum number of nonzero elements per row. This leads to

$$\|\delta A\| \le m\epsilon\|\,|A|\,\|.$$

Let β such that $\| \, |A| \, \| = \beta \|A\|$; then

$$\|\delta A\| \le m\epsilon\beta\|A\|, \quad \epsilon \le 1.01u.$$

Paige then moved on to analyze the unnormalized Lanczos algorithm in exact arithmetic. He gave new proofs of results by Kaniel [102], whose proofs were not completely correct. We have described these results in Chapter 1. They use Chebyshev polynomials to obtain bounds on the distances between eigenvalues of A and Ritz values. Then Paige gave some new results about the work of Lehmann (1963, 1966), who obtained ways to compute optimal intervals containing eigenvalues of A. The second part of Paige's thesis considered the Lanczos process without normalization and reorthogonalization. In Paige's notations the Lanczos algorithm is written as

$$\beta_{k+1} v^{k+1} = Av^k - \gamma_k v^k - \delta_k v^{k-1},$$

with

$$\gamma_k = \frac{(v^k)^T A v^k}{(v^k)^T v^k},$$

since the vectors v^k are not normalized and

$$\delta_k = \frac{(v^{k-1})^T A v^k}{(v^{k-1})^T v^{k-1}} \quad \text{or} \quad \frac{\beta_k (v^k)^T v^k}{(v^{k-1})^T v^{k-1}}.$$

This δ_k is not to be confused with the last pivot function we used before. Paige denoted the first formula for δ_k as (A1) and the second as (A2). He stated that (A2) has better properties in finite precision arithmetic. In the rounding error analysis he supposed that $(2n+1)\epsilon$ and $m\beta\epsilon$ are smaller than 0.01 and that there is no normalization; hence $\beta_{k+1} = 1$. If $fl(Av^k) = A_k v^k$ with $A_k = A + \delta A_k$, then $\|\delta A_k\| \le m\beta\epsilon\|A\|$, and the hypothesis gives $\|A_k\| \le 1.01\|A\|$. Moreover, $fl((v^k)^T A v^k) = (v^k)^T A_k v^k + n\epsilon |(v^k)^T| \, |A_k v^k|$, which gives

$$\gamma_k \le 1.03\|A\|.$$

In finite precision we have

$$v^{k+1} = Av^k - \gamma_k v^k - \delta_k v^{k-1} - \delta v^k.$$

At the first step

$$\delta v^1 = [2\gamma_1 D(\epsilon) - D(\epsilon) A_1] v^1.$$

Therefore, $\|\delta v^1\| \le 3.02\epsilon\|A_1 v^1\| \le 3.06\epsilon\|A\| \, \|v^1\|$. More generally

$$\delta v^k = 2\delta_k D(\epsilon) v^{k-1} + [3\gamma_k D(\epsilon) - 2D(\epsilon) A_k] v^k,$$

and the bound depends on whether we are using (A1) or (A2). With (A1) $\|\delta v^k\| < 7.2\epsilon\|A\| \, \|v^k\|$, while with (A2) $\|\delta v^k\| < 2\epsilon\delta_k\|v^{k-1}\| + 5.03\epsilon\|A_k v^k\| < 2\epsilon\delta_k\|v^{k-1}\| + 5.09\epsilon\|Av^k\|$. The choice between (A1) and (A2) has an influence on the local orthogonality between v^k and v^{k+1}. Let θ_i be defined such that

$$\theta_1 = (v^1)^T v^2, \quad \theta_k = (v^k)^T v^{k+1} + \delta_k (v^{k-1})^T v^k.$$

This leads to a recurrence for the θ_i's. The values of the θ_i's can be bounded by

$$|\theta_1| < 2.05(n+2)\epsilon \|A\| \, \|v^1\|^2,$$

and for $k > 1$ and (A1)

$$|\theta_k| < 2.05(n+4)\epsilon \|A\| \, \|v^k\|^2,$$

while for (A2)

$$|\theta_k| < 2[\delta_k \|v^{k-1}\| + 1.01(n+3)\|A_k v^k\|]\epsilon \|v^k\|.$$

The recurrence for the local inner products can be solved and

$$(v^k)^T v^{k+1} = \sum_{i=1}^{k} (-1)^{k-i} \theta_i \prod_{r=i+1}^{k} \delta_r.$$

The product term is taken equal to 1 when it does not make sense. Using (A1) if $\theta = 2.05(n+4)\epsilon\|A\|$ we have $|\theta_k| < \theta\|v^k\|^2$. If it would be true that (as in exact arithmetic) $\delta_i = \|v^i\|^2/\|v^{i-1}\|^2$, then $|(v^k)^T v^{k+1}| < k\theta\|v^k\|^2$, giving local orthogonality. Unfortunately, the previous relation is not true in finite precision. If we suppose that everything is known until the step $k-1$ without any error, then at step k instead of v^k we compute $u^k = v^k + w^k$ and it can be shown that $\|w^k\| \leq 7\epsilon\|Av^{k-1}\|$. An approximate value $\bar{\delta}_k$ is computed such that

$$\bar{\delta}_k = \delta_k + \frac{(v^{k-1})^T A w^k}{(v^{k-1})^T v^{k-1}},$$

and therefore

$$|\bar{\delta}_k - \delta_k| \leq 7\epsilon\|A\|^2.$$

If δ_k approaches $O(\epsilon^2)\|A\|^2$, the relative error can be large. However, for variant (A2) the relation $\delta_i = \|v^i\|^2/\|v^{i-1}\|^2$ holds approximately and therefore $|(v^k)^T v^{k+1}| < k\theta\|v^k\|^2$. From this, Paige concluded that (A2) is a better variant of the Lanczos algorithm than (A1). These results were complemented in 1972 in [133], where Paige studied some more variants of the Lanczos algorithm. This time, he considered the normalized Lanczos algorithm that we write as

$$\beta_{k+1} v^{k+1} = Av^k - \gamma_k v^k - \delta_k v^{k-1},$$

to keep the same notations as before, although those of [133] are different. The value β_{k+1} is designed to give v^{k+1} with norm 1. Let $u^1 = Av^1$; then the possible choices for the algorithm are

$$\begin{aligned}
\gamma_k &= (v^k)^T A v^k, && (1)\\
\text{or } \gamma_k &= (v^k)^T u^k, && (2)\\
w^k &= u^k - \gamma_k v^k, && (3)\\
\beta_{k+1} &= \sqrt{(w^k)^T w^k}, && (4)\\
v^{k+1} &= \frac{w^k}{\beta_{k+1}}, && (5)\\
\delta_{k+1} &= (v^k)^T A v^{k+1}, && (6)\\
\text{or } \delta_{k+1} &= \beta_{k+1}, && (7)\\
u^{k+1} &= Av^{k+1} - \delta_{k+1} v^k. && (8)
\end{aligned}$$

The choices are between (1) and (2) on one hand and (6) and (7) on the other hand. From this point, Paige performed a rounding error analysis and numerical experiments tending to show that (1,7) and (2,7) are the best algorithms. The one which is most in use today is (2,7).

In his Ph.D. thesis Paige continued to analyze the unnormalized Lanczos algorithm using (A2). Since there was no normalization, the generated tridiagonal matrix was not symmetric. However, Paige noticed that it can be symmetrized (as any tridiagonal matrix with nonvanishing lower and upper diagonals). Then he proved some of the results that we have already mentioned in Chapter 1 about cofactors of entries of tridiagonal matrices to get expressions for the entries of the eigenvectors of the Lanczos matrix.

To avoid square roots in the symmetrized tridiagonal matrix Paige changed notations. Since, when using (A2) δ_k is always positive, it is now denoted by δ_k^2. Doing this the tridiagonal matrix is similar to a symmetric one denoted by C_k having the γ_k's on the diagonal and the δ_k's on the subdiagonal. Its eigenvalues are (in our notation) the Ritz values $\theta_i^{(k)}$. Paige denoted them as $\mu_i^{(k)}$ ordering from the largest to the smallest, and the matrix of the (orthonormal) eigenvectors is denoted by $Y^{(k)}$ with columns $y_i^{(k)}$. Notice that these notations are different from ours.

For $j < k$, if we apply C_k to the rth eigenvector of C_j completed with zeros, we have

$$C_k \begin{pmatrix} y_r^{(j)} \\ 0 \\ 0 \\ \vdots \\ 0 \end{pmatrix} = \begin{pmatrix} \mu_r^{(j)} y_r^{(j)} \\ \delta_{j+1} y_{j,r}^{(j)} \\ 0 \\ \vdots \\ 0 \end{pmatrix}.$$

Concerning the algorithm in exact arithmetic, by moving the term $\mu_r^{(j)} y_r^{(j)}$ to the left-hand side and computing the residual, we obtain

$$\min_i |\mu_i^{(k)} - \mu_r^{(j)}| \le \delta_{j+1} |y_{j,r}^{(j)}|.$$

This expresses the fact we have mentioned in Chapter 1, that once at step j, δ_{j+1} times the last element of a normalized eigenvector $y_r^{(j)}$ of the Lanczos tridiagonal matrix is small, then on subsequent iterations there is always a Ritz value within this distance of $\mu_r^{(j)}$. By applying C_k to $(\, y_t^{(j)} \quad \cdots \quad y_{t+s}^{(j)} \,)$ Paige showed that there exist $s+1$ integers $i_0, \dots, i_s \le k$ such that

$$\sum_{r=0}^{s} (\mu_{i_r}^{(k)} - \mu_{t+r}^{(j)})^2 \le \delta_{j+1}^2 \sum_{r=0}^{s} (y_{j,t+r}^{(j)})^2.$$

Thus, if a group of $s+1$ eigenvalues is such that the right-hand side is small (i.e., the last components of the corresponding eigenvectors are small), then $s+1$ eigenvalues have converged to a certain accuracy. Two other expressions can be obtained for $j < k$:

$$(\delta_{j+1} y_{j,r}^{(j)})^2 \sum_{i=1}^{k} \left[\frac{y_{j+1,i}^{(k)}}{\mu_i^{(k)} - \mu_r^{(j)}} \right]^2 = 1,$$

$$\sum_{t=1}^{j} (y_{t,i}^{(k)})^2 = (\delta_{j+1} y_{j+1,i}^{(k)})^2 \sum_{r=1}^{j} \left[\frac{y_{j,r}^{(j)}}{\mu_i^{(k)} - \mu_r^{(j)}} \right]^2.$$

Doing a change of variable ($w^j = v^j/(\delta_2 \cdots \delta_j)$) to be able to use the symmetric matrix C_k the Lanczos relation in finite precision is written

$$AW_k = W_k C_k + \delta_{k+1} w^{k+1}(e^k)^T - G_k.$$

With these notations $\|w^1\| = 1$ and $\|w^k\| = \alpha^{n+(k-1)/2}$. Moreover, the norms of the columns g^i of G_k are less than $1.01(9 + m\beta)\epsilon\|A\|$. Let U_k be a strictly upper triangular matrix defined by $(0 \quad (w^1)^T w^2 \quad \cdots \quad W_{k-1}^T w^k)$; then

$$W_k^T W_k = U_k^T + D(\alpha^{2n+k-1}) + U_k.$$

By multiplying the Lanczos relation by W_k^T and subtracting the transpose we have

$$C_k(U_k^T + U_k) - (U_k^T + U_k)C_k = \delta_{k+1}(W_k^T w^{k+1}(e^k)^T - e^k(w^{k+1})^T W_k)$$
$$+ W_k^T G_k - G_k^T W_k + \mathrm{diag}(W_k^T W_k)C_k - C_k \mathrm{diag}(W_k^T W_k).$$

Let $u_{i,r} = (w^i)^T w^r$; then

$$\mathrm{diag}(C_k U_k^T - U_k^T C_k) = \mathrm{diag}\,(\,\delta_2 u_{1,2}, \quad \delta_3 u_{2,3} - \delta_2 u_{1,2}, \quad \ldots, \quad -\delta_k u_{k-1,k}\,)\,.$$

We have $|\delta_{i+1} u_{i,i+1}| < 2.2(n+4)i\epsilon\|A\|$. Writing the upper triangular part of the previous relation Paige obtained a fundamental relation describing the orthogonality in the algorithm,

$$C_k U_k - U_k C_k = \delta_{k+1} W_k^T w^{k+1}(e^k)^T - H_k,$$

where H_k is upper triangular with elements $h_{i,r}$,

$$h_{1,1} = \delta_2 u_{1,2}, \quad h_{i,i} = \delta_{i+1} u_{i,i+1} - \delta_i u_{i-1,i}, \quad i = 2, \ldots, k,$$
$$h_{i-1,i} = -(w^{i-1})^T g^i + (g^{i-1})^T w^i + \delta_i((w^i)^T w^i - (w^{i-1})^T w^{i-1}),$$
$$h_{r,i} = -(w^r)^T g^i + (g^r)^T w^i, \quad r = 1, \ldots, i-2,$$

where the g^j are the columns of G_k. The elements of H_k can be bounded by

$$|h_{i,i}| < 4.4(n+4)i\epsilon\|A\|,$$

$$|h_{i-1,i}| < 1.01(3.2n + 2m\beta + 20)\epsilon\|A\|,$$
$$|h_{r,i}| < 2.02(9 + m\beta)\epsilon\|A\|, \quad i = 1, \ldots, r-1.$$

This shows that the entries of H_k are small and gives a bound on the Frobenius norm of H_k

$$\|H_k\|_F < [2.6(n+4)(k+1)^{3/2} + 1.5(m\beta + 10)k]\epsilon\|A\|.$$

Using the spectral decomposition of C_k with Y_k being the matrix of eigenvectors, denoting by M_k the diagonal matrix of the eigenvalues $\mu_i^{(k)}$ and $Z_k = W_k Y_k$, and multiplying by Y_k^T on the left and Y_k on the right,

$$M_k Y_k^T U_k Y_k - Y_k^T U_k Y_k M_k = \delta_{k+1} Z_k^T w^{k+1}(e^k)^T Y_k - Y_k^T H_k Y_k.$$

Denoting by $\epsilon_{i,r}$ the elements of $Y_k^T H_k Y_k$, writing the diagonal elements of the last relation, and noticing that the diagonal elements of the left-hand side are zero gives the important relation (which is the most famous result of Paige's thesis),

$$(z^i)^T w^{k+1} = \frac{\epsilon_{i,i}}{\delta_{k+1} y_{k,i}}.$$

$y_{k,i}$ is the last element of y_i. The approximate eigenvector z^i at step k is not far from being orthogonal to the normalized Lanczos vector w^{k+1} unless z^i has converged with an accuracy close to $|\epsilon_{i,i}|$. This is the justification of what is generally expressed as "loss of orthogonality goes hand in hand with convergence." Other relations were developed by Paige. Writing the previous relation for $i = 1, \ldots, k$ gives

$$\delta_{k+1} W_k^T w^{k+1} = Y_k b^k,$$

where b^k is a vector with components $b_i^k = \epsilon_{i,i}/y_{k,i}$. By looking at some entries, we have, for instance,

$$\delta_{k+1}(w^k)^T w^{k+1} = \sum_{i=1}^k \epsilon_{i,i} = \sum_{i=1}^k h_{i,i}$$

and

$$\delta_{k+1}(w^{k-1})^T w^{k+1} = \sum_{i=1}^k \epsilon_{i,i} \frac{y_{k-1,i}}{y_{k,i}} = \frac{\sum_{i=1}^k (\mu_i - \gamma_k)\epsilon_{i,i}}{\delta_k}.$$

Another relation is

$$(\mu_i - \mu_r)(y^i)^T U_k y_r = \epsilon_{i,i} \frac{y_{k,r}}{y_{k,i}} - \epsilon_{i,r}.$$

This gives

$$(\mu_i - \mu_r)(y_i)^T (U_k^T + U_k) y_r = \epsilon_{i,i} \frac{y_{k,r}}{y_{k,i}} - \epsilon_{r,r} \frac{y_{k,i}}{y_{k,r}} + \epsilon_{r,i} - \epsilon_{i,r}.$$

We also have

$$(z^i)^T A z^r = \mu_r (z^i)^T z^r + \delta_{k+1} y_{k,r} (z^i)^T w^{k+1} + (y_i)^T W_k^T G_k y_r.$$

This leads to

$$(z^i)^T z^r = \frac{\delta_{k+1}(y_{k,r} z^i - y_{k,i} z^r)^T w^{k+1} + f_{i,r}}{\mu_i - \mu_r},$$

with $f_{i,r} = (y^i)^T (W_k^T G_k - G_k^T W_k) y^r$. Finally,

$$(z^i)^T z^r = \frac{\epsilon_{i,i} y_{k,r}/y_{k,i} - \epsilon_{r,r} y_{k,i}/y_{k,r} + f_{i,r}}{\mu_i - \mu_r},$$

giving an expression for the inner product of two approximate eigenvectors (which should be equal to 0 in exact arithmetic when $i \neq r$). Then Paige studied the question of convergence of the eigenvalues in finite precision arithmetic. If y^i is a normalized eigenvector of C_k with eigenvalue μ_i and z^i the Ritz vector, then

$$Az^i = \mu_i z^i + \delta_{k+1} y_{k,i} w^{k+1} G_k y^i,$$

and therefore there exists an eigenvalue λ_s of A such that

$$|\mu_i - \lambda_s| < 1.01 \frac{\delta_{k+1}|y_{k,i}| + g_k}{\|z^i\|},$$

with $g_k = \sqrt{k}(9+m\beta)\epsilon\|A\|$. This means that if $\delta_{k+1}|y_{k,i}|$ is small, as long as $\|z^i\|$ is not too small, μ_i is a good approximation of λ_s. From this result, it is interesting to look for bounds on $\|z^i\|$. Unfortunately, it is possible to have the norm of the Ritz vector $\|z^i\|$ very different from unity, and Paige gave a simple example illustrating this. The study of the norm of the approximate eigenvector is really clever but too technical to be repeated here. However, if μ_i is a well-separated eigenvalue, Paige showed that $0.8 < \|z^i\| < 1.2$. If there are $s+1$ eigenvalues from μ_t to μ_{t+s} which are close together, if they are well separated from the rest of the spectrum almost in the same sense as before, it can be shown that

$$0.65(s+1) < \sum_{i=t}^{t+s} (z^i)^T z^i < 1.35(s+1).$$

Considering the convergence of eigenvalues Paige first proved that at least one eigenvalue of the matrices C_k converges in finite precision arithmetic by iteration n. We have that $(w^i)^T w^i = \alpha^{2n+i-1}$, so if we define D_k as a diagonal matrix with nonzero elements $1/\|w^i\|$, then $D_k W_k^T W_k D_k$ is positive semidefinite with eigenvalues π_i ordered in ascending order and

$$\|W_k D_k\|_F^2 = \sum_{i=1}^{k} \pi_i = k.$$

Let $Q_k^T \Pi_k Q_k$ be the spectral decomposition of $D_k W_k^T W_k D_k$. Since $D_k W_k^T W_k D_k = I + D_k(U_k^T + U_k)D_k$, we have, π_i being the diagonal elements of Π_k,

$$\sum_{i=1}^{k} (\pi_i - 1)^2 = \|\Pi_k Q_k - Q_k\|_F^2 = 2\|D_k U_k D_k\|_F^2 \le \alpha^{4n+2k}\|U_k\|_F^2 \le 2.041\|U_k\|_F^2$$

if we assume that $(2n+k)\epsilon < 0.01$. If the number of Lanczos steps k is larger than n, the dimension of the problem, then the columns of W_k are linearly dependent and $D_k W_k^T W_k D_k$ must have at least $k-n$ zero eigenvalues. Suppose there are r zero eigenvalues $\pi_1 = \cdots = \pi_r = 0$; this gives

$$\sum_{i=1}^{k} (\pi_i - 1)^2 = r + \sum_{i=r+1}^{k} (\pi_i - 1)^2$$

and

$$\left[\sum_{i=r+1}^{k} (\pi_i - 1)\right]^2 \le (k-r) \sum_{i=r+1}^{k} (\pi_i - 1)^2,$$

with equality holding only if π_i, $i = r+1, \dots, k$, are constant equal to $k/(k-r)$. Hence,

$$2.041\|U_k\|_F^2 \ge r + \frac{(k-k+r)^2}{k-r} = \frac{kr}{k-r}.$$

The matrix $W_k^T W_k$ must be singular for some $k \le n+1$, so $r = 1$. If

$$a = \max_{i \le j < k} |(z^i_{(j)})^T w^{j+1}|,$$

$$0.49 < \|U_k\|_F^2 = \sum_{j=1}^{k-1} \|W_j^T w^{j+1}\|^2 = \sum_{j=1}^{k-1} \|Z_{(j)}^T w^{j+1}\|^2 \le \frac{a^2 k(k-1)}{2}.$$

This shows that $a > 0.98/k$. Therefore, for some $k \le n+1$ and some $i \le k$,

$$|(z^i_{(k)})^T w^{k+1}| = \left| \frac{\epsilon^{(k)}_{i,i}}{\delta_{k+1} y^{(k)}_{k,i}} \right| > \frac{0.98}{k},$$

which shows that there exists an eigenvalue μ of C_m, $m \ge k$, such that

$$|\mu - \mu_i^{(k)}| \le \delta_{k+1} |y^{(k)}_{k,i}| < 1.03k^2(4.4n + 1.5m\beta + 32)\epsilon \|A\|.$$

Some previous results show that if $\mu_i^{(k)}$ is well separated from the other Ritz values in such a way that $|\mu_i^{(k)} - \mu_j^{(k)}| > k^3(8.8n + 3m\beta + 64)\epsilon\|A\|$, then there exists an eigenvalue λ of A such that

$$|\lambda - \mu_i^{(k)}| < 1.3[k^2(4.4n + 1.5m\beta + 32) + k^{1/2}(9 + m\beta)]\epsilon \|A\|,$$

which proves the convergence. If the separation hypothesis is not fulfilled, Paige proved that nevertheless

$$|\lambda - \mu_i^{(k)}| < 1.1[k^3(4.4n + 1.5m\beta + 32) + k^{3/2}(9 + m\beta)]\epsilon \|A\|.$$

It is unfortunate that these nice and illuminating results were not published in mathematical journals soon after Paige's thesis defense. In [134], published in 1976, Paige studied the algorithm (2,7) we have seen before with the Lanczos vectors normalized to 1. The notations are a little bit different from those in Paige's Ph.D. thesis, but the hypotheses on the finite precision arithmetic are more or less the same. The results were gathered in a theorem which contains most of the results everyone should know about Lanczos algorithm in finite precision arithmetic. In this result, ϵ is less than the machine precision.

Theorem 3.3. *Let* $\epsilon_0 = (n+4)\epsilon$, $\epsilon_1 = (7 + m\beta)\epsilon$. *Then,*

$$AV_k = V_k T_k + \delta_{k+1} v^{k+1} (e^k)^T + \delta V_k.$$

For $j = 1, 2, \dots, k$

$$|(v^{j+1})^T v^{j+1} - 1| \le \epsilon_0,$$

$$\|\delta v^j\| \le \epsilon_1 \|A\|,$$

$$\delta_{j+1} |(v^j)^T v^{j+1}| \le 2\epsilon_0 \|A\|,$$

$$|\delta_j^2 + \alpha_j^2 + \delta_{j+1}^2 - \|Av^j\|^2| \le 4j(3\epsilon_0 + \epsilon_1)\|A\|^2.$$

If R_k is the strictly upper triangular part of $V_k^T V_k$, then

$$T_k R_k - R_k T_k = \delta_{k+1} V_k^T v^{k+1} (e^k)^T + H_k,$$

where H_k is upper triangular with elements such that $|h_{1,1}| \leq 2\epsilon_0 \|A\|$, and for $j = 2, 3, \dots, k$

$$|h_{j,j}| \leq 4\epsilon_0 \|A\|, \quad |h_{j-1,j}| \leq 2(\epsilon_0 + \epsilon_1)\|A\|,$$

$$|h_{i,j}| \leq 2\epsilon_1 \|A\|, \quad i = 1, 2, \dots, j-2.$$

The proof of this theorem relies on the fact that the computed u^k in algorithm (2,7) satisfies

$$u^k = Av^k - \delta_k v^{k-1} - \delta u^k,$$

with

$$\|u^k\| \leq (1 + 2k(7 + m\beta + 3(n+4))\epsilon)\|A\|,$$

and

$$\|\delta u^k\| \leq (1 + m\beta)\epsilon \|A\| + 2\|u^{k-1}\|\epsilon.$$

The reader must be careful that the denominations of the algorithms are not always the same in all of Paige's works. For instance, there are different (A1) and (A2) algorithms. The algorithm which is most recommended is (2,7).

The last paper whose results we are going to review was submitted in 1979 and published in 1980 in *Linear Algebra and its Applications* [135]. One of the main statements of this paper is that until an eigenvalue has fully converged giving a very small eigenvalue interval, the finite precision algorithm behaves remarkably like the Lanczos algorithm using full reorthogonalization. This paper shows that at least one very small interval containing an eigenvalue of A has been found by the nth step. It also states that it had not been proven (yet at that time) that all eigenvalues of A will eventually be given by the algorithm.

The paper starts by just recalling the theorem of [134] we have quoted before as Theorem 3.3, although the values of ϵ_0 and ϵ_1 are twice those of the theorem which increases the restriction on k and on the size of the problem. However, this allows us to have better bounds. The matrix H_k which is now denoted as δR_k is bounded by

$$\|\delta R_k\|_F^2 \leq 2[2(5k-4)\epsilon_0^2 + 4(k-1)\epsilon_0\epsilon_1 + k(k-1)\epsilon_1^2]\|A\|^2.$$

If we denote $\epsilon_2 = \sqrt{2}\max(6\epsilon_0, \epsilon_1)$, then

$$\|\delta R_k\|_F \leq k\epsilon_2 \|A\|.$$

Then the fundamental result relating the loss of orthogonality to eigenvalue convergence is proved,

$$(z_{(k)}^j)^T v^{k+1} = -\frac{\epsilon_{j,j}^{(k)}}{\delta_{k+1} y_{k,j}^{(k)}},$$

and hence, the Ritz vector $z_{(k)}^j$ is almost orthogonal to v^{k+1} if we have not obtained a small eigenvalue interval around $\mu_j^{(k)}$ and the eigenvector approximation norm is not too small.

Concerning the accuracy of the eigenvalues, we have

$$\min_i |\lambda_i - \mu_j^{(k)}| \le \frac{\delta_{k,j}(1+\epsilon_0) + \sqrt{k}\epsilon_1 \|A\|}{\|z_{(k)}^j\|}.$$

If $\|z_{(k)}^j\|$ is close to 1, we have an eigenpair to within about $\delta_{k,j} = \delta_{k+1} y_{i,j}^{(k)}$. For the Ritz values, Paige also proved that

$$\lambda_{min}(A) - k^{\frac{5}{2}} \|A\| \epsilon_2 \le \mu_j^{(k)} \le \lambda_{max}(A) + k^{\frac{5}{2}} \|A\| \epsilon_2.$$

This precise but thorough paper contains many other results from [132]. Paige also published in collaboration interesting papers on the sensitivity of the Lanczos algorithm to perturbations in the matrix A or the right-hand side; see [138], [139].

3.4 Illustration of the work of Chris Paige

In this section, we return to our own notations. We checked numerically that the version which works the best is computing η_{k+1} as the norm of the computed vector. Regarding Paige's variant for the other Lanczos coefficient, Figure 3.6 shows the $\log_{10}$ of $|(v^{k-1}, v^k)|$ as a function of k for the Strakos30 matrix. The dashed curve is the basic algorithm and the solid one Paige's variant. We see that the local orthogonality is better preserved with Paige's proposal. Figure 3.7 shows the same for $|(v^{k-2}, v^k)|$. Of course, the improvement is a little smaller. In Figure 3.8 we see the $\log_{10}$ of $|(v^1, v^k)|$, which measures the loss of orthogonality, and the $\log_{10}$ of $\lambda_n - \theta_k^{(k)}$ (dashed), which measures the first convergence of

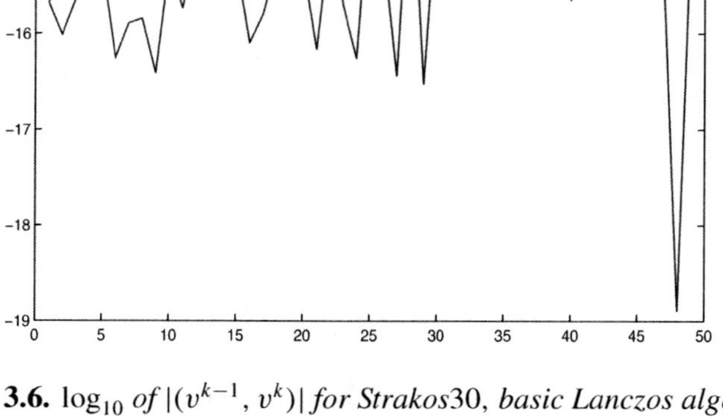

Figure 3.6. $\log_{10}$ *of* $|(v^{k-1}, v^k)|$ *for Strakos30, basic Lanczos algorithm (dashed) and Paige's variant (solid)*

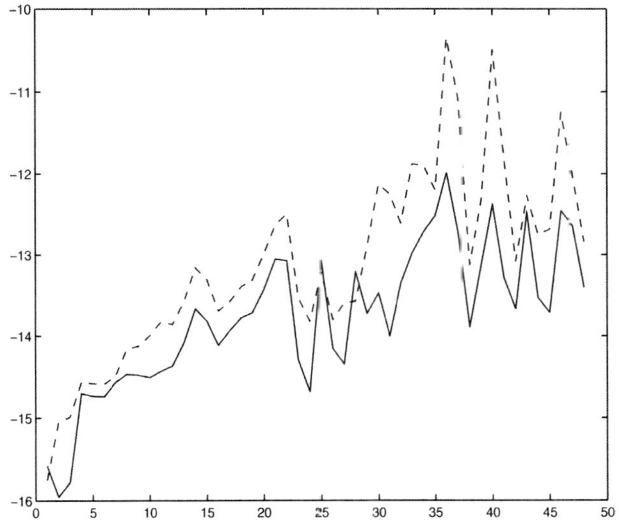

Figure 3.7. $\log_{10}$ *of* $|(v^{k-2}, v^k)|$ *for* Strakos30, *basic Lanczos algorithm (dashed) and Paige's variant (solid)*

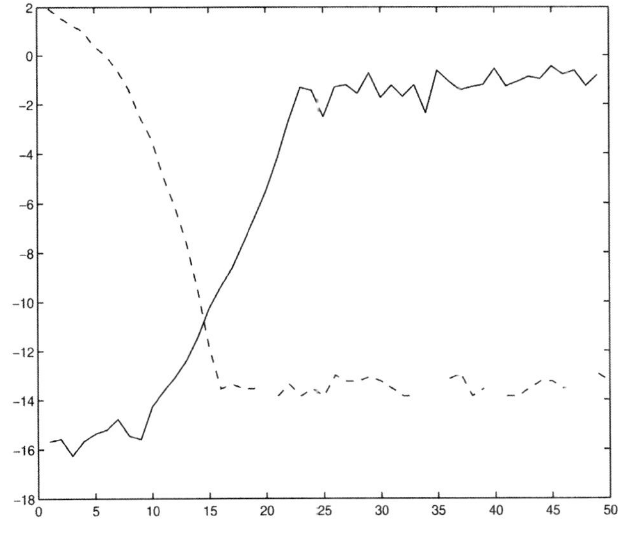

Figure 3.8. $\log_{10}$ *of* $|(v^1, v^k)|$ *(solid) and* $\lambda_n - \theta_k^{(k)}$ *(dashed) for* Strakos30

a Ritz value. Figure 3.9 shows the main ingredients of convergence, the last element of the eigenvector of T_k corresponding to the largest Ritz value z_k^k, the product $|\eta_{k+1} z_k^k \bar{v}_n^k|$, and $\lambda_n - \theta_k^{(k)}$. Figure 3.10 is another example with a tridiagonal matrix $(\, -1 \quad 4 \quad -1\,)$. We see the orthogonality with the first Lanczos vector (solid) and the convergence of the smallest Ritz value to the smallest eigenvalue of A (dashed). This helps us understanding why "loss of orthogonality goes hand in hand with eigenvalue convergence."

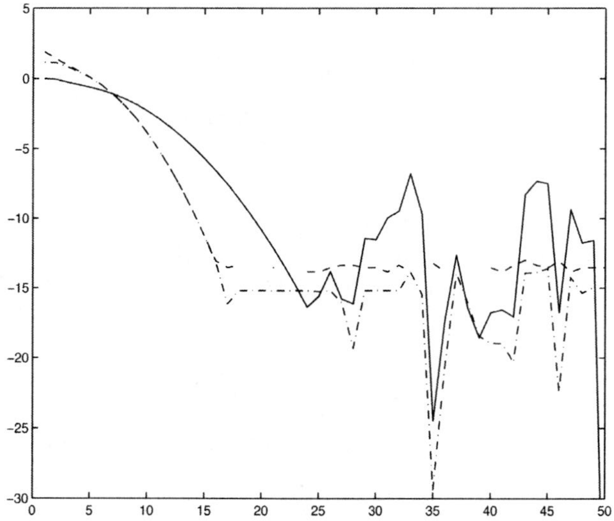

Figure 3.9. $\log_{10}$ *of* $|z_k^k|$ *(solid),* $|\eta_{k+1} z_k^k \bar{v}_n^k|$ *(dot-dashed) and* $\lambda_n - \theta_k^{(k)}$ *(dashed) for Strakos30*

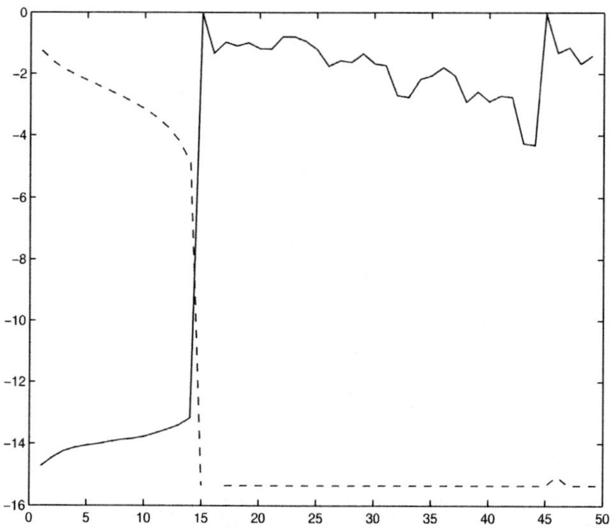

Figure 3.10. $\log_{10}$ *of* $|(v^1, v^k)|$ *(solid) and* $\theta_1^{(k)} - \lambda_1$ *(dashed) for* $trid(-1\ 4\ -1)$

3.5 The work of Joe Grcar

In his Ph.D. thesis (1981), which had not been published in journals, Grcar attempted a forward analysis of the errors in the Lanczos algorithm. Even though this fact was already implicit in the papers of Paige, Grcar clearly attributed the loss of orthogonality to instability of the recurrences of the algorithm, although he did not fully describe conditions for which these recurrences are stable or unstable. Below we summarize some of his results. This will exemplify the difficulties of a forward analysis of the rounding errors in the Lanczos algorithm. However, since we shall give solutions of three-term nonhomogeneous recurrences in Chapter 4, we shall skip most details.

Doing a forward analysis, Grcar introduced a notation for the exact and computed results. He denoted the computed quantities with a tilde. As in most other works the computed matrix recurrence is denoted as

$$A\tilde{V}_k = \tilde{V}_k\tilde{T}_k + \tilde{\eta}_{k+1}\tilde{v}^{k+1}(e^k)^T - F_k,$$

where F_k is a matrix which accounts for the local errors. For the computed basis vectors this is written as

$$\tilde{\eta}_{k+1}\tilde{v}^{k+1} = A\tilde{v}^k - \tilde{\alpha}_k\tilde{v}^k - \tilde{\eta}_k\tilde{v}^{k-1} + f^{k+1}.$$

Grcar wrote, "it is evident that cancellation and rounding errors by themselves do not cause the vast loss of orthogonality which typifies the Lanczos algorithm" [73, p. 17]. He showed the results of a single precision experiment where after step 10 the algorithm was computed in double precision and where the loss of orthogonality still grows. We shall show later on some experiments in the same spirit but looking at the components of the projections of the Lanczos basis vectors on the eigenvectors of A. We shall also report some numerical results showing that, unfortunately, double precision results do not always well represent the exact arithmetic results. The conclusion of Grcar was that the loss of orthogonality is caused by the growth of the local errors through the instability of the recurrences, but notice that this conclusion was already given in the work of Paige.

Grcar identified two properties on which his analysis is based, although there is no proof of them. He called them the projection and uncoupling properties. The projection property is that the global error $u^j = \tilde{v}^j - v^j$ is orthogonal to v^k for $k \geq j-1$. A (sufficient) condition for this property to occur is that $\|u^j\|$ should not be larger than $\sqrt{u}$. This is equivalent to the projections of the computed vectors on the exact vectors to be zero for $k \geq j-1$. The uncoupling property is that the recurrence coefficients of the Lanczos algorithm are accurate as long as the global error vectors are smaller than $\sqrt{u}$, although this was not given a very precise meaning. Grcar added that "it should be understood that the projection property and the uncoupling property are not absolute truths."

Then he derived three-term recurrences for the error vector u^k:

$$u^{k+1} = \frac{Au^k - \alpha_k u^k - \eta_k u^{k-1}}{\eta_{k+1}} + g^{k+1}.$$

The nonhomogeneous term is given by

$$g^{k+1} = h_{k+1}^{k+1} - \frac{\tilde{\eta}_{k+1} - \eta_{k+1}}{\eta_{k+1}} v^{k+1} - \frac{\tilde{\alpha}_k - \alpha_k}{\eta_{k+1}} v^k - \frac{\tilde{\eta}_k - \eta_k}{\eta_{k+1}} v^{k-1},$$

where

$$h_{k+1}^{k+1} = \frac{f^{k+1} - (\tilde{\eta}_{k+1} - \eta_{k+1})u^{k+1} - (\tilde{\alpha}_k - \alpha_k)u^k - (\tilde{\eta}_k - \eta_k)u^{k-1}}{\eta_{k+1}}.$$

We note that h_{k+1}^{k+1} depends on the error vector. The solution of the three-term recurrence for the error is given in the next theorem; see also Chapter 4.

Theorem 3.4. *Let h_j^i the vectors produced by the recurrences*

$$h_{j+1}^j = \frac{Ah_j^j - \alpha_j h_j^j}{\eta_{j+1}}, \quad h_{k+1}^j = \frac{Ah_k^j - \alpha_k h_k^j - \eta_k h_{k-1}^j}{\eta_{k+1}}, \quad k > j+1.$$

Let t_j^i be the vectors produced by the same recurrences starting from $t_j^j = g^j - h_j^j$. Then,

$$u^l = \sum_{j=1}^{l} (h_l^j + t_l^j).$$

This result is given in terms of the exact Lanczos vectors v^k and the differences in the coefficients. We can see that at the beginning of the iterations a sequence h_j^i begins at h_j^j where the local error f^j/η_{j+1} dominates a second order term which is a nonlinear relation between the error vectors. Grcar stated that "the linear recurrence formulas alone are the dominant factor in the growth of the error in the early stages of the algorithm when the projection property and the uncoupling property evidence some structure in the global errors." Grcar was much interested in deriving relations for the projections of the global error on the exact Lanczos vectors. In Chapter 4 we shall be more interested in projections on the eigenvectors of A. He proved the following result.

Proposition 3.5. *Let t^j lies in the span of $(v^1, v^2, \dots, v^j)$ and t^k, $k \geq j$, be defined by*

$$t^{j+1} = \frac{At^j - \alpha_j t^j}{\eta_{j+1}},$$

$$t^k = \frac{At^{k-1} - \alpha_{k-1} t^{k-1} - \eta_{k-1} t^{k-2}}{\eta_k}, \quad k > j+1.$$

Then t^k lies in the span of $(v^1, v^2, \dots, v^k)$ and

$$(t^k)^T v^k = (t^j)^T v^j, \quad (t^k)^T v^{k-1} = \frac{\eta_j}{\eta_k} (t^j)^T v^{j-1}.$$

This leads to the proof of the next theorem for the projections.

Theorem 3.6. *If $l > k$,*

$$(v^l)^T u^k = \sum_{j=1}^{k} (v^l)^T h_k^j,$$

$$(v^k)^T u^k = \sum_{j=1}^{k} (v^k)^T h_k^j - \sum_{j=1}^{k} \frac{\tilde{\eta}_j - \eta_j}{\eta_j},$$

$$(v^{k-1})^T u^k = \sum_{j=1}^{k} (v^{k-1})^T h_k^j - \sum_{j=1}^{k-1} \frac{\tilde{\alpha}_j - \alpha_j}{\eta_k}.$$

If the projections $(v^l)^T u^k$ are small for $l \geq k-1$, this is because the algorithm should dampen the terms $(v^l)^T h_k^1, (v^l)^T h_k^2, \dots, (v^l)^T h_k^k$.

Grcar studied the errors obtained in normalizing the basis vectors. The error in the norm is influenced significantly only by the square of the norm of the vector's error, provided the vector is orthogonal to its own error. In these conditions if the relative error in the vector is less than $\sqrt{u}$, then the computed norm is almost error free and bounds can be obtained for $\tilde{\eta}_k$. In the notations of the Lanczos algorithm, Grcar proved that

$$|\, \|\tilde{v}^k\| - 1 \,| \leq \left(n - \frac{17}{4}\right) u + O(u^{\frac{3}{2}}).$$

Concerning the projections of the error, under some minor conditions,

$$|(\tilde{v}^k)^T u^k| \leq \left(n + \frac{15}{2}\right) u + O(u^{\frac{3}{2}}).$$

Grcar obtained bounds on the (relative) differences of the Lanczos coefficients in exact and finite precision arithmetics. All the previous analysis tends to prove that the growth of the errors in the Lanczos vectors or the coefficients are caused by the recurrences. Therefore, Gcrar moved on to study the Lanczos recurrence, that is, with the exact coefficients. We have already studied some of these properties in previous chapters, notably those about the Lanczos polynomials and the eigenvectors of the tridiagonal matrices. Then Grcar was concerned about the growth of the projections. He defined for $j \leq k$,

$$M(j,k,l) = \max_{s^j \neq 0} \frac{|(s^k)^T v^l|}{\|s^j\|},$$

where s^k is obtained from the Lanczos recurrence starting from s^j at step j,

$$s^{j+1} = \frac{As^j - \alpha_j s^j}{\eta_{j+1}},$$

$$s^k = \frac{As^{k-1} - \alpha_{k-1} s^{k-1} - \eta_{k-1} s^{k-2}}{\eta_k}.$$

It is easy to prove the following result.

Theorem 3.7.

$$M(j,k,l) = \|p_{j,k}(A)v^l\|.$$

The numerical experiments in [73] show that with j and l fixed, $M(j,k,l)$ as a function of k is relatively flat when $k \le l+1$. The question arises to understand why this is so. Grcar introduced yet another definition with even more indices. For $1 \le j \le k \le m$, where m is the last step, let

$$\mu(j,k,l,i) = (s^k)^T v^i,$$

where s^k is obtained by starting the Lanczos recurrence at step j with v^l. Then,

$$M(j,k,l)^2 = \sum_{i=1}^{m} \mu(j,k,l,i)^2.$$

For each i, $|\mu(j,k,l,i)|$ is a lower bound for $M(j,k,l)$. Recurrences can be exhibited between different values of $\mu(i,k,l,i)$. This leads us to prove

$$\mu(j,k,l,l+k-j) = \frac{\eta_{l+1}\eta_{l+2}\cdots\eta_{l+k-j}}{\eta_{j+1}\eta_{j+2}\cdots\eta_k}.$$

It gives a lower bound for $M(j,k,l)$ for $l > k$. Moreover, it can be shown that $\mu(j,k,l,i) = 0$ when $|l-i| > k-j$. Finally, we have the following result.

Proposition 3.8. *If $k > l+1 > j$, then when $k \ge 2(l+1)-j$*

$$M(j,k,l) \ge \frac{\eta_j}{\eta_k}\frac{\eta_j\eta_{j+1}\cdots\eta_l}{\eta_{j+k-l}\eta_{j+k-l+1}\cdots\eta_k},$$

or when $k < 2(l+1)-j$

$$M(j,k,l) \ge \frac{\eta_j}{\eta_k}\frac{\eta_j\eta_{j+1}\cdots\eta_{j+k-l+1}}{\eta_{l+1}\eta_{l+2}\cdots\eta_{k-1}}.$$

These results show that if the η's decrease, then $M(j,k,l)$ is large when $k > l+1 > j$. It is much more difficult to obtain upper bounds for $M(j,k,l)$ than lower bounds. These projections monitor the growth of the error sequences. Since decreasing η are common in practice, Grcar concluded that "it seems that stable instances of the Lanczos algorithm are rare." Of course, we know that loss of orthogonality goes with convergence of Ritz values. So, with this definition, only cases where all eigenvalues and eigenvectors converge at the last step preserve orthogonality and are stable.

Finally, Grcar considered ways to stabilize the Lanczos algorithm. It is well known that full reorthogonalization is a good way to cure the problems, even though some nice properties of the Lanczos algorithm are then lost. An interesting question is what is the action of reorthogonalization. If $\tilde{v}^k$ is reorthogonalized against $\tilde{v}^j$, $j < k$, then $(\tilde{v}^k)^T v^j = (u^k)^T v^j$

changes to

$$[\tilde{v}^k - ((\tilde{v}^k)^T \tilde{v}^j)\tilde{v}^j]^T v^j = (u^k)^T v^j - ((u^k)^T v^j + (v^k)^T u^j + (u^k)^T u^j)(1 + (u^j)^T v^j)$$
$$= (v^k)^T u^j + O(u),$$

provided $\|u^k\| \leq \sqrt{u}$ and $\|u^j\| \leq \sqrt{u}$. Therefore an error component of size $|(u^k)^T v^j|$ is traded for one of size $|(u^j)^T v^k|$ which is small when $j < k$.

Proposition 3.9. *If $\hat{v}^k$ is the result of reorthogonalizing $\tilde{v}^k$ against the previous Lanczos vectors, then*

$$\hat{v}^k = \tilde{v}^k - \sum_{j=1}^{k-1} ((\tilde{v}^j)^T \tilde{v}^k)\tilde{v}^j.$$

If the global errors are bounded by $\sqrt{u}$, then

$$\hat{v}^k - v^k = r - \sum_{j=1}^{k-1} ((u^j)^T v^k) v^j + t,$$

where $\|t\| \leq (k-1)(3+\sqrt{u})u$ and r is the projection of u^k into the orthogonal complement of $span(V_{k-1})$.

Grcar advocated periodic reorthogonalization. A criterion is set to monitor the orthogonality and when it is reached a full reorthogonalization with all the previous vectors is done and the criterion is reset. Suppose we orthogonalize at iteration j and then s^{j-1} and s^j are set, respectively, to zero and a vector y^j with only ones in the first j positions and zero elsewhere. The Lanczos recurrence is run with $\tilde{T}_k$ instead of A and an additional nonhomogeneous term y^k. When $\|s^l\|/\|s^j\|$ is too large (that is, $> \sqrt{u}$) a new reorthogonalization is done. The criterion is set in this way using the tridiagonal matrix to avoid additional matrix-vector multiplications with A.

3.6 Illustration of the work of Joe Grcar

Figure 3.11 exemplifies the projection property. The elements $(\tilde{V}_{30}^T V_{30})_{i,j}$ are of the order of the roundoff unit for $j > i$. The dashed curve in Figure 3.12 is the coefficient α_k for Lanczos with double reorthogonalization which we consider as the "exact" result for this example. The solid curve is for the standard Lanczos algorithm. The difference between the coefficients with and without reorthogonalization is shown in Figure 3.13. It is at the level of 10^{-14}; that is, $u\|A\|$ for the first 17 iterations and then it grows very rapidly. Remember that the largest eigenvalue converges up to roundoff at iteration 18. When the coefficients started to differ the finite precision one has a very erratic behavior. The coefficient η_k and the relative difference are shown in Figures 3.14 and 3.15. The coefficients in finite precision are the "exact" ones (up to rounding errors) for quite a while and then they differ quite dramatically after the first convergence of an eigenvalue when orthogonality is lost.

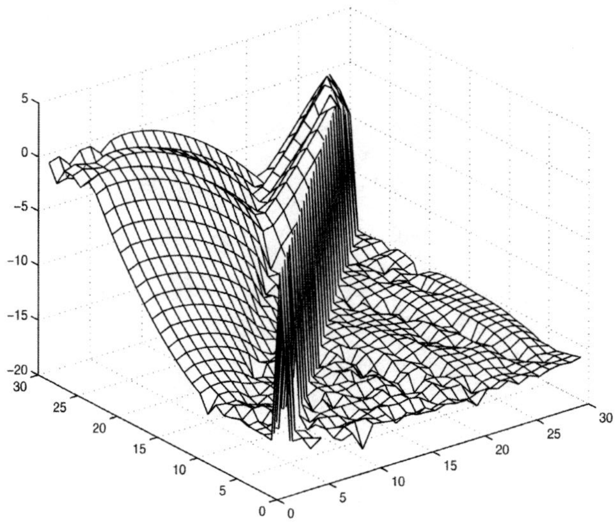

Figure 3.11. *The projection property:* $\log_{10}(|\tilde{V}_{30}^T V_{30}|)$ *for the Strakos30 matrix*

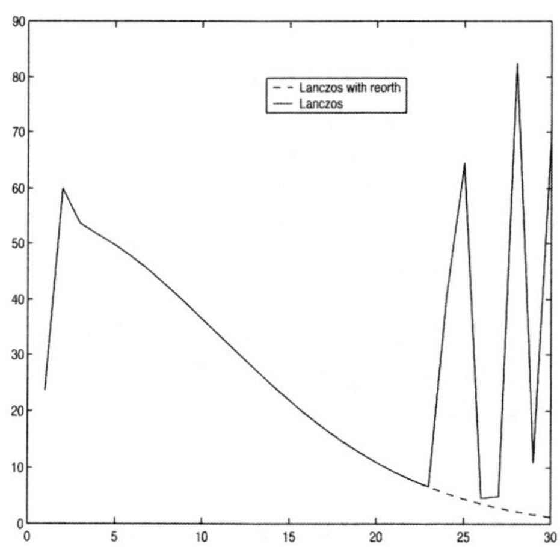

Figure 3.12. *Strakos30, coefficient* α_k

3.7 The work of Horst Simon

The Ph.D. thesis of Horst Simon [162] (1982) was concerned with solving symmetric linear systems with the Lanczos algorithm. Therefore, he was not directly interested in eigenvalue convergence. The main results of his work were subsequently published in two papers [163], [164]. The thesis started with a review of the Lanczos method in exact arithmetic and its relations to the CG method as well as other methods like MINRES and SYMMLQ. Simon moved on to the analysis of the algorithm in finite precision. It is of interest to quote the first

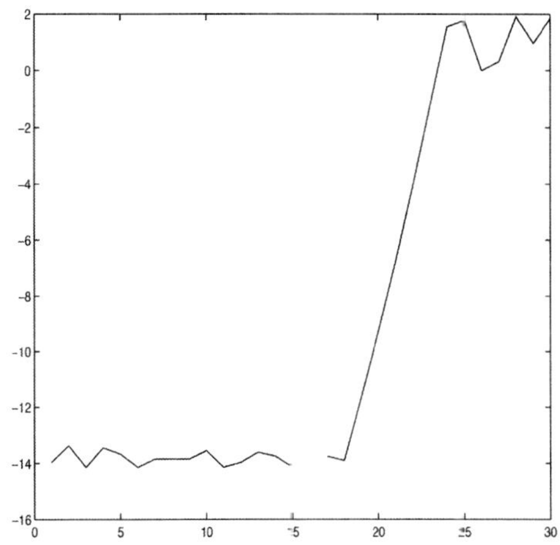

Figure 3.13. *Strakos30, uncoupling property:* $\log_{10}(|\tilde{\alpha}_k - \alpha_k|)$

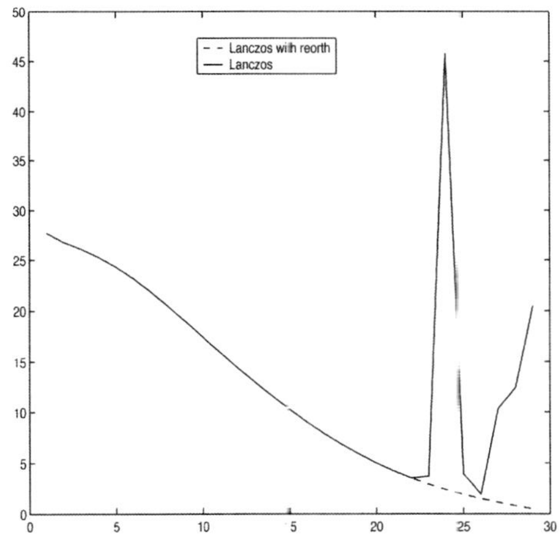

Figure 3.14. *Strakos30, coefficient* η_k

paragraph of section 2.1: "Most error analyses start out by making some assumptions on the roundoff errors which will occur when elementary operations like addition, etc... are carried out in floating point computation with relative precision ε. Based on these assumptions upper bounds on the errors in vector inner products, matrix-vector multiplication, etc... are derived or the reader is referred to Wilkinson. After providing these tools then finally the object of analysis itself is approached. Lengthy and complicated derivations finally yield error bounds which are rigorous but, in most cases, unrealistically pessimistic." According to this, Simon decided to make a few simplifying assumptions on the behavior of the Lanczos algorithm in finite precision. He did not consider the exact quantities, so he used the standard

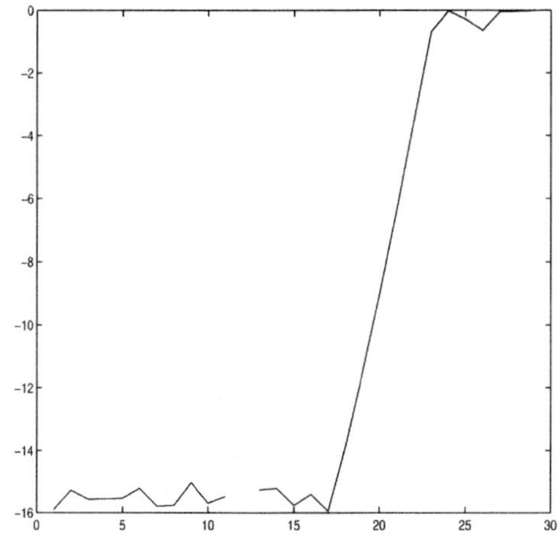

Figure 3.15. *Strakos30, uncoupling property:* $\log_{10}(|\tilde{\eta}_k - \eta_k|/\eta_k)$

notation for computed quantities which satisfy the following equation:

$$\eta_{k+1} v^{k+1} = A v^k - \alpha_k v^k - \eta_k v^{k-1} - f^k,$$

where f^k represents the local roundoff errors. In matrix form this translates into

$$A V_k - V_k T_k = \eta_{k+1} v^{k+1} (e^k)^T + F_k.$$

Simon supposed that $\|F_k\| \leq u \|A\|$. Of course, this is very close to what was obtained by Paige. Let the first k Lanczos vectors satisfy

$$|(v^i)^T v^j| \leq \omega_k, \quad i, j = 1, \dots, k, \quad i \neq j.$$

The smallest ω_k for which this holds is called the level of orthogonality. If $\omega_k = \sqrt{u}$, the vectors are said to be semiorthogonal. Simon assumed that the Lanczos vectors are exactly normalized $(v^j)^T v^j = 1, \ j = 1, \dots, k$, and that they are locally orthogonal $|(v^{j+1})^T v^j| \leq u_1, \ j = 1, \dots, k$, where $1 \gg u_1 > u$. The constant u_1 is supposed to be such that $u_1 \ll \sqrt{u}$. It is also assumed that no η_j ever becomes negligible. Let us also suppose in this section that A is positive definite. Quoting Simon, "the loss of orthogonality can be viewed as the result of an amplification of each local error after its introduction in the computation." The main result is the following.

Theorem 3.10. *Let $w_{i,j}$ be the elements of the matrix $W_k = V_k^T V_k$. They satisfy the following recurrence:*

$$w_{j,j} = 1, \quad j = 1, \dots, k, \quad w_{j,j-1} = u_j, \quad j = 2, \dots, k,$$

$$\eta_{j+1} w_{j+1,i} = \eta_{i+1} w_{j,i+1} + (\alpha_i - \alpha_j) w_{j,i} + \eta_i w_{j,i-1} - \eta_j w_{j-1,i} + (v^j)^T f^i - (v^i)^T f^j,$$

$1 \leq i < j$, *and* $w_{j,i+1} = w_{i+1,j}$. *Moreover,* $w_{j,0} = 0$ *and* $u_j = (v^j)^T v^{j-1}$.

Proof. The proof of this result is straightforward. Write the Lanczos equations for j and i and scalar multiply, respectively, by v^i and v^j before taking the difference. □

This result can also be written in a matrix form

$$\eta_{j+1} V_j^T v^{j+1} = T_j V_j^T v^j - \alpha_j V_j^T v^j - \eta_j V_j^T v^{j-1} + \eta_{j+1} e^j (v^{j+1})^T v^j + F_j^T v^j - V_j^T f^j.$$

Let R_j be the strictly upper triangular part of W_j and $\bar{r}^i$ its columns. If $w^{j+1} = V_j^T v^{j+1}$,

$$\eta_{j+1} w^{j+1} = T_j \bar{r}^j - \alpha_j \bar{r}^j - \eta_j \bar{r}^{j-1} + g^j,$$

where $g^j = F_j v^j - V_j f^j$. By using the fact proved by Paige that $\|T_j\| \le \|A\| + \|F_j\|$ and taking norms we obtain

$$\eta_{j+1} \|w^{j+1}\| \le 2\|A\| \max(\|\bar{r}^j\|, \|\bar{r}^{j-1}\|) + O(u\|A\|).$$

The growth of the level of orthogonality at each step is bounded by $2\|A\|/\eta_{j+1}$. The loss of orthogonality is initiated by the local error f^k, but its growth is determined by the Lanczos recurrence, that is, the computed coefficients α_k and η_k.

Now let $\theta_j^{(k)}$ be the Ritz values, that is, the exact eigenvalues of the computed T_k and $z_{(k)}^j$ or z^j the corresponding eigenvectors. We denote by $y^j = V_k z^j$ the corresponding approximations of eigenvectors (notice this is different from Paige's notations). From the equation giving w^{j+1} we have in matrix form

$$\eta_{j+1} w^{j+1} (e^j)^T = T_j R_j - R_j T_j + G_j,$$

where G_j is the strictly upper triangular part of $F_j^T V_j - V_j^T F_j$. This is similar to the equation that was derived by Paige. Then if we multiply on the left by $(z^i)^T$ and on the right by z^i and we denote by $\sigma_{j,i}$ the bottom element of the ith eigenvector z^i,

$$\eta_{j+1} (z^i)^T w^{j+1} \sigma_{j,i} = \theta_i^{(j)} (z^i)^T R_j z^i - (z^i)^T R_j z^i \theta_i^{(j)} + (z^i)^T G_j z^i.$$

Therefore,

$$\eta_{j+1} (y^i)^T v^{j+1} \sigma_{j,i} = (z^i)^T G_j z^i = \gamma_{i,i},$$

and we recover Paige's result (with different notations),

$$(y^i)^T v^{j+1} = \frac{\gamma_{i,i}}{\eta_{j+1} \sigma_{j,i}}.$$

As we have seen before, this shows that the way by which $(y^i)^T v^{j+1}$ can become large is by having $\sigma_{j,i}$ small, that is, convergence of one eigenvalue.

Simon provided an interesting small example showing that even if two matrices have close elements this does not imply that the Lanczos vectors computed with these matrices are close when starting from the same vector. The first matrix is

$$A = \begin{pmatrix} 1 & \eta & 0 \\ \eta & 2 & 3 \\ 0 & 3 & 4 \end{pmatrix}.$$

If η is small, this matrix is close to

$$\bar{A} = \begin{pmatrix} 1 & 0 & \eta \\ 0 & 2 & 3 \\ \eta & 3 & 4 \end{pmatrix}.$$

However, if $v^1 = (1 \quad 0 \quad 0)^T$, computing with A gives $v^2 = (0 \quad 1 \quad 0)^T$. Using the matrix $\bar{A}$ one obtains $\bar{v}^2 = (0 \quad 0 \quad 1)^T$, the values of α_1 being the same—equal to 1. This shows that a backward analysis (for the Lanczos vectors) cannot always work. One of the most interesting result in Simon's thesis is the following.

Theorem 3.11. *For $k > 3$, if the level of orthogonality ω between the Lanczos vectors in V_{k+1} satisfies*

$$\omega \le \frac{1}{2(k-2)},$$

then the computed matrix T_k is similar to a matrix $\bar{T}_k$ which is a perturbation of the orthogonal projection A_P of A onto $span(V_k)$ and

$$\|A_P - \bar{T}_k\| \le 2\sqrt{k}\omega\eta_{k+1} + 2u\|A\| + O(u^2).$$

Proof. The condition on ω is used to prove that the Lanczos vectors are linearly independent with, of course, $k \le n$. The QR factorization of V_k is written as $V_k = N_k L_k^T$, where N_k is an $n \times k$ orthonormal matrix and L_k^T is a $k \times k$ upper triangular matrix with positive diagonal elements. Then $W_k = V_k^T V_k = L_k L_k^T$ and L_k is the Cholesky factor of W_k. As we have seen before when looking at the QR factorization of the Krylov matrix, the first k columns of L_{k+1}^T are those of L_k^T completed by a zero in the last position and let l_{k+1} be the last column. Multiplying the Lanczos relation by V_k^T we obtain

$$V_k^T V_k T_k = V_k^T A V_k - \eta_{k+1} V_k^T v^{k+1} (e^k)^T - V_k F_k.$$

Using the QR factorization and multiplying by L_k^{-1} on the left and L_k^{-T} on the right,

$$L_k^T T_k L_k^{-T} = N_k^T A N_k - \eta_{k+1} N_k^T N_{k+1} l_{k+1} (e^k)^T L_k^{-T} - N_k^T F_k L_k^{-T}.$$

Let $\bar{T}_k = L_k^T T_k L_k^{-T}$. This matrix, which is similar to T_k, is a perturbation of $A_P = N_k^T A N_k$, the orthogonal projection of A onto the span of V_k. The norm of the perturbation can be bounded by

$$\|A_P - \bar{T}_k\| \le \eta_{k+1} \|N_k^T N_{k+1} l_{k+1} (e^k)^T L_k^{-T}\| + \|N_k^T F_k L_k^{-T}\|.$$

But $N_k^T N_{k+1} = (I_k \quad 0)$ and $L_k^{-1} e^k = l_{k,k}^{-1} e^k$, where $l_{k,k}$ is the bottom element of l_k, that is, the last element of the diagonal of L_k. Without any hypothesis on ω we have

$$\|A_P - \bar{T}_k\| \le \eta_{k+1} l_{k,k}^{-1} \|\bar{l}_{k+1}\| + u\|A\| \, \|L_k^{-T}\|,$$

where $\bar{l}_{k+1}$ is made of the k upper elements of the vector l_{k+1} which is the last column of the Cholesky factor L_{k+1}. It remains to study the matrix W_k and its Cholesky factors. By

Simon's hypothesis the diagonal elements of W_k are 1 and the absolute values of the other elements are less than 1. By Gerschgorin's theorem

$$\lambda_{min}(W_k) \geq 1 - (k-1)\omega, \quad \lambda_{max}(W_k) \leq 1 + (k-1)\omega.$$

Therefore, W_k is positive definite if $\omega \leq 1/(k-1)$. If $\omega \leq 1/2(k-2)$ and $k > 2$,

$$\|L_k\| = \sqrt{\lambda_{max}(L_k L_k^T)} = \sqrt{\lambda_{max}(W_k)} \leq \sqrt{1 + (k-1)\omega} \leq \sqrt{2}.$$

Similarly $\|L_k^{-T}\| \leq \sqrt{2}$. It remains to bound the (k,k) element of the Cholesky factor of W_k. This was done in the paper [164]. Let $\delta_{i,j}$ be the elements of L_k in the Cholesky factorization of W_k. If $\omega \leq 1/2(k-2)$, then

$$\delta_{i,j} = w_{i,j} + O(k\omega^2), \quad 1 \leq j < i \leq k,$$

$$\delta_{i,i} = 1 + O(k\omega^2), \quad 1 \leq i \leq k.$$

This is proved by induction. Using this result for L_{k+1},

$$\|\bar{l}_{k+1}\|^2 = \sum_{j=1}^{k} [w_{k+1,j} + O(k\omega^2)]^2 \leq k\omega^2 + O(k^2\omega^3).$$

Thus, we have bounded all the terms in the upper bound of $\|A_P - \bar{T}_k\|$. □

When ω is larger the conclusion no longer holds since the asymptotic analysis can no longer be done. However, interesting results can still be obtained as long as $l_{k,k}^{-1}$, $\|\bar{l}_{k+1}\|$ and $\|L_k^{-T}\|$ are reasonably bounded if the Cholesky factorization of W_k exists, that is, when W_k is nonsingular. Another interesting result is the following theorem.

Theorem 3.12. *If $\omega \leq \sqrt{u/k}$, then*

$$N_k^T A N_k = T_k + H_k,$$

where the elements of H_k are of order $O(u\|A\|)$.

Proof. We have

$$N_k^T A N_k = L_k^{-1} T_k L_k + \eta_{k+1} L_k^{-1} e^k \bar{l}_{k+1}^T + L_k^{-1} F_k^T N_k.$$

By induction it is enough to prove that the last columns of $N_k^T A N_k$ and T_k differ only by terms of order $O(u\|A\|)$. But

$$N_k^T A N_k e^k = L_k^{-1} T_k L_k e^k + \eta_{k+1} L_k^{-1} e^k \bar{l}_{k+1}^T e^k + L_k^{-1} F_k^T N_k e^k.$$

$L_k^{-1} T_k L_k$ is a lower Hessenberg matrix. Only the last two elements of the vector $L_k^{-1} T_k L_k e^k$ are nonzero. The elements of L_k are denoted by $\delta_{i,j}$, so the elements in the bottom right corner of L_k are

$$\begin{pmatrix} \delta_{k-1,k-1} & 0 \\ \delta_{k,k-1} & \delta_{k,k} \end{pmatrix}.$$

Let the corresponding elements of L_k^{-1} be

$$\begin{pmatrix} \delta_{k-1,k-1}^{-1} & 0 \\ \delta_{k,k-1}^{(-1)} & \delta_{k,k}^{-1} \end{pmatrix},$$

with

$$\delta_{k,k-1}^{(-1)} = -\frac{\delta_{k,k-1}}{\delta_{k,k}\delta_{k-1,k-1}}.$$

Then,

$$\begin{aligned} L_k^{-1} T_k L_k e^k &= L_k^{-1} T_k (\delta_{k,k} e^k) = \delta_{k,k} L_k^{-1} (\alpha_k e^k + \eta_k e^{k-1}) \\ &= \delta_{k,k} \alpha_k \delta_{k,k}^{-1} e^k + \delta_{k,k} \eta_k (\delta_{k-1,k-1}^{-1} e^{k-1} + \delta_{k,k-1}^{(-1)} e^k). \end{aligned}$$

For the second term on the right-hand side of the equation for $N_k^T A N_k e^k$ we have

$$\eta_{k+1} L_k^{-1} e^k l_{k+1}^T e^k = \eta_{k+1} \delta_{k,k}^{-1} \delta_{k+1,k} e^k.$$

We have $\delta_{k-1,k-1} = 1 + O(u)$, $\delta_{k,k} = 1 + O(u)$. Hence, the same is true for their reciprocals. Also, $\delta_{k,k-1} = w_{k,k-1} + O(u)$ and $\delta_{k+1,k} = w_{k+1,k} + O(u)$. This shows that $\delta_{k,k-1}^{(-1)} = -w_{k,k-1} + O(u)$. Altogether we have

$$N_k^T A N_k e^k = \alpha_k e^k + \eta_k e^{k-1} + (\eta_{k+1} w_{k+1,k} - \eta_k w_{k,k-1}) e^k + O(u)\|A\|.$$

This uses the fact that

$$\eta_k + \eta_{k+1} \le \|T_k\| \le (1+u)\|A\|.$$

The proof is concluded by using the hypothesis on local orthogonality

$$\eta_{k+1} (v^{k+1})^T v^k = O(u)\|A\|.$$

Finally,

$$N_k^T A N_k e^k = \alpha_k e^k + \eta_k e^{k-1} + O(u)\|A\|. \quad \square$$

Simon studied different strategies to maintain semiorthogonality. Let $\omega_0 = \sqrt{u/k}$. At some step of the Lanczos algorithm we have

$$\eta'_{k+1} \tilde{v}_{k+1} = A v^k - \alpha_k v^k - \eta_k v^{k-1} - (f^k)'.$$

If $|(\tilde{v}_{k+1})^T v^j| > \omega_0$ for some j, one chooses $k-1$ numbers $\xi_1, \dots, \xi_{k-1}$ to compute

$$\eta_{k+1} v_{k+1} = \eta'_{k+1} \tilde{v}_{k+1} - \sum_{j=1}^{k-1} \xi_j v^j - f^k.$$

A semiorthogonalization strategy is defined by choosing the ξ_j such that $|(v^j)^T v^{k+1}| \le \omega_0$ and (by redefining f^k)

$$\eta_{k+1} v_{k+1} = A v^k - \alpha_k v^k - \eta_k v^{k-1} - \sum_{j=1}^{k-1} \xi_j v^j - f^k,$$

with $\|f^k\| \le u\|A\|$. Of course, semiorthogonalization preserves local orthogonality.

Lemma 3.13. *With semiorthogonalization,*

$$\eta_{k+1}(v^{k+1})^T v^k = O(u)\|A\|.$$

Simon proved that with semiorthogonalization the property on $N_k^T A N_k$ holds without any condition.

Theorem 3.14. *With semiorthogonalization,*

$$N_k^T A N_k = T_k + H_k,$$

where H_k is of order $u\|A\|$.

Simon noted that if theoretically a level of orthogonality of $\sqrt{u/k}$ is required, practically a level of $\sqrt{u}$ is enough to obtain the properties of the theorem. He proved that full reorthogonalization and the periodic reorthogonalization of Grcar are semiorthogonalization methods. This is also true for the more complicated selective orthogonalization algorithm of Parlett and Scott [142] and also [141], although this is much more difficult to prove. Based on the previous results Simon proposed a simpler strategy called partial reorthogonalization (PRO). Details are given in [162], [163]. Formally, one iteration of PRO is the following:

1. Perform a regular Lanczos step
$$w^{k+1} = Av^k - \alpha_k v^k - \eta_k v^{k-1} - f^k.$$
2. Update $\omega_{k+1,j}$, $j = 1, \dots, k$.
3. Based on this information find a set of indices $L(k)$ smaller or equal to k (eventually empty) and compute
$$\eta_{k+1} v_{k+1} = w^{k+1} - \sum_{j\in L(k)} [(w^{k+1})^T v^j] v^j - f^k.$$

However, some details have to be given about how to compute the level of orthogonality and how to choose the set of indices $L(k)$. The problem with the computation of the level of orthogonality is that the local roundoff terms f^k are unknown. The terms $(v^k)^T f^j - (v^j)^T f^k$ are important only as long as $\omega_{k,j}$ is smaller than $\sqrt{u}$. Simon proposed to replace these terms by random values chosen from appropriate distributions. The recurrence relations for $\omega_{k,j}$ are now

$$\omega_{j,j} = 1, \quad j = 1, \dots, k, \quad \omega_{j,j-1} = \psi_j, \quad j = 2, \dots, k,$$

$$\omega_{k+1,j} = \frac{1}{\eta_{k+1}}[\eta_{k+1}\omega_{k,j+1} + (\alpha_j - \alpha_k)\omega_{k,j} + \eta_j\omega_{k,j-1} - \eta_k\omega_{k-1,j}] + \theta_{k,j}.$$

Based on statistical studies of the roundoff terms, Simon proposed to use

$$\theta_{k,j} = u(\eta_{j+1} + \eta_{k+1})\Theta, \quad \Theta \in N(0, 0.3),$$

where $N(0, 0.3)$ represents normally distributed random numbers with mean 0 and standard deviation 0.3. The other term is chosen as

$$\psi_j = un \frac{\eta_2}{\eta_{k+1}} \Psi, \qquad \Psi \in N(0, 0.6).$$

One has also to reset the $\omega_{k,j}$ after a reorthogonalization. Simon's proposal is $\omega_{k+1,j} \in N(0, 1.5)u$. These values were chosen to overestimate the true level of orthogonality. Concerning reorthogonalization, one can see that if we decide, for instance, to orthogonalize v^{k+1} against all the previous v^j, then it will also be necessary to reorthogonalize v^{k+2} against all the previous vectors. This is because by looking at the recurrence formulas the inner product $(v^{k+2})^T v^j$ involves a term $(v^k)^T v^j$ that must have been close to $\sqrt{u}$ since there exists at least a j for which $|(v^{k+1})^T v^j| \geq \sqrt{u}$. Reorthogonalizing twice in a row solves this problem. It remains to choose the set of indices $L(k)$. Of course, it is not enough to orthogonalize against the offending vector. Simon's strategy if $|\omega_{k+1,j}| \geq \sqrt{u}$ is to check the neighbors until $\omega_{k+1,j-s}$ and $\omega_{k+1,j+r}$ are found such that

$$|\omega_{k+1,j-s}| \leq \eta, \qquad |\omega_{k+1,j+r}| \leq \eta.$$

Then v^{k+1} is orthogonalized against v^{j-s} to v^{j+r}. At the next step v^{k+2} is orthogonalized against v^{j-s+1} to v^{j+r-1}. The value $\eta = u^{\frac{3}{4}}$ was found by experiments.

3.8 Illustration of the work of Horst Simon

We use the Strakos30 example. Figure 3.16 shows the logarithm of the level of orthogonality ω. The dashed horizontal line is the ($\log_{10}$ of the) square root of the roundoff unit u and the dot-dashed curve is $\sqrt{u/k}$. The level of orthogonality reaches $\sqrt{u}$ at iteration 18. Figure 3.17 gives the same information and in addition the left dashed curve starts at the same roundoff level with a multiplicative factor of $2\|A\|/\eta_{k+1}$ at each iteration. Remember that this was established as an upper bound on the level of orthogonality. Of course the values are much larger than ω, but the slope is right, as we can see by looking at the right dashed curve, which is the same but with a shift of 10 iterations. For this example, after an initial phase, when the level of orthogonality starts to rapidly grow it does it at this rate of $2\|A\|/\eta_{k+1}$. In Figure 3.18 the solid curve is $\log_{10}(\|A_P - \bar{T}_k\|)$ and the dashed curve is the first term in the bound $\log_{10}(\sqrt{k}\omega\eta_{k+1})$. Figure 3.19 shows the same curves (although the solid one is hidden under another solid curve) plus the dot-dashed curve, which is $\log_{10}(u\|A\|\ \|L_k^{-T}\|)$, the second term in the bound, and the solid curve, which is $\eta_{k+1} l_{k,k}^{-1} \|\bar{l}_{k+1}\|$, the first term in the bound. In this example we see that the first term dominates the second one. The norm of L_k^{-T} starts increasing significantly at iteration 24. In Figure 3.20 the solid curve is again the level of orthogonality. The dashed curve is $\log_{10}(\|V_k^T A V_k - T_k\|)$. The dot-dashed curve is $\log_{10}(\|N_k^T A N_k - T_k\|)$. The upper dot-dashed "horizontal" curve is $\sqrt{u/k}$. The lower horizontal dot-dashed line is $\log_{10}(u\|A\|)$. We see that $\|V_k^T A V_k - T_k\|$ grows from the beginning with approximately the same slope as the level of orthogonality. $\|N_k^T A N_k - T_k\|$ is almost constant at a level of a small multiple of $u\|A\|$ until iteration 18, which is the one at which the level of orthogonality becomes larger than $\sqrt{u}$.

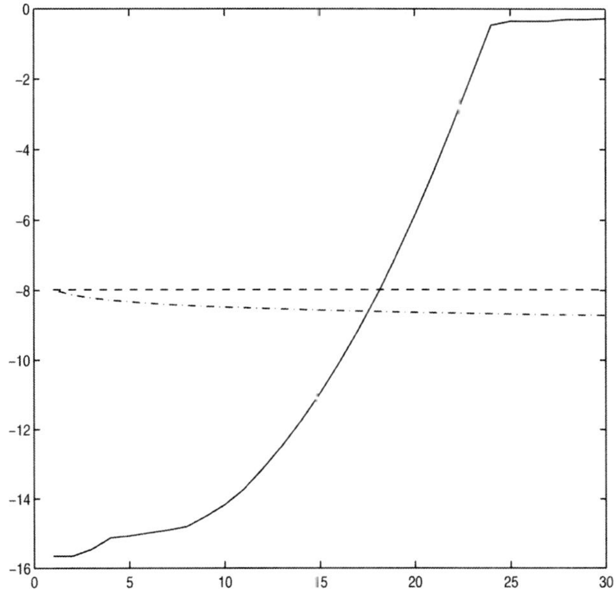

Figure 3.16. *Strakos*30, $\log_{10}$ *of the level of orthogonality* ω

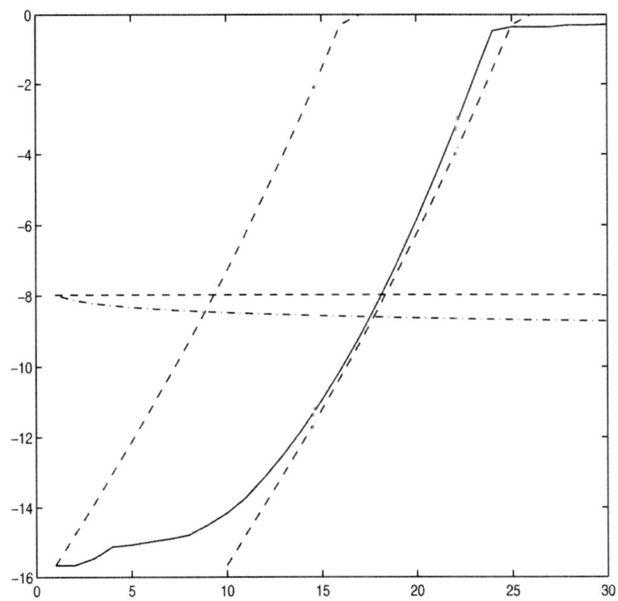

Figure 3.17. *Strakos*30, $\log_{10}$ *of the level of orthogonality* ω *and bound*

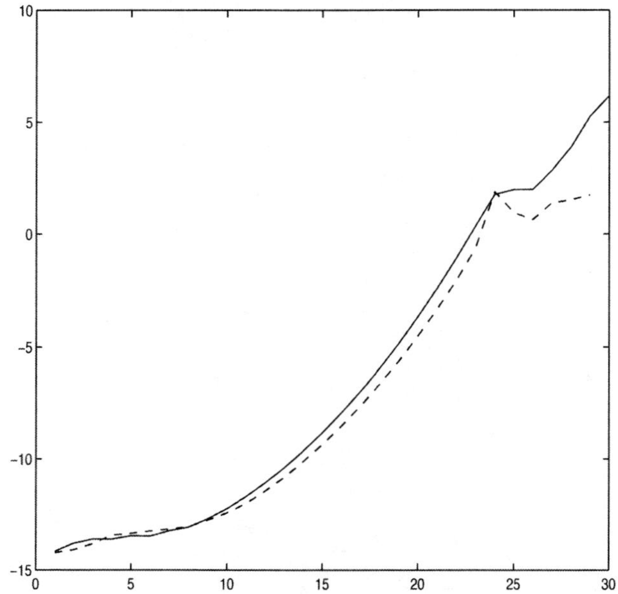

Figure 3.18. *Strakos30*, $\log_{10}(\|A_P - \bar{T}_k\|)$ *and* $\log_{10}(\sqrt{k}\omega\eta_{k+1})$

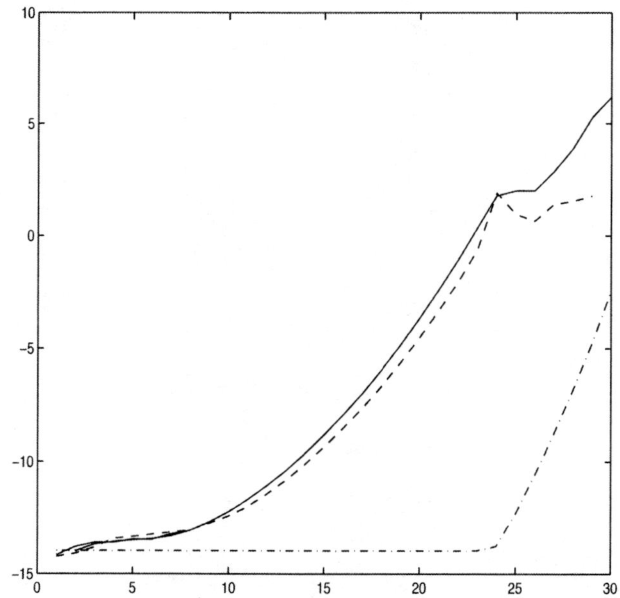

Figure 3.19. *Strakos30, same as Figure* 3.18 *plus the terms in the bound*

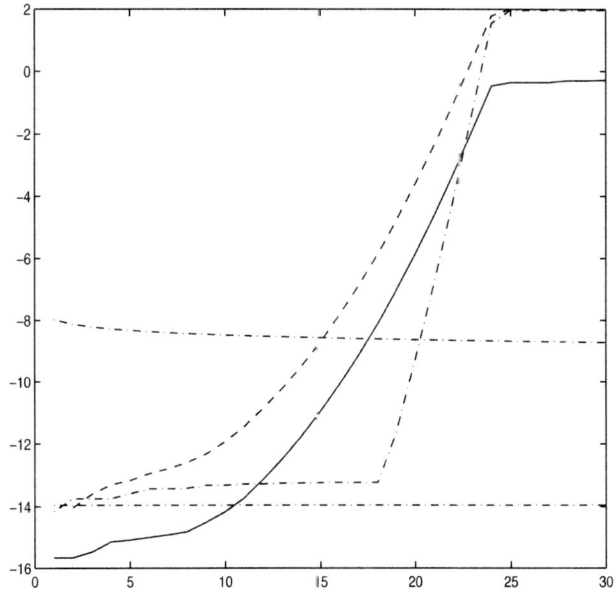

Figure 3.20. *Strakos*30, $\log_{10}(\|N_k^T A N_k - T_k\|)$ *and others*

3.9 The work of Anne Greenbaum and Zdenek Strakoš

The work of Anne Greenbaum gives results about the Lanczos and CG algorithms. In [76] she studied perturbed Lanczos recurrence like

$$AV_k = V_k T_k + w^k (e^k)^T + F_k.$$

However, she used a different view point than in other works we have already examined. She showed that the matrix T_k generated at a given step k of a perturbed Lanczos recurrence is the same as the one generated by an exact Lanczos recurrence applied to a matrix of larger dimension than A whose eigenvalues are all close to those of A. This can be considered as a backward error analysis in a sense different from the usual one. For instance, for Gaussian elimination analysis shows that the finite precision computation corresponds to a matrix $A + \delta A$ and bounds are given on δA. Here we have a larger matrix whose eigenvalues are perturbations around the eigenvalues of A. This is done by showing that at iteration k, T_k can be extended to a matrix

$$T_{k+M} = \begin{pmatrix} T_k & \eta_{k+1} & & & & \\ \eta_{k+1} & \alpha_{k+1} & \eta_{k+2} & & & \\ & \eta_{k+2} & \alpha_{k+2} & \eta_{k+3} & & \\ & & \ddots & \ddots & \ddots & \\ & & & \eta_{k+M-1} & \alpha_{k+M-1} & \eta_{k+M} \\ & & & & \eta_{k+M} & \alpha_{k+M} \end{pmatrix},$$

with $\eta_{k+M+1} = 0$ whose eigenvalues are all close to those of A. The result is obtained by applying the Lanczos algorithm in exact arithmetic to T_{k+M} using e^1 as the initial vector. If this construction can be done,

$$AV_{k+M} = V_{k+M}T_{k+M} + \tilde{F}_{k+M},$$

where $\tilde{F}_{k+M} = (f^1, \dots, f^{k-1}, \tilde{f}^k, \dots, \tilde{f}^{k+M})$, where the first $k-1$ columns are the perturbations in the finite precision Lanczos steps and the other ones are perturbations arising from the construction explained below. One way to construct such an extension is to continue with the Lanczos recurrence after step k, making small additional perturbations to the recurrence if necessary, in order to generate a zero coefficient $\eta_j = 0$ at or before step $k + M + 1$ by orthogonalizing the new vectors again the previous ones as in exact arithmetic. Assuming this happens at step $k + M + 1$ and denoting by Z_{k+M} the matrix of the eigenvectors z^j of T_{k+M}, Θ_{k+M} the matrix of the eigenvalues, and $Y_{k+M} = V_{k+M}Z_{k+M}$ with columns y^j, we have

$$AY_{k+M} = Y_{k+M}\Theta_{k+M} + \tilde{F}_{k+M}Z_{k+M}.$$

If we suppose that for the $k-1$ first steps $\|f^j\| \le \epsilon\|A\|$, by looking at the residuals of the vectors y^j considered as approximations to the eigenvectors we obtain that

$$|\theta_j - \lambda_i| \le \frac{1}{\|y^j\|}\left[(k-1)(\epsilon\|A\|)^2 + \sum_{j=k}^{k+M}\|\tilde{f}^j\|^2\right]^{\frac{1}{2}}.$$

Greenbaum applied results of Paige that we have reviewed before to show that every eigenvalue of T_{k+M} lies within $\sigma(k+M)^3\zeta\|A\|$ of an eigenvalue of A if the perturbation terms satisfy

$$\max\{\|f^1\|, \dots, \|f^{k-1}\|, \|\tilde{f}^k\|, \dots, \|\tilde{f}^{k+M}\|\} \le \zeta\|A\|,$$

and σ is a constant independent of ζ, n, k, M, and $\|A\|$. This leads Greenbaum to the following theorem [76].

Theorem 3.15. *The matrix T_k generated at step k of a perturbed Lanczos recurrence is equal to that generated by an exact Lanczos recurrence applied to a matrix whose eigenvalues lie within*

$$\sigma(k+M)^3\max\{\epsilon\|A\|, \|\tilde{f}^k\|, \dots, \|\tilde{f}^{k+M}\|\}$$

of eigenvalues of A where $\tilde{f}^k, \dots, \tilde{f}^{k+M}$ are the smallest perturbations that will cause a coefficient η_j to be zero at or before step $k + M + 1$ and ϵ is of the order of the machine epsilon ϵ_M.

Since according to Greenbaum's own words "the details of the theorems and proofs are gory even though the basic ideas are not," we shall just show how to construct the extensions of T_k and the consequences that can be obtain for CG. Most of the difficulties in those proofs come from proving that the additional perturbation terms are small. The construction of the extension of T_k starts by considering the already converged Ritz vectors at iteration k, but they have to be chosen carefully. This can be explained by introducing a few definitions. A Ritz value θ_i is said to be well separated from the other Ritz values if

$$\min_{k \neq i} \frac{|\theta_k - \theta_i|}{\|A\ |} > \mu.$$

It is said to be part of a cluster if

$$\min_{k \neq i} \frac{|\theta_k - \theta_i|}{\|A\|} \leq \Delta.$$

The numbers μ and Δ have to be chosen such that all Ritz values fall into one of these two categories. A Ritz vector corresponding to well-separated Ritz value is considered to be converged if

$$\eta_{k+1,i} \leq \delta$$

and unconverged otherwise where $\eta_{k+1,i}$ is the product of η_{k+1} and the absolute value of the last component z_k^i of the eigenvector of T_k. A linear combination

$$y^C = \frac{1}{w^C} \sum_{l \in C} z_k^l y^l, \quad (w^C)^2 = \sum_{l \in C} (z_k^l)^2,$$

of Ritz vectors y^l corresponding to a cluster of Ritz values is considered to be converged if

$$\eta_{k+1} w^C \leq \delta$$

and unconverged otherwise. The value

$$\theta_C = \frac{1}{2} \left(\max_{l \in C} \theta_l + \min_{l \in C} \theta_l \right)$$

is the cluster value.

With these definitions we can proceed to the construction of the extension of T_k. We shall give most of the results without proof. For details, see [76]. Let $T_k = Z_k \Theta_k Z_k^T$ be the spectral decomposition of T_k and m be the number of converged Ritz vectors. This identifies a subset of $k - m$ vectors in span$(v^1, \ldots, v^k)$ such that the chosen vectors are mutually orthogonal and v^{k+1} is orthogonal to all of them. Define $\hat{\Theta}_{k-m}$ to be the diagonal matrix whose diagonal elements are the well-separated Ritz values that correspond to unconverged Ritz vectors and cluster values that correspond to unconverged cluster vectors. Let $\hat{Y}_{k-m}$ be the matrix whose columns are the corresponding unconverged Ritz vectors or unconverged cluster vectors and $\hat{Z}_{k-m}$ be the same for the eigenvectors of T_k. By multiplying the Lanczos matrix relation to the right by $\hat{Z}_{k-m}$, we have

$$A\hat{Y}_{k-m} = \hat{Y}_{k-m}\hat{\Theta}_{k-m} + w^k (e^k)^T \hat{Z}_{k-m} + F_k \hat{Z}_{k-m} + G,$$

$$G = V_k (T_k \hat{Z}_{k-m} - \hat{Z}_{k-m} \hat{\Theta}_{k-m}).$$

Now, according to Paige, the vector w^k is approximately orthogonal to the columns of $\hat{Y}_{k-m}$. So exactly orthogonalizing w^k against the span of the column of $\hat{Y}_{k-m}$ resulting in $w^{k'}$ will give a small perturbation to the recurrence. If $v^{k+1} = w^{k'} / \|w^{k'}\|$ and the next vectors

$v^{k+2}, \ldots, v^{k+j}$ are constructed to be orthogonal to each other and to the columns of $\hat{Y}_{k-m}$, the next vector in the recurrence

$$w^{k+j} = Av^{k+j} - \alpha_{k+j}v^{k+j} - \eta_{k+j}v^{k+j-1}$$

will be approximately orthogonal to the columns of $\hat{Y}_{k-m}$. It is also true that w^{k+j} is approximately orthogonal to $v^{k+1}, \ldots, v^{k+j}$. Hence one can make a small perturbation to the recurrence to exactly orthogonalize w^{k+j} against these vectors and continue. Because of the orthogonality properties that are enforced after step k, the extended recurrence will reach a step before $k+M$, where $\eta_j = 0$. This is because the set $(\hat{Y}_{k-m}, v^{k+1}, \ldots, v^{k+M})$, where $M = n+m-k$, is a set of n orthogonal vectors in a space of dimension n leading to $\eta_{k+M+1} = 0$. This construction is justified by the next two lemmas proved by Greenbaum that we state without proof. Then we shall provide details on how the recurrence is continued.

Lemma 3.16. *The columns of G satisfy $g^j = 0$ if $\hat{z}^k$ is an eigenvector of T_k and*

$$\|g^j\| \leq \frac{(c_{max} - 1)\Delta\|A\|}{2}\sqrt{c_{\max}}$$

if $\hat{z}^j$ is a linear combination of eigenvectors where $c_{\max}$ is the maximum number of elements in an unconverged cluster or 1 *if there is no unconverged clusters. Moreover,*

$$\|\hat{Y}_{k-m}^T v^k\| \leq \|v^k\| \frac{k\gamma\epsilon}{\delta/\|A\|}$$

and

$$\|\eta_{k+1}v^k - \sum_{t=1}^{k-m} \eta_{k+1}w_t\hat{y}^t\| \leq m\delta,$$

where $w_t = z_k^t$ if $\hat{y}^t$ is a Ritz vector or $w_t = w_C$ if $\hat{y}^t$ is a cluster vector and γ is a quantity intervening in Paige's bounds.

Concerning the approximate orthogonality of the columns of $\hat{Y}_{k-m}$, Greenbaum obtained the following bound.

Lemma 3.17.

$$\|I - \hat{Y}_{k-m}^T\hat{Y}_{k-m}\| \leq \max\left\{\frac{k(k-1)\gamma\epsilon}{\mu}, \frac{(c_{\max}-1)k\phi_1\epsilon}{(\delta/\|A\|)^2\mu}\right\}$$
$$+(k-m-1)\max\left\{\frac{\phi_2\epsilon}{(\delta/\|A\|)\mu}, \frac{(c_{\max}-1)\phi_3\epsilon}{(\delta/\|A\|)\mu}\right\},$$

with $\phi_1 = 3\gamma + \nu$, $\phi_2 = 2\gamma + (\delta/\|A\|)\nu$, $\phi_3 = 4\gamma + \sqrt{2}(\delta/\|A\|)\nu$, and γ and ν are quantities intervening in Paige's bounds; see [76].

Suppose now that ϵ is small enough to have $\hat{Y}_{k-m}^T\hat{Y}_{k-m}$ nonsingular. We now show more details on how the Lanczos recurrence is continued. Using the QR factorization, the matrix of the unconverged Ritz vectors can be written

$$\hat{Y}_{k-m} = \bar{Y}_{k-m}R,$$

where the columns of $\bar{Y}_{k-m}$ are orthonormal and R is upper triangular. The first step in the extension process is orthogonalizing w^k, resulting in

$$w^{k'} = w^k - \bar{Y}_{k-m}\bar{Y}_{k-m}^T w^k.$$

The new vector satisfies

$$w^{k'} = Av^k - \alpha_k v^k - \eta_k v^{k-1} + \tilde{f}^k, \qquad \tilde{f}^k = f^k + \bar{Y}_{k-m}\bar{Y}_{k-m}^T w^k.$$

The next "Lanczos" vector is

$$v^{k+1} = \frac{w^{k'}}{\eta'_{k+1}}, \qquad \eta'_{k+1} = \|w^{k'}\|,$$

and hence,

$$\bar{Y}_{k-m}^T v^{k+1} = 0, \qquad \|v^{k+1}\| = 1.$$

The other vectors w^{k+j}, $j = 1, \dots,$ are defined by

$$w^{k+j} = Av^{k+j} - \alpha_{k+j}v^{k+j} - \eta_{k+j}v^{k+j-1} - \tilde{f}^{k+j},$$

$$\tilde{f}^{k+j} = (\bar{Y}_{k-m}, v^{k+1}, \dots, v^{k+j})(\bar{Y}_{k-m}, v^{k+1}, \dots, v^{k+j})^T (Av^{k+j} - \alpha_{k+j}v^{k+j} - \eta_{k+j}v^{k+j-1}),$$

making w^{k+j} orthogonal to the previous vectors. The vectors w^{k+j} are exactly normalized to give

$$v^{k+j+1} = \frac{w^{k+j}}{\eta_{k+j+1}}, \qquad \eta_{k+j+1} = \|w^{k+j}\|.$$

The perturbation terms are bounded in the next lemma; see [76].

Lemma 3.18.

$$\|\tilde{f}^k\| \le \epsilon\|A\| + \frac{2k\gamma\epsilon\|A\|}{\delta/\|A\|},$$

$$\|\tilde{f}^{k+1}\| \le \frac{2k(k-1)\gamma\epsilon\|A\|}{(\delta/\|A\|)\mu} + 2\sqrt{k}\epsilon\|A\| + (c_{\max} - 1)\sqrt{c_{\max}}\sqrt{k-m}\Delta\|A\|,$$

$$\|\tilde{f}^{k+j}\| \le 2\sqrt{k}\epsilon\|A\| + (c_{\max} - 1)\sqrt{c_{\max}}\Delta\|A\| + m\delta, \quad j = 2, \dots, M.$$

It remains to decide what should be the values of μ, δ, and Δ. If $\delta = \sqrt{\epsilon}\|A\|$ and the Ritz values with unconverged Ritz vectors are well separated (which implies Δ must be of the order $\sqrt{\epsilon}$), then $c_{\max} = 1$. If μ is taken to be the minimum relative separation between unconverged Ritz vectors, then all the bounds are of the order $\sqrt{\epsilon}$. However, if there are not well-separated Ritz values with unconverged Ritz vectors, then $c_{\max}$ is larger than 1 and some terms involving $c_{\max}$ in the bounds can be large. In this case a value $\delta = \epsilon^{\frac{1}{4}}\|A\|$ can be used. If $\Delta \le \epsilon^{\frac{1}{4}}$ and $\mu \ge \epsilon^{\frac{1}{4}}$, then the bounds are of the order $\epsilon^{\frac{1}{4}}$, which is small if ϵ is small enough. This leads Greenbaum [76] to the following result.

Theorem 3.19. *The tridiagonal matrix T_k generated at step k of a perturbed Lanczos recurrence with A is equal to that generated by an exact Lanczos recurrence applied to a larger matrix whose eigenvalues lie within*

$$\sigma(k+M)^3 O(k^2\gamma\sqrt{\epsilon}\|A\|) \quad or \quad \sigma(k+M)^3 O(c_{\max}^{\frac{3}{2}} k \epsilon^{\frac{1}{4}}\|A\|)$$

of eigenvalues of A.

Concerning CG, Greenbaum [76] wrote the perturbed relations as

$$x^k = x^{k-1} + \gamma_{k-1} p^{k-1} + \delta_{x^k},$$

$$r^k = r^{k-1} - \gamma_{k-1} A p^{k-1} + \delta_{r^k},$$

$$p^k = r^k + \beta_k p^{k-1} + \delta_{p^k}.$$

Denoting

$$u^k = (-1)^k \frac{r^k}{\|r^k\|}, \qquad U_k = (\, u^0 \quad \dots \quad u^{k-1}\,),$$

and eliminating the vectors p^k in the finite precision CG equations gives

$$AU_k = U_k T_k + \eta_{k+1} u^k (e^k)^T + G_k,$$

where the elements of the tridiagonal matrix T_k are computed from the CG coefficients in the same way as for the exact CG and Lanczos algorithms. The columns of the perturbation matrix $G_k = (\, g^0 \quad \dots \quad g^{k-1}\,)$ are

$$g^j = \frac{(-1)^{j+1}}{\|r^j\|} \left[\frac{\beta_j}{\gamma_{j-2}} \delta_{r^j} - A\delta_{p^j} + \frac{1}{\gamma_{j-1}} \delta_{r^{j+1}} \right].$$

Greenbaum, defining the error as $\varepsilon^k = A^{-1} r^k$ and using her results for Lanczos recurrences, proved a result relating the norm of the error in finite precision and a norm related to the larger matrix $\bar{A}$.

Theorem 3.20. *Let k be given and $\bar{\varepsilon}^j = \bar{A}^{-1}\bar{r}^j$. Let $\tau \le \sigma(k+M)^3 \epsilon \|A\|$. If the initial error satisfies $\|\bar{\varepsilon}^0\|_{\bar{A}} = \|\varepsilon^0\|_A$ and $\lambda_1 - \tau > 0$, then*

$$\left| 1 - \frac{\|\bar{\varepsilon}^k\|_{\bar{A}}}{\|\varepsilon^k\|_A} \right| \le (\bar{\kappa}^3 + \bar{\kappa}^2) \frac{\tau}{\|A\|} + \bar{\kappa} \sum_{i=1}^{n} \left| (Q^T u^k)_i^2 - \sum_{l \in C_i} (\bar{Q}^T \bar{u}^k)_l^2 \right|$$

$$+ \bar{\kappa}^2 \sum_{i=1}^{n} \left| (Q^T u^0)_i^2 - \sum_{l \in C_i} (\bar{Q}^T \bar{u}^0)_l^2 \right|,$$

where

$$\bar{\kappa} = \frac{\lambda_n + \tau}{\lambda_1 + \tau}.$$

The columns of the matrices Q and $\bar{Q}$ are the eigenvectors of A and $\bar{A}$.

By using results she proved about components of the Lanczos vectors, Greenbaum obtained the simpler following result.

Theorem 3.21. *Under the same conditions as in Theorem* 3.20,

$$\left| 1 - \frac{\|\bar{\varepsilon}^k\|_{\Theta_{k+M}}}{\|\varepsilon^k\|_A} \right| \leq (\bar{\kappa}^3 + \bar{\kappa}^2)\frac{\bar{\tau}}{\|A\|} + (\bar{\kappa}^2 + \bar{\kappa})O((k+M)^2\epsilon).$$

We remark that the value m depends on the given k and, in fact, the larger matrix $\bar{A}$ could depend also on the iteration number. We also remark that the widths of the intervals where the eigenvalues of $\bar{A}$ lie depend on k. Concerning the dependence of $\bar{A}$ on k in her construction, we quote Greenbaum [80]: "There are many matrices $\bar{A}$ such that finite precision CG for A behaves like exact CG for $\bar{A}$ for the first k steps. For some, but not all of these matrices $\bar{A}$, it is also true that finite precision CG for A behaves like exact CG for $\bar{A}$ for the first $k+1$ steps. And for some of these the analogy holds for the first $k+2$ steps, etc.... So if you have a bound, say, K on the number of steps you will run (and assuming, of course, that $K \ll 1/u$), the same matrix $\bar{A}$ can be used to describe what happens at all steps of the computation."

This was illustrated in [81], in which Greenbaum and Strakoš described numerical experiments related to the previous results. They used the Strakoš matrix with $n = 24$, $\lambda_1 = 0.1$, $\lambda_n = 100$, and $\rho = (0.4 \quad 0.6 \quad 0.8 \quad 0.9 \quad 1)$. The eigenvalues of these matrices are depicted in Figure 3.21. The A-norms of the errors obtained with CG are in Figure 3.22 and the norms of the residuals in Figure 3.23. The solid curve for $\rho = 1$ is the one

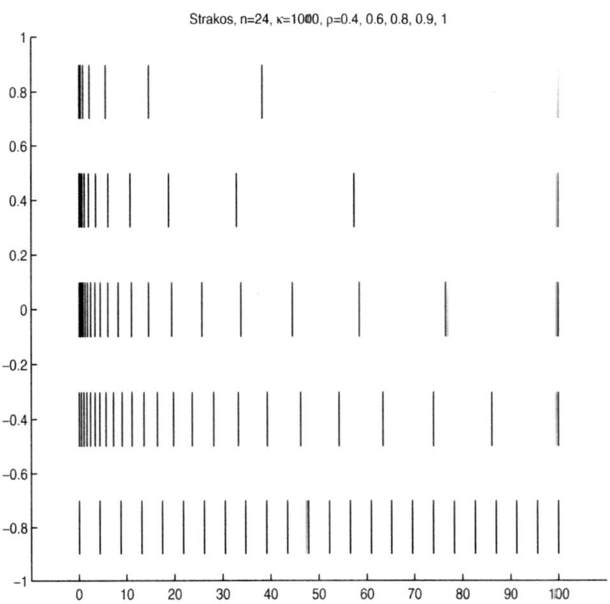

Figure 3.21. *Eigenvalues of the Strakoš matrix with $n = 24$, $\rho = 0.4$ at the top, $\rho = 1$ at bottom*

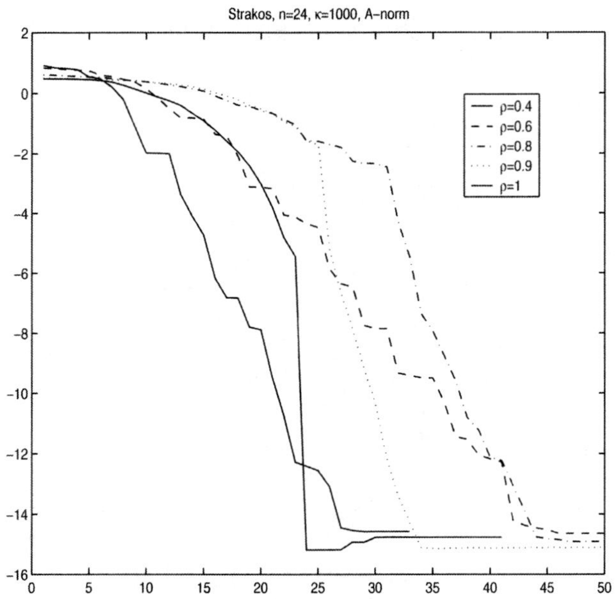

Figure 3.22. *A-norms of the errors $x - x^k$ for the Strakoš matrix with $n = 24$*

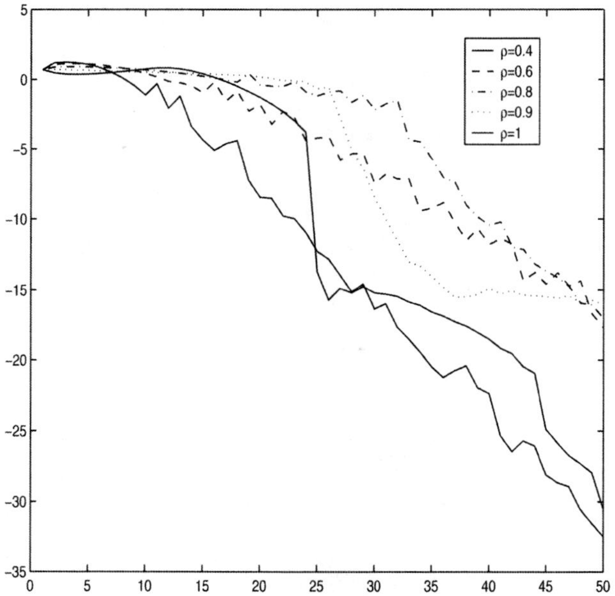

Figure 3.23. *l_2 norms of the iterative residuals r^k for the Strakoš matrix with $n = 24$*

with a rapid drop at iteration 24. For CG convergence, the worst case is the dot-dashed curve corresponding to $\rho = 0.8$. For the uniform distribution corresponding to $\rho = 1$ we obtain convergence to the roundoff level in 24 iterations. They also provided results for matrices constructed a priori by spreading many eigenvalues in very small intervals around the eigenvalues of A. The convergence curves for these matrices using full reorthogonalization reproduce quite well the behavior observed for A in finite precision arithmetic. A complete mathematical justification of this approach has not been provided yet.

More results are described in Strakoš and Greenbaum [185]. In [184] Strakoš introduced the family of matrices we are using quite intensively as examples in this work and reported some numerical results. He also proved the following results. Writing the perturbed Lanczos relation as

$$AV_k = V_k T_k + \eta_{k+1} v^{k+1} (e^k)^T + \delta V_k$$

and using the spectral decompositions of $A = Q\Lambda Q^T$ and $T_k = Z_k \Theta_k Z_k^T$ we have

$$\Lambda Q^T V_k Z_k = Q^T V_k Z_k \Theta_k + Q^T [\eta_{k+1} v^{k+1} (e^k)^T + \delta V_k] Z_k.$$

Denoting $W_k = Q^T V_k Z_k$ and $\Omega_k = Q^T [\eta_{k+1} v^{k+1} (e^k)^T + \delta V_k] Z_k$, we can write

$$\Lambda W_k = W_k \Theta_k + \Omega_k.$$

This is the equivalent of what we have seen in exact arithmetic except that Ω_k (with entries $\omega_{i,j}^{(k)}$) is different owing to the perturbation term δV_k. Looking at the elements of this matrix equality we have

$$(\lambda_i - \theta_j^{(k)}) w_{i,j}^{(k)} = \omega_{i,j}^{(k)}.$$

The norm of Ω_k can be bounded using Paige's results:

$$\|\Omega_k\| \leq \eta_{k+1} \left(1 + \frac{\epsilon_0}{2}\right)^{\frac{1}{2}} + \sqrt{k} \|A\| \epsilon_1.$$

This leads to the next theorem.

Theorem 3.22. *If the eigenvector q^i has a nonzero component in the initial Lanczos vector, there exists a Ritz value $\theta_j^{(k)}$ such that*

$$|\lambda_i - \theta_j^{(k)}| \leq \frac{\sqrt{k}(\eta_{k+1} + \sqrt{k}\epsilon_1 \|A\|)}{|(q^i, v^1)|} [1 + o(\epsilon)].$$

Proof. The proof given in [184] is the following. Consider $\lambda_i - \theta_j^{(k)}$ to be the minimum of the distances. Then,

$$|\lambda_i - \theta_j^{(k)}| \, |w_{i,t}^{(k)}| \leq \|\omega_{i,t}^{(k)}\| \leq \|\Omega_k\|.$$

Let ϕ_t be a set of real numbers such that $\sum_{t=1}^k \phi_t^2 = 1$,

$$|\lambda_i - \theta_j^{(k)}| \left| \sum_{t=1}^k \phi_t w_{i,t}^{(k)} \right| \leq \|\Omega_k\| \sum_{t=1}^k |\phi_t| \leq \sqrt{k} \|\Omega_k\|.$$

But

$$\sum_{t=1}^{k} \phi_t w_{i,t}^{(k)} = (q^i)^T V_k \left(\sum_{t=1}^{k} \phi_t z^t \right),$$

where q^i is an eigenvector of A and z^i an eigenvector of T_k. We can choose the numbers ϕ_t such that

$$\sum_{t=1}^{k} \phi_t z^t = e^1. \quad \square$$

From this result we have the following conclusion for the Greenbaum construction of the extension T_{k+M}.

Corollary 3.23. *Under the same hypothesis,*

$$|\lambda_i - \bar{\theta}_j^{(k+M)}| \leq \frac{(k+M)\zeta \|A\|}{|(q^i, v^1)|}[1 + o(\epsilon)].$$

3.10 The work of J. Cullum and R. Willoughby

The main thesis of the book by Cullum and Willoughby [27] is that the Lanczos algorithm could be used in finite precision without any form of reorthogonalization to compute all or a part of the spectrum of A. This could be at the expense of doing many more iterations than the order of the matrix. Of course, in this case the Lanczos vectors are linearly dependent and we have more Ritz values than the number of distinct eigenvalues of A. To motivate this strategy the authors used the equivalence between Lanczos and CG algorithms and gave a method to decide which Ritz values should be discarded.

Let us concentrate on Chapter 3 of the first volume of [27]. The authors recall how the equivalence of CG and Lanczos algorithms is obtained in exact arithmetic. Then they want to extend these results to finite precision. So they show, using Paige's works, that from the vectors and coefficients computed by the Lanczos algorithm, one can generate vectors and coefficients that approximately verify the CG recurrences, up to quantities multiplied by u the roundoff unit. From this, they interpret these approximate CG relations as an algorithm to approximately minimize a function which is proportional to the A-norm of the error. They gave arguments trying to show that the values of the function at the CG iterates go to zero. However, as we shall see this is doubtful with their definition of the error since there is an iterate after which the norms of the error $x - x^k$ stagnate whence the norm of the computed iterative residual goes to zero. In [27] this decrease was supposed to prove that we shall eventually be able to compute all the eigenvalues of A if we iterate long enough, a property the authors denoted as the Lanczos phenomenon.

The identification test for eigenvalues proposed by Cullum and Willoughby [27] is the following: any Ritz value of T_k which is also an eigenvalue of $T_{2,k}$ is labeled as "spurious" and discarded from the list of computed eigenvalues. All the other eigenvalues are accepted.

3.11 The work of V. Druskin and L. Knizhnerman

In a series of papers from 1989 to 1999 Druskin and Knizhnerman studied methods based on the Lanczos algorithm to compute approximations of vectors $f(A)\phi$, where ϕ is a given vector and f is a smooth function; see Druskin and Knizhnerman [37], [38], [40], Knizhnerman [103], [105], Druskin, Greenbaum, and Knizhnerman [42], and Greenbaum, Druskin, and Knizhnerman [82]. Some of these papers also give results about the convergence of Ritz values in finite precision arithmetic. Let us review some of these results in chronological order.

The paper [37] introduced the techniques used in most of their papers. One method which was proposed to approximate $f(A)\phi$ with $\|\phi\| = 1$ was to run the Lanczos algorithm obtaining (in exact arithmetic)

$$AV_k = V_k T_k + \eta_{k+1} v^{k+1} (e^k)^T$$

and then take $V_k f(T_k) e^1$ as the approximation to $f(A)\phi$. In fact if p is a polynomial of degree less than or equal to $k-1$, then $p(A)\phi = V_k p(T_k) e^1$. In several of these papers the technique which is used is to map the eigenvalue interval $[\lambda_1, \lambda_n]$ to $[-1, 1]$, setting

$$S_k = \frac{\lambda_n + \lambda_1}{\lambda_n - \lambda_1} I - \frac{2}{\lambda_n - \lambda_1} T_k,$$

which gives $\|S_k\| \leq 1$, and then to use Chebyshev polynomials to obtain bounds on the norm of the errors. For instance, if

$$h(x) = f\left(\frac{\lambda_n + \lambda_1 - (\lambda_n - \lambda_1)x}{2}\right),$$

where $h(x) = \sum_{j=0}^{\infty} h_j C_j(x)$ is the Chebyshev series of h, then if the series is absolutely convergent, we have

$$\|f(A)\phi - V_k f(T_k) e^1\| \leq 2 \sum_{j=k}^{\infty} |h_j|.$$

In the case where $f(x) = 1/x$, that is, when we look for the solution of the linear system $Au = \phi$, Druskin and Knizhnerman [37] proved that

$$\|A^{-1}\phi - V_k T_k^{-1} e^1\| \leq \frac{2\epsilon}{1-\epsilon} \|A^{-1}\phi\|,$$

where

$$\epsilon = \frac{1}{C_k\left(\frac{\lambda_n + \lambda_1}{\lambda_n - \lambda_1}\right)}.$$

We notice this is a bound of the l_2 norm of the error in solving (in exact arithmetic) the linear system with the Lanczos algorithm with a zero initial guess since $\|\phi\| = 1$.

The 1991 paper [38] considered the same algorithms in finite precision arithmetic. This work uses Paige's results, mainly the inequality for Ritz values

$$\lambda_1 - \eta \leq \theta_i^{(k)} \leq \lambda_n + \eta,$$

with

$$\eta = \sqrt{k} k^2 \|A\| \epsilon_2, \quad \epsilon_2 = \sqrt{2} \max[12(n+4)u, \epsilon_1], \quad \epsilon_1 = (7 + m\beta)u.$$

The eigenvalue interval to consider is $[\lambda_1 - \eta, \lambda_n + \eta]$. Let

$$B = \frac{\lambda_n + \lambda_1}{\lambda_n - \lambda_1 + 2\eta} I - \frac{2}{\lambda_n - \lambda_1 + 2\eta} A, \quad S_k = \frac{\lambda_n + \lambda_1}{\lambda_n - \lambda_1 + 2\eta} I - \frac{2}{\lambda_n - \lambda_1 + 2\eta} T_k,$$

and

$$h(x) = f\left(\frac{\lambda_n + \lambda_1 - (\lambda_n - \lambda_1 + 2\eta)x}{2}\right).$$

The paper studied a nonhomogeneous scalar Chebyshev recurrence relation:

$$w_0 = b, w_1 = c, \quad w_{k+2} - 2xw_{k+1} + w_k = a_k, \quad k \geq 1,$$

where b, c, a_k are given real numbers. Druskin and Knizhnerman proved that

$$|w_k| \leq k\left(|b| + |c| + \sum_{l=1}^{k-2} |a_l|\right).$$

For a vector relation

$$w_0 = b,\ w_1 = c, \quad w_{k+2} - 2Bw_{k+1} + w_k = a_k, \quad k \geq 1,$$

for a matrix B with $\|B\| \leq 1$ and vectors b, c, and a_k, we have

$$\|w_k\|^2 \leq k^2(k+1)\left(\|b\|^2 + \|c\|^2 + \sum_{l=1}^{k-2} \|a_l\|^2\right).$$

If $\|B\| < 1$,

$$\|w_k\|^2 \leq \frac{k+1}{1 - \|B\|^2}\left(\|b\|^2 + \|c\|^2 + \sum_{l=1}^{k-2} \|a_l\|^2\right).$$

This is used in proving the following result:

$$\|f(A)\phi - \|\phi\| V_k f(T_k)e^1\| \leq 4\sqrt{\frac{2}{\pi}}\|h\|\,\|\phi\|\frac{\|A\|}{\lambda_n - \lambda_1}k^3\epsilon_1 + \frac{10}{\sqrt{24}}\|\phi\|\sqrt{k}\sum_{j=k}^{\infty}|g_j|,$$

with $\|h\|$ in $]-1, 1[$ defined with a weight $1/\sqrt{1-x^2}$. Then Druskin and Knizhnerman gave some results about Ritz values convergence in the finite precision Lanczos algorithm. They consider an eigenvalue of A, $\lambda_r = 0$, which can always be obtained with a shift of the spectrum.

Theorem 3.24. *Suppose that $\|A\| \leq 0.9$ and some restrictive conditions on k relative to u. Then there exists an index i such that*

$$|\theta_i^{(k)}| \leq \frac{1}{2}\left[\left(\frac{10\sqrt{k}}{\sqrt{6}|(v^1, q^r)|}\right)^{\frac{2}{k-2}} - 1\right],$$

where q^r is the eigenvector of A corresponding to $\lambda_r = 0$.

Proof. The proof of this theorem [38] uses the previous results with a function h which is given by

$$h(x) = \frac{C_{k'}\left(\frac{1+p^2-2x^2}{1-p^2}\right)}{C_{k'}\left(\frac{1+p^2}{1-p^2}\right)},$$

where $k' = [(k-1)/2]$, $l < p < 1$, with

$$l = \frac{1}{2}\left[\left(\frac{10\sqrt{k}}{\sqrt{6}|(v^1, q^r)|}\right)^{\frac{2}{k-2}} - 1\right].$$

An example of such a function is given in Figure 3.24 (with parameters not satisfying these requirements). The dashed lines are for $x = \pm p$. This function is such that

$$|h(x)| \le \begin{cases} 1, & x \in [-1, 1], \\ \left[C_{k'}(\frac{1+p^2}{1-p^2})\right]^{-1}, & x \in [-1, 1] \setminus [-p, p]. \end{cases}$$

This is illustrated in Figure 3.24, where the dot-dashed horizontal line gives the maximum value outside $[-p, p]$. The proof of the result is obtained by showing that the maximum of the absolute value of the function g over the Ritz values satisfies

$$\max_{\theta_i^{(k)}} |h(\theta)| \ge w$$

for some w. Then it is proved that

$$C_{k'}\left(\frac{1+p^2}{1-p^2}\right) \ge \frac{1}{w}.$$

This implies that

$$\max_{\theta_i^{(k)}} \left| C_{k'}\left(\frac{1+p^2-2(\theta_i^{(k)})^2}{1-p^2}\right)\right| > 1$$

and therefore there exists a $\theta_i^{(k)}$ giving the maximum for which $|\theta_i^{(k)}| < p$. Letting p tending to l gives the result. □

If

$$\frac{1}{k-2}\log\frac{10\sqrt{k}}{\sqrt{6}|(v^1, q^r)|} \ll 1,$$

the right-hand side of the inequality in the last theorem behaves like

$$l' = \frac{1}{k}\log\frac{10\sqrt{k}}{\sqrt{6}|(v^1, q^r)|}.$$

This is illustrated in Figure 3.25, where the solid curve is the bound l as a function of k and the dashed curve shows the values of l'. These results were obtained with a vector v^1 having all components equal. We see from the figure that the drawback of this interesting result besides the restriction $\|A\| \le 0.9$ (which is insignificant since A can always be scaled such that this bound holds) is the fact that the bound is only very slowly decreasing with k, implying that it could take a very large number of iterations to be sure that we get a given eigenvalue with an acceptable precision. Slightly better bounds were obtained in [38] when allowing for separation of the eigenvalue 0 from the other ones. However, the

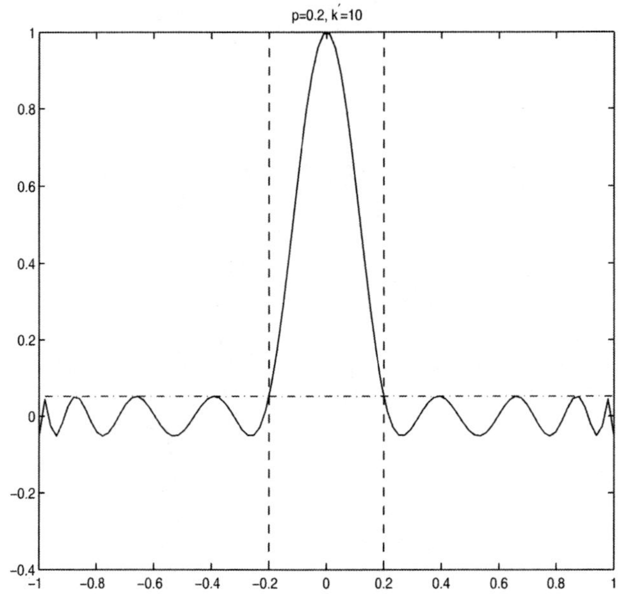

Figure 3.24. *Function h*

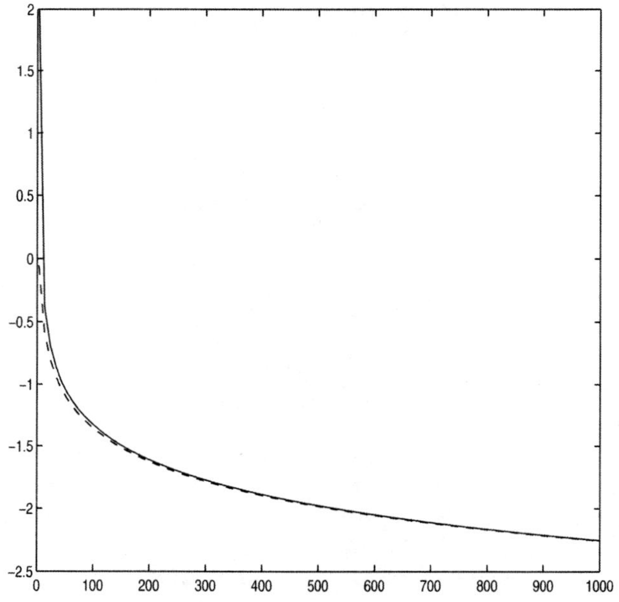

Figure 3.25. *Bounds*

previous result solves the important theoretical question of knowing if all the eigenvalues of A are eventually found by the Lanczos algorithm. Even though the answer is positive, it is unfortunately of little practical interest since the number of iterations to obtain a given accuracy can eventually be very large.

In the paper [103] Knizhnerman improved some of Greenbaum's results about the clustering of the Ritz values using different techniques. The proofs are rather technical, so we will not repeat them here. However, we shall quote some of the interesting results. The notations are the same as before except that $\epsilon_1 = 2\epsilon_1$. We denote by $Sp(A)$ the set of eigenvalues of A. The c_i's are some constants independent of k.

Lemma 3.25. *Let $l_0 = c_1 k\epsilon_2\|A\|$ and $\delta = c_2 k^3\epsilon_2\|A\|$. Then for any pair of indices $1 \le i < j \le k$, for $l = \theta_j^{(k)} - \theta_i^{(k)} \ge l_0$, we have*

$$[\theta_i^{(k)} - \delta, \theta_j^{(k)} + \delta] \cap Sp(A) \neq \emptyset.$$

Therefore, up to a small quantity proportional to u, there is at least one eigenvalue of A between any two Ritz values. This result is a finite precision equivalent of what we know from the interlacing theorem in exact arithmetic (since then, there is an iteration m for which the eigenvalues of T_m are among those of A).

Theorem 3.26. *Let λ_r be a given eigenvalue of A with $\phi_r = (v^1, q^r) \neq 0$ and $\gamma = \min_{i\neq r}|\lambda_r - \lambda_i| > 0$. Let also p be a polynomial of degree less than or equal to $k-3 \ge 0$ such that $p(\lambda_r) = 1$ and*

$$|p(x)| \le \begin{cases} 1, & x \in [\lambda_1 - \eta, \lambda_n + \eta], \\ \mu, & x \in [\lambda_1 - \eta, \lambda_n + \eta]\backslash]\lambda_{r-1} + \delta, \lambda_{r+1} - \delta[. \end{cases}$$

Suppose the usual restrictions on k and n relative to u apply as well as $D \le c_3\phi_r$ with $D = \sqrt{k}\mu + \Delta$, $\Delta = k^3\epsilon_2\|A\|(\lambda_n - \lambda_1)^{-1}$ and $k\epsilon_2 \le c_4\gamma\|A\|^{-1}$. Then

$$\|A\|^{-1}\min_i|\theta_i^{(k)} - \lambda_r| = O(\max[\sqrt{k}\phi_r^{-2}(k\sqrt{k}\epsilon_2\|A\|\gamma^{-1} + D), \phi_r^{-1}(k\sqrt{k}\epsilon_2 + D)\|A\|\gamma^{-1}]).$$

A polynomial satisfying the assumptions of the last theorem can be constructed in the same way as for the function h before. The constant γ measures the isolation of the eigenvalue λ_r.

Theorem 3.27. *Let $K = Sp(T_k)\cap]\lambda_r + \delta, \lambda_{r+1} - \delta[$ and $d = dist([\min K, \max K], Sp(A))$. Then if conditions $\eta \le c_5\|A\|$ and $d \ge \max(c_5k^3\epsilon_2, c_7k\epsilon_2^{\frac{1}{3}})\|A\|$ are satisfied, $card(K) \le 1$.*

This means that there is at most one Ritz value in K.

Theorem 3.28. *Let $K = Sp(T_k)\cap]\lambda_r + \delta, \lambda_{r+1} - \delta[$ and $d = dist([\min K, \max K], Sp(A))$, the eigenvalue of A closest to the interval $[\min K, \max K]$ being that with separation γ. Then if conditions $\eta \le c_5\|A\|$ and $d \ge c_8k^3\epsilon_2\|A\|$ as well as*

$$d \ge \max[c_9\gamma^{-1}(k\epsilon_2\|A\|)^2, c_{10}\sqrt{\gamma^{-1}k^3\epsilon_2\|A\|^3}],$$

are satisfied, $card(K) \le 1$.

The intervals which are obtained in the results of [103] are shorter than those in Greenbaum's work. The paper [40] by Druskin and Knizhnerman is mainly a summary of results of previous papers that were published as translations of Russian papers. The paper [42] by Druskin, Greenbaum, and Knizhnerman studies the residuals in different problems using the Lanczos algorithm. When solving linear systems we have

$$r^k = r^0 - AV_k y^k, \quad T_k y^k = \|r^0\| e^1.$$

In finite precision this gives

$$r^k = -\|r^0\| \left(\eta_{k+1} v^{k+1} (e^k)^T T_k^{-1} e^1 + F_k \frac{y^k}{\|r^0\|} \right).$$

Taking norms and using Paige's results it follows that

$$\sqrt{1 - 2\epsilon_0} \eta_{k+1} |(e^k, T_k^{-1} e^1)| - \sqrt{k} \epsilon_1 \|A\| \frac{\|y^k\|}{\|r^0\|} \leq \frac{\|r^k\|}{\|r^0\|},$$

$$\frac{\|r^k\|}{\|r^0\|} \leq \sqrt{1 + 2\epsilon_0} \eta_{k+1} |(e^k, T_k^{-1} e^1)| + \sqrt{k} \epsilon_1 \|A\| \frac{\|y^k\|}{\|r^0\|}.$$

Therefore, if $\sqrt{k}\epsilon_1 \|A\| \, \|y^k\| / \|r^0\|$ is not too large, we have

$$\frac{\|r^k\|}{\|r^0\|} \simeq \eta_{k+1} |(e^k, T_k^{-1} e^1)|,$$

which is similar to exact arithmetic (of course with a different T_k). Using these results and Paige's results about location of the Ritz values, the authors obtained the following result.

Theorem 3.29. *Assume that A is positive definite and $\lambda_1 - (k+1)^2 \sqrt{k+1}\epsilon_2 \|A\| > 0$. Then,*

$$\frac{\|r^k\|}{\|r^0\|} \leq \sqrt{1 + 2\epsilon_0} 2\sqrt{\tilde{\kappa}} \left(\frac{\sqrt{\tilde{\kappa}} - 1}{\sqrt{\tilde{\kappa}} + 1} \right)^k + \sqrt{k} \tilde{\kappa} \epsilon_1,$$

with

$$\tilde{\kappa} = \frac{\lambda_n + (k+1)^2 \sqrt{k+1}\epsilon_2 \|A\|}{\lambda_1 - (k+1)^2 \sqrt{k+1}\epsilon_2 \|A\|}$$

as an equivalent of the condition number.

Unfortunately, this result cannot be used to show CG convergence in finite precision because of the presence of the second term in the right-hand side which is growing with k.

3.12 Recent related results

During the writing of the present work the author became aware of the Ph.D. thesis of Zemke [205], who considered Krylov methods in finite precision. Some of the facts and results to be presented in Chapter 4 can also be found in [205], although they are considered from a different perspective. The work of Zemke has a quite ambitious goal: the study of most known Krylov methods in finite precision arithmetic. Most of the results that are developed in this work are based on writing solutions of perturbed Krylov recurrences like

$$AV_k - V_k C_k = M_k - F_k,$$

where $M_k = c_{k+1,k} v^{k+1} (e^k)^T$ and F_k stands for the rounding errors. One of the techniques used in [205] is to consider these relations as Sylvester equations for V_k and to convert them to standard linear systems for $vec(V_k)$ which is obtained by putting all the columns of V_k in sequence in a vector. Formulas for the solution are given in [205, pp. 145–148]. Studies of the projections of the columns of V_k on the eigenvectors of A are given in [205, pp. 172–176]. Even though the techniques are different, the results are somehow similar to what will be done in Chapter 4, so we do not repeat them here. A study of some terms in the solution is also done in the thesis and examples are given. The Ph.D. thesis of Zemke is a recommended reading.

The appearance of multiple copies of eigenvalues and the formation of tight clusters of Ritz values pose many technical difficulties in some of the results we have reviewed so far. Ritz values can stabilize only close to an eigenvalue of A. If the stabilized Ritz value is well separated, then the norm of the Ritz vector cannot significantly differ from unity. When a Ritz value is a part of a tight cluster, then some Ritz pairs corresponding to the cluster can have strange properties.

In [185] several conjectures concerning these clusters of Ritz values have been formulated by Strakoš and Greenbaum:

- Does any well separated cluster consisting of at least two Ritz values approximate an eigenvalue of A?
- Is any Ritz value in a well-separated cluster stabilized to within a small δ? In [185] it was conjectured that the answer is positive and δ is proportional to the square root of the size of the cluster interval divided by the square root of the separation of the cluster from the other Ritz values.
- If Ritz values in a well-separated cluster closely approximate some eigenvalue λ_i of A, does the sum of weights of these Ritz values in the corresponding Riemann–Stieltjes integral closely approximate the weight of the original eigenvalue λ_i?

These questions were investigated by Wülling in [202] and [203] with the following answers:

- Every tight well separated cluster of at least two Ritz values must stabilize; i.e., the answer is positive.
- There are tight well-separated clusters of Ritz values in which none of the Ritz values is stabilized to within a small δ; i.e., the answer is negative. However, in cases for which there are stabilized Ritz values, the conjecture about δ is right.
- The weights in the Riemann–Stieltjes integral corresponding to the kth Gauss quadrature of the Riemann–Stieltjes integral determined by A and v^1 must stabilize; i.e., the answer is positive.

The proofs of Wülling are cleverly based on the use of the residue theorem from complex analysis.

Chapter 4

The Lanczos algorithm in finite precision

From the works we reviewed in Chapter 3, particularly those of Paige and Greenbaum, we already know many things about the behavior of the Lanczos algorithm in finite precision arithmetic. We know that the loss of orthogonality implies convergence of Ritz values. We also have some results showing that the deviation from orthogonality arises from the growth of local roundoff errors through the Lanczos algorithm recurrences.

In this chapter we are going to study the Lanczos algorithm in finite precision arithmetic from another perspective than what we have seen in Chapter 3. We shall consider the behavior of the components of the Lanczos vectors on the eigenvectors of A. We shall see that because of the rounding errors these components have a very interesting structure. This leads to the mathematical problem of obtaining expressions for the solutions of three-term nonhomogeneous scalar recurrences. We shall first derive a solution using orthogonal polynomials and associated polynomials. Then we shall study the growth or decrease of these polynomials. This will give us insights about the origin of the growth of the roundoff errors and why and when multiple copies of the eigenvalues appear. However, we shall be able to do only a partial analysis of the polynomial solution and thus we shall derive another way to express the solution involving the Lanczos matrix T_k which will allow us to obtain bounds for the perturbation terms caused by rounding errors.

4.1 Numerical examples

Still using the Strakos30 example, we would like to show some more facts about the Lanczos algorithm. First, we want to show experimentally what is the level of local roundoff. Since this cannot be easily done with the basic MATLAB software, we use the Symbolic Toolbox, which is based on a Maple kernel. In Maple there is a variable called *Digits* which allows one to change the number of decimal digits with which the computations are done. We can estimate the roundoff errors by doing a computation with, say, 16 decimal digits and another one with twice that, computing the difference (with 32 digits) to evaluate the roundoff level. From these computations, we see in Figures 4.1 to 4.3 that the roundoff is of order $u\|A\|$. All the computations in this chapter use an initial vector with equal components. For comments on extended precision computations, see Higham [94].

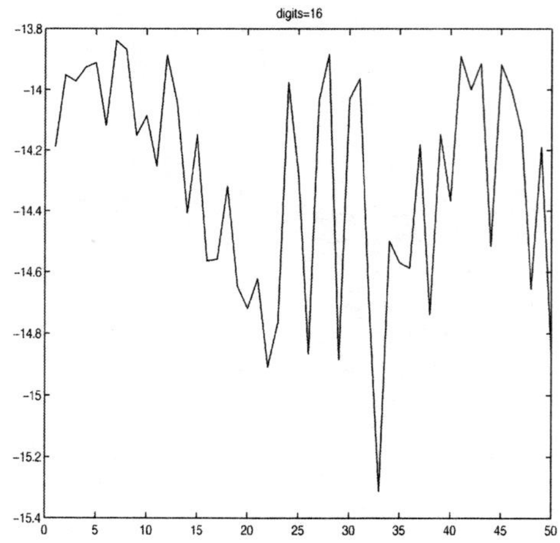

Figure 4.1. *Strakos*30, $\log_{10}(\|f^k\|)$ *with* 16 *digits* (16–32)

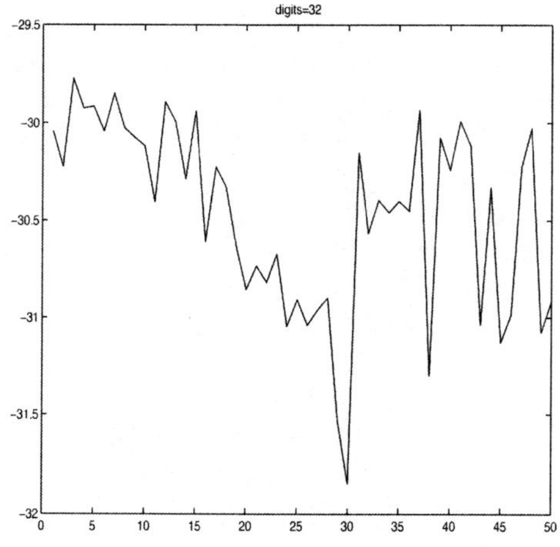

Figure 4.2. *Strakos*30, $\log_{10}(\|f^k\|)$ *with* 32 *digits* (32–64)

One may argue that the local roundoff errors are minimum in this example since the Strakos30 matrix is diagonal and there are only 30 multiplications in a matrix-vector product. Therefore, we constructed a dense matrix with the same eigenvalues, which we denote as Strakos30b. We computed the orthonormal matrix Q_2 of the eigenvectors of the tridiagonal matrix of the one-dimensional Poisson equation $(-1, 2, -1)$ and defined $A_b = Q_2^T A Q_2$. The norms of the local errors are shown in Figure 4.4. The solid curve is the same as before and the dashed curve is the norm of the local error with A_b. We see that it is larger but not by a large factor. In average, the ratio of both norms is not much larger than the dimension of the matrix (30) and for most iterations much smaller than that.

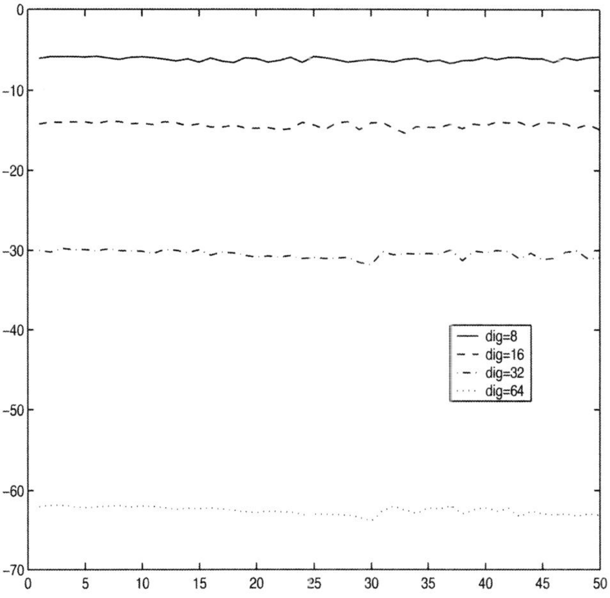

Figure 4.3. *Strakos*30, $\log_{10}$ *of the norm of the local error with different values of digits (*8, 16, 32, *and* 64*)*

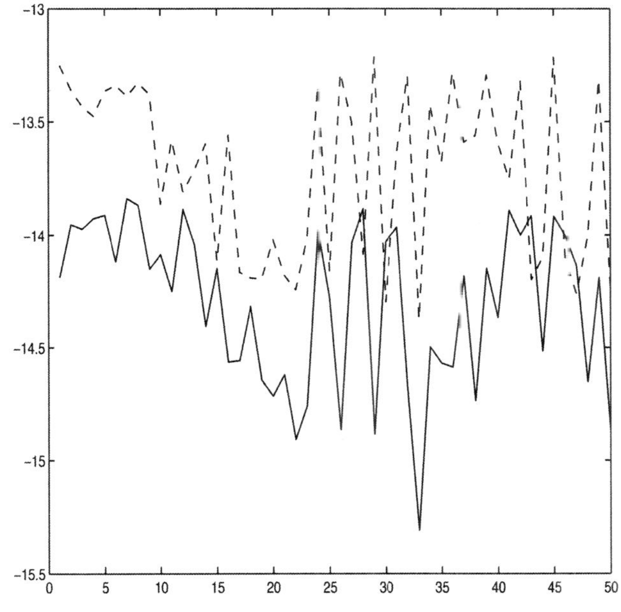

Figure 4.4. $\log_{10}$ *of the norm of the local error with* 16 *digits, Strakos*30 *(solid) and Strakos*30*b (dashed)*

For the Strakos30 example which has distinct eigenvalues (even though some of them are very close) the computation with double reorthogonalization of the new Lanczos vector against all the preceding ones at each iteration represents quite well the extended precision results and we shall use it as the "exact arithmetic" result. This is shown in Figure 4.5, where we have the result of the computation of the last component of the Lanczos vectors with double reorthogonalization (dashed) and a variable precision computation with $Digits = 64$ (solid). Of course the component v_{30}^k with reorthogonalization could not decrease beyond the roundoff unit level $u \approx 1.16\ 10^{-16}$. With reorthogonalization the Lanczos vectors are orthogonal up to machine precision and all eigenvalues are found by iteration 30 with a good accuracy.

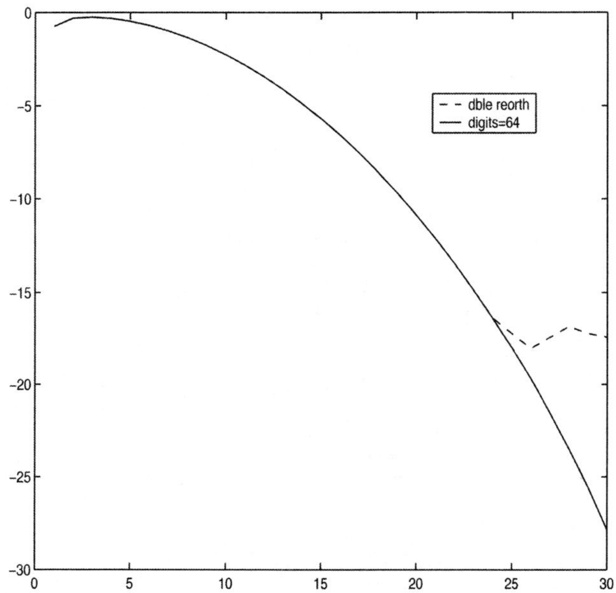

Figure 4.5. *Strakos*30, $\log_{10}(|v_{30}^k|)$ *with double reorthogonalization (dashed) and with* 64 *digits (solid)*

In this example, since the matrix A is diagonal, the matrix of its eigenvectors is the identity matrix, $Q = Q^T = I$. The Lanczos vectors also represent the projections of the Lanczos vectors on the eigenvectors of A. This allows us to look directly at the components of the Lanczos vectors on eigenvectors for which a Ritz value is converging to the corresponding eigenvalue of A without introducing any further rounding errors. For our example the first eigenvalue to converge is the largest one λ_n since the largest eigenvalues are well separated. So, we have to look at the last component v_n^k of the Lanczos vectors as a function of k. In theory, it must decrease because, in exact arithmetic, it is equal to the value of the Lanczos polynomial times the initial component $p_k(\lambda)v_n^1$ at $\lambda = \lambda_n$. It is proportional to

$$\prod_i (\lambda_n - \theta_i^{(k)}).$$

As we have seen before, since the largest Ritz value "converges" monotonically to the largest eigenvalue, the value of the Lanczos polynomial must also decrease sooner or later. This is what we see in Figure 4.5.

Figure 4.6 shows the $\log_{10}$ of the (absolute value of the) last component of v^k both with double reorthogonalization and without. The dashed curve is with reorthogonalization, the solid curve is the Lanczos algorithm in finite precision, and the dotted curve is the difference of the two. The dot-dashed horizontal line is $\sqrt{u}$. We see that the difference is increasing almost from the start and that when it reaches $\sqrt{u}$, instead of still decreasing like it should in theory, the last component in finite precision starts increasing up to $O(1)$ and then decreases again whence the last component computed with reorthogonalization decreases to roundoff level. These results are complemented by Figure 4.7, where in addition to what is on the previous figure the other solid curve is the $\log_{10}$ of the (absolute value of the) last component of the sequence computed in finite precision arithmetic but with the "exact" coefficients α_k and η_k computed with double reorthogonalization. The other dotted curve is the difference with the solution with orthogonalization. We see that these two curves are very close to the finite precision computation until the last component reaches $O(\sqrt{u})$ because the Lanczos coefficients in both computations are still not too far away. After this stage, the last components in exact and finite precision are completely different. Because of the normalization of the Lanczos vectors, the finite precision component cannot grow larger than 1 whence the component in finite precision but with "exact" coefficients continues to increase since the vectors are then unnormalized. This shows that the difference between exact and finite precision results cannot be explained only by the difference in the computation of

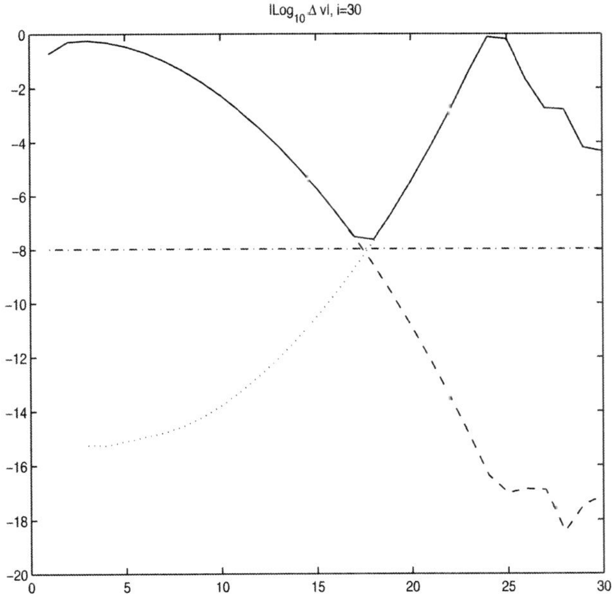

Figure 4.6. *Strakos30,* $\log_{10}(|v_{30}^k|)$ *with (dashed), without reorthogonalization (solid), and their difference (dotted)*

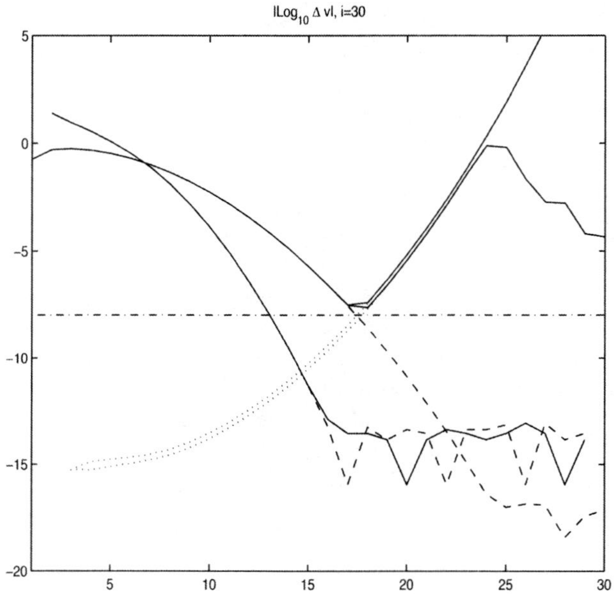

Figure 4.7. *Strakos30,* $\log_{10}(|v_{30}^k|)$ *with (dashed), without reorthogonalization (solid), and more*

the coefficients. Using the "right" coefficients is not enough. In Figure 4.7 the solid and dashed curves starting above 0 give the $\log_{10}$ of the distances of the maximum Ritz values to the maximum eigenvalue of A in finite precision and with double reorthogonalization. The precision with which the largest eigenvalue is obtained is around $\lambda_n u \simeq 10^{-14}$. It is reached at the same time the last component of (the projection of) the Lanczos vector is around $\sqrt{u}$.

Figure 4.8 shows the $\log_{10}$ of the (absolute value of the) last component with reorthogonalization (dashed) and without (solid) but with 100 iterations for the latter. This is a very interesting pattern. We see that after going up to $O(1)$ the finite precision last component decreases again to the level of $\sqrt{u}$ and then goes up again. This happens almost periodically. To complement this, Figure 4.9 shows the distance of the second largest Ritz value to the largest eigenvalue of A as a solid curve. Figure 4.10 shows the same for the fifth-largest Ritz value. We see that each time the last component reaches $O(1)$ we get convergence of a new copy of the largest eigenvalue of A. The precision with which it is obtained is more or less the same each time, as shown in Figure 4.11, where we show the distances of λ_n to the first 10 Ritz values.

From these figures, it seems that $\log_{10}(|v_n^k)|$ and $\log_{10}(|\tilde{v}_n^k - v_n^k|)$ are almost symmetric around $\log_{10}(\sqrt{u})$ until $|\tilde{v}_n^k|$ reaches $O(1)$. This means that

$$\log_{10}(|v_n^k|) + \log_{10}(|\tilde{v}_n^k - v_n^k|) \simeq 2\log_{10}(\sqrt{u})$$

and therefore we should have

$$|v_n^k|\,|\tilde{v}_n^k - v_n^k| \simeq O(u).$$

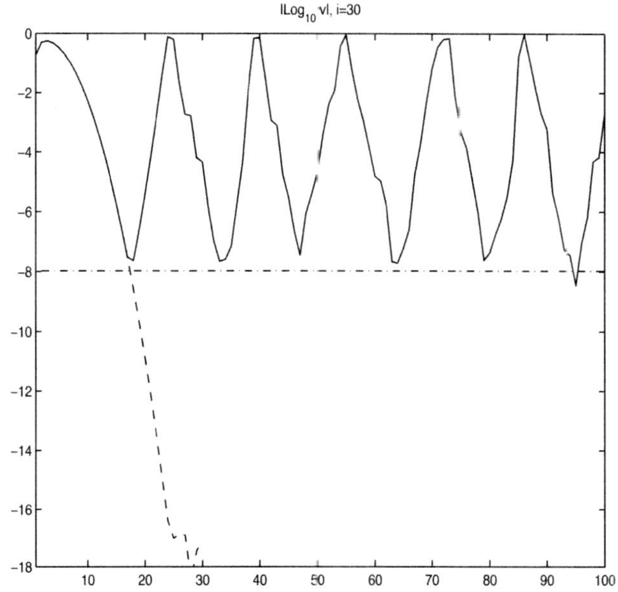

Figure 4.8. *Strakos*30, $\log_{10}(|v_{30}^k|)$ *with (dashed) and without reorthogonalization (solid)*

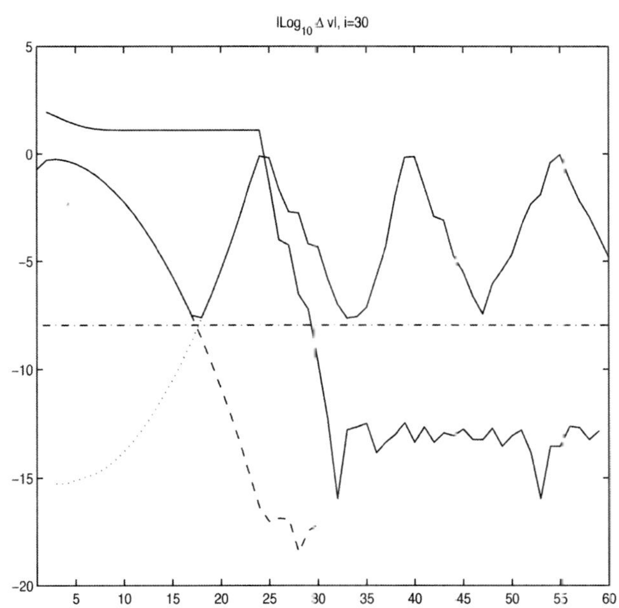

Figure 4.9. *Strakos*30, $\log_{10}(|v_{30}^k|)$ *and distance to the second-largest Ritz value*

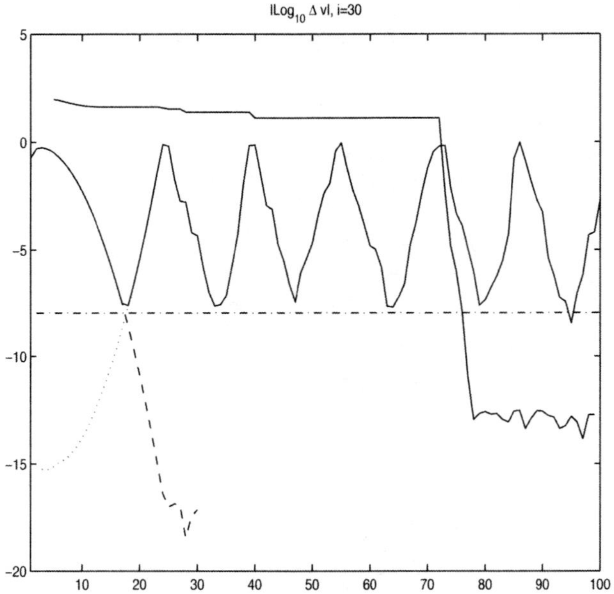

Figure 4.10. *Strakos*30, $\log_{10}(|v_{30}^k|)$ *and distance to the fifth-largest Ritz value*

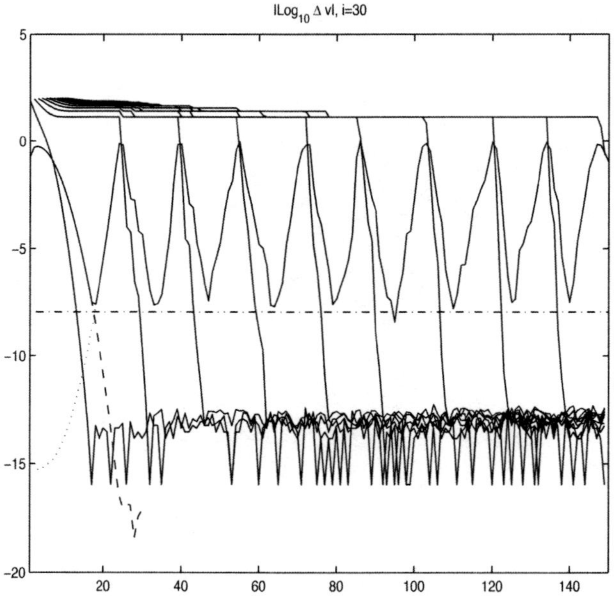

Figure 4.11. *Strakos*30, $\log_{10}(|v_{30}^k|)$ *and the distances to the* 10 *largest Ritz values*

This phenomenon arises for all the components of $\tilde{v}^k$, as shown in Figure 4.12, where $\log_{10}(|\tilde{v}^k_{20}|)$ is plotted together with the distances of λ_{20} to the closest Ritz values when they are smaller than 0.1 represented by black points. We see that each time the given component reaches $O(1)$ a new copy of the corresponding eigenvalue appears.

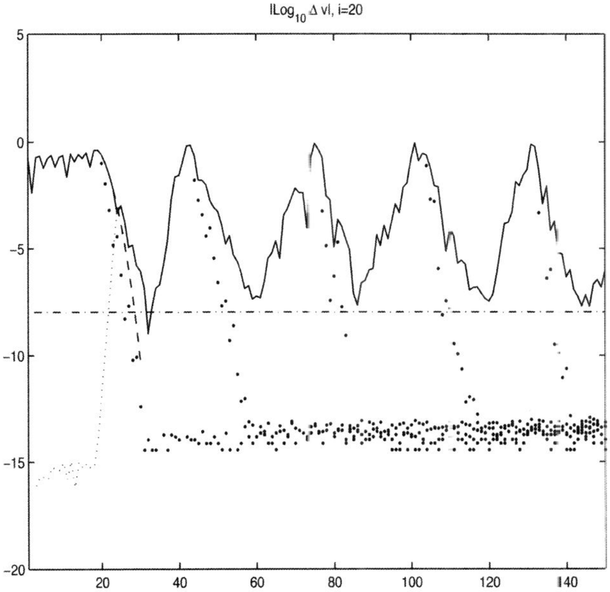

Figure 4.12. *Strakos*30, $\log_{10}(|v^k_{20}|)$ *and the distances to closest Ritz value*

What we have described before can be considered as the "generic" behavior of the Lanczos algorithm, at least when there are well-separated eigenvalues. However, one can find some examples whose behavior may appear different from the generic one. Let us consider an example proposed by Wülling [202]. The matrix A is diagonal of order 23 with eigenvalues λ_i defined by

$$\lambda_1 = -100, \quad \lambda_{i+1} = \lambda_i + 0.1, \quad i = 1, \ldots, 10,$$

$$\lambda_{12} = 0, \quad \lambda_{12+i} = -\lambda_{12-i}, \quad i = 1, \ldots, 11.$$

Therefore, $\lambda_{12} = 0$ is an isolated eigenvalue and provided the initial vector is not chosen in a weird way, it is the first to converge. Figure 4.13 shows the corresponding component of the Lanczos vectors when we choose an initial vector with equal components. The value $|v^k_{12}|$ oscillates and every second iteration there are two Ritz values close to zero. This example shows that the number of Ritz values in a cluster is not always increasing. The behavior of the component of the Lanczos vector seems to agree with the generic behavior when the Lanczos vectors are no more linearly independent after iteration 23.

Even though this example may seem to show that something different from what we have described before may happen, this is caused by the very particular choice of the initial vector. If we choose a random initial vector, then we obtain what is shown in Figure 4.14.

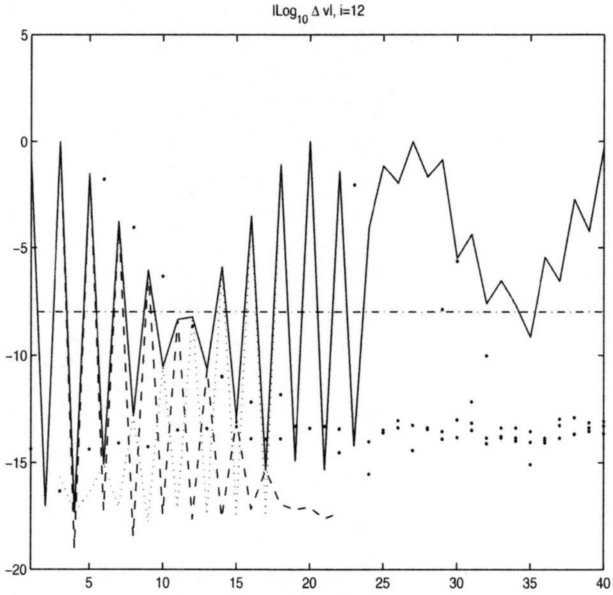

Figure 4.13. *Wülling's example,* $\log_{10}(|v_{12}^k|)$ *and the distances to closest Ritz value,* $v^1 = e/\|e\|$

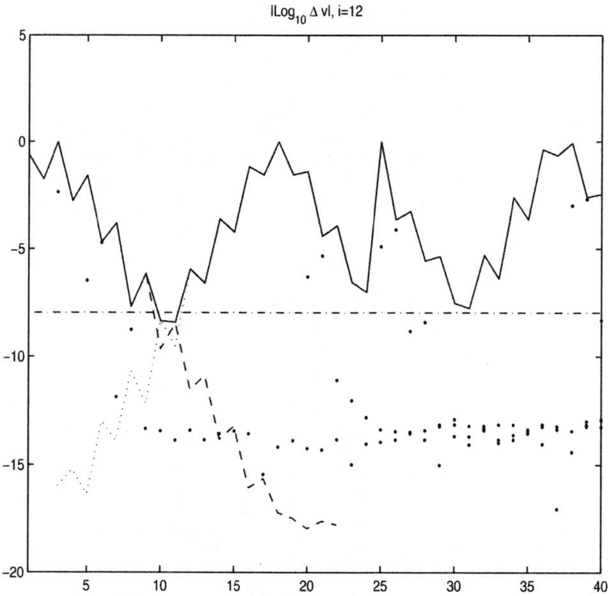

Figure 4.14. *Wülling's example,* $\log_{10}(|v_{12}^k|)$ *and the distances to closest Ritz value,* v^1 *random*

Despite having small oscillations the 12th component of the Lanczos vectors follows the generic behavior. Of course, having such examples shows that it may not be possible to prove very general theorems about the generic behavior.

4.2 Solution of three-term recurrences

Considering the previous numerical experiments, it is interesting to obtain expressions for the components of the Lanczos vectors on the eigenvectors of A in finite precision. In this section, we shall give a solution in terms of polynomials and then study their properties. These components are given by three-term scalar nonhomogeneous recurrence. For the sake of simplicity and generality we shall use some new notations. Let s_1 be a real number and suppose we have three sets of given coefficients τ_i and ζ_i and perturbations f_i, $i = 1, 2, \dots$. We define the sequence $s_i(\lambda)$, s_1 and λ being two given real numbers, as

$$\zeta_2 s_2 = (\lambda - \tau_1)s_1 + f_1,$$

$$\zeta_{k+1}s_{k+1} = (\lambda - \tau_k)s_k - \zeta_k s_{k-1} + f_k, \quad k = 2, \dots.$$

As we said before, the solution of such a second order recurrence with nonconstant coefficients is known; see, for instance, Mallik [114]. His result is summarized in the next theorem.

Theorem 4.1. *Let initial conditions w_1 and w_2 be given as well as coefficients $a_{k,i}$, $i = 1, 2$, $k = 1, \dots,$ and nonhomogeneous terms h_k, $k = 3, \dots$, and*

$$w_{k+2} = a_{k,1}w_{k+1} + a_{k,2}w_k + h_{k+2}, \quad k = 1, \dots.$$

The solution is given by

$$w_{k+2} = d_{k,2}w_2 + d_{k,1}w_1 + (d_{k,2-k}h_{k+1} + d_{k,3-k}h_k + \dots - d_{k,0}h_3) - h_{k+2}, \quad k = 1, \dots,$$

where

$$d_{k,j} = \sum_{r=1}^{k+j-1} \sum_{\substack{(l_1,\dots,l_r) \\ 1 \le l_i \le 2 \\ l_r \ge j \\ l_1+\dots+l_r = k+j-1}} \prod_{m=1}^{r} a_{k+l_m-\sum_{n=1}^{m} l_n, l_m}.$$

It is difficult to exploit this result for our purposes since the product of the recurrence coefficients has no particular meaning for the Lanczos algorithm. Therefore, we shall look for other ways to obtain the solution which are better suited to the Lanczos algorithm. We rewrite the second order (three-term) recurrence as a block first order recurrence, the initial condition being

$$\begin{pmatrix} s_2 \\ s_1 \end{pmatrix} = \begin{pmatrix} \frac{\lambda-\tau_1}{\zeta_2} & -\frac{\zeta}{\zeta_2} \\ 1 & 0 \end{pmatrix} \begin{pmatrix} s_1 \\ 0 \end{pmatrix} + \begin{pmatrix} \frac{f_1}{\zeta_2} \\ 0 \end{pmatrix},$$

because of the starting vector, which has a second component $s_0 = 0$, the value of ζ can be arbitrary. Let

$$y^1 = \begin{pmatrix} s_1 \\ s_0 \end{pmatrix}, \quad B_1 = \begin{pmatrix} \frac{\lambda-\tau_1}{\zeta_2} & -\frac{\zeta}{\zeta_2} \\ 1 & 0 \end{pmatrix}, \quad g^1 = \begin{pmatrix} \frac{f}{\zeta_2} \\ 0 \end{pmatrix}.$$

The block recurrence is

$$\begin{pmatrix} s_{k+1} \\ s_k \end{pmatrix} = \begin{pmatrix} \frac{\lambda-\tau_k}{\zeta_{k+1}} & -\frac{\zeta_k}{\zeta_{k+1}} \\ 1 & 0 \end{pmatrix} \begin{pmatrix} s_k \\ s_{k-1} \end{pmatrix} + \begin{pmatrix} \frac{f_k}{\zeta_{k+1}} \\ 0 \end{pmatrix}.$$

By denoting

$$y^{k+1} = \begin{pmatrix} s_{k+1} \\ s_k \end{pmatrix}, \quad B_k = \begin{pmatrix} \frac{\lambda-\tau_k}{\zeta_{k+1}} & -\frac{\zeta_k}{\zeta_{k+1}} \\ 1 & 0 \end{pmatrix}, \quad g^k = \begin{pmatrix} \frac{f_k}{\zeta_{k+1}} \\ 0 \end{pmatrix},$$

this is a first order block recurrence

$$y^{k+1} = B_k y^k + g^k,$$

with y^1 given. It is easy to write the solution of this recurrence.

Lemma 4.2. *The solution of the recurrence $y^{k+1} = B_k y^k + g^k$ starting from y^1 is given by*

$$y^{k+1} = B_k \cdots B_1 y^1 + \sum_{l=1}^{k-1} B_k \cdots B_{l+1} g^l + g^k, \quad k = 1, 2, \ldots .$$

Proof. The proof is straightforward by induction. □

Of course, we are interested in the behavior of the term arising from the nonhomogeneous perturbations g^l. The first idea we may have is to try to bound the perturbation terms. To bound these terms we can compute the norms of the matrices B_l. Unfortunately, we remark that since

$$\max_{i,j} |(B_l)_{i,j}| \le \|B_l\|,$$

the norm of B_l is always larger than 1. Therefore, we cannot use this to prove that the perturbation terms are small and, in fact, as we have already seen in the numerical experiments, they may not be small. But, eventually one can obtain bounds of their growth in this way. However, this involves only the coefficients of the recurrence and we are more interested in expressions involving the Ritz values. Of course, we are concerned only with the first component of the solution y^k. We can express the solution in terms of polynomials and give the first main result of this chapter.

Theorem 4.3. *Let j be given and $p_{j,k}$ be the polynomial determined by*

$$p_{j,j-1} = 0, \quad p_{j,j} = 1,$$

$$\zeta_{k+1} p_{j,k+1}(\lambda) = (\lambda - \tau_k) p_{j,k}(\lambda) - \zeta_k p_{j,k-1}(\lambda), \quad k = j, \ldots .$$

The solution of the perturbed recurrence

$$\zeta_{k+1} s_{k+1} = (\lambda - \tau_k) s_k - \zeta_k s_{k-1} + f_k$$

starting from $s_0 = 0$ and s_1 is given by

$$s_{k+1} = p_{1,k+1}(\lambda) s_1 + \sum_{l=1}^{k} p_{l+1,k+1}(\lambda) \frac{f_l}{\zeta_{l+1}}.$$

Proof. We are interested in the first component of y^{k+1} given in Lemma 4.2. The proof can be obtained by induction on k. Consider $B_k \cdots B_{l+1} g^l$ for $l < k$, and let $t^l = g^l$ and $t^{l+1} = B_{l+1} t^l$; then

$$t^{l+1} = \begin{pmatrix} \frac{\lambda - \tau_{l+1}}{\zeta_{l+2}} & -\frac{\zeta_{l+1}}{\zeta_{l+2}} \\ 1 & 0 \end{pmatrix} \begin{pmatrix} t_1^l \\ t_2^l \end{pmatrix} = \begin{pmatrix} \frac{(\lambda - \tau_{l-1}) t_1^l}{\zeta_{l+2}} - \frac{\zeta_{l+1} t_2^l}{\zeta_{l+2}} \\ t_1^l \end{pmatrix}.$$

Since $t_2^l = 0$, we have

$$t_1^{l+1} = \frac{\lambda - \tau_{l+1}}{\zeta_{l+2}} t_1^l = p_{l+1,l+2}(\lambda) g_1^l.$$

At the next step, since $t_2^{l+1} = t_1^l$ we have

$$t_1^{l+2} = \frac{\lambda - \tau_{l+2}}{\zeta_{l+3}} t_1^{l+1} - \frac{\zeta_{l+2}}{\zeta_{l+3}} t_1^l.$$

Hence,

$$t_1^{l+2} = \frac{\lambda - \tau_{l+2}}{\zeta_{l+3}} p_{l+1,l+2} g_1^l - \frac{\zeta_{l+2}}{\zeta_{l+3}} p_{l+1,l+1} g_1^l = p_{l+1,l+3}(\lambda) g_1^l.$$

The proof is ended by induction. □

The result of Theorem 4.3 is also given in Cullum and Willoughby [27, Lemma 4.4.1, p. 119]. Unfortunately, no real consequences were obtained from this result in this book. One can also find similar expressions in the Ph.D. thesis of Zemke [205]. The polynomials $p_{j,k}$ are usually called the associated polynomials. Some properties of these polynomials are recalled in papers by Skrzipek [170], [171]. The associated polynomials $p_{j,k}$ are orthogonal with respect to a measure $\omega^{(j)}$ that depends on j. Some results are also given in a paper by Belmehdi [7], from which we obtained the following result.

Proposition 4.4. *Let $j > 1$, $k \geq j$,*

$$p_{j,k+1} p_{1,k} - p_{j,k} p_{1,k+1} = \frac{\zeta_j}{\zeta_{k+1}} p_{1,j-1}.$$

Dividing by $p_{1,k} p_{j,k}$, which we suppose to be nonzero, we obtain

$$p_{1,k} p_{j,k} = \frac{\zeta_j}{\zeta_{k+1}} \frac{p_{1,j-1}}{\frac{p_{j,k+1}}{p_{j,k}} - \frac{p_{1,k+1}}{p_{1,k}}}.$$

We shall use this relation later on. The expression of $p_{j,k}$ using eigenvalues of principal submatrices of T_k has been given by Grcar [73] (notice there is a typo in the proof of Lemma 8 of [73]; in the last two lines β_j should be β_{j+1}).

Lemma 4.5. *The associated polynomial $p_{j,k}$, $k \geq j$, is given by*

$$p_{j,k}(\lambda) = (-1)^{k-j} \frac{\chi_{j,k-1}(\lambda)}{\zeta_{j+1} \cdots \zeta_k},$$

where $\chi_{j,k}(\lambda)$ is the determinant of $T_{j,k} - \lambda I$, $T_{j,k}$ being the tridiagonal matrix obtained from the coefficients of the second order recurrence from step j to step k, that is, discarding the $j - 1$ first rows and columns of T_k.

4.3 The Lanczos vectors in finite precision

In the previous section we exhibited a solution of the nonhomogeneous scalar three-term recurrence using associated polynomials. In this section we shall apply these results to the Lanczos algorithm. The previous results are interesting since they show that the possible growth of the local roundoff perturbations is linked to the matrices $\tilde{T}_{j,k} - \lambda I$ for all $j \le k$, more precisely, to the values of their determinants at the eigenvalues of A.

How could we use these results? There are different sequences of vectors we may want to consider:

- The Lanczos vectors in exact arithmetic v^k given by the exact coefficients α_k and η_k. In this case, the perturbation terms are zero and the polynomials $p_{1,k}$ are the orthogonal Lanczos polynomials we studied in Chapter 1.

- The Lanczos vectors in finite precision $\tilde{v}^k$ with coefficients $\tilde{\alpha}_k$ and $\tilde{\eta}_k$; then the perturbation terms correspond to the local roundoff and the polynomials $\tilde{p}_{1,k}$ are no longer orthogonal for the same measure as $p_{1,k}$. Moreover, the first term in the solution is not the solution in exact arithmetic (even though it is very close to it until the first Ritz value has converged) but the solution with no roundoff and with the finite precision coefficients.

- The Lanczos vectors in finite precision $\hat{v}^k$ but computed with the exact coefficients α_k, η_k.

- The Lanczos vectors in exact arithmetic $\check{v}^k$ but with the finite precision coefficients $\tilde{\alpha}_k, \tilde{\eta}_k$.

Moreover, what we are really interested in here are the projections of these Lanczos vectors on the eigenvectors of A. In exact arithmetic we have $v^{k+1} = p_{1,k+1}(A)v^1$. Concerning the vectors $\tilde{v}^k$ in finite precision and applying the previous results, we have the following result.

Theorem 4.6. *Let j be given and $\tilde{p}_{j,k}$ be the polynomial determined by*

$$\tilde{p}_{j,j-1} = 0, \quad \tilde{p}_{j,j} = 1,$$

$$\tilde{\eta}_{k+1}\tilde{p}_{j,k+1}(\lambda) = (\lambda - \tilde{\alpha}_k)\tilde{p}_{j,k}(\lambda) - \tilde{\eta}_k\tilde{p}_{j,k-1}(\lambda), \quad k = j, \dots .$$

Then,

$$\tilde{v}^{k+1} = \tilde{p}_{1,k+1}(A)\tilde{v}^1 + \sum_{l=1}^{k} \tilde{p}_{l+1,k+1}(A)\frac{f^l}{\tilde{\eta}_{l+1}}.$$

The first term on the right-hand side is $\check{v}^{k+1} = \tilde{p}_{1,k+1}(A)\tilde{v}^1$.

Proof. We have

$$Q^T\tilde{v}^{k+1} = \frac{1}{\tilde{\eta}_{k+1}}((\Lambda - \tilde{\alpha}_k I)Q^T\tilde{v}^k - \tilde{\eta}_k Q^T\tilde{v}^{k-1} + Q^T f^k).$$

Therefore,

$$(Q^T \tilde{v}^{k+1})_i = \tilde{p}_{1,k+1}(\lambda_i)(Q^T \tilde{v}^1)_i + \sum_{l=1}^{k} \tilde{p}_{l+1,k+1}(\lambda_i) \frac{(Q^T f^l)_i}{\tilde{\eta}_{l+1}}.$$

Since $A = Q \Lambda Q^T$ and $Q^T Q = Q Q^T = I$ we obtain the result. □

We notice that the first term $\check{v}^{k+1} = \tilde{p}_{1,k+1}(A)\tilde{v}^1$ is different from what we have in exact arithmetic since the coefficients of the polynomial are the ones computed in finite precision. An elegant proof of a result similar to the previous one was given by Zemke in [207]. The characterization of the polynomial $\tilde{p}_{j,k}$ was given in Lemma 4.5. If we denote by $\tilde{\theta}_i^{(j,k)}$ the eigenvalues of $\tilde{T}_{j,k}$, we have

$$\tilde{p}_{j,k}(\lambda) = (-1)^{k-j} \frac{\prod_i (\tilde{\theta}_i^{(j,k-1)} - \lambda)}{\tilde{\eta}_{j+1} \cdots \tilde{\eta}_k}.$$

The values of the components of the projections of the computed Lanczos vectors on the eigenvectors of A depend on how close the eigenvalues of A are to the eigenvalues of all the submatrices $\tilde{T}_{j,k}$, $j = 1, \ldots, k$. This means that we have to look closely at these eigenvalues and their evolution when k increases for a given j.

For the Strakos30 example the eigenvalues of $\tilde{T}_{2,k}$ are shown in Figure 4.15 from $k = 2$ at the top to $k = 50$ at the bottom. The (small) stars on the x-axis are the eigenvalues of A. This figure shows that an eigenvalue very close to $\lambda_{max} = 100$ appears at iteration 25. This is precisely the time when the last component of the Lanczos vector reaches almost 1. It is also the time when we start having a second copy of λ_{max} in the set of Ritz values. This is also shown in Figure 4.16, where we see that the distance of the largest eigenvalue of $\tilde{T}_{2,k}$ to λ_{max} is more or less constant after iteration 7 and starts decreasing at iteration 25 and that it decreases very fast (six iterations) to the roundoff level. The appearance of this eigenvalue close to $\lambda_{30} = 100$ for $\tilde{T}_{2,k}$ is caused by the interlacing of the eigenvalues of $\tilde{T}_{1,k}$ and $\tilde{T}_{2,k}$ because the two largest eigenvalues of $\tilde{T}_{1,k}$ are very close to λ_{30} after iteration 25. Figure 4.17 shows that the same phenomenon happens for $\tilde{T}_{3,k}$. In fact, if we look closely at matrices $\tilde{T}_{j,k}$ we see a Ritz value close to λ_{30} appearing at iteration 25 for $j = 2, \ldots, 24$. The matrices $\tilde{T}_{25,k}$, $k > 25$, have an eigenvalue close to λ_{30} appearing for $k = 40$. For $k \geq 40$, the matrices $\tilde{T}_{j,k}$, $j < 25$ have at least two eigenvalues close to λ_{30}.

This is not directly implied by the Cauchy interlacing theorem. However, we can use Theorem 1.9 to explain this phenomenon. Consider, for instance, the matrix $\tilde{T}_{3,k}$ for $k \geq 25$; using the notations of Theorem 1.9 we have $j = 1$ and $\tilde{T}_1 = (\tilde{\alpha}_1)$. Now, we have to consider the intervals defined by the eigenvalues of $\tilde{T}_{3,k}$, $\tilde{\alpha}_1$, and $\pm\infty$. If $\tilde{\alpha}_1$ is reasonably far from λ_{30} (which should be the case except for some special choices of v^1), say $\tilde{\alpha}_1 < \tilde{\theta}_{k-1}^{(3,k)}$, then Theorem 1.9 implies that the two close eigenvalues of $\tilde{T}_k$, $\tilde{\theta}_k^{(k)}$ and $\tilde{\theta}_{k-1}^{(k)}$ are such that

$$\tilde{\theta}_{k-3}^{(3,k)} < \tilde{\theta}_{k-1}^{(k)} < \tilde{\theta}_{k-2}^{(3,k)} < \tilde{\theta}_k^{(k)},$$

showing that the largest Ritz value $\tilde{\theta}_{k-2}^{(3,k)}$ is close to λ_{30} This is also true for $\tilde{T}_j$ and $\tilde{T}_{j+2,k}$, $j > 1$, as long as the eigenvalues of $\tilde{T}_j$ are far enough from λ_{30}. A Ritz value

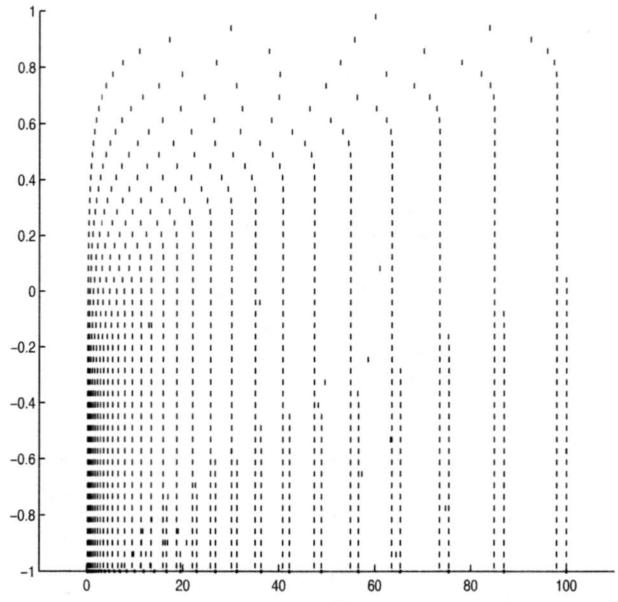

Figure 4.15. *Strakos30, eigenvalues of* $\tilde{T}_{2,k}$

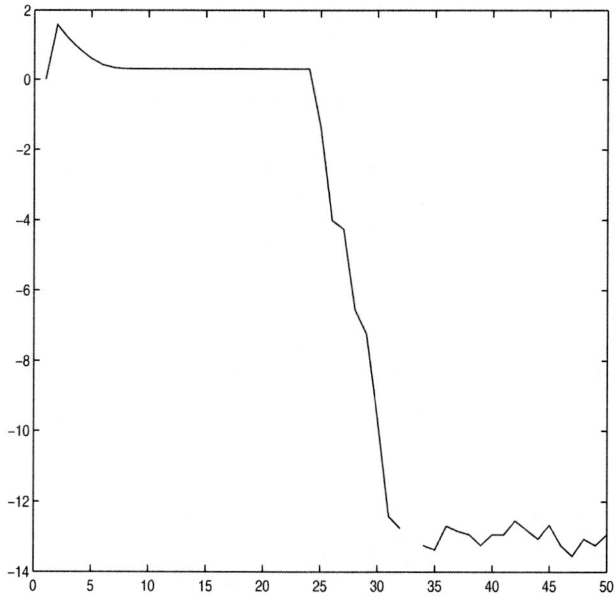

Figure 4.16. *Strakos30,* $\log_{10}$ *of the distance of the largest eigenvalue of* $\tilde{T}_{2,k}$ *to* 100

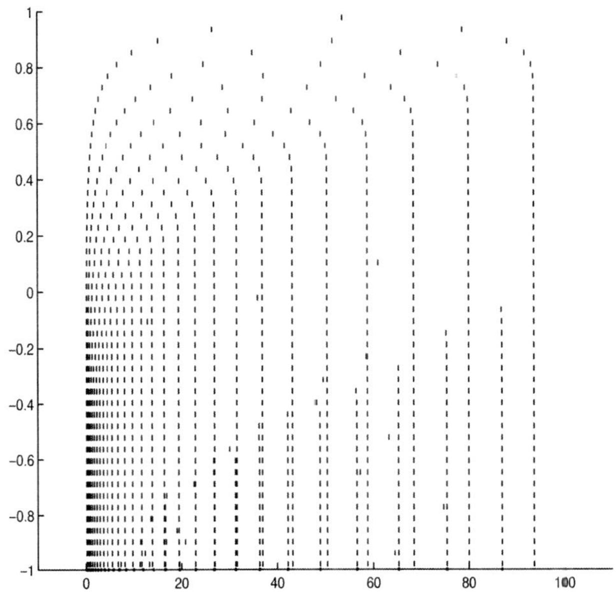

Figure 4.17. *Strakos30, eigenvalues of* $\tilde{T}_{3,k}$

converges to λ_{30} to full accuracy at iteration 24. After iteration 24 and before iteration 40, there is one eigenvalue of $\tilde{T}_j$ close to λ_{30}. After iteration 40, a new eigenvalue close to λ_{30} appears for $\tilde{T}_{25,k}$ because there are three eigenvalues in the cluster for $\tilde{T}_k$. Even though we do not give a formal proof, the same thing happens later for the other eigenvalues, as we can see in Figures 4.15 and 4.17.

What we have obtained so far enables us to look at the behavior of the components of the projections of the Lanczos vectors on the eigenvectors of A, which is our main goal in this chapter. If we are interested in a genuine forward analysis (which can be done only up to $k \le n$), we are not completely finished yet with the Lanczos vector $\tilde{v}^{k+1}$, if we want to compare it to v^{k+1}, since the first term in the solution $\tilde{v}^{k+1} = \tilde{p}_{1,k+1}(A)v^1$ is not what we would like to see, that is, the value in exact arithmetic $v^{k+1} = p_{1,k+1}(A)v^1$. Looking at the three-term recurrences for both polynomials we have

$$\eta_{k+1} p_{1,k+1}(\lambda) = (\lambda - \alpha_k) p_{1,k}(\lambda) - \eta_k p_{1,k-1}(\lambda),$$

$$\tilde{\eta}_{k+1} \tilde{p}_{1,k+1}(\lambda) = (\lambda - \tilde{\alpha}_k) \tilde{p}_{1,k}(\lambda) - \tilde{\eta}_k \tilde{p}_{1,k-1}(\lambda).$$

Setting $\Delta p_k = p_{1,k}(\lambda) - \tilde{p}_{1,k}(\lambda)$ we have

$$\tilde{\eta}_{k+1} \Delta p_{k+1} = (\lambda - \tilde{\alpha}_k) \Delta p_k - \tilde{\eta}_k \Delta p_{k-1} + g_k(\lambda),$$

with

$$g_k(\lambda) = (\tilde{\eta}_{k+1} - \eta_{k+1}) p_{1,k+1}(\lambda) + (\tilde{\alpha}_k - \alpha_k) p_{1,k}(\lambda) + (\tilde{\eta}_k - \eta_k) p_{1,k-1}(\lambda).$$

We see that g_k is a polynomial of the same order as $p_{1,k+1}$. We also have

$$\eta_{k+1}\Delta p_{k+1} = (\lambda - \alpha_k)\Delta p_k - \eta_k \Delta p_{k-1} + \tilde{g}_k(\lambda),$$

with

$$\tilde{g}_k(\lambda) = (\tilde{\eta}_{k+1} - \eta_{k+1})\tilde{p}_{1,k+1}(\lambda) + (\tilde{\alpha}_k - \alpha_k)\tilde{p}_{1,k}(\lambda) + (\tilde{\eta}_k - \eta_k)\tilde{p}_{1,k-1}(\lambda).$$

These recurrences for Δp_k are of the same form as the ones we studied in Theorem 4.3. Hence, we can easily write the solution.

Proposition 4.7. *We have*

$$p_{1,k+1}(\lambda) - \tilde{p}_{1,k+1}(\lambda) = \Delta p_{k+1} = \sum_{l=1}^{k} \tilde{p}_{l+1,k+1}(\lambda)\frac{g_l(\lambda)}{\tilde{\eta}_{l+1}}$$

and

$$\Delta p_{k+1} = \sum_{l=1}^{k} p_{l+1,k+1}(\lambda)\frac{\tilde{g}_l(\lambda)}{\eta_{l+1}}.$$

Proof. By noticing that $\Delta p_{1,1} = p_{1,1} - \tilde{p}_{1,1} = 0$ and applying the previous results we obtain the proof. □

Therefore, $p_{1,k}$ (resp., $\tilde{p}_{1,k}$) can be expressed as a combination of the $\tilde{p}_{j,k}$ (resp., $p_{j,k}$). Proposition 4.7 shows us two things. First, we see that if the differences in the coefficients are small (which is true at the beginning of the computation), then $\tilde{p}_{1,k}$ is not much different from $p_{1,k}$.

Proposition 4.8. *If* $|\alpha_l - \tilde{\alpha}_l| \le C_\alpha$ *and* $|\eta_l - \tilde{\eta}_l| \le C_\eta$, $l = 1, \dots, k$,

$$|\Delta p_{k+1}(\lambda_i)(Q^T v^1)_i| \le (2C_\eta + C_\alpha)\sum_{l=1}^{k}\frac{|\tilde{p}_{l+1,k+1}(\lambda_i)|}{\tilde{\eta}_{l+1}}.$$

Proof. With our hypothesis the perturbation term satisfies

$$|g_l(\lambda_i)(Q^T v^1)_i| \le 2C_\eta + C_\alpha,$$

since $|p_{1,l}(\lambda_i)(Q^T v^1)_i| \le 1$. This proves the result. □

As long as $2C_\eta + C_\alpha \le \sqrt{u}$ and $|\tilde{p}_{l+1,k+1}| \le \tilde{\eta}_{l+1}\sqrt{u}$, then $|\Delta p_{k+1}(\lambda_i)(Q^T v^1)_i| \simeq ku$. The second fact is that we can obtain a complete expression for the difference of v^k and $\tilde{v}^k$, as long as it makes sense, in particular $k < n$.

Theorem 4.9.

$$\tilde{v}^{k+1} = v^{k+1} + \sum_{l=1}^{k} \tilde{p}_{l+1,k+1}(A) g_l(A)\frac{v^1}{\tilde{\eta}_{l+1}} + \sum_{l=1}^{k} \tilde{p}_{l+1,k+1}(A)\frac{f^l}{\tilde{\eta}_{l+1}}.$$

Proof. This is the straightforward application of the previous results which give

$$(Q^T\tilde{v}^{k+1})_i = (Q^T v^{k+1})_i + \sum_{l=1}^{k} \tilde{p}_{l+1,k+1}(\lambda_i)g_l(\lambda_i)\frac{(Q^T\tilde{v}^1)_i}{\tilde{\eta}_{l+1}}$$
$$+ \sum_{l=1}^{k} \tilde{p}_{l+1,k+1}(\lambda_i)\frac{(Q^T f^l)_i}{\tilde{\eta}_{l+1}}.$$

We note that an analogous expression can be obtained using $\tilde{g}_l$. □

This result is essentially similar to what was obtained in Grcar's Ph.D. thesis [73] except that he was not specifically interested in the projections on the eigenvectors of A but in the projections of the computed vectors on the exact ones. It shows that the difference between the exact and the finite precision vectors comes from two sources. One is the difference between the coefficients and the other is arising from the growth of the local errors. Of course, they are closely related except at the beginning of the computation since then the coefficients in exact arithmetic and finite precision arithmetic are the same up to rounding errors and the terms involving g_l are small. The question we must now consider is what are the behavior and the possible growth of the terms in the right-hand side of $Q^T(\tilde{v}^{k+1} - v^{k+1})$. When do they grow and when do they stay small? As we shall see, it is difficult to answer this question, and this is a strong limitation of a forward analysis. However, what is much more interesting is to understand the behavior of $Q^T\tilde{v}^{k+1}$ since this reveals why and when we obtain multiple copies of the eigenvalues.

Nevertheless, denoting $\bar{v}^k = Q^T\tilde{v}^k$, let us say that we are interested in looking at

$$(\bar{v}^{k+1})_i = (Q^T v^{k+1})_i + \sum_{l=1}^{k} \tilde{p}_{l+1,k+1}(\lambda_i)g_l(\lambda_i)\frac{(\bar{v}^1)_i}{\tilde{\eta}_{l-1}}$$
$$+ \sum_{l=1}^{k} \tilde{p}_{l+1,k+1}(\lambda_i)\frac{(Q^T f^l)_i}{\tilde{\eta}_{l+1}}.$$

It seems difficult to study the general behavior of such an expression. We remark that, so far, we have not supposed that the perturbation terms f^l are small. The main difficulty with this analysis is that we do not have any analytic formula for terms like $(Q^T f^l)_i$; we just have bounds and we do not even know their sign. Hence, we choose to look at the absolute values of the terms of the sums. Of course this gives only a bound of the difference $|(\bar{v}^{k+1})_i - (Q^T v^{k+1})_i|$ and the fact that an upper bound is large does not prove that the expression is large. Even though there can eventually be some cancellations between several terms, we consider that if the absolute values of some terms are large, then it is likely than the sum is large. In any case, it will reveal the influence of each local perturbation. Moreover, we shall see that the absolute values of the terms in the sum all behave in the same way.

To somehow explain the behavior of the components (on the eigenvectors of A) of the Lanczos vectors we must look at the polynomials $\tilde{p}_{j,k}$ and therefore at $\tilde{\chi}_{j,k}$ for different values of j. There are at least two ways to do this.

For the first one we use the pivot functions and recall that we have $\tilde{\chi}_{1,k}(\lambda) = \tilde{\delta}_1(\lambda)\cdots\tilde{\delta}_k(\lambda)$, where $\tilde{\delta}_i(\lambda) = \tilde{\delta}_{1,i}(\lambda)$, $i = 1, \dots,$ are the last pivot functions, that is, the diagonal elements of the Cholesky decomposition of $\tilde{T}_k - \lambda I$. Similarly for $\tilde{\chi}_{j,k}(\lambda)$ we can introduce the diagonal elements $\tilde{\delta}_{j,i}(\lambda)$ of the Cholesky decomposition of $\tilde{T}_{j,k} - \lambda I$ starting at index j given by

$$\tilde{\delta}_{j,j}(\lambda) = \tilde{\alpha}_j - \lambda, \quad \tilde{\delta}_{j,i}(\lambda) = \tilde{\alpha}_i - \lambda - \frac{\tilde{\eta}_i^2}{\tilde{\delta}_{j,i-1}(\lambda)}, \qquad i = j+1, \dots,$$

provided $\tilde{\delta}_{j,i-1}(\lambda) \neq 0$. Then $\tilde{\chi}_{j,k} = \tilde{\delta}_{j,j}\cdots\tilde{\delta}_{j,k}$. Note we can also define $\delta_{j,i}$ with similar formulas by removing the tildes. To compare $\tilde{\chi}_{1,k}$ to $\tilde{\chi}_{j,k}$ we would like to compare $\tilde{\delta}_{1,k}(\lambda)$ and $\tilde{\delta}_{j,k}(\lambda)$. This is done in the next proposition.

Proposition 4.10. *For $k > j > 1$, if $\tilde{\delta}_{1,i}(\lambda) \neq 0$, $i = 1, \dots, k-1$, and $\tilde{\delta}_{j,i}(\lambda) \neq 0$, $i = j, \dots, k-1$, then*

$$\tilde{\delta}_{1,k}(\lambda) = \tilde{\delta}_{j,k}(\lambda) - \frac{(\tilde{\eta}_j\cdots\tilde{\eta}_k)^2}{(\tilde{\delta}_{j,j}(\lambda)\cdots\tilde{\delta}_{j,k-1}(\lambda))(\tilde{\delta}_{1,j-1}(\lambda)\cdots\tilde{\delta}_{1,k-1}(\lambda))}.$$

Proof. For the sake of simplicity we drop the arguments λ. Since

$$\tilde{\delta}_{1,j} = \tilde{\alpha}_j - \lambda - \frac{\tilde{\eta}_j^2}{\tilde{\delta}_{1,j-1}}, \qquad \tilde{\delta}_{j,j} = \tilde{\alpha}_j - \lambda,$$

we have

$$\tilde{\delta}_{1,j} = \tilde{\delta}_{j,j} - \frac{\tilde{\eta}_j^2}{\tilde{\delta}_{1,j-1}}.$$

At the next step we have

$$\tilde{\delta}_{1,j+1} = \tilde{\alpha}_{j+1} - \lambda - \frac{\tilde{\eta}_{j+1}^2}{\tilde{\delta}_{1,j}} = \tilde{\delta}_{j,j+1} + \tilde{\eta}_{j+1}^2\left(\frac{1}{\tilde{\delta}_{j,j}} - \frac{1}{\tilde{\delta}_{1,j}}\right) = \tilde{\delta}_{j,j+1} + \tilde{\eta}_{j+1}^2\left(\frac{\tilde{\delta}_{1,j} - \tilde{\delta}_{j,j}}{\tilde{\delta}_{1,j}\tilde{\delta}_{j,j}}\right).$$

Therefore,

$$\tilde{\delta}_{1,j+1} = \tilde{\delta}_{j,j+1} - \frac{(\tilde{\eta}_j\tilde{\eta}_{j+1})^2}{\tilde{\delta}_{j,j}\tilde{\delta}_{1,j}\tilde{\delta}_{1,j-1}}.$$

The proof is ended by induction on the second index. $\square$

This result leads to a simple determinant equality.

Corollary 4.11. *If $k > j > 1$ and for all λ such that $\tilde{\delta}_{j,k}(\lambda) - \tilde{\delta}_{1,k}(\lambda) \neq 0$, if we set $\tilde{\chi}_{1,0}(\lambda) = 1$, we have*

$$\tilde{\chi}_{1,k-1}(\lambda)\tilde{\chi}_{j,k-1}(\lambda) = (\tilde{\eta}_j\cdots\tilde{\eta}_k)^2\frac{\tilde{\chi}_{1,j-2}(\lambda)}{\tilde{\delta}_{j,k}(\lambda) - \tilde{\delta}_{1,k}(\lambda)}.$$

Proof. To obtain $\tilde{\chi}_{1,k-1}$ from $\tilde{\delta}_{j,k} - \tilde{\delta}_{1,k}$ we multiply the numerator and the denominator by $\tilde{\delta}_{1,1} \cdots \tilde{\delta}_{1,j-2}$. □

What we are interested in are the absolute values of the polynomials $\tilde{p}_{1,k}$ and $\tilde{p}_{j,k}$. But since

$$|\tilde{p}_{j,k}(\lambda)| = \frac{|\tilde{\chi}_{j,k-1}(\lambda)|}{\tilde{\eta}_{j+1} \cdots \tilde{\eta}_k},$$

we have the following result.

Theorem 4.12. *For $k > j > 1$ if $\tilde{\delta}_{j,k}(\lambda) - \tilde{\delta}_{1,k}(\lambda) \neq 0$, we have*

$$|\tilde{p}_{1,k}(\lambda)|\,|\tilde{p}_{j,k}(\lambda)| = \frac{\tilde{\eta}_j\,|\tilde{p}_{1,j-1}(\lambda)|}{|\tilde{\delta}_{j,k}(\lambda) - \tilde{\delta}_{1,k}(\lambda)|}.$$

We have a similar expression for $|p_{1,k}(\lambda)|\,|p_{j,k}(\lambda)|$ by omitting the tildes.

Proof. This is a straightforward consequence of the previous results. It can also be seen as a consequence of Proposition 4.4. □

This is what we can prove rigorously about the product of these polynomial values. If we are able to show that the right-hand side in Theorem 4.12 is more or less constant as a function of k, this will show that $|\tilde{p}_{j,k}|$ is increasing when $|\tilde{p}_{1,k}|$ is decreasing. It is now of interest to consider the behavior of $\tilde{p}_{1,k}(\lambda_i)$. Although we shall not give any formal proof, we are going to explain what happens computationally. To do this we start by considering the exact polynomial $p_{1,k}$. We already know that $|p_{1,k}(\lambda_i)\bar{v}_i^1| \leq 1$ and $|p_{1,k}(\lambda_i)|$ decreases when k is large and increasing (even though there exist cases for which it is more or less constant as a function of k and zero at the last step). In fact, $p_{1,l}(\lambda_i) = 0, l \geq n$. To compare $p_{1,k}$ with $p_{j,k}$ we use

$$|p_{1,k}(\lambda_i)|\,|p_{j,k}(\lambda_i)| = \frac{\eta_j\,|p_{1,j-1}(\lambda_i)|}{|\delta_{j,k}(\lambda_i) - \delta_{1,k}(\lambda_i)|}.$$

The zeros of the last pivot function $\delta_{1,k}(\lambda)$ are the Ritz values $\theta_j^{(k)}$. In exact arithmetic $\delta_{1,k}(\lambda_i)$ and $\delta_{j,k}(\lambda_i)$ cannot be both zero and cannot be equal by the interlacing property and $\delta_{1,k}(\lambda_i)$ tends to 0 or at least $\delta_n(\lambda_i) = 0$.

This is illustrated in Figures 4.18 and 4.19 for $\lambda = \lambda_n$ and $j = 2$ for which the computations were done in extended precision with *Digits* $= 64$ as well as for the next figures. The denominator is bounded when k increases and the numerator decreases when j increases. Hence the term on the right-hand side is bounded from above and from below and more or less $O(1)$.

This implies that when $|p_{1,k}|$ decreases, $|p_{j,k}|$ increases and, as we have said before, there is a symmetry around the starting value. This is what we see in Figure 4.20 for $p_{1,k}(\lambda_n)$ and $p_{2,k}(\lambda_n)$. A picture of the first $|p_{j,k}(\lambda_n)|$ starting from $j = 2$ is given in Figure 4.21. They are all increasing as a function of k, but remember that the polynomials for $j > 1$ are not involved in "exact" computations.

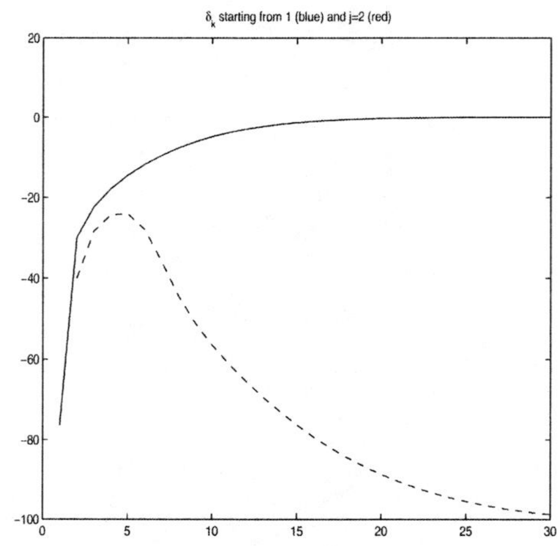

Figure 4.18. *Strakos30, $\delta_{1,k}(\lambda_n)$ (solid) and $\delta_{2,k}(\lambda_n)$ (dashed)*

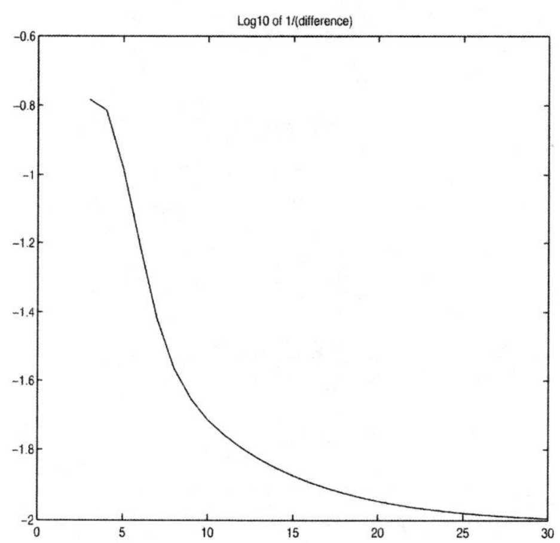

Figure 4.19. *Strakos30, $\log_{10}$ of $1/|\delta_{1,k}(\lambda_n) - \delta_{2,k}(\lambda_n)|$*

This helps us understand the behavior of $\tilde{p}_{1,k}(\lambda_i)$. From Proposition 4.7 we know that

$$\tilde{p}_{1,k+1}(\lambda_i) = p_{1,k+1}(\lambda_i) - \sum_{l=1}^{k} p_{l+1,k+1}(\lambda_i)\frac{\tilde{g}_l(\lambda_i)}{\eta_{l+1}}.$$

When a Ritz value is converging to λ_i, $|p_{1,k+1}(\lambda_i)|$ is decreasing to 0 and simultaneously $|p_{l+1,k+1}(\lambda_i)|$ is increasing. This applies to the generic case for which there is a move of a Ritz value towards an eigenvalue of A. At the beginning of the iterations $\tilde{g}_l$ is small, of the order of u (or at maximum $O(\lambda_n u)$). Hence, $|\tilde{p}_{1,k+1}(\lambda_i)|$ starts by decreasing like $|p_{1,k+1}(\lambda_i)|$, until for a sufficiently large k a term $|p_{l+1,k+1}(\lambda_i)\tilde{g}_l|$ becomes larger than $|p_{1,k+1}(\lambda_i)|$. This happens for two reasons: because $|\tilde{g}_l|$ may no longer be small (because the

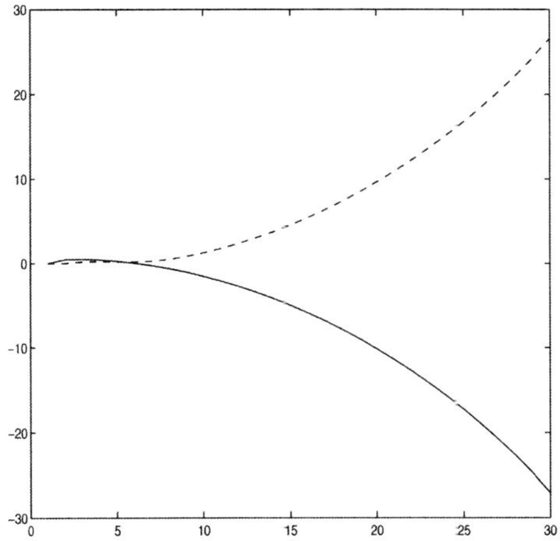

Figure 4.20. *Strakos30,* $\log_{10}$ *of* $|p_{1,k}(\lambda_n)|$ *(solid) and* $|p_{2,k}(\lambda_n)|$ *(dashed)*

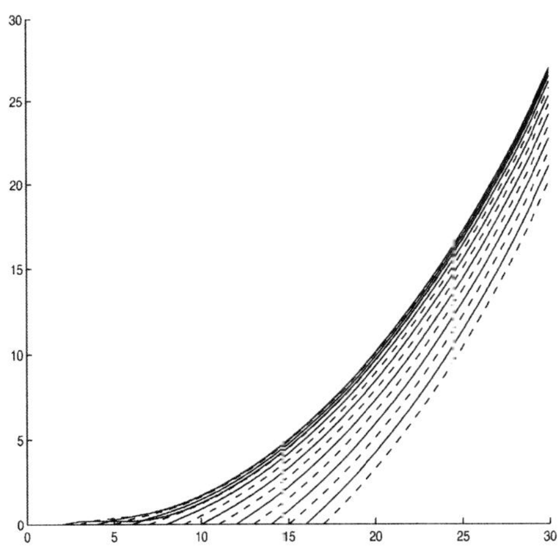

Figure 4.21. *Strakos30,* $\log_{10}$ *of the first* $|p_{j,k}(\lambda_n)|$, $j > 1$

exact and computed coefficients may be different) and because $|p_{l+1,k+1}(\lambda_i)|$ is increasing for all k. Notice that for $\lambda = \lambda_n$, the values of the polynomials at λ_n are positive. The value $\tilde{p}_{1,k+1}(\lambda_i)$ must first decrease (eventually with some oscillations) when a Ritz value starts converging and then increase when the perturbation terms are large enough. Now let us consider the product $|\tilde{p}_{1,k}|\,|\tilde{p}_{j,k}|$. The easiest case to analyze is $j = 2$, for which we have the simpler expression

$$|\tilde{p}_{1,k}(\lambda)|\,|\tilde{p}_{2,k}(\lambda)| = \frac{\tilde{\eta}_2}{|\tilde{\delta}_{2,k}(\lambda) - \tilde{\delta}_{1,k}(\lambda)|}.$$

If we are interested, for instance, in the largest eigenvalue λ_n, the absolute value of the product of $\tilde{p}_{1,k}(\lambda_n)(Q^T v^1)_n$ and the first term in the sum of the local perturbation terms $\tilde{p}_{2,k}(\lambda_n)(Q^T f^2)_n/\tilde{\eta}_2$ is

$$\frac{\tilde{\eta}_2}{|\tilde{\delta}_{2,k}(\lambda_n) - \tilde{\delta}_{1,k}(\lambda_n)|} |(Q^T \tilde{v}^1)_n| \frac{|(Q^T f^2)_n|}{\tilde{\eta}_2}.$$

The terms f_i^2 are of the order at most $\lambda_n u$ and unless $(Q^T v^1)_n$ is too small the term $|(Q^T v^1)_n|\,|(Q^T f^2)_n|$ is of order u. The value of the product depends mainly on

$$\frac{1}{|\tilde{\delta}_{2,k}(\lambda_n) - \tilde{\delta}_{1,k}(\lambda_n)|}.$$

In theory $\tilde{\delta}_{1,k}(\lambda)$ and $\tilde{\delta}_{2,k}(\lambda)$ cannot be both zero by the interlacing property. However, if in exact arithmetic $\delta_{1,k}(\lambda_n)$ is closer and closer to 0, this is not what happens in finite precision. This is what is shown in Figures 4.22 and 4.23, where we see that $\tilde{\delta}_{1,k}(\lambda_n)$ seems to converge to 0 until iteration 18 when there is a start of rapid change in the coefficients $\tilde{\alpha}_k$ and $\tilde{\eta}_k$ and $\tilde{\delta}_{1,k}(\lambda_n)$ seems to converge to the same values as $\tilde{\delta}_{2,k}(\lambda_n)$. Therefore, $1/|\tilde{\delta}_{2,k}(\lambda_n) - \tilde{\delta}_{1,k}(\lambda_n)|$ which was $O(1)$ starts increasing very fast. This can be explained as follows. In exact arithmetic we always have $\theta_k^{(k)} \leq \lambda_n$. When $\theta_k^{(k)}$ converges to λ_n the function $\delta_{1,k}(\lambda)$ is almost vertical close to the poles which are $\theta_j^{(k-1)}$ and like $\alpha_k - \lambda$ in between; see Figure 4.24. The value of $\delta_{1,k}$ at λ_n is then computed on the rightmost branch of the function and is close to 0. In finite precision, it may happen that even though $\tilde{\theta}_{k-1}^{(k-1)}$ is very close to λ_n, it is greater and the value at λ_n may be computed on the next to last branch of the function and then is almost $\alpha_k - \lambda_n$ which is large and negative because the value of α_k is close to smaller eigenvalues at that moment of the computation.

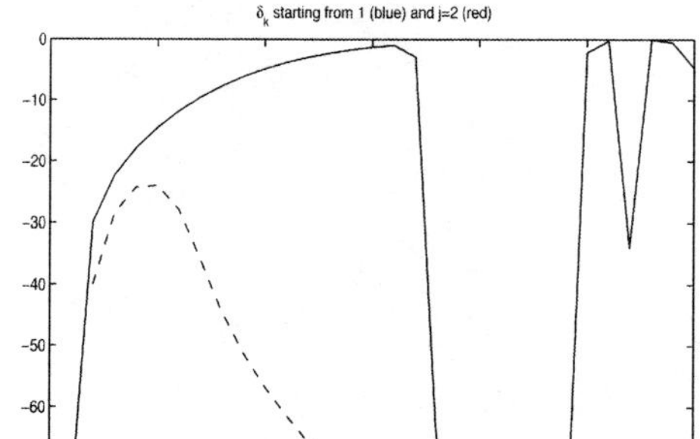

Figure 4.22. *Strakos30, $\tilde{\delta}_{1,k}(\lambda_n)$ (solid) and $\tilde{\delta}_{2,k}(\lambda_n)$ (dashed)*

At least until a Ritz value has first converged to a good accuracy $1/(|\tilde{\delta}_{2,k}(\lambda_n) - \tilde{\delta}_{1,k}(\lambda_n)|)$ is of order 1 and $|\tilde{p}_{j,k}(\lambda_n)|$ increases when $|\tilde{p}_{1,k}(\lambda_n)|$ decreases. Therefore, the perturbation terms arising from the local errors also increase when there is convergence of

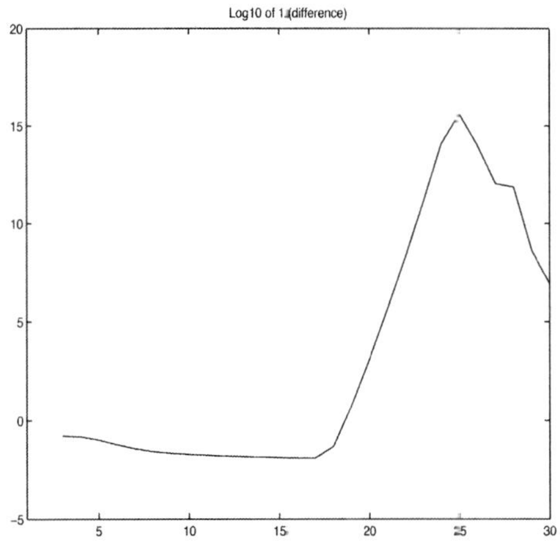

Figure 4.23. *Strakos*30, $\log_{10}$ *of* $1/|\tilde{\delta}_{1,k}(\lambda_n) - \tilde{\delta}_{2,k}(\lambda_n)|$

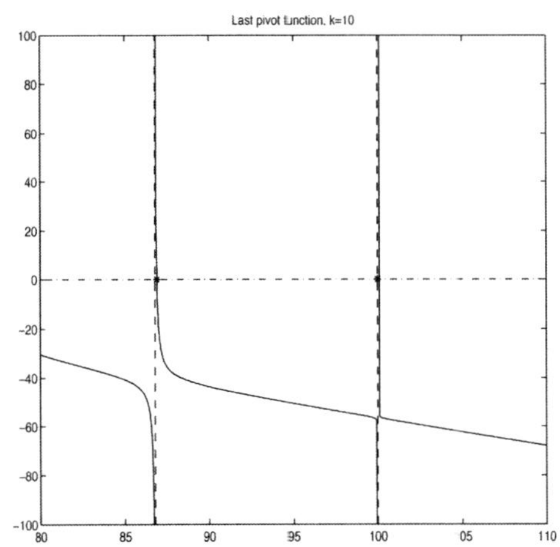

Figure 4.24. *Strakos*30, *function* $\delta_{1,k}$ *at iteration* 10

a Ritz value towards the given eigenvalue of A. For the other terms with $j > 2$ besides the difference of the $\tilde{\delta}$'s which behaves the same, there is another multiplying factor $\tilde{p}_{1,j-1}(\lambda_n)$ which grows, but it cannot be too large when j is small. When both j and k are large we cannot understand what is the behavior of $\tilde{p}_{j,k}$ from this formula. A more fruitful way to study the growth of the perturbation terms is to use the same technique as in Theorem 1.28.

Theorem 4.13. *Let l be such that*

$$\tilde{\theta}_{l-1}^{(j,k-1)} < \lambda_i < \tilde{\theta}_l^{(j,k-1)}.$$

If

$$|\lambda_i - \tilde{\theta}_l^{(j,k)}| > \eta_{k+1},$$

$|\tilde{p}_{j,k}(\lambda_i)|$ *is increasing as a function of* $k > j$.

Proof. The ratio $|\tilde{p}_{j,k+1}(\lambda_i)|/|\tilde{p}_{j,k}(\lambda_i)|$ is equal to

$$\frac{|\lambda_i - \tilde{\theta}_l^{(j,k)}|}{\tilde{\eta}_{k+1}} \frac{|\lambda_i - \tilde{\theta}_1^{(j,k)}|}{|\lambda_i - \tilde{\theta}_1^{(j,k-1)}|} \cdots \frac{|\lambda_i - \tilde{\theta}_{l-1}^{(j,k)}|}{|\lambda_i - \tilde{\theta}_{l-1}^{(j,k-1)}|} \frac{|\lambda_i - \tilde{\theta}_{l+1}^{(j,k)}|}{|\lambda_i - \tilde{\theta}_{l}^{(j,k-1)}|} \cdots \frac{|\lambda_i - \tilde{\theta}_{k-j+1}^{(j,k)}|}{|\lambda_i - \tilde{\theta}_{k-j}^{(j,k-1)}|}.$$

By the interlacing property all the ratio of distances of Ritz values to λ_i in the right-hand side are larger than 1 and the first ratio is larger than 1 by hypothesis, proving the result. □

Hence, if λ_i is far enough from the Ritz values $\tilde{\theta}_l^{(j,k)}$, the absolute value of the polynomial is increasing. We have seen that this is what happens at the beginning of the computation for the Strakos30 example for which the perturbation terms grow when the main term decreases. For the Lanczos vector projections that stay almost constant with Ritz values far from eigenvalues, there is no growth of the perturbations. This growth happens only when a Ritz value is converging to an eigenvalue, meaning that the corresponding component of the projections of the Lanczos vectors decreases and that the perturbations grow. When they reach $O(1)$ another copy of the same eigenvalue has to appear to stop the growth. When a perturbation term like

$$\left| \tilde{p}_{l+1,k+1}(\lambda_i) \frac{(Q^T f^l)_i}{\tilde{\eta}_{l+1}} \right|$$

starting from $O(u)$ has almost reached 1, the only way it can be stopped from growing unboundedly anymore when k increases is for $|\tilde{p}_{l,k+1}(\lambda_i)|$ to decrease since $\tilde{\eta}_{l+1}$ cannot be very large. The following result, which is almost a corollary of Theorem 4.13, shows that if the distances of the Ritz values to an eigenvalue are bounded from below, then the perturbation grows unboundedly.

Proposition 4.14. *If there exist numbers* $C_l > 1$ *independent of* k *such that*

$$|\lambda_i - \tilde{\theta}_j^{(l,k-1)}| > C_l \tilde{\eta}_{l+j}$$

for all $j, k > l$, $j = 1, \ldots, k - l$, *then*

$$|\tilde{p}_{l,k}(\lambda_i)| > C_l^{k-l}.$$

Proof. The proof is straightforward. □

Clearly if the condition of the proposition is fulfilled, the value of the polynomial at λ_i continues to grow. At some point the growth has to be stopped to ensure the normalization, so there must be one term $\lambda_i - \tilde{\theta}_j^{(l,k)}$ which is small for all l. A new copy of the same eigenvalue appears in all the matrices $T_{j,k}$ defined so far as we have seen before, stopping the growth of the perturbation terms. Then we are back in the situation we had in the beginning with a component $|(Q^T \bar{v}^k)_i|$ which is almost equal to 1. The convergence of the new copy being fast the corresponding term also decreases fast, compensating the additional ratios larger than 1 that appear in the product, and the value of the polynomial starts decreasing. When the distance of the Ritz value to the eigenvalue stabilizes to a multiple of the unit roundoff, $|\tilde{p}_{l,k}(\lambda_i)|$ starts increasing again because the small term is now almost constant and there are more new multiplicative terms larger than 1 in the product. The perturbation terms that start after the new copy has appeared are growing because there are no eigenvalues of the corresponding $\tilde{T}_{j,k}$ close to the eigenvalue of A. This behavior can go on forever (or at least until we switch off the computer!), producing additional copies of the eigenvalues of A.

The value of $\tilde{p}_{j,k}(\lambda_n)$ is shown for different values of j in Figure 4.25, the curves for j and $j+1$ being alternatively solid and dashed (by mistake $j = 22$ is missing). The solid curve below the others is for $j = 1$. It decreases at the beginning until the perturbation started at iteration 1 becomes large enough at iteration 18. It grows until iteration 25, when a decrease is started by the convergence of a new copy of λ_n. All the other polynomials which have started (up to $j = 24$) have the same behavior; they start decreasing at iteration 25. This is because these polynomials are multiples of determinants involving matrices $\tilde{T}_{j,k}$. As we have shown, they all have an eigenvalue close to the extra copy of the eigenvalue of $\tilde{T}_{1,k}$ that has just appeared.

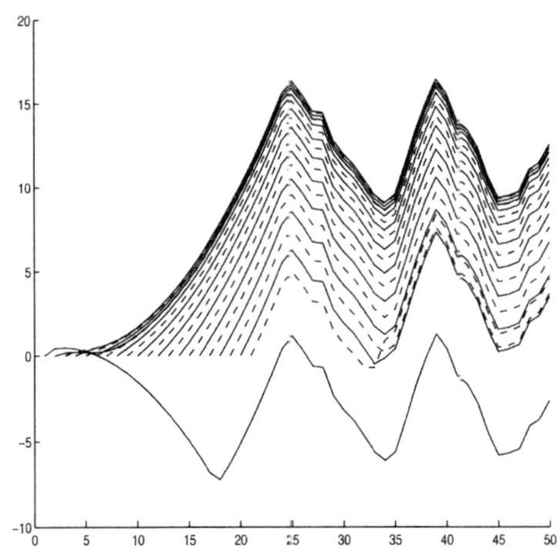

Figure 4.25. *Strakos*30, $\log_{10}(|\tilde{p}_{j,k}(\lambda_n)|)$ *for* $j = 1, \ldots, 21$

We see in Figure 4.26 that at the same time (iteration 25) there is a perturbation term that starts growing which will reach its maximum at iteration 39 which corresponds to the second "bump" in $\bar{v}^k_{30}$. At the same time all the other polynomials with $j > 24$ start decreasing. At iteration 40 another perturbation starts growing, etc., explaining the almost periodic behavior of the components of the projections of the Lanczos vectors. These computations were done in extended precision starting from the Lanczos coefficients computed in double precision. Figure 4.27 shows the same information for λ_{10} and Figure 4.28 gives the values

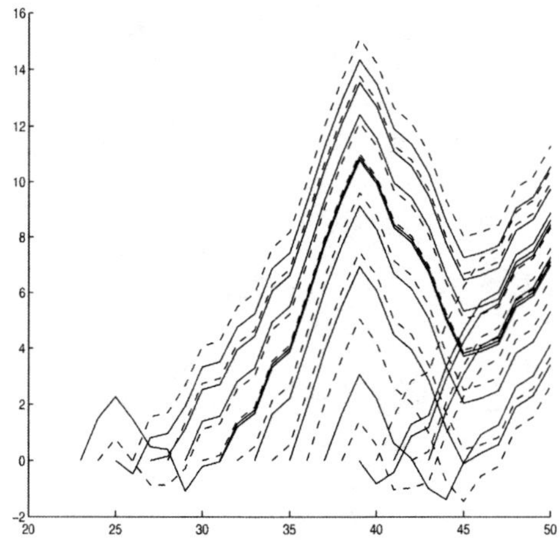

Figure 4.26. *Strakos*30, $\log_{10}(|\tilde{p}_{j,k}(\lambda_n)|)$ *for* $j = 23, \ldots, 43$

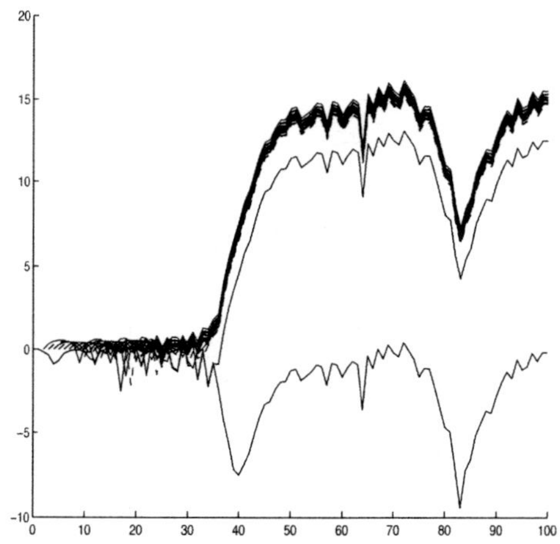

Figure 4.27. *Strakos*30, $\log_{10}(|\tilde{p}_{j,k}(\lambda_{10})|)$ *for* $j = 1, \ldots, 21$

of the polynomials $\tilde{p}_{j,k}$ for $j = 30$ to 40. The polynomials stay bounded until there is a Ritz value "converging" to λ_{10} around iteration 30.

4.4 Another solution to three-term recurrences

Considering polynomials for the Lanczos algorithm in finite precision gives an elegant solution to the three-term recurrences. However, this does not tell the whole story since we have been able to look only at the individual terms in the sums and not at the global behavior. We are now going to look at the solution of a three-term nonhomogeneous recurrence from

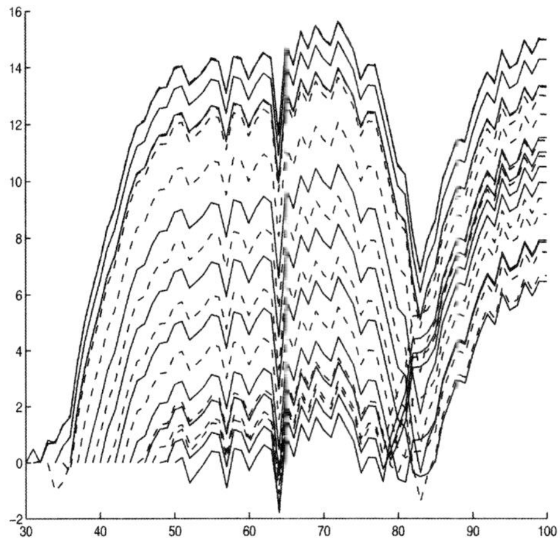

Figure 4.28. *Strakos30*, $\log_{10}(|\tilde{p}_{j,k}(\lambda_{10})|)$ *for* $j = 30, \ldots$

a different viewpoint. We consider the problem as a triangular solve and exhibit the inverse of the triangular matrix. We are looking again for the solution of the following recurrence with s_1 given

$$\zeta_2 s_2 = (\lambda - \tau_1)s_1 + f_1,$$

$$\zeta_{k+1}s_{k+1} = (\lambda - \tau_k)s_k - \zeta_k s_{k-1} + f_k, \quad k = 2, \ldots.$$

For the sake of simplicity we take $\lambda = 0$ and let

$$L_{k+1} = \begin{pmatrix} 1 & 0 & \cdots & \cdots & 0 & 0 \\ \tau_1 & \zeta_2 & 0 & \cdots & 0 & 0 \\ \zeta_2 & \tau_2 & \zeta_3 & & \vdots & \vdots \\ & \ddots & \ddots & \ddots & 0 & \vdots \\ & & \zeta_{k-1} & \tau_{k-1} & \zeta_k & 0 \\ & & & \zeta_k & \tau_k & \zeta_{k+1} \end{pmatrix}.$$

Denoting as usual by e^1 the first column of the identity matrix of dimension k, by e^k its last column, and by T_k the tridiagonal matrix given by the recurrence coefficients τ_i, ζ_i, we have

$$L_{k+1} = \begin{pmatrix} (e^1)^T & 0 \\ T_k & \zeta_{k+1}e^k \end{pmatrix}.$$

Let $s^{k+1} = (s_1 \quad \cdots \quad s_{k+1})^T$ be the vector of the solution at steps 1 to $k+1$ and $g = s_1$, $h = (f_1 \quad \cdots \quad f_k)^T$; then the nonhomogeneous recurrence at all stages up to $k+1$ can be written in matrix form as

$$L_{k+1}s^{k+1} = \begin{pmatrix} g \\ h \end{pmatrix}.$$

Later on we shall use this with $T_k - \lambda I$ instead of T_k to obtain the solution of the problem we are interested in for the Lanczos algorithm. The simplest way to obtain s_{k+1} is to solve this linear system analytically. It is relatively easy to obtain the following result.

Theorem 4.15. *The solution of the three-term recurrence (with $\lambda = 0$) can be written as*

$$s_{k+1} = -\frac{1}{\zeta_{k+1}(T_k^{-1})_{1,k}} s_1 + \frac{1}{\zeta_{k+1}(T_k^{-1})_{1,k}} \sum_{j=1}^{k} (T_k^{-1})_{j,1} f_j.$$

It is not particularly easy to obtain bounds of the perturbation terms using the previous result. Therefore we are going to look for another expression of the solution.

For the sake of generality, we are going to solve a problem corresponding to a long-term recurrence (where s_{k+1} depends on all s_i, $i = 1, \dots, k$) for which we have

$$L_{k+1} = \begin{pmatrix} (e^1)^T & 0 \\ H_k^T & \zeta_{k+1} e^k \end{pmatrix},$$

where H_k is an upper Hessenberg matrix (therefore H_k^T is a lower Hessenberg matrix). We suppose that all the matrices H_k are nonsingular. Our goal is now to find an expression for the inverse of L_{k+1} involving H_k. Even though this may seem a little strange we are first going to complicate the problem a bit. Let us write

$$L_{k+1} L_{k+1}^T (L_{k+1}^{-T} s^{k+1}) = \begin{pmatrix} g \\ h \end{pmatrix};$$

if $y^{k+1} = L_{k+1}^{-T} s^{k+1}$, the solution is obtained by solving $L_{k+1} L_{k+1}^T y^{k+1} = (g \quad h)^T$ and through $s^{k+1} = L_{k+1}^T y^{k+1}$. Let us now consider $L_{k+1} L_{k+1}^T$. We have

$$L_{k+1} L_{k+1}^T = \begin{pmatrix} 1 & (e^1)^T H_k \\ H_k^T e^1 & H_k^T H_k + \zeta_{k+1}^2 e^k (e^k)^T \end{pmatrix}.$$

For the sake of simplicity let us denote

$$L_{k+1} L_{k+1}^T = \begin{pmatrix} 1 & y^T \\ y & X \end{pmatrix}.$$

We want to solve

$$\begin{pmatrix} 1 & y^T \\ y & X \end{pmatrix} \begin{pmatrix} \xi \\ z \end{pmatrix} = \begin{pmatrix} g \\ h \end{pmatrix},$$

which will give us the inverse of the matrix.

Lemma 4.16. *Supposing X is nonsingular, the inverse of*

$$\begin{pmatrix} 1 & y^T \\ y & X \end{pmatrix}$$

is given by

$$\begin{pmatrix} \frac{1}{1-y^T X^{-1} y} & -\frac{1}{1-y^T X^{-1} y} y^T X^{-1} \\ -\frac{1}{1-y^T X^{-1} y} X^{-1} y & (X - yy^T)^{-1} \end{pmatrix}.$$

Proof. Writing the equations of the linear system we have

$$\xi + y^T z = g,$$

$$\xi y + X z = h.$$

By eliminating z with the second equation, $z = X^{-1}(h - \xi y)$ and

$$\xi = \frac{1}{1 - y^T X^{-1} y}(g - y^T X^{-1} h).$$

This gives us the first row of the inverse. By plugging the value of ξ into the second equation we obtain after some manipulations

$$z = \left[X^{-1} + \frac{1}{1 - y^T X^{-1} y} X^{-1} y y^T X^{-1} \right] h - \frac{1}{1 - y^T X^{-1} y} [X^{-1} y] g.$$

Using the Sherman–Morrison formula we can recognize that the first term within braces is nothing other but $(X - yy^T)^{-1}$, that is, the inverse of the Schur complement. This could have been seen directly by eliminating x using the first equation and is, in fact, an indirect proof of the Sherman–Morrison formula. □

Now we apply Lemma 4.16 to the matrix $L_{k+1} L_{k+1}^T$. Therefore, $y = H_k^T e^1$ and

$$X = H_k^T H_k + \zeta_{k+1}^2 e^k (e^k)^T.$$

We obtain X^{-1} with the Sherman–Morrison formula

$$X^{-1} = H_k^{-1} H_k^{-T} - \frac{1}{\omega_k} H_k^{-1} H_k^{-T} e^k (e^k)^T H_k^{-1} H_k^{-T},$$

with

$$\omega_k = \frac{1}{\zeta_{k+1}^2} + (e^k)^T H_k^{-1} H_k^{-T} e^k.$$

Let us compute $y^T X^{-1} y$; it is

$$(e^1)^T H_k \left[H_k^{-1} H_k^{-T} - \frac{1}{\omega_k} H_k^{-1} H_k^{-T} e^k (e^k)^T H_k^{-1} H_k^{-T} \right] H_k^T e^1,$$

and hence

$$y^T X^{-1} y = 1 - \frac{(H_k^{-T})_{1,k}^2}{\omega_k},$$

which gives

$$\frac{1}{1 - y^T X^{-1} y} = \frac{\omega_k}{(H_k^{-T})_{1,k}^2}.$$

We can compute the lower right $k \times k$ block term U of the inverse which after some manipulations is given by

$$U = H_k^{-1} H_k^{-T} + \frac{\omega_k}{(H_k^{-T})_{1,k}^2} H_k^{-1} e^1 (e^1)^T H_k^{-T}$$
$$- \frac{1}{(H_k^{-T})_{1,k}} [H_k^{-1} e^1 (e^k)^T H_k^{-1} H_k^{-T} - H_k^{-1} H_k^{-T} e^k (e^1)^T H_k^{-T}].$$

The lower left $k \times 1$ block term w of the inverse is

$$w = -\frac{\omega_k}{(H_k^{-T})_{1,k}^2} H_k^{-1} e^1 + \frac{1}{(H_k^{-T})_{1,k}} H_k^{-1} H_k^{-T} e^k.$$

The inverse we were looking for is

$$\begin{pmatrix} \frac{\omega_k}{(H_k^{-T})_{1,k}^2} & w^T \\ w & U \end{pmatrix}.$$

By multiplying from the left with L_{k+1}^T, we obtain L_{k+1}^{-1},

$$L_{k+1}^{-1} = \begin{pmatrix} \frac{\omega_k}{(H_k^{-T})_{1,k}^2} e^1 + H_k w & e^1 w^T + H_k U \\ \zeta_{k+1} (e^k)^T w & \zeta_{k+1} (e^k)^T U \end{pmatrix}.$$

Theorem 4.17. *The inverse of the triangular matrix*

$$L_{k+1} = \begin{pmatrix} (e^1)^T & 0 \\ H_k^T & \zeta_{k+1} e^k \end{pmatrix}$$

is given by

$$(L_{k+1}^{-1})_{(1:k,1)} = \frac{1}{(H_k^{-T})_{1,k}} H_k^{-T} e^k,$$

$$(L_{k+1}^{-1})_{(1:k,2:k+1)} = H_k^{-T} - \frac{1}{(H_k^{-T})_{1,k}} H_k^{-T} e^k (e^1)^T H_k^{-T},$$

$$(L_{k+1}^{-1})_{(k+1,1)} = \zeta_{k+1} (e^k)^T \left[-\frac{\omega_k}{(H_k^{-T})_{1,k}^2} H_k^{-1} e^1 + \frac{1}{(H_k^{-T})_{1,k}} H_k^{-1} H_k^{-T} e^k \right],$$

$$(L_{k+1}^{-1})_{(k+1,2:k+1)} = \zeta_{k+1} (e^k)^T H_k^{-1} \left[H_k^{-T} + \frac{\omega_k}{(H_k^{-T})_{1,k}^2} e^1 (e^1)^T H_k^{-T} \right.$$
$$\left. - \frac{1}{(H_k^{-T})_{1,k}} \left[e^1 (e^k)^T H_k^{-1} H_k^{-T} + H_k^{-T} e^k (e^1)^T H_k^{-T} \right] \right].$$

Proof. There are interesting cancellations in the expressions we have to consider. We have

$$\frac{\omega_k}{(H_k^{-T})_{1,k}^2} e^1 + H_k w = \frac{1}{(H_k^{-T})_{1,k}} H_k^{-T} e^k$$

and

$$e^1 w^T + H_k U = H_k^{-T} - \frac{1}{(H_k^{-T})_{1,k}} H_k^{-T} e^k (e^1)^T H_k^{-T}.$$

The lower left entry is $\zeta_{k+1} (e^k)^T w$, which is

$$\zeta_{k+1} (e^k)^T \left[-\frac{\omega_k}{(H_k^{-T})_{1,k}^2} H_k^{-1} e^1 + \frac{1}{(H_k^{-T})_{1,k}} H_k^{-1} H_k^{-T} e^k \right].$$

The lower right block is $\zeta_{k+1}(e^k)^T U$, which is

$$\zeta_{k+1}(e^k)^T H_k^{-1} \Bigg[H_k^{-T} + \frac{\omega_k}{(H_k^{-T})_{1,k}^2} e^1 (e^1)^T H_k^{-T} - \frac{1}{(H_k^{-T})_{1,k}} [e^1 (e^k)^T H_k^{-1} H_k^{-T} + H_k^{-T} e^k (e^1)^T H_k^{-T}] \Bigg]. \quad \square$$

It does not seem trivial that the matrix

$$H_k^{-T} - \frac{1}{(H_k^{-T})_{1,k}} H_k^{-T} e^k (e^1)^T H_k^{-T}$$

is strictly lower triangular if H_k is upper Hessenberg. But this can be easily verified if $H_k = H_k^T = T_k$ is a symmetric tridiagonal matrix in the following way. We use the following characterization of inverses of tridiagonal matrices; see, for instance, [25], [117].

Lemma 4.18. *If T_k is irreducible and tridiagonal, there exist two sequences $\{u_i\}$, $\{w_i\}$, $i = 1, \ldots, k$, with $u_1 = 1$ such that*

$$T_k^{-1} = \begin{pmatrix} u_1 w_1 & u_1 w_2 & u_1 w_3 & \cdots & u_1 w_k \\ u_1 w_2 & u_2 w_2 & u_2 w_3 & \cdots & u_2 w_k \\ u_1 w_3 & u_2 w_3 & u_3 w_3 & \cdots & u_3 w_k \\ \vdots & \vdots & \vdots & \ddots & \vdots \\ u_1 w_k & u_2 w_k & u_3 w_k & \cdots & u_k w_k \end{pmatrix}.$$

Using the first and last pivot functions at 0, the elements u_i and w_i are given by

$$w_1 = \frac{1}{d_1}, \quad w_i = (-1)^{i-1} \frac{\zeta_2 \cdots \zeta_i}{d_1 \cdots d_i}, \quad i = 2, \ldots, k,$$

$$u_k = \frac{1}{\delta_k w_k}, \quad u_{k-i} = (-1)^i \frac{\zeta_{k-i+1} \cdots \zeta_k}{\delta_{k-i} \cdots \delta_k w_k}, \quad i = 1, \ldots, k-1.$$

We can simplify the expression for u_i. Using the definition of w_k we have

$$u_i = (-1)^{i-1} \frac{d_1 \cdots d_k}{\delta_i \cdots \delta_k} \frac{1}{\zeta_2 \cdots \zeta_i}.$$

By noticing that $d_1 \cdots d_k = \delta_1 \cdots \delta_k = \det(T_k)$ this gives

$$u_i = (-1)^{i-1} \frac{\delta_1 \cdots \delta_{i-1}}{\zeta_2 \cdots \zeta_i}.$$

Then we remark that the matrix

$$T_k^{-1} - \frac{1}{(T_k^{-1})_{1,k}} T_k^{-1} e^k (e^1)^T T_k^{-1}$$

can be factored as

$$\left[I - \frac{1}{(T_k^{-1})_{1,k}} T_k^{-1} e^k (e^1)^T \right] T_k^{-1}.$$

The matrix $e^k(e^1)^T$ has only one nonzero element in position $(k, 1)$. Therefore, the matrix $T_k^{-1}e^k(e^1)^T$ has only nonzero elements in the first column which is equal to $T_k^{-1}e^k$. Since we divide this vector by its first element the structure of $U_k = I - \frac{1}{(T_k^{-1})_{1,k}} T_k^{-1} e^k (e^1)^T$ is the following:

$$\begin{pmatrix} 0 & 0 & \cdots & 0 \\ x & 1 & 0 & 0 \\ \vdots & & \ddots & \vdots \\ x & 0 & \cdots & 1 \end{pmatrix}.$$

The $(i, 1)$ element is

$$-\frac{(T_k^{-1})_{i,k}}{(T_k^{-1})_{1,k}}.$$

This matrix can also be written as

$$U_k = \begin{pmatrix} 0 & 0 & \cdots & \cdots & 0 \\ -u_2 & 1 & 0 & \cdots & 0 \\ -u_3 & 0 & 1 & & \vdots \\ \vdots & \vdots & & \ddots & 0 \\ -u_k & 0 & \cdots & \cdots & 1 \end{pmatrix}.$$

We multiply this matrix to the right by T_k^{-1}. The first row of the product is 0. For $i > 1$, the ith row of the product is the ith row of T_k^{-1} minus u_i times the first row of T_k^{-1}. By noticing that $u_1 = 1$, we see that all the elements of the strictly upper triangular part of the product vanish.

Theorem 4.19. *When T_k is symmetric tridiagonal,*

$$L_k^{-1} = \begin{pmatrix} 1 & 0 & \cdots & \cdots & 0 \\ u_2 & u_1 w_2 - u_2 w_1 & 0 & \cdots & 0 \\ u_3 & u_1 w_3 - u_3 w_1 & u_2 w_3 - u_3 w_2 & & \vdots \\ \vdots & \vdots & \vdots & \ddots & 0 \\ u_k & u_1 w_k - u_k w_1 & u_2 w_k - u_k w_2 & \cdots & u_{k-1} w_k - u_k w_{k-1} \end{pmatrix}.$$

Let us compute the elements $u_i w_j - u_j w_i,\ i < j,\ i = 1, \ldots, k-1,\ j = 2, \ldots, k$.

Proposition 4.20.

$$u_i w_j - u_j w_i = (-1)^{i+j} \left(\frac{\zeta_{i+1} \cdots \zeta_j}{\delta_i \cdots \delta_j} - \frac{1}{\delta_j} \frac{d_{i+1} \cdots d_j}{\zeta_{i+1} \cdots \zeta_j} \right).$$

The solution of our nonhomogeneous three-term recurrence relation can be written using the tridiagonal matrix T_k of the coefficients.

Theorem 4.21. *The k first elements of the solution of the three-term recurrence (with $\lambda = 0$) are given by*

$$s^k = \left(L_{k+1}^{-1}\begin{pmatrix} s_1 \\ h \end{pmatrix}\right)_{1:k} = \frac{s_1}{(T_k^{-1})_{1,k}} T_k^{-1} e^k + \left[I - \frac{1}{(T_k^{-1})_{1,k}} T_k^{-1} e^k (e^1)^T \right] T_k^{-1} h.$$

Moreover,

$$s_k = \frac{(T_k^{-1})_{k,k}}{(T_k^{-1})_{1,k}} s_1 + (T_k^{-1} h)_k - \frac{(T_k^{-1})_{k,k}}{(T_k^{-1})_{1,k}} (T_k^{-1} h)_1.$$

(notice that s_k does not depend on $h_k = f_k$). The element s_k can also be written as

$$s_k = u_k s_1 + \sum_{l=1}^{k-1} (u_l w_k - u_k w_l) f_l.$$

Moreover, we can check that the expressions given in the previous theorem are equivalent to what we have obtained by directly solving the triangular system. Let $\vartheta_k = (T_k^{-1})_{k,k}/(T_k^{-1})_{1,k}$. Then we consider

$$\sum_{j=1}^{k} [(T_{k+1}^{-1})_{j,k+1} - \vartheta_{k+1} (T_{k+1}^{-1})_{j,1}] f_j.$$

We can express the entries of the inverse of T_{k+1} using the inverse of T_k. We have seen that

$$\frac{(T_{k+1}^{-1})_{k+1,k+1}}{(T_{k+1}^{-1})_{1,k+1}} = \frac{1}{\delta_{k+1}(T_{k+1}^{-1})_{1,k+1}} = -\frac{1}{\zeta_{k+1}(T_k^{-1})_{1,k}};$$

therefore, we have

$$\vartheta_{k+1} = -\frac{1}{\zeta_{k+1}(T_k^{-1})_{1,k}} = -\frac{\delta_k}{\zeta_{k+1}} \vartheta_k.$$

The first k entries of the last column of the inverse of T_{k+1} are

$$(T_{k+1}^{-1} e^{k+1})_{1:k} = -\frac{\zeta_{k+1}}{\delta_{k+1}} (T_k^{-1} e^k)_{1:k}.$$

For the first column, we have

$$(T_{k+1}^{-1})_{j,1} = (T_k^{-1})_{j,1} + \frac{\zeta_{k+1}^2}{\delta_{k+1}} (T_k^{-1})_{j,k} (T_k^{-1})_{k,1}.$$

This gives

$$\vartheta_{k+1} (T_{k+1}^{-1})_{j,1} = -\frac{\delta_k}{\zeta_{k+1}} \vartheta_k (T_k^{-1})_{j,1} - \frac{\zeta_{k+1}}{\delta_{k+1}} (T_k^{-1})_{j,k}.$$

Then,

$$\sum_{j=1}^{k} [(T_{k+1}^{-1})_{j,k+1} - \vartheta_{k+1} (T_{k+1}^{-1})_{j,1}] f_j = \frac{\delta_k}{\zeta_{k+1}} \sum_{j=1}^{k} \vartheta_k (T_k^{-1})_{j,1} f_j.$$

It simplifies to

$$\sum_{j=1}^{k}[(T_{k+1}^{-1})_{j,k+1} - \vartheta_{k+1}(T_{k+1}^{-1})_{j,1}]f_j = \frac{1}{\zeta_{k+1}(T_k^{-1})_{1,k}}\sum_{j=1}^{k}(T_k^{-1})_{j,1}f_j,$$

and we recover the solution we obtained directly.

4.5 Bounds for the perturbation terms

Having a solution of the three-term recurrence involving a general symmetric matrix T_k for the coefficients, we can now apply the results of the previous section to the problem of the computation of the Lanczos vectors in finite precision. Let $\breve{T}_k = \tilde{T}_k - \lambda_i I$ for a given i, where $\tilde{T}_k$ is the computed Lanczos matrix. The solution of the nonhomogeneous recurrence satisfied by the ith component of the projection of the Lanczos vectors (slightly changing the notations) $\bar{v}_i^k = (Q^T \tilde{v}^k)_i$ is given by the previous theorems using $\breve{T}_k$ instead of T_k. We denote the perturbation term by $h^{(i)}$ with $h_j^{(i)} = (q^i)^T f^j$, $j = 1, \dots, k$. The first term in the solution at iteration k is

$$\frac{(\breve{T}_k^{-1})_{k,k}}{(\breve{T}_k^{-1})_{1,k}}\bar{v}_i^1 = -\frac{1}{\tilde{\eta}_k(\breve{T}_{k-1}^{-1})_{1,k-1}}\bar{v}_i^1 = \tilde{u}_k\bar{v}_i^1 = \tilde{p}_{1,k}(\lambda_i)\bar{v}_i^1,$$

as we already know. By using the expressions for u_j (adapted to $\breve{T}_k$) this can be written as

$$\tilde{u}_k\bar{v}_i^1 = (-1)^{k-1}\frac{\tilde{\delta}_1(\lambda_i)\cdots\tilde{\delta}_{k-1}(\lambda_i)}{\tilde{\eta}_2\cdots\tilde{\eta}_k}\bar{v}_i^1,$$

and we recover the expression we already have for $\tilde{p}_{1,k}$. We can bound the perturbation term corresponding to the solution of the recurrence for all iterations from 1 to k by

$$\|\tilde{U}_k\breve{T}_k^{-1}h^{(i)}\| \le \|\tilde{U}_k\|\,\|\breve{T}_k^{-1}\|\,\|h^{(i)}\|,$$

where $\tilde{U}_k = I - \frac{1}{(\breve{T}_k^{-1})_{1,k}}\breve{T}_k^{-1}e^k(e^1)^T$. But $\|\tilde{U}_k\|$ can be computed easily.

Lemma 4.22. *Let us denote*

$$\tilde{U}_k = \begin{pmatrix} 0 & 0 \\ x & I \end{pmatrix}.$$

Then,

$$\|\tilde{U}_k\| = \sqrt{1 + x^T x}.$$

Proof. The norm is the largest singular value, that is, the square root of the largest eigenvalue of $\tilde{U}_k^T\tilde{U}_k$. Therefore we are looking for eigenvalues μ and eigenvectors $(\,y \quad z\,)^T$ such that

$$\begin{pmatrix} x^T x & x^T \\ x & I \end{pmatrix}\begin{pmatrix} y \\ z \end{pmatrix} = \mu\begin{pmatrix} y \\ z \end{pmatrix}.$$

Writing the equations, if $\mu \neq 1$, we obtain

$$z = \frac{y}{\mu - 1} x.$$

Plugging this into the other equation gives, if $y \neq 0$,

$$\mu^2 - (1 + x^T x)\mu = 0.$$

Hence, we have two eigenvalues

$$\mu = 0, \quad \mu = 1 + x^T x.$$

If $\mu = 1$, then $y = 0$. We have a multiple eigenvalue equal to 1 with eigenvectors satisfying $x^T z = 0$. The matrix $\tilde{U}_k^T \tilde{U}_k$ has only three distinct eigenvalues $\mu = 0$, $\mu = 1$, and $\mu = 1 + x^T x$. □

In our application, we have

$$1 + x^T x = \sum_{l=1}^{k} (\tilde{u}_l)^2,$$

since $\tilde{u}_1 = 1$. The question now is to find a bound for $1 + x^T x$ that will give us a bound for $\|\tilde{U}_k\|$.

Lemma 4.23. *If $\bar{v}_i^1 \neq 0$, let $(\hat{v}_i^l)^2 = (\tilde{p}_{1,l}(\lambda_i)\bar{v}_i^1)^2$; then*

$$\sum_{l=1}^{k} (\tilde{u}_l)^2 = \frac{\sum_{l=1}^{k} (\hat{v}_i^l)^2}{(\bar{v}_i^1)^2}.$$

Moreover, there exists a constant C independent of k such that

$$\sqrt{\sum_{l=1}^{k} (\tilde{u}_l)^2} \leq \frac{\sqrt{k} C}{|\bar{v}_i^1|}.$$

Proof. We have $\tilde{u}_l = \tilde{p}_{1,l}(\lambda_i)$. Without perturbations (local errors) the projections of the Lanczos vectors are linked by $(\hat{v}_i^l)^2 = (\tilde{p}_l(\lambda_i)\bar{v}_i^1)^2$. The value $\hat{v}_i^l$ is what is obtained with the scalar three-term recurrence without rounding errors starting from $\bar{v}_i^1$ but using the computed coefficients. Hence,

$$\sum_{l=1}^{k} (\hat{v}_i^l)^2 = \left(\sum_{l=1}^{k} \tilde{p}_l(\lambda_i)^2 \right) (\bar{v}_i^1)^2,$$

and if $\bar{v}_i^1 \neq 0$, this implies

$$\sum_{l=1}^{k} (\tilde{u}_l)^2 = \frac{\sum_{l=1}^{k} (\hat{v}_i^l)^2}{(\bar{v}_i^1)^2}.$$

Since the normalizing coefficients $\tilde{\eta}_j$ were computed from vectors $\tilde{v}^l$, the norm of $\hat{v}^l$ is not exactly 1. But, from what we have established for the polynomials $\tilde{p}_{1,k}$, we have $|\hat{v}_i^l| \leq C$, where C is a constant independent of k, the growth of the polynomial being stopped by the multiple copies of the eigenvalues. It is likely that $C = 1$, but, unfortunately, we do not have a proof for that. Nevertheless, we obtain the last result of the lemma. □

When $\bar{v}_i^1 \neq 0$, the norm of $\tilde{U}_k$ is bounded by

$$\frac{\sqrt{k}C}{|\bar{v}_i^1|}.$$

Since

$$\bar{v}_i^k = \frac{(\breve{T}_k^{-1})_{k,k}}{(\breve{T}_k^{-1})_{1,k}}\bar{v}_i^1 + (e^k)^T \tilde{U}_k \breve{T}_k^{-1} h^{(i)},$$

where $h^{(i)} = (\, \bar{f}_i^1 \quad \cdots \quad \bar{f}_i^k \,)^T$, we have the following result.

Theorem 4.24. *The perturbation term is bounded by*

$$\|\tilde{U}_k \breve{T}_k^{-1} h^{(i)}\| \leq \frac{\sqrt{k}C}{|\bar{v}_i^1|} \frac{\|h^{(i)}\|}{\min_j |\theta_j^{(k)} - \lambda_i|}.$$

Proof. We have

$$\|\tilde{U}_k \breve{T}_k^{-1} h^{(i)}\| \leq \|\tilde{U}_k\| \frac{\|h^{(i)}\|}{\min_j |\theta_j^{(k)} - \lambda_i|}.$$

Using the bound for $\|\tilde{U}_k\|$ we obtain the result. □

We note that

$$\|h^{(i)}\|^2 = \sum_{j=1}^{k} (q^i, f^j)^2 \leq \sum_{j=1}^{k} \|f^j\|^2.$$

In fact, the term for $j = k$ can be omitted since by cancellation the term with $h_k^{(i)}$ does not appear in the solution. If $\bar{v}_i^1$ is large (close to 1), at the beginning the bound is small, but in this case there is a fast convergence of a Ritz value to λ_i. If the eigenvalue λ_i is not well approximated by a Ritz value, the perturbation term is small, provided $\bar{v}_i^1$ is not too small. The bound can be $O(1)$ if $\min_j |\theta_j^{(k)} - \lambda_i| = O(u)$. Of course, the fact that the bound is large does not prove that the perturbation term is large. But, here we have another proof of the fact that if there is no convergence of Ritz values, the perturbation terms stay small.

4.6 Some more examples

To complement our investigations we are going to look at some more numerical examples. In these experiments the initial vectors have equal components. The first figures are concerned with the Strakos30 example. Figure 4.29 shows for each i the minimum distance of Ritz values to the eigenvalue λ_i of A at iteration 30 with different values of *Digits* in variable precision and with double reorthogonalization. The matrices $\tilde{T}_{30}$ were computed in variable precision, but the Ritz values were computed using floating point double precision, which explains that the distance cannot be smaller than 10^{-16}. We see that up to $Digits = 32$ the smallest eigenvalues are not well approximated at iteration 30. In fact in this case we need $Digits = 35$ to get all the eigenvalues to double precision at iteration 30.

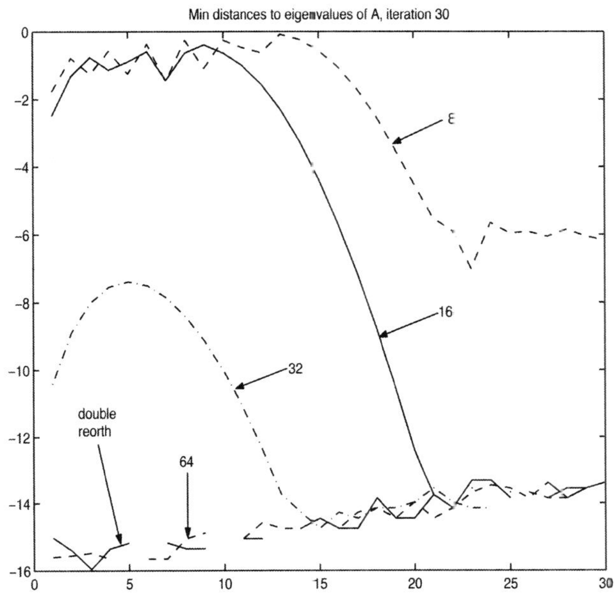

Figure 4.29. *Strakos*30, $\log_{10}$ *of min distance of Ritz values to the ith eigenvalue of A as a function of i, iteration* 30

Figure 4.30 shows $\log_{10}(|\bar{v}_{30}^k|)$ with different values of $Digits$. We see that $\log_{10}(|\bar{v}_{30}^k|)$ decreases to roughly $-Digits/2$ and then the growing perturbations are getting large enough to be the dominant part. The beginning of the computation is the same, whatever the precision. The period of the oscillations is, of course, different when changing the precision. The larger the number of decimal digits, the later we have the first increase of the component $\tilde{v}_{30}^k$. The multiple copies of the largest eigenvalue of A will appear at different iterations. When the precision is increasing, the number of iterations between two copies is increasing. So, for a given number of iterations, we get fewer and fewer copies. Figures 4.31 and 4.32 show the distances of the largest eigenvalue of A to Ritz values for $Digits = 32$ and 64 when they are smaller than a given threshold. For these figures the Ritz values were computed in extended precision. In Figure 4.32 the dashed curve is hidden behind the solid curve. Figure 4.33 is the same as Figure 4.30 but for $\log_{10}(|\bar{v}_1^k|)$, the first component of the projections of the Lanczos vectors. Whatever the precision of the computation, the same behavior is observed. It must be noticed that even when the precision is increased, sooner or later there will be the appearance of multiple copies of the eigenvalues if we iterate long enough. With reorthogonalization we have all the eigenvalues at iteration 30. If we continue the iterations, this is equivalent to restarting the computation.

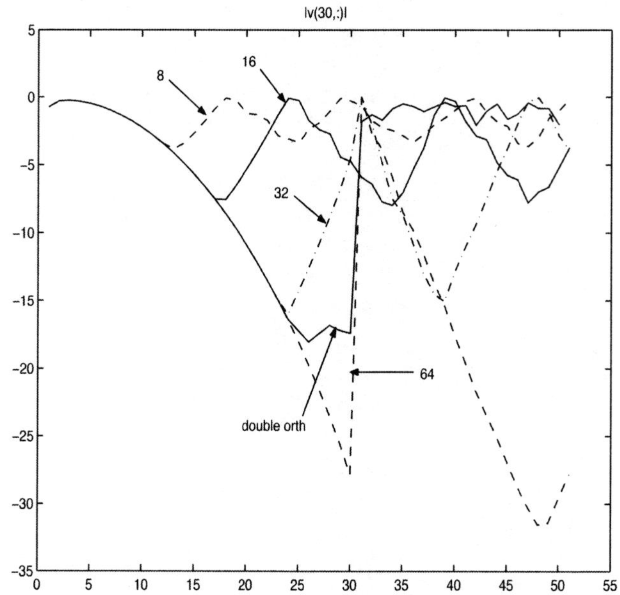

Figure 4.30. *Strakos*30, $\log_{10}(|\bar{v}_{30}^k|)$ *with different values of Digits*

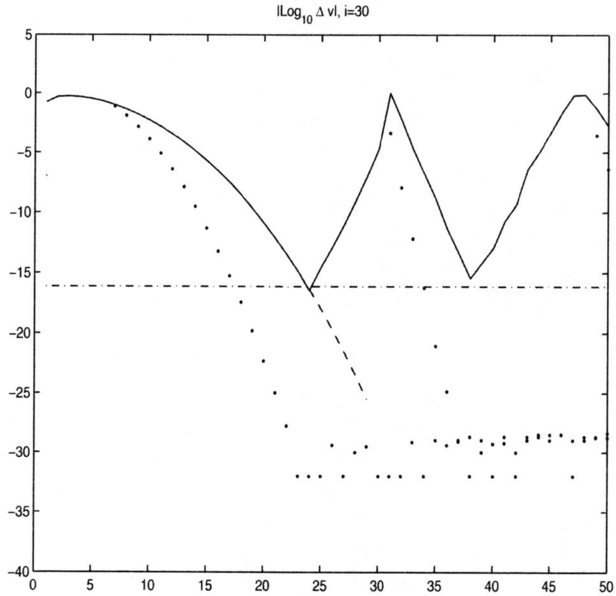

Figure 4.31. *Strakos*30, $\log_{10}(|\bar{v}_{30}^k|)$ *with* $Digits = 32$

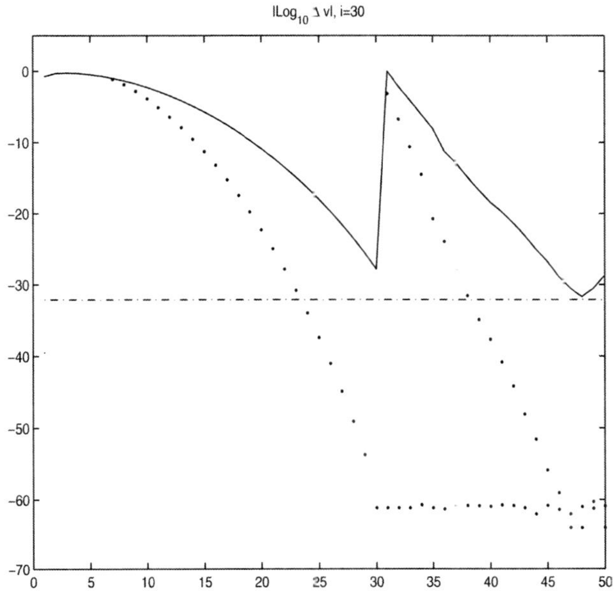

Figure 4.32. *Strakos*30, $\log_{10}(|\bar{v}_{30}^k|)$ *with* $Digits = 64$

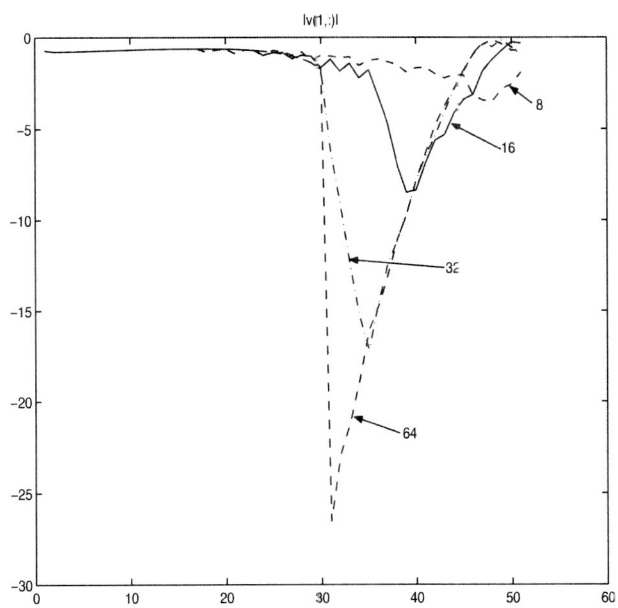

Figure 4.33. *Strakos*30, $\log_{10}(|\bar{v}_1^k|)$ *with different values of Digits*

Then we consider a case with multiple eigenvalues. We generate a diagonal Strakoš matrix with $\lambda_1 = 0.1$, $\lambda_n = 100$, $\rho = 0.9$, and $n = 20$. We duplicate the five largest eigenvalues to obtain a matrix of order 25 with five double eigenvalues. We denote this matrix as Strakos25d. Of course, as we said before, the Lanczos algorithm is unable to detect the double eigenvalues. It can only compute the 20 distinct eigenvalues. This is done in exact arithmetic at iteration 20. Figures 4.34 and 4.35 show the components 25 and 24 which correspond to the largest double eigenvalue. We see they are exactly the same. So, it seems that in this case the double eigenvalues do not cause any particular problem.

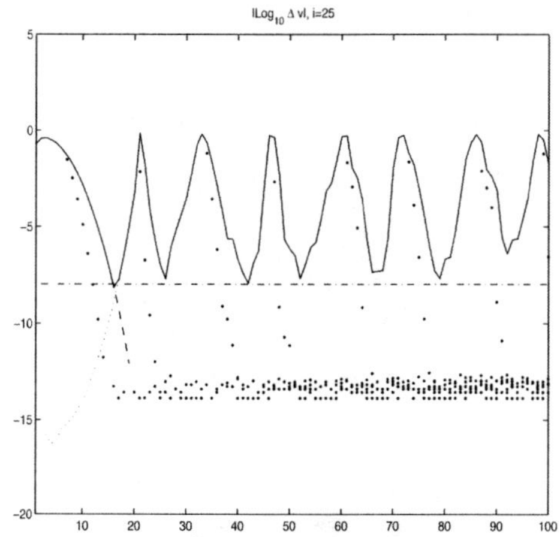

Figure 4.34. *Strakos25d, component* 25

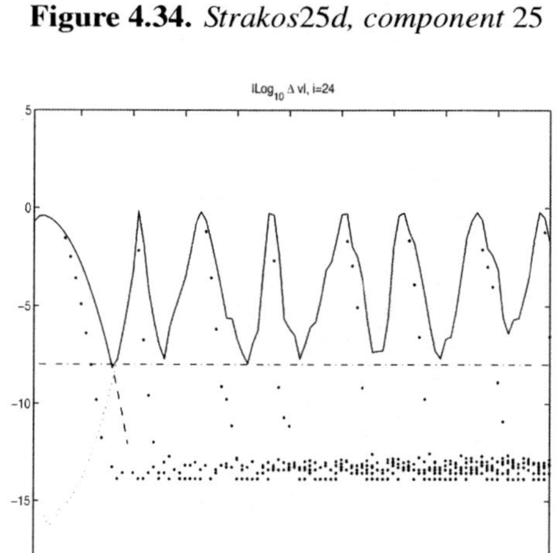

Figure 4.35. *Strakos25d, component* 24

The next example has a cluster of large eigenvalues. We first generate a diagonal Strakoš matrix with $\lambda_1 = 0.1$, $\lambda_n = 100$, $\rho = 0.9$, and $n = 20$ as before. Therefore, we have

$\lambda_{20} = 100$. The last 10 eigenvalues $\lambda_i, i = 21, \ldots, 30$, are $\lambda_i = 100 + (i - 20) * 0.001$. So, we have a cluster of 11 eigenvalues well separated from the rest of the spectrum. The first Ritz value to converge corresponds to λ_{19}, the largest eigenvalue outside the cluster. This is shown in Figure 4.36. The first component is given in Figure 4.37. Some components corresponding to the cluster are shown in Figures 4.38 to 4.41. Convergence is much more difficult for the eigenvalues in the cluster. We see that there are many small oscillations in the components of the Lanczos vectors. Ritz values stagnate for a few iterations and then start moving again. The global behavior is the same as what we have described before with multiple copies of each eigenvalue. All the eigenvalues in the cluster converge around iteration 60 whence all the other eigenvalues have converged by iteration 24.

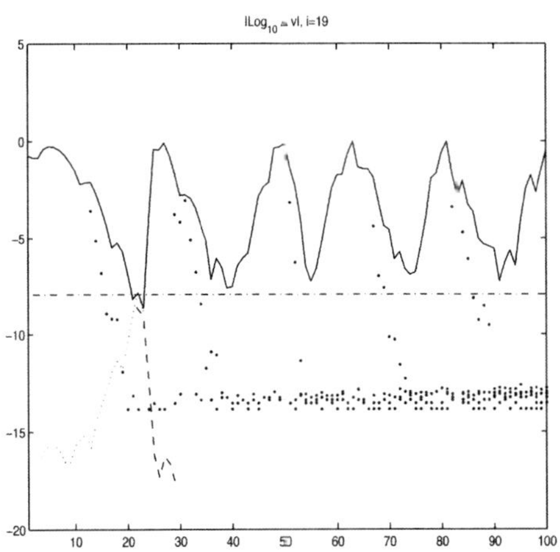

Figure 4.36. *Strakos with cluster,* $\log_{10}(|\bar{v}_{19}^k|)$

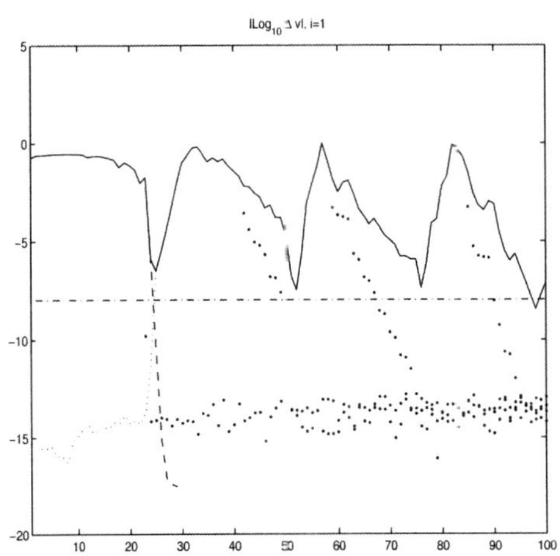

Figure 4.37. *Strakos with cluster,* $\log_{10}(|\bar{v}_1^k|)$

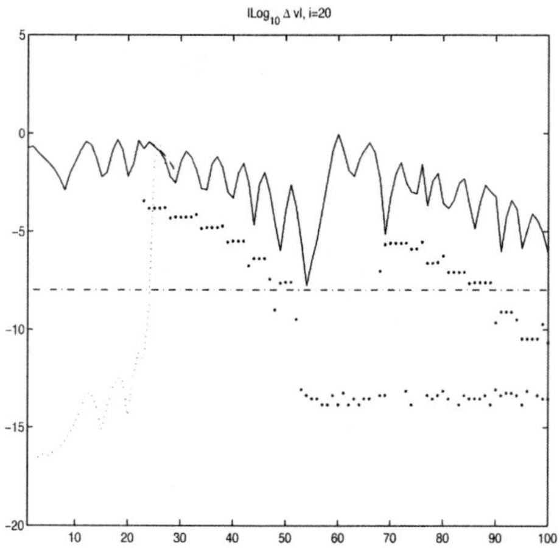

Figure 4.38. *Strakos with cluster,* $\log_{10}(|\bar{v}_{20}^k|)$

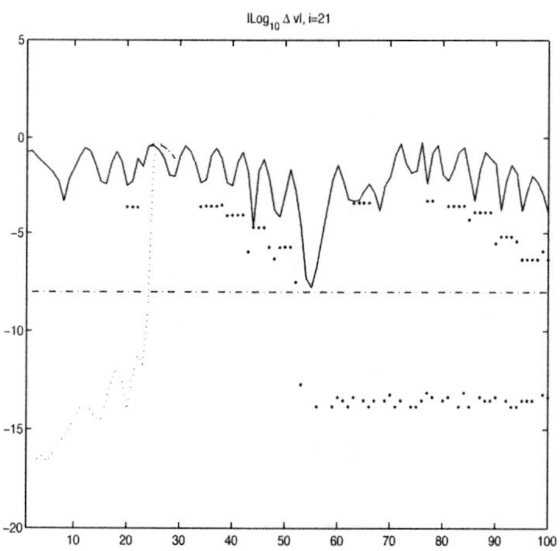

Figure 4.39. *Strakos with cluster,* $\log_{10}(|\bar{v}_{21}^k|)$

We now consider a matrix for which the Ritz values do not converge before step n. We have seen that bad convergence can be obtained by carefully selecting the initial vector for any matrix. However, some matrices also lead to slow convergence even with an initial vector with equal components on the eigenvectors of A. The matrix A arises from the discretization of the one-dimensional Poisson equation. This gives a tridiagonal matrix

$$A = \begin{pmatrix} 2 & -1 & & & \\ -1 & 2 & -1 & & \\ & \ddots & \ddots & \ddots & \\ & & -1 & 2 & -1 \\ & & & -1 & 2 \end{pmatrix}.$$

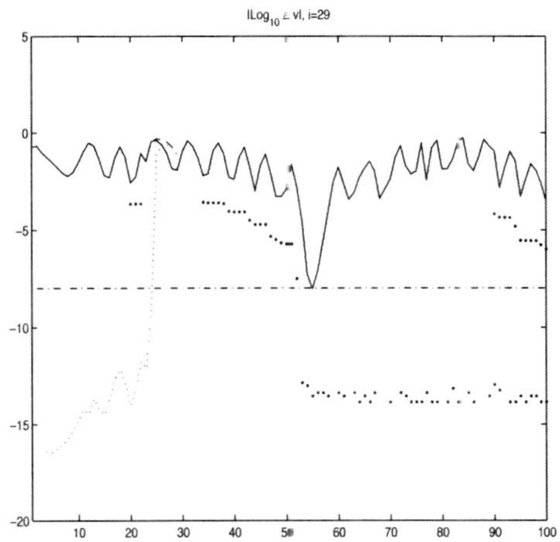

Figure 4.40. *Strakos with cluster,* $\log_{10}(|\bar{v}_{29}^k|)$

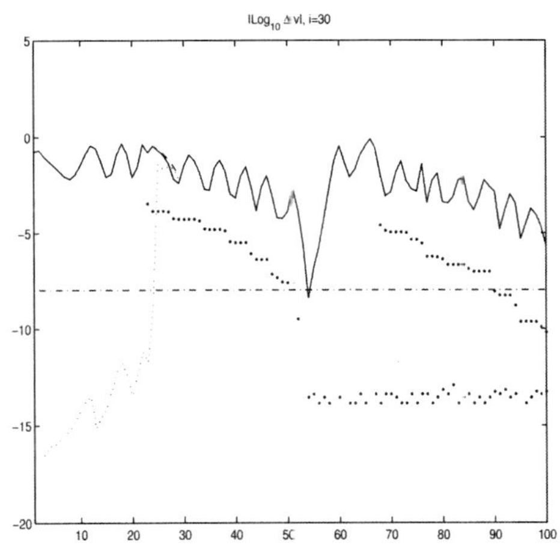

Figure 4.41. *Strakos with cluster,* $\log_{10}(|\bar{v}_{30}^k|)$

We denote this 20×20 matrix as Lap1d20. It may seem strange to apply the Lanczos algorithm to a tridiagonal matrix since it is a tridiagonalization algorithm. But, we should get the same results for any symmetric matrix having the same eigenvalue distribution as A. Figures 4.42 and 4.43 give the last and tenth components of the projections of the Lanczos vectors on the eigenvectors of A. The dashed curves for double reorthogonalization are behind the solid curves. We see that the differences (dotted) stay at roundoff level. The black dots show the distances of the closest Ritz value to the corresponding eigenvalue. Convergence (within roundoff) occurs at iteration 20. Figures 4.44 and 4.45 show the values of the polynomials $\tilde{p}_{j,k}(\lambda_i)$ for $j = 1, \ldots, 20$, $\lambda_i = \lambda_{20}$, and $\lambda_i = \lambda_{10}$. We see that they do not grow too much.

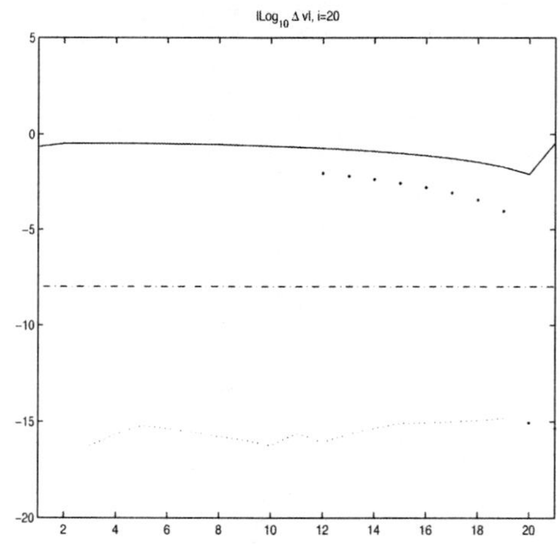

Figure 4.42. *Lap1d20, last component of the projection of Lanczos vectors*

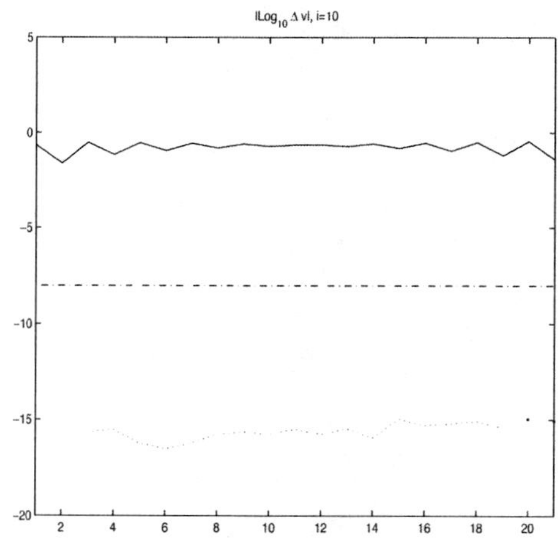

Figure 4.43. *Lap1d20 tenth component of the projection of Lanczos vectors*

4.7 Conclusions of this chapter

Based on the previous examples and what we have proved in this chapter and reviewed in Chapter 3, the generic scenario of how the Lanczos algorithm works in finite precision arithmetic is the following:

- The local roundoff is almost constant, being $O(u)$ (typically of order $u\|A\|$).

- The perturbation terms first start to grow when a Ritz value starts converging to an eigenvalue of A.

- Until one perturbation term is larger than $O(\sqrt{u})$ the finite precision coefficients are almost the same as the exact ones.

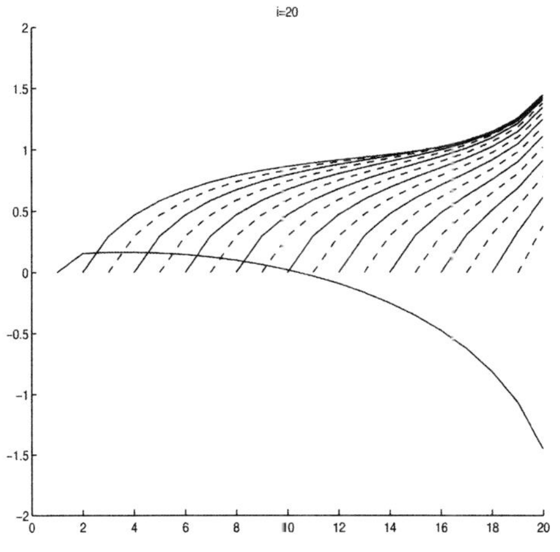

Figure 4.44. *Lap1d20,* $\tilde{p}_{j,k}(\lambda_{20})$ *for* $j = 1, \dots, 20$

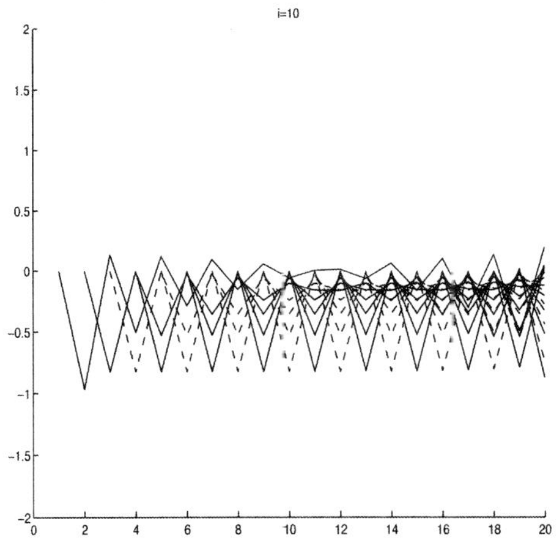

Figure 4.45. *Lap1d20,* $\tilde{p}_{j,k}(\lambda_{10})$ *for* $j = 1, \dots, 20$

- When the perturbation terms reach $O(\sqrt{u})$ for the first time the coefficients start to differ and the main term $|\tilde{p}_{1,k}\bar{v}_i^1|$ starts to increase instead of decreasing as in exact arithmetic, the perturbation terms generated at each iteration continue to grow.

- When the component of the projection of $\tilde{v}^k$ is close to 1, because of the normalization of the computed Lanczos vectors, the growth of the perturbations is stopped by generating a new copy of the eigenvalue of A. When the component is close to 1 the algorithm works again to compute the corresponding eigenvalue of A.

- There is a subsequent decrease of the component of $\tilde{v}^k$ and new perturbation terms generated during the last iterations are now growing and the process repeats again.

- When a Ritz value has converged once, the process of decrease and increase of the components of $\tilde{v}^k$ on the eigenvalues of A is more or less periodic. New copies of the Ritz values start to appear each time the component of the projection is close to 1 and then decreases to $O(\sqrt{u})$.

Of course, one can find specially designed matrices or initial vectors such that the behavior is slightly different. But in most practical cases, the above scenario is what happens in finite precision computations. In practical computations one does not know the eigenvectors of A and what the user sees is just the appearance of clusters of Ritz values.

Chapter 5

The CG algorithm in finite precision

This chapter considers the CG algorithm in finite precision arithmetic. We shall first look at some examples with different matrices from the Matrix Market [115] and the Strakos30 matrix. We shall study the problem of equivalence of the Lanczos and CG algorithms in finite precision arithmetic. CG is equivalent to the Lanczos algorithm but with different rounding error terms. From the results we have obtained for the Lanczos algorithm we shall easily obtain some results for CG. We shall consider three-term and two-term CG recurrences. Then we shall obtain results for the local orthogonality of the residuals and the local conjugacy of the descent directions. Finally, we shall give results on CG convergence showing that CG converges in finite precision arithmetic, provided the condition number of A is not too large relative to the roundoff unit u.

In this chapter we do not consider the exact arithmetic quantities. For simplicity we shall drop the tilde on the computed quantities.

5.1 Numerical examples

Let us start by looking at some examples. In all these computations, the right-hand side b is a random vector and the initial vector x^0 is zero. As before with the Lanczos algorithm we use the Strakos30 matrix. Figure 5.1 shows the loss of orthogonality of the computed CG residuals r^k. In Figure 5.2 we see the loss of conjugacy of the descent directions p^k. For the sake of clarity we have normalized the residuals as $r^k/\|r^k\|$ and the descent directions as $p^k/\sqrt{(p^k, Ap^k)}$. Both figures are more or less the same, showing that both vectors lose their orthogonality or conjugacy almost at the same time.

Bcsstk01 and Nos4 are matrices from the Matrix Market [115]. In Figures 5.3 to 5.6 we show the $\log_{10}$ of some norms. The dot-dashed curve is the l_2 norm of the error $x - x^k$, and the dashed curve is the A-norm of the error $(x - x^k, A(x - x^k))^{\frac{1}{2}}$. The "exact" solution x was computed using Gaussian elimination. The solid curve is the norm of the iteratively computed residual and the dotted curve is what is sometimes called the "true" residual $b - Ax^k$. We put the word "true" within quotes because we shall see that, in some examples, $b - Ax^k$ could be called the "wrong" residual. It is only the true residual regarding to the value of x^k and the computation of $b - Ax^k$ in finite precision. In what follows we

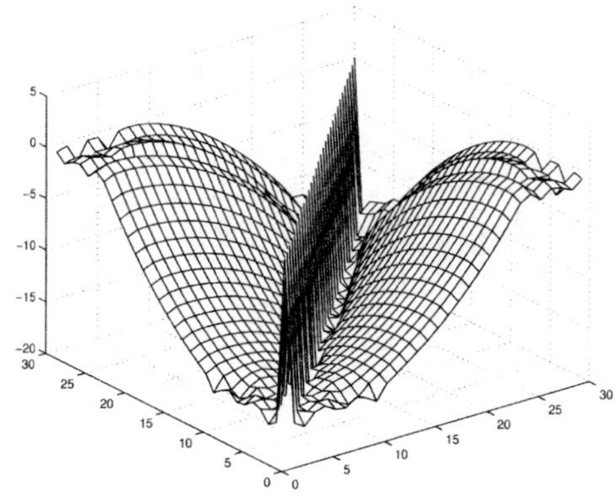

Figure 5.1. *Strakos*30, $\log_{10}(|(r^i, r^j)|)$ *normalized*

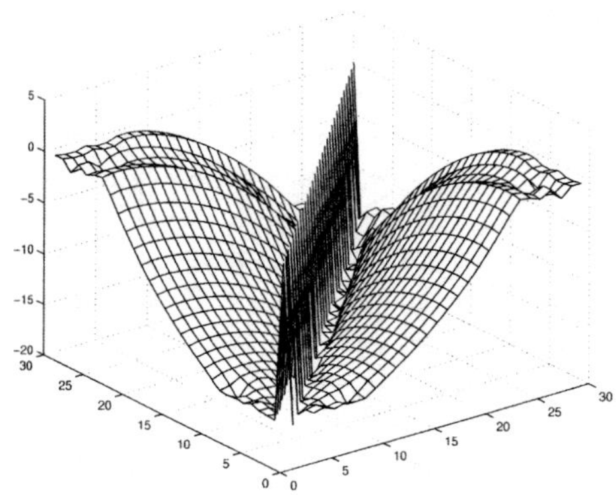

Figure 5.2. *Strakos*30, $\log_{10}(|(p^i, Ap^j)|)$ *normalized*

shall denote it as the x-residual and the one computed iteratively will be called the iterative residual or simply the residual. In fact what we could have access to numerically is not the exact value of the x-residual but $fl(b - Ax^k)$. We have seen before that

$$fl(b - Ax) = b - Ax + Cu + O(u^2), \quad |C| \leq (m+1)|A|\,|x| + |b|.$$

Even if x is the exact solution the floating point result is of order Cu.

We see from Figures 5.3 to 5.6 that we can have quite different behaviors. For the Strakos30 matrix all the norms are close to each other. The norm of the residual is sometimes larger and sometimes smaller than the l_2 norm of the error and it exhibits only small oscillations. The decrease of the error norms is monotone (except after stagnation), as the theory says it should be. We remark that even though the order of the matrix is 30 we have to do more iterations to reach the maximum attainable accuracy, which is the level where some norms stagnate. After iteration 45 the norms of the error and the x-residual stagnate, while the norm of the iterative residual continues to decrease.

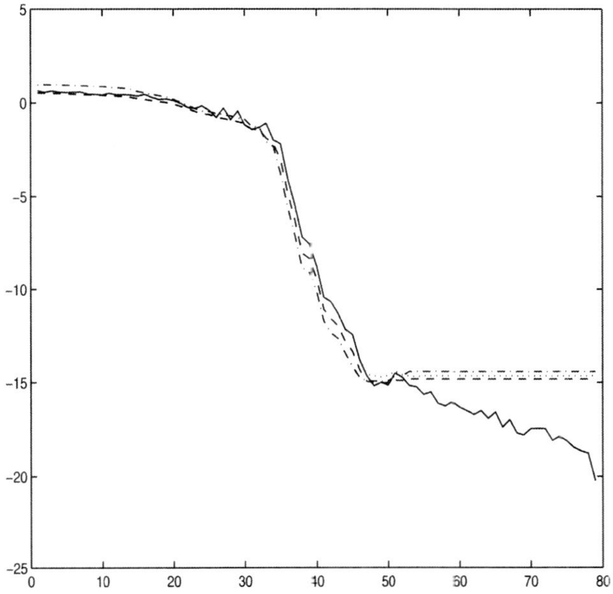

Figure 5.3. *Strakos30, norms of residuals and errors*

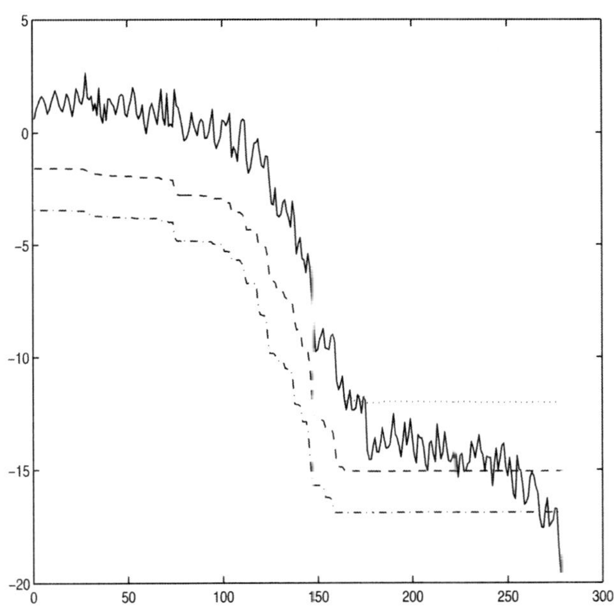

Figure 5.4. *Bcsstk01, norms of residuals and errors*

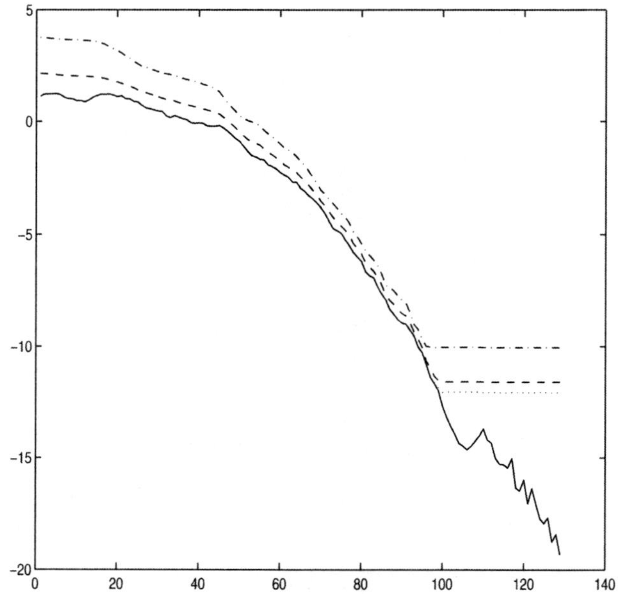

Figure 5.5. *Nos4, norms of residuals and errors*

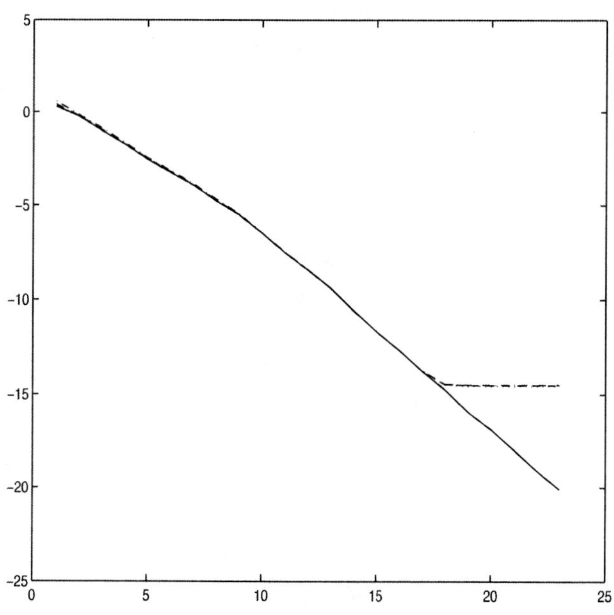

Figure 5.6. *Lap–ic, norms of residuals and errors*

For the matrix Bcsstk01 of order 48, the norm of the residual oscillates and is larger than the error norms. The norm of this matrix is 3 10^9 (which is the same as $\| \, |A| \, \|$) and the smallest eigenvalue is 3417, so the condition number is 8.8 10^5. To obtain small error or residual norms we have to do many more iterations than the order of the matrix. If we base the stopping criterion on the norm of the residual, we do more iterations than necessary since the norms of the error are smaller. We note also that the level of stagnation is different for the residual and the error norms.

Matrix Nos4 of order 100 is an example where the norm of the residual is smaller than the error norms. The norm of this matrix is 0.849 (the norm $\| \, |A| \, \|$ is 0.8632) and the smallest eigenvalue is 5.4 10^{-4}, so the condition number is 1.58 10^3.

All the previous examples are difficult cases where convergence is quite slow. Fortunately, this is not always the case, particularly when we are using a good preconditioner. The last example, which we denote by Lap–ic, is a full matrix of order 49 obtained by starting from the discretization of the Poisson equation in a square with a 7×7 Cartesian mesh using finite differences. We compute the incomplete Cholesky factorization of A with no fill-in (see, for example, [120]), which gives a symmetric positive definite matrix M. The matrix we consider is $M^{-\frac{1}{2}} A M^{-\frac{1}{2}}$. It has the same spectrum as the matrix A preconditioned by an incomplete Cholesky factorization. The norm is 1.2, the smallest eigenvalue is 0.368, the condition number is 3.28, and $\| \, |A| \, \| = 1.57$. We see in Figure 5.6 that all the curves for the norms of errors and residuals are almost the same until stagnation and the convergence is quite fast. Moreover, there are no oscillations of the residual norm.

We have seen that in exact arithmetic the Lanczos and CG algorithms are equivalent, the CG residuals being related to the Lanczos vectors. The questions related to finite precision arithmetic we would like to consider are the following:

- What happens to this relationship in finite precision?
- What is the origin of the differences between the iterative residuals and the x-residuals computed as $fl(b - Ax^k)$?
- How is the convergence delayed by rounding errors?
- What are the differences between the different versions of CG?

CG in finite precision has been mostly studied by Greenbaum (see [75], [76], [77], and her collaboration with Strakoš [81]). See also Bollen [11], [12] and Sleijpen, van der Vorst, and Modersitski [174].

5.2 Relations between CG and Lanczos in finite precision

Since we know the equivalence of CG and Lanczos algorithms in exact arithmetic, we start from CG and see what we obtain for the residuals since this is easier than the other way around—starting from the finite precision Lanczos algorithm. We write the CG relations in finite precision as

$$r^{k+1} = r^k - \gamma_k A p^k + \delta_r^k,$$

$$p^k = r^k + \beta_k p^{k-1} + \delta_p^{k-1}.$$

We have seen in Chapter 3 that we have

$$|\delta_r^k| \le u[(m+2)\gamma_k |A|\,|p^k| + |r^k|\,],$$

$$|\delta_p^{k-1}| \le u[\,|r^k| + 2\beta_k |p^{k-1}|\,].$$

The computed coefficients can be written as

$$\gamma_k = \frac{\|r^k\|^2}{(p^k, Ap^k)} + \delta_\gamma^k,$$

$$\beta_k = \frac{\|r^k\|^2}{\|r^{k-1}\|^2} + \delta_\beta^k,$$

with

$$|\delta_\gamma^k| \le \frac{\|r^k\|^2}{(p^k, Ap^k)} |C_\gamma| u + O(u^2), \quad |C_\gamma| \le (m+n)\frac{(|A|\,|p^k|, |p^k|)}{(Ap^k, p^k)} + n + 1,$$

$$|\delta_\beta^k| \le \frac{\|r^k\|^2}{\|r^{k-1}\|^2} |C_\beta| u + O(u^2), \quad |C_\beta| \le 1 + 2n.$$

Clearly these perturbations can be absorbed in δ_r^k and δ_p^{k-1}. Hence, we can consider the coefficients as computed exactly from the computed quantities. In this case we have to modify accordingly the bounds such that

$$|\delta_r^k| \le u[(m+2)\gamma_k |A|\,|p^k| + |r^k|] + |\delta_\gamma^k|\,|A|\,|p^k|,$$

$$|\delta_p^{k-1}| \le u[|r^k| + 2\beta_k |p^{k-1}|] + |\delta_\beta^k|\,|p^{k-1}|.$$

Doing this will be useful since in some proofs we have to compute the product of coefficients. This is much easier if we have the analytic expression without perturbations. It is also interesting to bound the norms of the perturbation terms.

Proposition 5.1.

$$\|\delta_p^{k-1}\| \le u\|r^k\| \left[1 + (3+2n)\frac{\|p^{k-1}\|}{\|r^{k-1}\|^2}\|r^k\|\right] + O(u^2),$$

$$\|\delta_r^k\| \le u\|r^k\| \left[1 + \bar{\kappa}(A)\frac{\|r^k\|}{\|p^k\|}\{(m+2) + (m+n)\bar{\kappa}(A) + n + 1\}\right] + O(u^2),$$

where $\bar{\kappa}(A) = \|\,|A|\,\|/\lambda_1$.

Proof. By inserting the values of the coefficients we have

$$\|\delta_r^k\| \le u\left[(m+2)\frac{\|r^k\|^2}{(p^k, Ap^k)}\|\,|A|\,\|\,\|p^k\| + \|r^k\| \right.$$
$$\left. + \frac{\|r^k\|^2}{(p^k, Ap^k)}\left\{(m+n)\frac{\|\,|A|\,\|\,\|p^k\|^2}{(p^k, Ap^k)} + n + 1\right\}\|\,|A|\,\|\,\|p^k\|\right].$$

We put $\|r^k\|$ in factor, introduce wherever needed ratios $\|r^k\|/\|p^k\|$, and use the fact that

$$\frac{(p^k, Ap^k)}{\|p^k\|^2} \geq \lambda_1.$$

For the other term, we have

$$\|\delta_p^{k-1}\| \leq u \left[\|r^k\| + 2\frac{\|r^k\|^2}{\|r^{k-1}\|^2}\|p^{k-1}\| \right.$$
$$\left. + \ (1+2n)\frac{\|r^k\|^2}{\|r^{k-1}\|^2}\|p^{k-1}\| \right] + O(u^2),$$

which proves the result. □

This result shows that the perturbation bounds can be large if $\|r^k\|$ is large or if $\bar{\kappa}(A)$ is large relative to u. For further reference we denote

$$\|\delta_p^{k-1}\| \leq u\|r^k\|(1 + C_p^{k-1}) + O(u^2),$$
$$\|\delta_r^k\| \leq u\|r^k\|(1 + C_r^k) + O(u^2).$$

The important thing to notice is that there is a factor $u\|r^k\|$ in the bounds for the norms of the local roundoff terms. The coefficients above are bounded provided the ratios $\|p^k\|/\|r^k\|$ and $\|r^k\|/\|p^k\|$ are bounded as well as the ratios of residual norms. Let us look at the local errors for the residual and the descent directions for the Strakos30 example. This is shown in Figure 5.7, where the solid curve is the local error for the residual and the dashed curve the local error for the descent direction. As before for the Lanczos algorithm, one may argue that the local error can be larger if we do not have a diagonal matrix. When the Strakoš diagonal matrix is symmetrically multiplied by an orthogonal matrix which gives a full matrix with the same spectrum, the norms are a little larger but not by much. The decrease of the norm of the local error is directly linked to the decrease of the residual norm, as shown in Figure 5.8, where we have the $\log_{10}$ of the local error (solid) and the residual norm multiplied by 10^{-14} (dotted). The computations of local errors have been done in variable precision with $Digits = 16$ and 32.

We can eliminate p^k from the CG relations to obtain a three-term recurrence for r^k. To do this we derive expressions for Ap^k using both equations:

$$Ap^k = \frac{r^k - r^{k+1}}{\gamma_k} + \frac{1}{\gamma_k}\delta_r^k,$$
$$Ap^k = Ar^k + \beta_k Ap^{k-1} + A\delta_p^{k-1}.$$

Equating these two expressions and replacing Ap^{k-1} by its value from the first equation at iteration k, we obtain

$$r^{k+1} = -\gamma_k Ar^k + \left(1 + \frac{\gamma_k \beta_k}{\gamma_{k-1}}\right) r^k - \frac{\gamma_k \beta_k}{\gamma_{k-1}} r^{k-1} + \delta_r^k - \frac{\gamma_k \beta_k}{\gamma_{k-1}} \delta_r^{k-1} - \gamma_k A \delta_p^{k-1}.$$

The initial conditions are r^0 given (as $fl(b - Ax^0)$) and $p^0 = r^0$, which is an exact relation. Therefore

$$r^1 = r^0 - \gamma_0 Ar^0 + \delta_r^0.$$

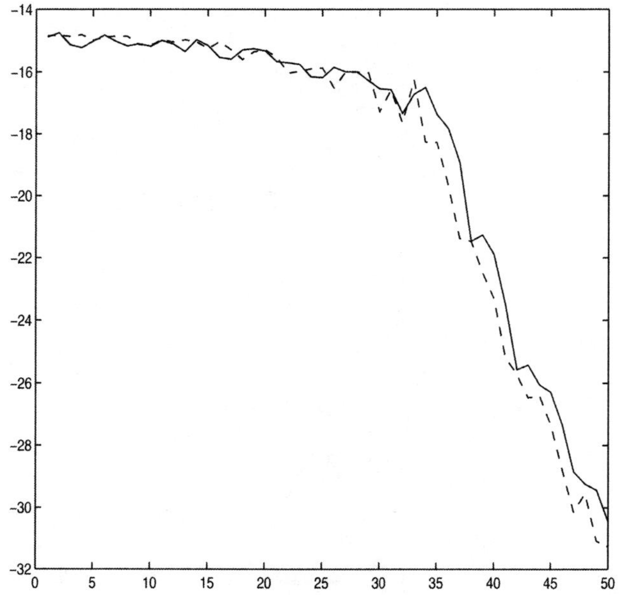

Figure 5.7. *Strakos*30, $\log_{10}$ *of norms of local errors for* r^k *(solid) and* p^k *(dashed)*

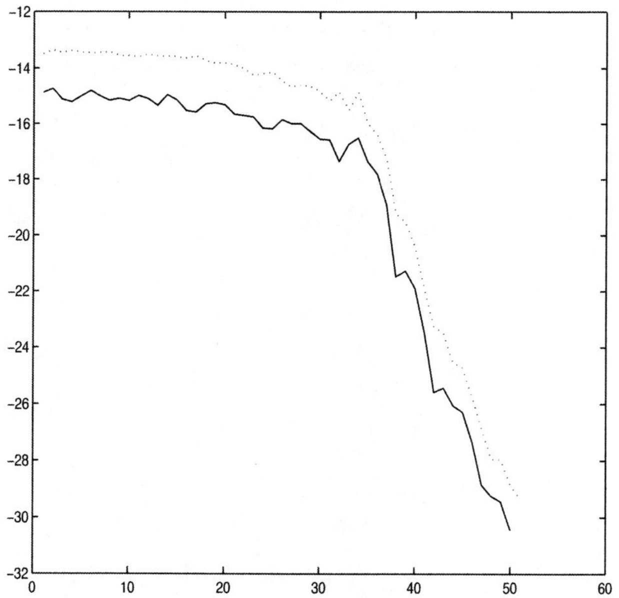

Figure 5.8. *Strakos*30, $\log_{10}$ *of norm of local error for* r^k *(solid) and* 10^{-14} *the norm of the residual (dotted)*

Proposition 5.2. *Let*

$$\bar{\delta}_r^k = \delta_r^k - \frac{\gamma_k \beta_k}{\gamma_{k-1}} \delta_r^{k-1} - \gamma_k A \delta_p^{k-1}, \quad k \geq 1, \quad \bar{\delta}_r^0 = \delta_r^0.$$

The equivalent three-term recurrence for CG in finite precision is written as

$$r^{k+1} = -\gamma_k A r^k + \left(1 + \frac{\gamma_k \beta_k}{\gamma_{k-1}}\right) r^k - \frac{\gamma_k \beta_k}{\gamma_{k-1}} r^{k-1} + \bar{\delta}_r^k, \quad k \geq 1,$$

$$r^1 = r^0 - \gamma_0 A r^0 + \bar{\delta}_r^0.$$

The perturbation term can be bounded as

$$\|\bar{\delta}_r^k\| \leq u \|r^k\| \left[1 + C_r^k + \frac{\|r^k\|^3}{\|r^{k-1}\|^3} \frac{\|p^{k-1}\|^2}{\|p^k\|^2} \kappa(A)(1 + C_r^{k-1}) + \frac{\|r^k\|^2}{\|p^k\|^2} \kappa(A)(1 + C_p^{k-1})\right].$$

Figure 5.9 shows the $\log_{10}$ of the norm of $\bar{\delta}_r^k$ as a function of k. It has more or less the same behavior as the local errors for r^k and p^k, except with more pronounced peaks between iterations 30 and 35. These peaks are probably caused by oscillations in the coefficients γ_k and β_k (particularly β_k) around iteration 30.

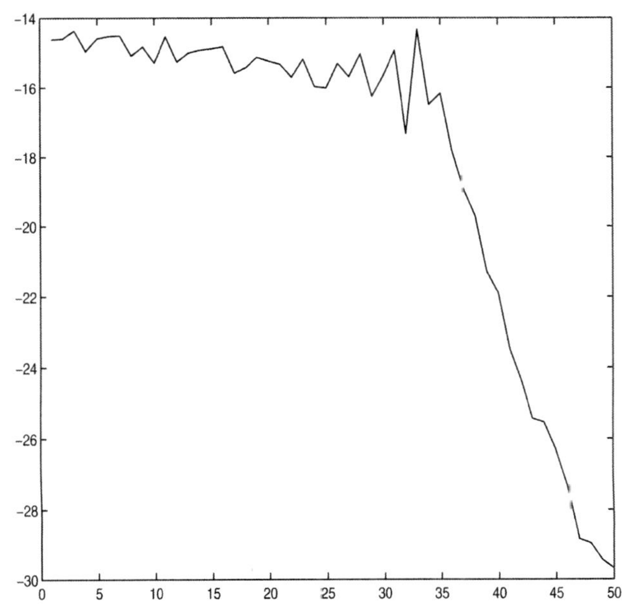

Figure 5.9. *Strakos30*, $\log_{10}$ *of norm of* $\bar{\delta}_r^k$

From this we can obtain the equivalent of the Lanczos vectors that we call the CG-Lanczos vectors and we denote as w^k,

$$w^{k+1} = (-1)^k \frac{r^k}{\|r^k\|}.$$

We introduce this notation since these vectors may be different from the Lanczos vectors because the rounding errors are different. Although the notations are different, the next result is the same as that obtained by Greenbaum [76], [79]; see Chapter 3.

Theorem 5.3. *The equivalent three-term recurrence for the CG-Lanczos vectors is*

$$\eta_{k+1} w^{k+1} = A w^k - \alpha_k w^k - \eta_k w^{k-1} + (-1)^k \frac{\bar{\delta}_r^{k-1}}{\gamma_{k-1} \|r^{k-1}\|},$$

$$\eta_{k+1} = \frac{\sqrt{\beta_k}}{\gamma_{k-1}}, \quad \alpha_k = \frac{1}{\gamma_{k-1}} + \frac{\beta_{k-1}}{\gamma_{k-2}}, \quad \alpha_1 = \frac{1}{\gamma_0},$$

with

$$\gamma_k = \frac{\|r^k\|^2}{(Ap^k, p^k)}, \quad \beta_k = \frac{\|r^k\|^2}{\|r^{k-1}\|^2}.$$

The initial vector is given as $w^1 = r^0 / \|r^0\|$ *and*

$$\eta_2 w^2 = A w^1 - \alpha_1 w^1 - \frac{\bar{\delta}_r^0}{\gamma_0 \|r^0\|}.$$

If we want to refer to the computed coefficients, we still have the same relationship, but we have to modify the perturbation term as

$$(-1)^k \frac{\bar{\delta}_r^{k-1}}{\gamma_{k-1} \|r^{k-1}\|} + (-1)^{k+1} \frac{\delta_\eta^{k+1} r^k}{\|r^k\|} - \delta_\alpha^k w^k - \delta_\eta^k w^{k-1},$$

where $\bar{\delta}_r^{k-1}$ *does not involve the terms arising from the coefficients and*

$$\delta_\eta^{k+1} = -\frac{1}{2} \frac{\delta_\beta^k}{\gamma_{k-1}} + \frac{\sqrt{\beta_k} \delta_\gamma^{k-1}}{\gamma_{k-1}^2} + \frac{1}{2} \frac{\delta_\beta^k \delta_\gamma^{k-1}}{\gamma_{k-1}^2},$$

$$\delta_\alpha^k = \frac{\beta_{k-1} \delta_\gamma^{k-2}}{\gamma_{k-2}^2} - \frac{\delta_\beta^{k-1}}{\gamma_{k-2}} - \frac{\delta_\beta^{k-1} \delta_\gamma^{k-2}}{\gamma_{k-2}^2}, \quad k \geq 2.$$

Proof. We write the recurrence for r^k,

$$r^k = -\gamma_{k-1} A r^{k-1} + \left(1 + \frac{\gamma_{k-1} \beta_{k-1}}{\gamma_{k-2}}\right) r^{k-1} - \frac{\gamma_{k-1} \beta_{k-1}}{\gamma_{k-2}} r^{k-2} + \bar{\delta}_r^{k-1}.$$

This translates into

$$(-1)^k \|r^k\| w^{k+1} = (-1)^k \|r^{k-1}\| \gamma_{k-1} A w^k + \left(1 + \frac{\gamma_{k-1} \beta_{k-1}}{\gamma_{k-2}}\right) (-1)^{k-1} \|r^{k-1}\| w^k$$
$$+ \frac{\gamma_{k-1} \beta_{k-1}}{\gamma_{k-2}} (-1)^{k-1} \|r^{k-2}\| w^{k-1} + \bar{\delta}_r^{k-1}.$$

Dividing by $(-1)^k \|r^k\|$,

$$w^{k+1} = \frac{\|r^{k-1}\|}{\|r^k\|}\gamma_{k-1} A w^k - \left(1 + \frac{\gamma_{k-1}\beta_{k-1}}{\gamma_{k-2}}\right)\frac{\|r^{k-1}\|}{\|r^k\|} w^k - \frac{\gamma_{k-1}\beta_{k-1}}{\gamma_{k-2}}\frac{\|r^{k-2}\|}{\|r^k\|} w^{k-1} + (-1)^k \frac{\bar{\delta}_r^{k-1}}{\|r^k\|}.$$

Remember that we have

$$\frac{1}{\sqrt{\beta_k}} = \frac{\|r^{k-1}\|}{\|r^k\|},$$

since the roundoff errors have been put in the δ term. By defining

$$\eta_{k+1} = \frac{\sqrt{\beta_k}}{\gamma_{k-1}}, \quad \alpha_k = \frac{1}{\gamma_{k-1}} + \frac{\beta_{k-1}}{\gamma_{k-2}},$$

we have

$$\eta_{k+1} w^{k+1} = A w^k - \alpha_k w^k - \eta_k w^{k-1} + (-1)^k \frac{\bar{\delta}_r^{k-1}}{\gamma_{k-1}\|r^{k-1}\|}. \quad \square$$

As in the Lanczos algorithm, we shall denote by T_k the matrix of the CG-Lanczos coefficients. Conversely the Lanczos algorithm is equivalent to CG with different rounding errors. Figure 5.10 shows (dashed curve) the $\log_{10}$ of the roundoff error associated with the three-term recurrence for w^{k+1}. It is interesting to compare this with the roundoff error in

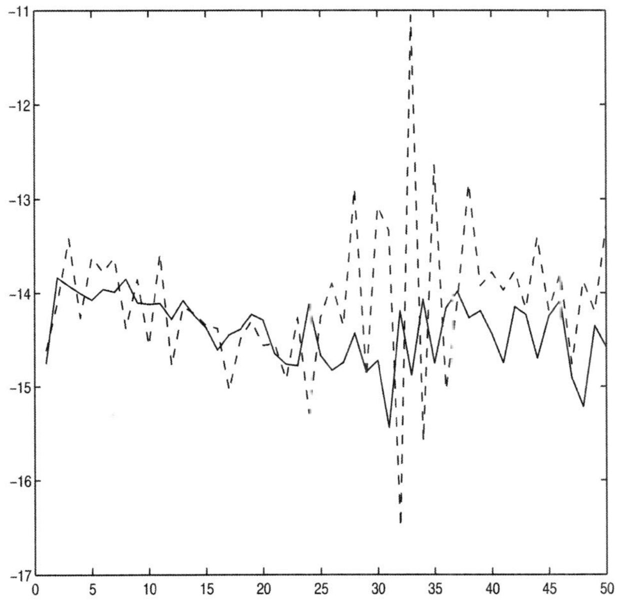

Figure 5.10. *Strakos30,* $\log_{10}$ *of norms of* $\frac{\bar{\delta}_r^{k-1}}{\gamma_{k-1}\|r^{k-1}\|}$ *(dashed) and the Lanczos algorithm local error (solid)*

the Lanczos algorithm, which is the solid curve. We see that we have different rounding errors, but the magnitudes are the same except for large peaks around iteration 30.

We see that the equation we obtain for the CG-Lanczos vectors is formally the same as the one for the Lanczos vectors. However, the roundoff terms are different. It is likely that the overall behavior will be the same with slight variations depending on the values of the roundoff terms.

5.3 Study of the CG three-term recurrences

We have at least three possibilities for studying CG in finite precision. We can analyze the equivalent three-term recurrence for w^k or the one for r^k (recurrences similar to this one have been studied by Gutknecht and Strakoš [86]), and finally we may consider the two-term recurrences for r^k and p^k.

For the three-term recurrence of w^k we can directly apply the results we have proved (Theorem 4.6) for the Lanczos recurrences. We consider the components of the projections of the CG-Lanczos vectors w^k on the eigenvectors of A that we denote by $\bar{w}^k$.

Theorem 5.4. *Let*

$$\tilde{\delta}^k = Q^T\left((-1)^k \frac{\delta_r^{k-1}}{\gamma_{k-1}\|r^{k-1}\|}\right),$$

j given, and $p_{j,k}$ be the polynomial determined by

$$p_{j,j-1} = 0, \quad p_{j,j} = 1,$$

$$\eta_{k+1} p_{j,k+1}(\lambda) = (\lambda - \alpha_k) p_{j,k}(\lambda) - \eta_k p_{j,k-1}(\lambda), \quad k = j, \dots.$$

The solution of the perturbed recurrence

$$\eta_{k+1}\bar{w}_i^{k+1} = (\lambda_i - \alpha_k)\bar{w}_i^k - \eta_k \bar{w}_i^{k-1} + \tilde{\delta}_i^k$$

starting from $\bar{w}_i^0 = 0$ and $\bar{w}_i^1$ is given by

$$\bar{w}_i^{k+1} = p_{1,k+1}(\lambda_i)\bar{w}_i^1 + \sum_{l=1}^{k} p_{l+1,k+1}(\lambda_i)\frac{\tilde{\delta}_i^l}{\eta_{l+1}}.$$

We could have as well used the results we proved when formulating the recurrence as a triangular system showing that there is no growth of any perturbation term as long as there is no Ritz value convergence.

Figure 5.11 shows $\bar{w}_{30}^k$ for CG (solid) and $\bar{v}_{30}^k$ for the Lanczos algorithm, dashed for Paige's version, and dot-dashed for the basic formula. Notice that what we see in the figure is not w^k but the computed w^k which is obtained by normalizing r^{k-1}. Of course this introduces some small errors which are not significant in this context.

We see that as long as the perturbation terms are negligible (iteration 18) the curves are almost the same. Since the local roundoff errors are different, the curves then start to differ, although the phenomenology is the same. The oscillations are not exactly in phase, leading

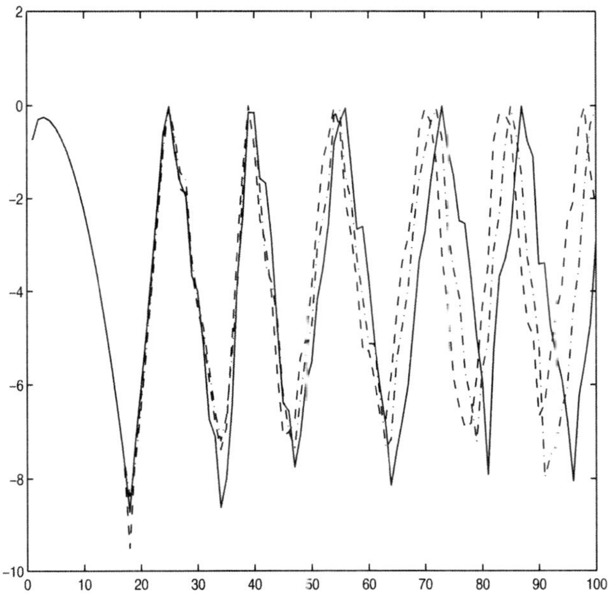

Figure 5.11. *Strakos30*, $\log_{10}$ *of* $|v_{30}^k|$ *for CG (solid), Paige's Lanczos (dashed), and Lanczos (dot-dashed)*

to eventually obtaining multiple copies of the largest eigenvalue at different iterations. The nonzero elements of the matrix T_{30} are quite different for these three computations. Nevertheless the Ritz values (which are the important quantities) are almost the same. Theorem 5.4 leads to the following result.

Theorem 5.5. *Let*

$$\bar{\delta}^k = (-1)^k \frac{\bar{\delta}_r^{k-1}}{\gamma_{k-1}\|r^{k-1}\|},$$

$$w^{k+1} = p_{1,k+1}(A)w^1 + \sum_{l=1}^{k} p_{l+1,k+1}(A)\frac{\bar{\delta}^l}{\eta_{l+1}}.$$

If we refer to the computed coefficients, we have to modify accordingly the perturbation term. The study of the behavior of the solution w^k is exactly the same as what we have done for the Lanczos algorithm except that the roundoff terms are different. We have seen in Figure 5.10 that they can sometimes be larger. When we first have convergence of a Ritz value to an eigenvalue of A, $\bar{w}_i^k$, the corresponding component of w^k decreases to $\sqrt{u}$ and then oscillates, producing multiple copies of that eigenvalue. From Theorem 5.5 we can obtain the solution of the recurrence for r^k, since by definition

$$w^{k+1} = (-1)^k \frac{r^k}{\|r^k\|}.$$

Theorem 5.6.

$$r^k = (-1)^k \frac{\|r^k\|}{\|r^0\|} p_{1,k+1}(A) r^0 + (-1)^k \|r^k\| \sum_{l=1}^{k} p_{l+1,k+1}(A) \frac{\bar{\delta}^l}{\eta_{l+1}},$$

where the polynomials are defined in Theorem 5.4.

If we go back to the definition of the perturbation term, we have

$$r^k = (-1)^k \frac{\|r^k\|}{\|r^0\|} p_{1,k+1}(A) r^0 + (-1)^k \sum_{l=1}^{k} p_{l+1,k+1}(A) \frac{\bar{\delta}_r^{l-1} \|r^k\|}{\|r^l\|}.$$

The first term in the solution is what we have without the local roundoff perturbation (although in exact arithmetic the coefficients are different), as we have seen when studying the three-term form of CG in Chapter 2. It is interesting to note that even though there are terms $1/\|r^l\|$ in the sum, there is a $\|r^k\|$ in front of the sum. After simplification the term factor of the polynomial in the sum is

$$\gamma_{l-1} \frac{\|r^k\|}{\|r^l\|} \left(\delta_r^{l-1} - \frac{\beta_{l-1}}{\gamma_{l-2}} \delta_r^{l-2} - A\delta_p^{l-2} \right).$$

Its absolute value is bounded by

$$u\gamma_{l-1} \frac{\|r^k\|}{\|r^l\|} \left[\|r^l\|(1 + C_r^{l-1}) + \frac{\|r^{l-1}\|^2}{\|r^{l-2}\|} \frac{\|p^{l-2}\|^2}{\|r^{l-2}\|^2} \|A\|(1 + C_r^{l-2}) + \|A\| \, \|r^{l-1}\|(1 + C_p^{l-2}) \right].$$

Remember that the C_r^k and C_p^k terms are bounded by ratios of norms of p^k and r^k at different iterations. We are also interested in the components of r^k on the eigenvectors of A. We have

$$(Q^T r^k)_i = \bar{r}_i^k = (-1)^k \frac{\|r^k\|}{\|r^0\|} p_{1,k+1}(\lambda_i) \bar{r}_i^0 + (-1)^k \|r^k\| \sum_{l=1}^{k} p_{l+1,k+1}(\lambda_i) \frac{\tilde{\delta}_i^l}{\eta_{l+1}}.$$

The second term in the right-hand side can be written as

$$(-1)^k \sum_{l=1}^{k} p_{l+1,k+1}(\lambda_i) \gamma_{l-1} \frac{\|r^k\|}{\|r^l\|} \left(Q^T(\delta_r^{l-1})_i - \frac{\beta_{l-1}}{\gamma_{l-2}} Q^T(\delta_r^{l-2})_i - \lambda_i Q^T(\delta_p^{l-2})_i \right).$$

The behavior of $\bar{r}_i^k$ is the same as the one of $\bar{w}_i^k$ (and of the corresponding component of the projection of the Lanczos vector) except that there is a $\|r^k\|$ in front. The roots of the polynomial $p_{1,k+1}$ are the CG Ritz values. The last component $\bar{r}_{30}^k$ for the Strakos30 example is shown in Figure 5.12. We see the oscillations of the components of the residual (the large oscillations are caused by the convergence of Ritz values to λ_{30}) but also the general trend of decrease. The dashed curve is the norm of the residual. It is not small for the 40 first iterations because of the components of the projections of the residual corresponding to the

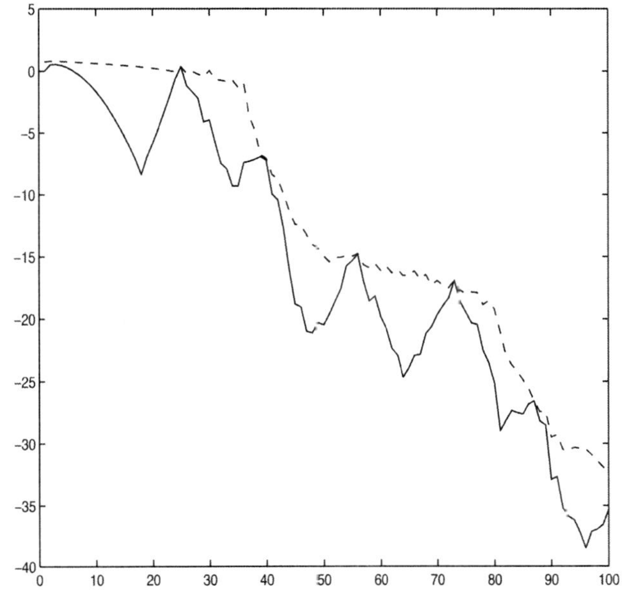

Figure 5.12. *Strakos*30, $\log_{10}(|\bar{r}^k_{30}|)$ *(solid) and* $\log_{10}(\|r^k\|)$ *(dashed)*

smallest eigenvalues. This allows the 30th component to go up after the initial decrease. Remember that since Q is an orthogonal matrix, we have

$$\|r^k\|^2 = \sum_{i=1}^{n} (\bar{r}^k_i)^2.$$

In exact arithmetic after a Ritz value has converged, the corresponding components of the projection of the residual and the error vanish. This is not the case in finite precision arithmetic since after decreasing the component of the residual may come back to contribute to the norm until a new copy of the eigenvalue appears. This can delay convergence. This can happen only if the norm of the residual has not decreased too much in the meantime, as we see in Figure 5.12. Figure 5.13 shows all the components of $\bar{r}^k$ as a function of k, $\bar{r}^k_{30}$ in the foreground and $\bar{r}^k_1$ in the background. We see that the last components oscillate while the first ones stay more or less constant before decreasing rapidly a little before iteration 40. The first convergence of a Ritz value to λ_1 occurs at iteration 30 in exact arithmetic and a little before iteration 40 in finite precision. In this example, the level of $\|r^k\|$ is mainly driven by the first components of the projection of the residual, as we can see in Figure 5.14. In the Lanczos algorithm we have obtained at least one copy of each eigenvalue by iteration 40. This corresponds to the rapid decrease of the residual norm. Remember that the components of $\bar{w}^k$ have large oscillations when we have convergence of Ritz values. We see that the small oscillations that we have in the norm of the residual correspond to peaks in the (last, in the first iterations) components of the residual. This can also be seen for the matrix Bcsstk01 of size 48. In Figure 5.15 we have $\log_{10}(\|r^k\|)$ (solid) and two components of $\bar{r}^k$. In this

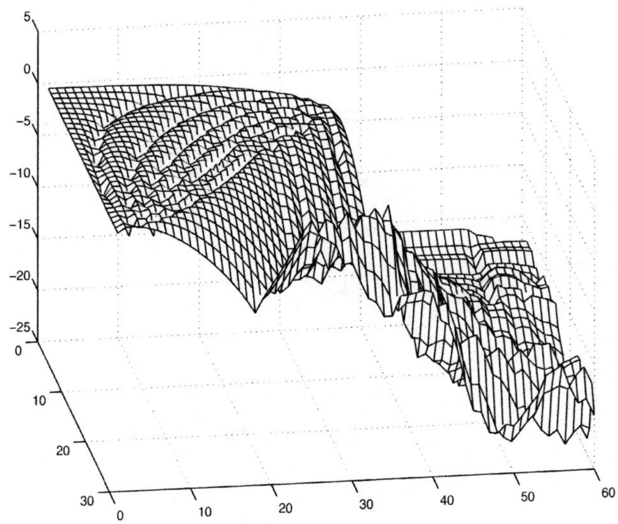

Figure 5.13. *Strakos*30, $\log_{10}(|\bar{r}^k|)$, $k = 1, \ldots, 60$

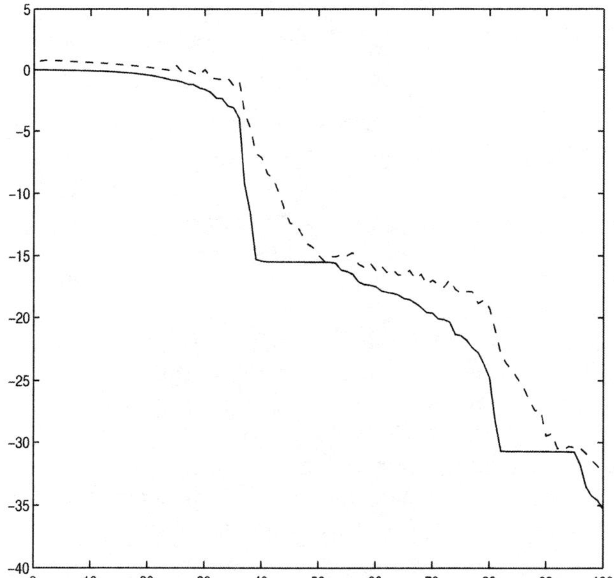

Figure 5.14. *Strakos*30, $\log_{10}(|\bar{r}_1^k|)$ *(solid) and* $\log_{10}(\|r^k\|)$ *(dashed)*

example the level of the norm of the residual is given by the first components of $\bar{r}^k$ and the peaks by the last ones for which the Ritz values converge first.

The result in Theorem 5.6 is interesting since it shows that the perturbation term decreases with $\|r^k\|$, as we have seen in the example. The expression for r^k involves $\|r^k\|$ in the right-hand side and it is interesting to be able to find another expression of the solution

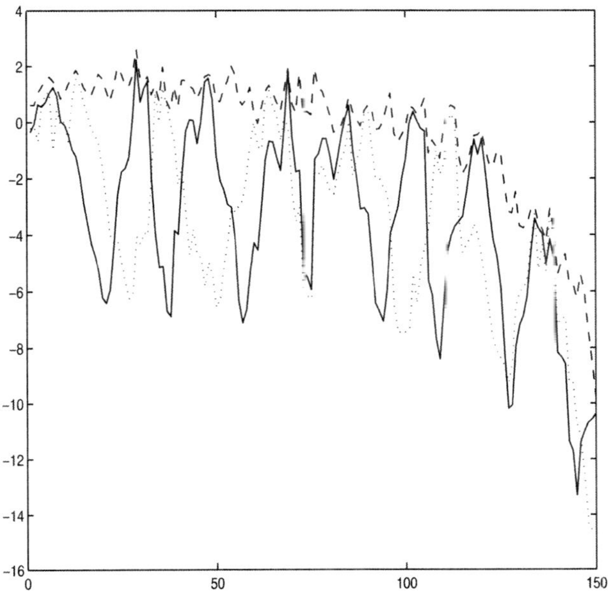

Figure 5.15. *Bcsstk01*, $\log_{10}(|\bar{r}^k_{48}|)$ *(solid)*, $\log_{10}(|\bar{r}^k_{45}|)$ *(dotted), and* $\log_{10}(\|r^k\|)$ *(dashed)*

for r^k. Therefore it is useful to directly study the equation for r^k. From Proposition 5.2 we have

$$r^{k+1} = -\gamma_k A r^k + \left(1 + \frac{\gamma_k \beta_k}{\gamma_{k-1}}\right) r^k - \frac{\gamma_k \beta_k}{\gamma_{k-1}} r^{k-1} + \bar{\delta}^k_r.$$

Now, we are interested in the components of r^k on the eigenvectors of A, $\bar{r}^k = Q^T r^k$ and let $\tilde{\delta_r}^k = Q^T \bar{\delta}^k_r$,

$$\bar{r}^{k+1}_i = -\gamma_k \lambda_i \bar{r}^k_i + \left(1 + \frac{\gamma_k \beta_k}{\gamma_{k-1}}\right) \bar{r}^k_i - \frac{\gamma_k \beta_k}{\gamma_{k-1}} \bar{r}^{k-1}_i + (\tilde{\delta_r}^k)_i,$$

$$\bar{r}^1_i = (1 - \gamma_0 \lambda_i)\bar{r}^0_i + (\tilde{\delta_r}^0)_i.$$

This recurrence is not exactly in the form that was considered in Theorem 4.3. But, nothing in the proof of Theorem 4.3 imposes to have a similar coefficient for the terms of order $k+1$ and $k-1$ as well as a 1 in front of the λ term. We have also to take care of the fact that the indices are different as well as the initial conditions. With the same kind of proof we have the following result.

Theorem 5.7. *Let j given and $p^r_{j,k}$ be the polynomial determined by*

$$p^r_{j,j-1} = 0, \quad p^r_{j,j} = 1,$$

$$p^r_{j,k+2}(\lambda) = \left(\left(1 + \frac{\gamma_k \beta_k}{\gamma_{k-1}}\right) - \gamma_k \lambda\right) p^r_{j,k+1}(\lambda) - \frac{\gamma_k \beta_k}{\gamma_{k-1}} p^r_{j,k}(\lambda), \quad k = j-1, \ldots,$$

$$p^r_{1,2}(\lambda) = 1 - \gamma_0 \lambda.$$

The solution of the perturbed recurrence

$$s_{k+1} = \left(\left(1 + \frac{\gamma_k \beta_k}{\gamma_{k-1}}\right) - \gamma_k \lambda\right) s_k - \frac{\gamma_k \beta_k}{\gamma_{k-1}} s_{k-1} + f_k, \quad k \geq 1,$$

$$s_1 = (1 - \gamma_0 \lambda) s_0 + f_0,$$

starting from s_0 *is given by*

$$s_k = p^r_{1,k+1}(\lambda) s_0 + \sum_{l=1}^{k} p^r_{l+1,k+1}(\lambda) f_{l-1}.$$

We apply this result to $\bar{r}^k$ to get the following.

Theorem 5.8.

$$\bar{r}^k_i = p^r_{1,k+1}(\lambda_i)\bar{r}^0_i + \sum_{l=1}^{k} p^r_{l+1,k+1}(\lambda_i)(\tilde{\delta}^{l-1}_r)_i,$$

$$r^k = p^r_{1,k+1}(A) r^0 + \sum_{l=1}^{k} p^r_{l+1,k+1}(A)\tilde{\delta}^{l-1}_r.$$

We shall study later on the properties of these polynomials in terms of determinants of tridiagonal matrices. Examples of polynomials $p^r_{j,k}$ are given in Figures 5.16 and 5.17. In Figure 5.16 we have $p^r_{1,k}(\lambda_n)$ as a function of k as the decreasing solid curve. It first decreases and then increases when the largest Ritz value converges. All the other polynomials $p^r_{j,k}$ at $\lambda_n = 100$, alternatively solid and dashed curves, are first increasing and then decreasing in average even though there are some local increases when copies of the maximum eigenvalue have converged. Remember that the values of these polynomials for $j > 1$ are multiplied with terms of order u. These computations were done with double precision CG coefficients and computation of the polynomial values in variable precision with $Digits = 32$. Figure 5.17 shows the same information at $\lambda_1 = 0.1$. The first polynomial decreases and all the others are almost constant at the beginning and then decrease. Solution of three-term recurrences like those in Theorem 5.7 were also studied by Gutknecht and Strakoš in [86].

It is also interesting to derive expressions for the error. However, we have to be careful about how to define the error. Ideally the error is $\epsilon^k = x - x^k$, where x is the exact solution and it is related to the residual by $A\epsilon^k = r^k$. But this is true only if the residual is $b - Ax^k$. In CG we have seen that the iterative residual can be different from $b - Ax^k$. We can no longer use the relation between the error and the residual. From what we have obtained before it seems reasonable to consider $(A^{-1}r^k, r^k) = \|A^{-\frac{1}{2}}r^k\|^2$ as a measure of the error. We denote this "error" by a new symbol $\varepsilon^k = A^{-1}r^k$, where r^k is the iterative residual. Concerning the components of ε^k on the eigenvectors of A we have $\bar{\varepsilon}^k_i = \bar{r}^k_i/\lambda_i$. This leads to

$$\bar{\varepsilon}^k_i = p^r_{1,k+1}(\lambda_i)\frac{\bar{r}^0_i}{\lambda_i} + \sum_{l=1}^{k} p^r_{l+1,k+1}(\lambda_i)\frac{(\tilde{\delta}^{l-1}_r)_i}{\lambda_i},$$

$$\varepsilon^k = p^r_{1,k+1}(A)\varepsilon^0 + \sum_{l=1}^{k} A^{-1} p^r_{l+1,k+1}(A)\tilde{\delta}^{l-1}_r.$$

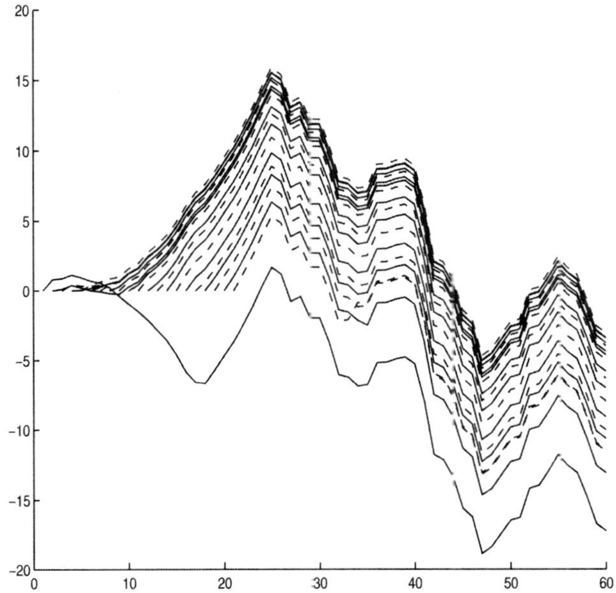

Figure 5.16. *Strakos*30, $\log_{10}$ *of polynomials* $|p^r_{j,k}(\lambda_n)|$

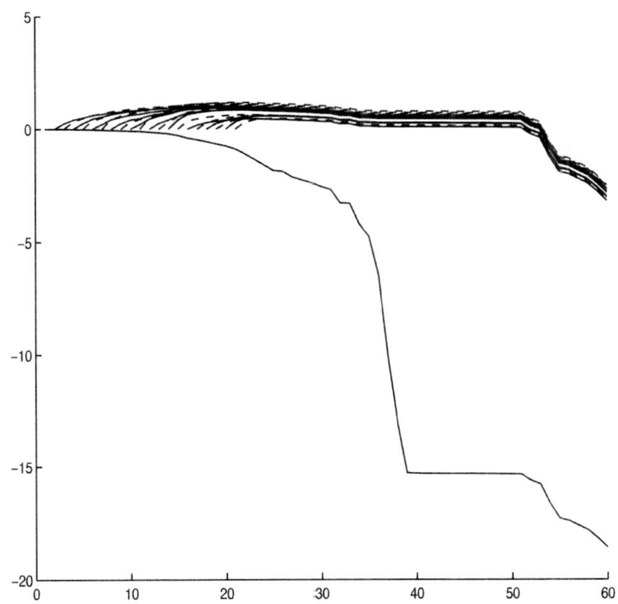

Figure 5.17. *Strakos*30, $\log_{10}$ *of polynomials* $|p^r_{j,k}(\lambda_1)|$

The components of the error have a $1/\lambda$ factor. Although it depends on the components of the initial residual, it is likely that the largest terms and perturbations in the error are obtained for the smallest eigenvalues, provided that the smallest Ritz values have not converged yet. Moreover,

$$\|\varepsilon^k\|^2 = \sum_{i=1}^{n} \left(\frac{\bar{r}_i^k}{\lambda_i} \right)^2, \qquad \|\varepsilon^k\|_A^2 = (A\varepsilon^k, \varepsilon^k) = \sum_{i=1}^{n} \frac{(\bar{r}_i^k)^2}{\lambda_i}.$$

Examples of components of the error are given in Figures 5.18 and 5.19 for the Strakos30 matrix. Unless the corresponding components of the residual are very small it is likely that the terms with small λ_i are the most important in the expression of the error

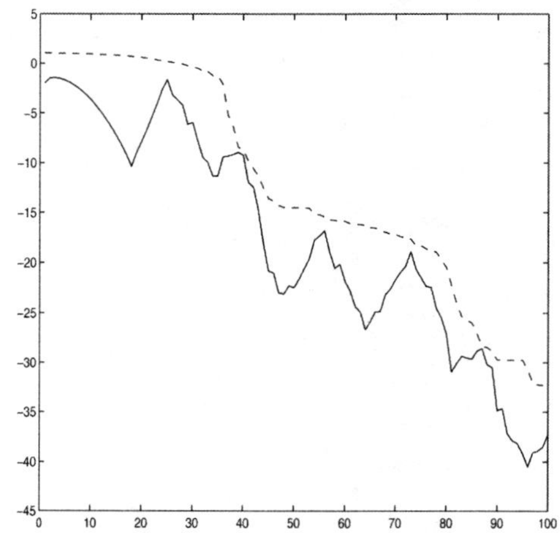

Figure 5.18. *Strakos30*, $\log_{10}(|\bar{\varepsilon}_{30}^k|)$ *(solid) and* $\log_{10}(\|\varepsilon^k\|_A)$ *(dashed)*

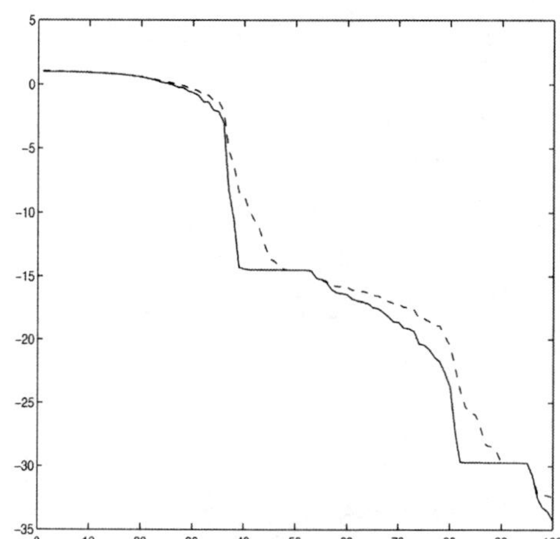

Figure 5.19. *Strakos30*, $\log_{10}(|\bar{\varepsilon}_1^k|)$ *(solid) and* $\log_{10}(\|\varepsilon^k\|_A)$ *(dashed)*

norms. It is remarkable that, even though the components of the error can oscillate, the A- and l_2 norms are monotonically decreasing.

To relate the polynomials $p^r_{j,k}$ to determinants it is useful to write the CG three-term recurrence in matrix form. If we denote $R_k = (r^0 \quad r^1 \quad \cdots \quad r^{k-1})$, then writing the recurrence relations for r^k in matrix form, we have

$$AR_k = R_k \bar{T}_k + \frac{1}{\gamma_{k-1}} r^k (e^k)^T + \bar{F}_k,$$

with $\bar{F}_k = (\bar{\delta}^0_r/\gamma_0 \quad \cdots \quad \bar{\delta}^{k-1}_r/\gamma_{k-1})$. The matrix $\bar{T}_k$ is tridiagonal

$$\bar{T}_k = \begin{pmatrix} \frac{1}{\gamma_0} & -\frac{\beta_1}{\gamma_0} & & & \\ -\frac{1}{\gamma_0} & \frac{1}{\gamma_1}+\frac{\beta_1}{\gamma_0} & -\frac{\beta_2}{\gamma_1} & & \\ & \ddots & \ddots & \ddots & \\ & & -\frac{1}{\gamma_{k-3}} & \frac{1}{\gamma_{k-2}}+\frac{\beta_{k-2}}{\gamma_{k-3}} & -\frac{\beta_{k-1}}{\gamma_{k-2}} \\ & & & -\frac{1}{\gamma_{k-2}} & \frac{1}{\gamma_{k-1}}+\frac{\beta_{k-1}}{\gamma_{k-2}} \end{pmatrix}.$$

Note that the matrix $\bar{T}_k$ is not symmetric, but, as every tridiagonal matrix with nonvanishing lower and upper diagonals, it can be symmetrized. Let M be a diagonal matrix with nonzero diagonal coefficients μ_i. Then we consider $T_M = M\bar{T}_k M^{-1}$. If we choose μ_i, $i = 1, \dots, k$, such that

$$\frac{\mu_{i-1}}{\mu_i}\frac{\beta_{i-1}}{\gamma_{i-2}} = \frac{\mu_i}{\mu_{i-1}}\frac{1}{\gamma_{i-2}},$$

the matrix T_M is symmetric. This leads to

$$\frac{\mu_i}{\mu_{i-1}} = \sqrt{\beta_{i-1}},$$

and we see that the nondiagonal terms become

$$-\frac{\sqrt{\beta_i}}{\gamma_{i-1}}.$$

Thus, up to a change of the signs of the nondiagonal coefficients (which does not matter for the eigenvalues), the matrix T_M is nothing other than the matrix of the CG-Lanczos coefficients. Hence $\bar{T}_k$ has the same eigenvalues as the CG-Lanczos matrix T_k, that is, Ritz values $\theta^{(k)}_j$ (which are not exactly the ones from the Lanczos algorithm). We remark that

$$\mu_1 = 1, \quad \mu_2 = \frac{\|r^1\|}{\|r^0\|}, \dots, \mu_k = \frac{\|r^{k-1}\|}{\|r^0\|}.$$

These facts could have also been shown by relating the residual vectors to the CG-Lanczos vectors. If

$$W_k = (w^1 \quad w^2 \quad \cdots \quad w^k) = \left(\frac{r^0}{\|r^0\|} \quad -\frac{r^1}{\|r^1\|} \quad \cdots \quad (-1)^{k-1}\frac{r^{k-1}}{\|r^{k-1}\|}\right) = R_k D_k,$$

where D_k is a diagonal matrix with diagonal elements plus or minus the inverses of the norms of the residual vectors, then $R_k = W_k D_k^{-1}$ and the matrix relation is written as

$$AW_k D_k^{-1} = W_k D_k^{-1}\bar{T}_k - \frac{(-1)^k\|r^k\|}{\gamma_{k-1}} w^{k+1}(e^k)^T + \bar{F}_k.$$

Multiplying to the right by D_k we have

$$AW_k = W_k D_k^{-1} \bar{T}_k D_k - \frac{(-1)^k \|r^k\|}{\gamma_{k-1}} w^{k+1} (e^k)^T D_k + \bar{F}_k D_k.$$

The matrix $D_k^{-1} \bar{T}_k D_k$ is the matrix T_k of the CG-Lanczos coefficients. It is similar to $\bar{T}_k$. For the second term of the right-hand side we see that $(e^k)^T D_k$ is a row vector whose coefficients are zero except for the last one, which is $(-1)^{k-1}/\|r^{k-1}\|$. We obtain

$$\frac{1}{\gamma_{k-1}} \frac{\|r^k\|}{\|r^{k-1}\|} w^{k+1} (e^k)^T.$$

Hence this term is $\eta_{k+1} w^{k+1} (e^k)^T$. The perturbation term is $\bar{F}_k D_k$.

We can now look at finding an expression for the polynomials $p^r_{j,k}$. We denote by $\bar{T}_{j,k}$ the matrix obtained by deleting the first $j-1$ rows and columns of $\bar{T}_k$. It is similar to $T_{j,k}$. We have a recurrence relation for the determinants of these matrices,

$$\det(\bar{T}_{j,k} - \lambda I) = \left(\frac{1}{\gamma_{k-1}} + \frac{\beta_{k-1}}{\gamma_{k-2}} - \lambda \right) \det(\bar{T}_{j,k-1} - \lambda I) - \frac{\beta_{k-1}}{(\gamma_{k-2})^2} \det(\bar{T}_{j,k-2} - \lambda I).$$

Since the matrices are similar we have $\det(\bar{T}_{j,k} - \lambda I) = \chi_{j,k}(\lambda)$, the determinant of $T_{j,k} - \lambda I$. By comparing to the recurrence for the polynomials $p^r_{j,k}$ we obtain the following results.

Proposition 5.9. *Let $k > j \geq 1$,*

$$p^r_{j,k}(\lambda) = \gamma_{j-1} \cdots \gamma_{k-2} \chi_{j,k-1}(\lambda).$$

The roots of $p^r_{j,k}$ are the eigenvalues of $T_{j,k-1}$. In particular, the roots of $p^r_{1,k}$ are the Ritz values. When a Ritz value converges the absolute value of the polynomial at the corresponding eigenvalue decreases but the perturbation terms may increase.

Proposition 5.10. *Let $k \geq j$,*

$$p^r_{j,k}(\lambda) = (-1)^{k-j} \frac{\|r^{k-1}\|}{\|r^{j-1}\|} p_{j,k}(\lambda),$$

where $p_{j,k}$ is the polynomial of the CG-Lanczos recurrence.

Proof. Use the relation between the CG coefficients and the CG-Lanczos coefficients. □

From this we conclude that

$$\frac{\|r^k\|}{\|r^{k-1}\|} = \frac{|p^r_{k,k+1}(\lambda)|}{|p_{k,k+1}(\lambda)|} \quad \forall \lambda \text{ such that } p_{k,k+1}(\lambda) \neq 0.$$

The polynomial $p^r_{j,k}$ has the same oscillations as $p_{j,k}$ but damped by $\|r^{k-1}\|/\|r^{j-1}\|$. If $j \ll k$, this factor is likely to be small.

As we have done for the Lanczos algorithm we can also write the three-term recurrence for $\bar r_i^k$ as a triangular system $Lz = h$ with a vector of unknowns $z^T = (\bar r_i^0 \quad \bar r_i^1 \quad \cdots \quad \bar r_i^k)^T$, a right-hand side

$$h = \begin{pmatrix} \bar r_i^0 & \frac{\bar\delta^0}{\gamma_0} & \cdots & \frac{\bar\delta^{k-1}}{\gamma_{k-1}} \end{pmatrix}^T,$$

and a matrix (denoting $\lambda = \lambda_i$)

$$L = \begin{pmatrix} (e^1)^T & 0 \\ \lambda I - \bar T_k^T & \frac{1}{\gamma_{k-1}} e^k \end{pmatrix}.$$

As we have seen before the k first elements of the first column of L^{-1} are

$$\frac{(\lambda I - \bar T_k^T)^{-1} e^k}{(\lambda I - \bar T_k^T)^{-1}_{1,k}}.$$

If $T_k = Z_k \Theta_k Z_k^T$, we have $(\lambda I - \bar T_k^T)^{-1} e^k = D_k^{-1} Z_k (\lambda I - \Theta_k)^{-1} Z_k^T D_k e^k$. The numerator is

$$\sum_{j=1}^k \frac{(z_k^j)^2}{\lambda - \theta_j^{(k)}}.$$

The denominator is

$$(-1)^{k-1} \frac{\|r^0\|}{\|r^{k-1}\|} \sum_{j=1}^k \frac{z_k^j z_1^j}{\lambda - \theta_j^{(k)}}.$$

The k first elements of the other columns are given by those of the matrix

$$\left[I - \frac{1}{(\lambda I - \bar T_k^T)^{-1}_{1,k}} (\lambda I - \bar T_k^T)^{-1} e^k (e^1)^T \right] (\lambda I - \bar T_k^T)^{-1},$$

which is the factor of the perturbation term. We can now relate these expressions to the symmetric CG-Lanczos matrix T_k. We have $\bar T_k = D_k T_k D_k^{-1}$ and if z is an eigenvector of T_k, the corresponding unnormalized eigenvector of $\bar T_k$ is $y = D_k z$. It follows that

$$(\lambda I - \bar T_k^T)^{-1} = D_k^{-1} (\lambda I - T_k)^{-1} D_k$$

and

$$(\lambda I - \bar T_k^T)^{-1}_{1,k} = (-1)^{k-1} \frac{\|r^0\|}{\|r^{k-1}\|} (\lambda I - T_k)^{-1}_{1,k}.$$

As we have seen the kth element of the first column of the inverse (factor of $\bar r_i^0$) is

$$(-1)^{k-1} \frac{\|r^{k-1}\|}{\|r^0\|} \frac{(\lambda_i I - T_k)^{-1}_{k,k}}{(\lambda_i I - T_k)^{-1}_{1,k}} = p^r_{1,k}(\lambda_i).$$

Let

$$\bar U_k = I - \frac{1}{(\lambda I - \bar T_k^T)^{-1}_{1,k}} (\lambda I - \bar T_k^T)^{-1} e^k (e^1)^T, \quad U_k = I - \frac{1}{(\lambda I - T_k)^{-1}_{1,k}} (\lambda I - T_k)^{-1} e^k (e^1)^T.$$

Then,

$$\bar U_k (\lambda I - \bar T_k^T)^{-1} = D_k^{-1} U_k (\lambda I - T_k)^{-1} D_k.$$

By using the spectral decomposition of $T_k = Z_k \Theta_k Z_k^T$ and $h = (\bar{r}_i^0 \, \bar{h})^T$,

$$\bar{U}_k(\lambda I - \bar{T}_k^T)^{-1}\bar{h} = D_k^{-1} U_k Z_k (\lambda I - \Theta_k)^{-1} Z_k^T D_k \bar{h},$$

where $\bar{h}$ is the vector of local errors. We are interested in the last component of this vector. Using the same notations as for the Lanczos algorithm, this component is

$$(-1)^{k-1} \|r^{k-1}\| \sum_{j=1}^{k} \frac{(z_k^j - u_k z_1^j)}{\lambda - \theta_j^{(k)}} (Z_k^T D_k \bar{h})_j.$$

We remark that

$$\frac{(z_k^j - u_k z_1^j)}{\lambda - \theta_j^{(k)}} = \frac{p_k(\theta_j^{(k)}) - p_k(\lambda)}{\lambda - \theta_j^{(k)}} z_1^j,$$

where p_k is the polynomial associated with the CG-Lanczos matrix T_k. A term like this one can be large only if $\lambda - \theta_j^{(k)}$ is small, that is, if a Ritz value converges.

5.4 Study of the CG two-term recurrences

Another possible way to look at the problem of rounding errors in CG is to directly consider the two-term recurrences. In this section we shall derive the solution of these recurrences using polynomials.

If we write the spectral decomposition of A with the matrix Q of the eigenvectors and denote $\bar{r}^k = Q^T r^k$ and $\bar{p}^k = Q^T p^k$, denoting $\check{\delta}_r^k = Q^T \delta_r^k$ and $\check{\delta}_p^k = Q^T \delta_p^k$, we have

$$\bar{r}_i^{k+1} = \bar{r}_i^k - \gamma_k \lambda_i \bar{p}_i^k + (\check{\delta}_r^k)_i,$$

$$\bar{p}_i^{k+1} = \bar{r}_i^{k+1} + \beta_{k+1} \bar{p}_i^k + (\check{\delta}_p^k) = \bar{r}_i^k + (\beta_{k+1} - \gamma_k \lambda_i) \bar{p}_i^k + (\check{\delta}_r^k)_i + (\check{\delta}_p^k)_i.$$

In matrix form this is written as

$$y^{k+1} = B_k y^k + g^k,$$

where

$$y^k = \begin{pmatrix} \bar{r}_i^k \\ \bar{p}_i^k \end{pmatrix}, \quad B_k = \begin{pmatrix} 1 & -\gamma_k \lambda_i \\ 1 & \beta_{k+1} - \gamma_k \lambda_i \end{pmatrix}, \quad g^k = \begin{pmatrix} (\check{\delta}_r^k)_i \\ (\check{\delta}_r^k)_i + (\check{\delta}_p^k)_i \end{pmatrix}.$$

The recurrence starts from y^0. The solution is given by slightly modifying Lemma 4.2.

Lemma 5.11. *Let $k \geq 1$,*

$$\begin{pmatrix} \bar{r}_i^{k+1} \\ \bar{p}_i^{k+1} \end{pmatrix} = B_k \cdots B_0 \begin{pmatrix} \bar{r}_i^0 \\ \bar{p}_i^0 \end{pmatrix} + \sum_{l=0}^{k-1} B_k \cdots B_{l+1} \begin{pmatrix} (\check{\delta}_r^l)_i \\ (\check{\delta}_r^l)_i + (\check{\delta}_p^l)_i \end{pmatrix} + \begin{pmatrix} (\check{\delta}_r^k)_i \\ (\check{\delta}_r^k)_i + (\check{\delta}_p^k)_i \end{pmatrix}.$$

We cannot use the norm of the matrices B_k to efficiently bound the norms of r^k and p^k since $\|B_k\| > 1$. The solution of Lemma 5.11 can also be written in terms of polynomials.

We first define two sets of polynomials ϕ_k and ψ_k by

$$\begin{aligned}\phi_{k+1}(\lambda) &= \phi_k(\lambda) - \gamma_k \lambda \psi_k(\lambda),\\ \psi_{k+1}(\lambda) &= \phi_k(\lambda) + (\beta_{k+1} - \gamma_k \lambda)\psi_k(\lambda),\\ \phi_0(\lambda) &= 1, \quad \psi_0(\lambda) = 1.\end{aligned}$$

This is also written in matrix form as

$$\begin{pmatrix} \phi_{k+1}(\lambda) \\ \psi_{k+1}(\lambda) \end{pmatrix} = B_k \begin{pmatrix} \phi_k(\lambda) \\ \psi_k(\lambda) \end{pmatrix}.$$

Then, since $\bar{p}_i^0 = \bar{r}_i^0$,

$$B_k \cdots B_0 \begin{pmatrix} \bar{r}_i^0 \\ \bar{p}_i^0 \end{pmatrix} = \bar{r}_i^0 B_k \cdots B_0 \begin{pmatrix} 1 \\ 1 \end{pmatrix} = \bar{r}_i^0 \begin{pmatrix} \phi_{k+1}(\lambda_i) \\ \psi_{k+1}(\lambda_i) \end{pmatrix}.$$

Now we have to deal with the term

$$\begin{pmatrix} (\check{\delta}_r^l)_i \\ (\check{\delta}_r^l)_i + (\check{\delta}_p^l)_i \end{pmatrix} = (\check{\delta}_r^l)_i \begin{pmatrix} 1 \\ 1 \end{pmatrix} + (\check{\delta}_p^l)_i \begin{pmatrix} 0 \\ 1 \end{pmatrix}.$$

We define

$$\begin{pmatrix} \phi_{j,k+1}(\lambda) \\ \psi_{j,k+1}(\lambda) \end{pmatrix} = B_k \begin{pmatrix} \phi_{j,k}(\lambda) \\ \psi_{j,k}(\lambda) \end{pmatrix},$$
$$\phi_{j,j}(\lambda) = 1, \quad \psi_{j,j}(\lambda) = 1,$$

and

$$\begin{pmatrix} \bar{\phi}_{j,k+1}(\lambda) \\ \bar{\psi}_{j,k+1}(\lambda) \end{pmatrix} = B_k \begin{pmatrix} \bar{\phi}_{j,k}(\lambda) \\ \bar{\psi}_{j,k}(\lambda) \end{pmatrix},$$
$$\bar{\phi}_{j,j}(\lambda) = 0, \quad \bar{\psi}_{j,j}(\lambda) = 1.$$

Notice that only the initial conditions are different. With these notations we have $\phi_{0,k} = \phi_k$ and $\psi_{0,k} = \psi_k$.

Theorem 5.12.

$$\bar{r}_i^{k+1} = \phi_{k+1}(\lambda_i)\bar{r}_i^0 + \sum_{l=0}^{k} \phi_{l+1,k+1}(\lambda_i)(\check{\delta}_r^l)_i + \sum_{l=0}^{k} \bar{\phi}_{l+1,k+1}(\lambda_i)(\check{\delta}_p^l)_i,$$

$$\bar{p}_i^{k+1} = \psi_{k+1}(\lambda_i)\bar{r}_i^0 + \sum_{l=0}^{k} \psi_{l+1,k+1}(\lambda_i)(\check{\delta}_r^l)_i) + \sum_{l=0}^{k} \bar{\psi}_{l+1,k+1}(\lambda_i)(\check{\delta}_p^l)_i).$$

Proof. The proof is obvious from the definition of the polynomials. □

The interest of the last theorem is to clearly separate the influence of the roundoff from r^k and p^k. Examples of polynomials $\phi_{j,k}$ and $\psi_{j,k}$ at λ_n as a function of k are given in Figures 5.20 and 5.21. Figures 5.22 and 5.23 display the values of polynomials $\bar{\phi}_{j,k}$ and $\bar{\psi}_{j,k}$ at λ_n. They all have the same behavior.

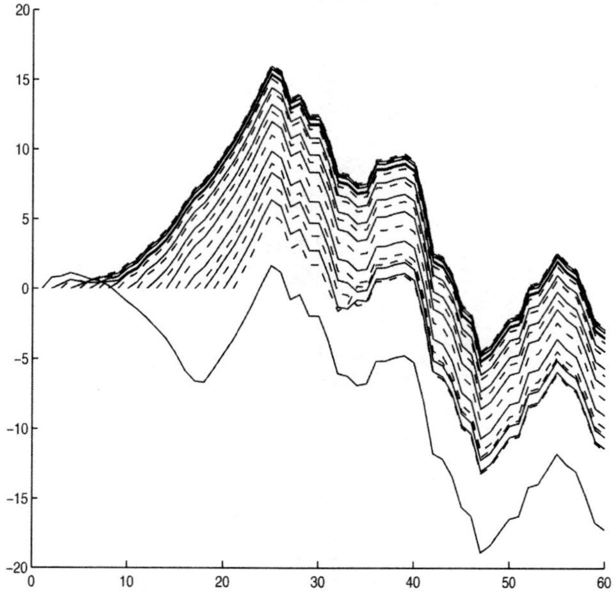

Figure 5.20. *Strakos*30, $\log_{10}$ *of polynomials* $|\phi_{j,k}(\lambda_n)|$

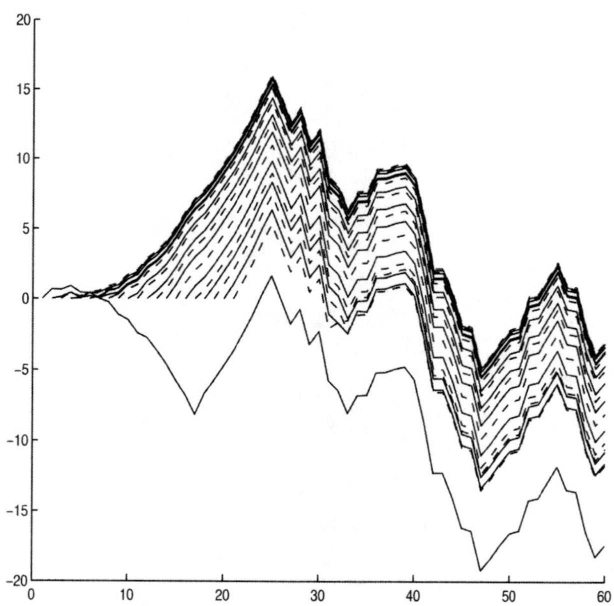

Figure 5.21. *Strakos*30, $\log_{10}$ *of polynomials* $|\psi_{j,k}(\lambda_n)|$

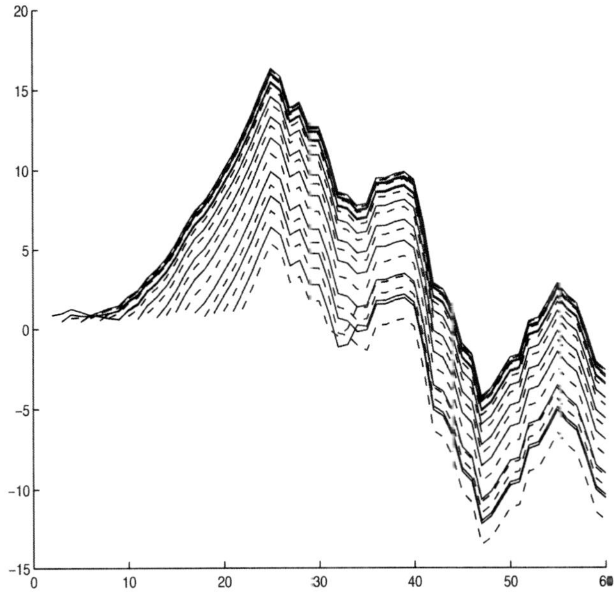

Figure 5.22. *Strakos*30, $\log_{10}$ *of polynomials* $|\bar{\phi}_{j,k}(\lambda_n)|$

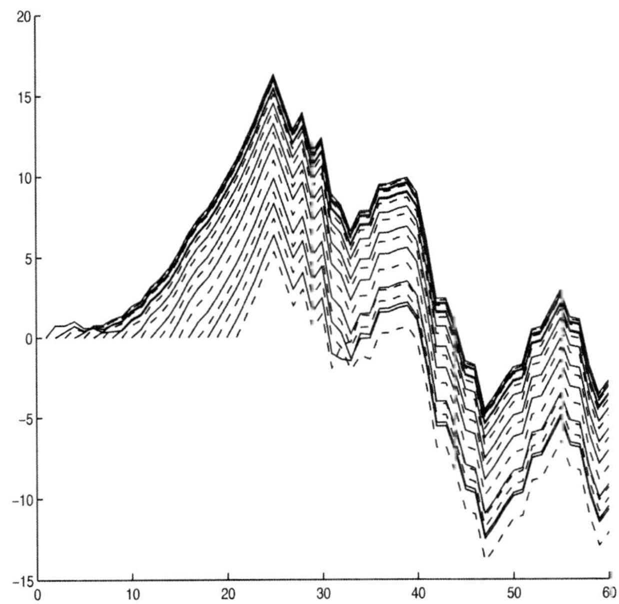

Figure 5.23. *Strakos*30, $\log_{10}$ *of polynomials* $|\bar{\psi}_{j,k}(\lambda_n)|$

It could be interesting to characterize all these polynomials. To study the properties of ϕ_k we can eliminate the polynomial ψ_k to obtain a three-term recurrence:

$$\frac{1}{\gamma_k}\phi_{k+1} + \lambda\phi_k - \left(\frac{1}{\gamma_k} + \frac{\beta_k}{\gamma_{k-1}}\right)\phi_k + \frac{\beta_k}{\gamma_{k-1}}\phi_{k-1} = 0,$$

with the initial condition $\phi_1 = 1 - \gamma_0\lambda$.

Proposition 5.13.

$$\phi_k(\lambda) = p^r_{1,k+1}(\lambda).$$

Unfortunately, it is not so simple to relate $\phi_{j,k}$ to determinants of matrices $\bar{T}_{j,k}$ because we now have the initial condition $\phi_{j,j} = 1$, $\psi_{j,j} = 1$, and

$$\phi_{j,j+1} = (1 - \gamma_j\lambda)\phi_{j,j}.$$

However, all the polynomials which are involved in these relations can be characterized up to a multiplicative factor as determinants of some tridiagonal matrices. It is difficult to analyze the behavior of these polynomials when k increases since this is also linked to the way the norm of the residual decreases.

5.5 Relations between p^k and r^k

In exact arithmetic we obtained a relation between p^k and r^k which shows that when we have convergence of CG or at least when $\|r^k\|$ is small, p^k behaves like r^k. In this section we are going to see what happens to this relationship in finite precision arithmetic.

Theorem 5.14.

$$p^k = r^k + \|r^k\|^2\left(\frac{r^{k-1}}{\|r^{k-1}\|^2} + \cdots + \frac{r^0}{\|r^0\|^2}\right) + uC\|r^k\|,$$

with a vector C such that

$$|C| \leq 1 + C_p^{k-1} + \frac{\|r^k\|}{\|r^{k-1}\|}(1 + C_p^{k-2}) + \cdots + \frac{\|r^k\|}{\|r^1\|}(1 + C_p^0) + O(u^2).$$

Proof. We have

$$p^k = r^k + \beta_k p^{k-1} + \delta_p^{k-1}.$$

By iterating this relation, we obtain for $k \geq 2$

$$p^k = r^k + \beta_k r^{k-1} + \cdots + \beta_k \cdots \beta_1 r^0 + \delta_p^{k-1} + \beta_k\delta_p^{k-2} + \cdots + \beta_k \cdots \beta_2\delta_p^0.$$

Simplifying this expression depends on the hypothesis we have done for β_k. We suppose we have put the error term on β_k in the local roundoff error; then we can directly compute the products $\beta_k \cdots \beta_j$. We have

$$p^k = r^k + \|r^k\|^2 \left(\frac{r^{k-1}}{\|r^{k-1}\|^2} + \cdots + \frac{r^0}{\|r^0\|^2} \right)$$

$$+ \delta_p^{k-1} + \frac{\|r^k\|^2}{\|r^{k-1}\|^2} \delta_p^{k-2} + \cdots + \frac{\|r^k\|^2}{\|r^1\|^2} \delta_p^0.$$

But, we know that

$$\|\delta_p^{j-1}\| \leq u\|r^j\|(1 + C_p^{j-1}) + O(u^2).$$

Using this bound for $\|\delta_p^{j-1}\|$, the perturbation term can be written as $Cu\|r^k\|$, where

$$|C| \leq 1 + C_p^{k-1} + \frac{\|r^k\|}{\|r^{k-1}\|}(1 + C_p^{k-2}) + \cdots + \frac{\|r^k\|}{\|r^1\|}(1 + C_p^0) + O(u^2). \quad \square$$

Since δ_p^j is a linear combination of r^{j+1} and p^j, like in exact arithmetic $span\{p^i,\ i = 0, \ldots, k\} \equiv span\{r^i,\ i = 0, \ldots, k\}$ despite the perturbations. Using the previous proposition we have a bound for the norm of p^k.

Proposition 5.15.

$$\|p^k\| \leq \|r^k\| \left\{ 1 + \frac{\|r^k\|}{\|r^{k-1}\|} + \cdots + \frac{\|r^k\|}{\|r^0\|} \right.$$

$$\left. + u\left[1 + C_p^{k-1} + \frac{\|r^k\|}{\|r^{k-1}\|}(1 + C_p^{k-2}) + \cdots + \frac{\|r^k\|}{\|r^0\|}(1 + C_p^0) \right] \right\} + O(u^2).$$

From this result we note that the bound on the norm of p^k depends on the ratio of the norms of residuals and there is a factor $\|r^k\|$ in front of the perturbation terms. The norm $\|p^k\|$ can be much different from $\|r^k\|$ only if there are large ratios of norms of residuals. If the variations are reasonably bounded, $\|p^k\|$ is of the same order as $\|r^k\|$. We shall see some other relations between $\|p^k\|$ and $\|r^k\|$ later on.

5.6 Local orthogonality in finite precision

Before considering the CG convergence problem we are going to look at local orthogonality. In this section we derive some expressions for inner products of residuals and descent directions that we shall use later. Results on local orthogonality can also be found in Strakoš and Tichý [186].

Proposition 5.16. *Let* $\delta_{rp}^j = (r^j, \delta_p^{j-1}) + (p^j, \delta_r^j)$*. If* C_j *is such that*

$$|\delta_{rp}^j| \leq u\|r^j\|^2 C_j,$$

we have

$$(r^{k+1}, p^k) = (r^k, r^k)uC_{rp}^k,$$

with

$$|C_{rp}^k| \leq C_0 + C_1 + \cdots + C_k.$$

Proof. Supposing that the CG coefficients are computed exactly, that is, their errors are in δ_r and δ_p, we have

$$(r^{k+1}, p^k) = (r^k - \gamma_k A p^k + \delta_r^k, p^k) = (r^k, p^k) - \gamma_k (Ap^k, p^k) + (\delta_r^k, p^k)$$
$$= (r^k, p^k) - (r^k, r^k) + (\delta_r^k, p^k).$$

But,

$$(r^k, p^k) = (r^k, r^k) + \beta_k (r^k, p^{k-1}) + (r^k, \delta_p^{k-1}).$$

By denoting $\delta_{rp}^k = (r^k, \delta_p^{k-1}) + (p^k, \delta_r^k)$, we have

$$(r^{k+1}, p^k) = \beta_k (r^k, p^{k-1}) + \delta_{rp}^k.$$

By iterating this formula, we have

$$(r^{k+1}, p^k) = \beta_k \cdots \beta_1 (r^1, p^0) + \beta_k \cdots \beta_2 \delta_{rp}^1 + \cdots + \beta_k \delta_{rp}^{k-1} + \delta_{rp}^k.$$

Since we have supposed that $\beta_k = (r^k, r^k)/(r^{k-1}, r^{k-1})$,

$$(r^{k+1}, p^k) = \|r^k\|^2 \left[\frac{(p^0, r^1)}{\|r^0\|^2} + \frac{\delta_{rp}^1}{\|r^1\|^2} + \cdots + \frac{\delta_{rp}^k}{\|r^k\|^2} \right].$$

Going back to the definition of δ_{rp}^j, we have

$$|\delta_{rp}^j| \le u \|r^j\|^2 (1 + C_p^{j-1}) + u \|p^j\| \, \|r^j\| (1 + C_r^j) + O(u^2).$$

By factoring $\|r^j\|^2$, we have

$$|\delta_{rp}^j| \le u \|r^j\|^2 \left(1 + \frac{\|p^j\|}{\|r^j\|} + C_p^{j-1} + \frac{\|p^j\|}{\|r^j\|} C_r^j \right).$$

By looking at the definition of C_r^j we see that the last term is bounded independently of j by an expression involving $\bar{\kappa}(A)$, m, and n. However, C_p^{j-1} depends on ratios $\|p^{k-1}\|/\|r^{k-1}\|$ and $\|r^k\|/\|r^{k-1}\|$. Nevertheless we can write

$$|\delta_{rp}^j| \le u \|r^j\|^2 C_j,$$

where C_j depends on ratios of norms of residuals and direction vectors. Hence

$$\frac{|\delta_{rp}^j|}{\|r^j\|^2} \le C_j u.$$

Since we have $p^0 = r^0$ and

$$r^1 = r^0 - \gamma_0 A r^0 + \delta_r^0,$$

we have

$$(r^1, p^0) = (r^1, r^0) = (\delta_r^0, r^0).$$

Therefore,

$$|(r^1, p^0)| \leq u\|r^0\|^2(1 + \| \, |A| \, \|\gamma_0[(m+2) + \delta_\gamma^0]) = u\| r^0\|^2 C_0.$$

We can write the term in factor of $\|r^k\|^2$ in (r^{k+1}, p^k) as uC_{rp}^k, with

$$|C_{rp}^k| \leq C_0 + C_1 + \cdots + C_k. \quad \square$$

We notice that the inner product (r^{k+1}, p^k) is not the computed value of this quantity. It is the exact result given from the computed vectors r^{k+1} and p^k. To obtain $fl((r^{k+1}, p^k))$ from this, we must use the results about the computation of inner products. The next thing to do is to consider (p^k, r^k).

Proposition 5.17.

$$(p^k, r^k) = (r^k, r^k)(1 + uC_{pr}^k),$$

with

$$|C_{pr}^k| \leq |C_{rp}^k| + |C_p^{k-1}|,$$

where C_{rp}^k is defined in the proof of Proposition 5.16 and

$$|(r^k, \delta_p^{k-1})| \leq u(r^k, r^k)(1 + C_p^{k-1}).$$

Proof. This is obtained by using the previous proposition. $\square$

Proposition 5.18. *Provided $1 + uC_{rp}^k > 0$ and $1 + uC_{rp}^k - u(1 + C_p^{k-1}) > 0$ we have*

$$\frac{\|r^k\|}{\|p^k\|} \leq \frac{1}{1 + uC_{rp}^k - u(1 + C_p^{k-1})} \leq 1 + u(1 + C_p^{k-1} - C_{rp}^k) + O(u^2).$$

Proof. Since we have

$$(p^k, r^k) = (r^k, r^k)(1 + uC_{rp}^k) + (r^k, \delta_p^{k-1}),$$

using the Cauchy–Schwarz inequality, we can write

$$\|r^k\| \, (1 + uC_{rp}^k) \leq \|p^k\| + \|\delta_p^{k-1}\|.$$

Looking back at the definitions,

$$\frac{\|\delta_p^{k-1}\|}{\|p^k\|} \leq u \frac{\|r^k\|}{\|p^k\|}(1 + C_p^{k-1}).$$

Hence,

$$\frac{\|r^k\|}{\|p^k\|}(1 + uC_{rp}^k - u(1 + C_p^{k-1})) \leq 1.$$

Provided $1 + uC_{rp}^k - u(1 + C_p^{k-1}) > 0$ we obtain

$$\frac{\|r^k\|}{\|p^k\|} \leq \frac{1}{1 + uC_{rp}^k - u(1 + C_p^{k-1})} \leq 1 + u(1 + C_p^{k-1} - C_{rp}^k) + O(u^2).$$

This may seem to be a circular argument since C^k_{rp} contains a term $u\|p^k\|/\|r^k\|$, but the previous relation implies that $\|r^k\|/\|p^k\|$ is bounded from above if we suppose that the same is true at previous iterations and that the ratios $\|r^k\|/\|r^j\|$ are bounded. □

Next we consider (r^{k+1}, r^k). We have

$$(r^{k+1}, r^k) = (r^k, r^k)\left(1 - \frac{(Ap^k, r^k)}{(Ap^k, p^k)}\right) + (\delta^k_r, r^k).$$

In exact arithmetic the second parenthesis of the right-hand side and δ^k_r are zero.

Proposition 5.19.

$$(r^k, r^{k-1}) = u\|r^{k-1}\|^2 C_{k-1}, \quad (Ap^k, p^{k-1}) = u\|r^{k-1}\|^2 C^{k-1}_{Ap}.$$

Proof. The proof is by induction. We have already seen that we can write (eventually changing the sign of C_0)

$$(r^1, r^0) = u\|r^0\|^2 C_0.$$

Moreover, since we have

$$(p^1, r^1) = (r^1, r^1) + uC_0(r^1, r^1) + (\delta^0_p, r^1)$$

and

$$(r^0, p^1) = u\|r^0\|^2 C_0 + (r^1, r^1) + (\delta^0_p, r^0),$$

it gives

$$(r^1, p^1) - (r^0, p^1) = uC_0(\|r^1\|^2 - \|r^0\|^2) + (\delta^0_p, r^1 - r^0).$$

Then,

$$\gamma_0(Ap^0, p^1) = (r^0 - r^1, p^1) + (\delta^1_r, p^1).$$

Therefore,

$$\gamma_0(Ap^0, p^1) = uC_0(\|r^0\|^2 - \|r^1\|^2) + (\delta^0_p, r^0 - r^1) + (\delta^1_r, p^1).$$

Since

$$\frac{1}{\gamma_0} = \frac{(Ar^0, r^0)}{(r^0, r^0)} \le \|A\|,$$

$$|(Ap^0, p^1)| \le u\|A\|\,[|C_0|\,(\|r^0\|^2 + \|r^1\|^2) + \|r^1\|(1 + C^0_p)(\|r^0\| + \|r^1\|) + \|r^1\|(1 + C^1_r)\|p^1\|].$$

This can be written as

$$(Ap^1, p^0) = (Ap^0, p^1) = u\|r^0\|^2 C^0_{Ap},$$

with C^0_{Ap} having a factor $\|A\|$. We make the induction hypothesis,

$$(r^j, r^{j-1}) = u\|r^{j-1}\|^2 C_{j-1}, \quad (Ap^j, p^{j-1}) = u\|r^{j-1}\|^2 C^{j-1}_{Ap}, \quad j = 1, \dots, k-1.$$

Then we have

$$(Ap^{k-1}, p^{k-1}) = (Ap^{k-1}, r^{k-1} + \beta_{k-1} p^{k-2} + \delta^{k-2}_p).$$

With the definition of β_{k-1} we obtain

$$(Ap^{k-1}, p^{k-1}) = (Ap^{k-1}, r^{k-1}) + (r^{k-1}, r^{k-1})uC_{Ap}^{k-2} + (Ap^{k-1}, \delta_p^{k-2}).$$

Therefore

$$\begin{aligned}
(r^k, r^{k-1}) &= \|r^{k-1}\|^2 - \frac{\|r^{k-1}\|^2}{(Ap^{k-1}, p^{k-1})}(Ap^{k-1}, r^{k-1}) + (\delta_r^{k-1}, r^{k-1}) \\
&= \|r^{k-1}\|^2 - \frac{\|r^{k-1}\|^2}{(Ap^{k-1}, p^{k-1})}[(Ap^{k-1}, p^{k-1}) - u\|r^{k-1}\|^2 C_{Ap}^{k-2} \\
&\quad -(Ap^{k-1}, \delta_p^{k-2})] + (\delta_r^{k-1}, r^{k-1}) \\
&= \|r^{k-1}\|^2 \left[\gamma_{k-1} u C_{Ap}^{k-2} + \frac{(Ap^{k-1}, \delta_p^{k-2})}{(Ap^{k-1}, p^{k-1})} \right] + (\delta_r^{k-1}, r^{k-1}).
\end{aligned}$$

This can be written as

$$(r^k, r^{k-1}) = u\|r^{k-1}\|^2 C_{k-2},$$

with

$$|C_{k-2}| \le \gamma_{k-1}|C_{Ap}^{k-2}| + \kappa(A)\frac{\|r^{k-1}\|}{\|p^{k-1}\|}(1 + C_p^{k-2}) + 1 + C_p^{k-1} + O(u^2).$$

We now consider $(Ap^k, p^{k-1}) = (p^k, Ap^{k-1})$ for which

$$\gamma_{k-1}(Ap^{k-1}, p^k) = (r^{k-1}, p^k) - (r^k, p^k) + (\delta_r^{k-1}, p^k).$$

We have

$$(r^k, p^k) = \|r^k\|^2(1 + uC_{rp}^k) + (r^k, \delta_p^{k-1})$$

and

$$\begin{aligned}
(r^{k-1}, p^k) &= (r^{k-1}, r^k + \beta_k p^{k-1} + \delta_p^{k-1}) \\
&= u\|r^{k-1}\|^2 C_{k-2} + \frac{\|r^k\|^2}{\|r^{k-1}\|^2}[\|r^{k-1}\|^2(1 + uC_{rp}^{k-1}) + (r^{k-1}, \delta_p^{k-2})] + (r^{k-1}, \delta_p^{k-1}).
\end{aligned}$$

Therefore

$$\begin{aligned}
(r^{k-1}, p^k) - (r^k, p^k) &= u\|r^{k-1}\|^2 C_{k-2} + u\|r^k\|^2(C_{rp}^{k-1} - C_{rp}^k) \\
&\quad + \frac{\|r^k\|^2(r^{k-1}, \delta_p^{k-2})}{\|r^{k-1}\|^2} + (r^{k-1} - r^k, \delta_p^{k-1}).
\end{aligned}$$

Noticing that

$$\frac{1}{\gamma_{k-1}} \le \|A\| \frac{\|p^{k-1}\|^2}{\|r^{k-1}\|^2},$$

we can write

$$(Ap^{k-1}, p^k) = u\|r^{k-1}\|^2 C_{Ap}^{k-1},$$

with

$$|C_{Ap}^{k-1}| \le \|A\| \frac{\|p^{k-1}\|^2}{\|r^{k-1}\|^2} \left[|C_{k-2}| + \frac{\|r^k\|^2}{\|r^{k-1}\|^2} (|C_{rp}^{k-1}| + |C_{rp}^k| + 1 + C_p^{k-2}) \right.$$
$$\left. + \frac{\|p^k\|}{\|r^{k-1}\|}(1 + C_r^{k-1}) + \frac{\|r^k\|}{\|r^{k-1}\|^2}(\|r^k\| + \|r^{k-1}\|)(1 + C_p^{k-1}) \right].$$

This proves that the induction hypotheses are satisfied at iteration k. □

The scalar (p^k, r^{k+1}) for the matrix Strakos30 is shown in Figure 5.24. Figures 5.25 to 5.27 depict (r^k, r^{k+1}), (p^k, Ap^{k+1}), and $(r^k, r^k) - (p^k, r^k)$. All these curves look the same. There is a large decrease when $\|r^k\|$ is rapidly decreasing.

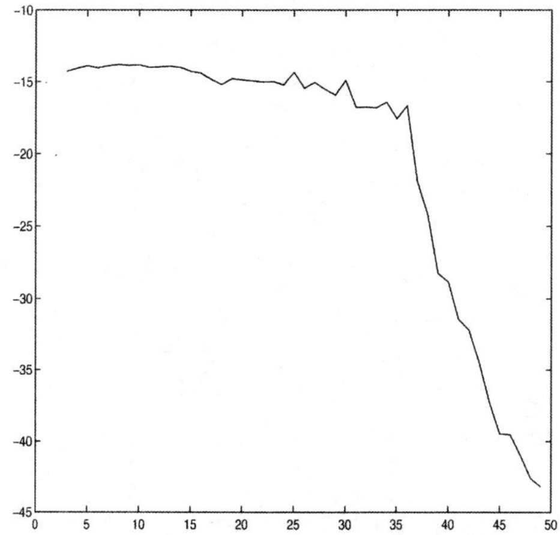

Figure 5.24. *Strakos*30, *local orth.*, $\log_{10}(|(p^k, r^{k+1})|)$

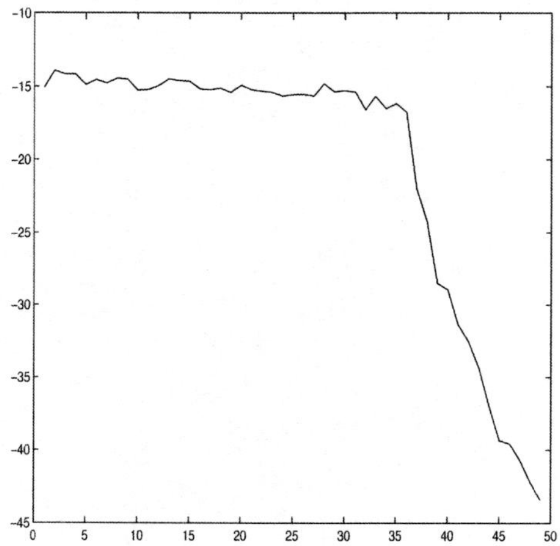

Figure 5.25. *Strakos*30, *local orth.*, $\log_{10}(|(r^k, r^{k+1})|)$

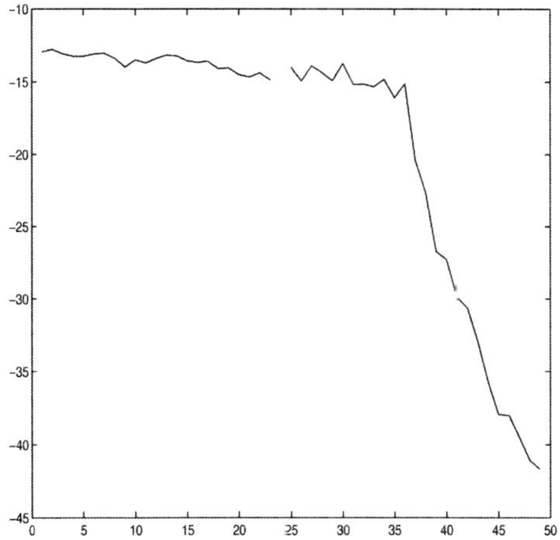

Figure 5.26. *Strakos30, local orth.,* $\log_{10}(|(p^k, Ap^{k+1})|)$

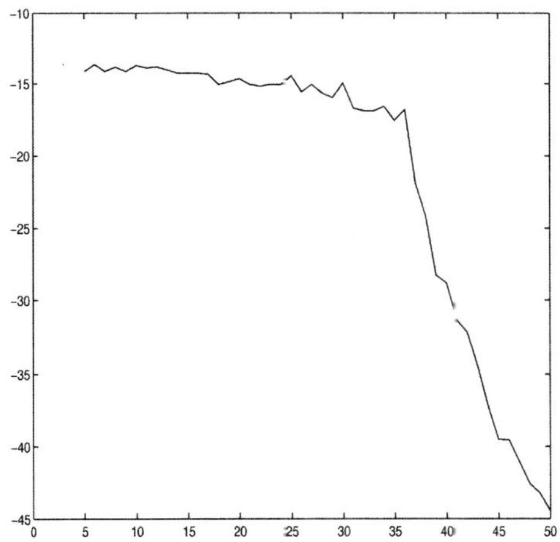

Figure 5.27. *Strakos30, local orth.,* $\log_{10}(|(r^k, r^k) - (p^k, r^k)|)$

Proposition 5.20.

$$\frac{(Ap^k, p^k)}{(Ar^k, r^k)} = 1 - \theta_k^2 + u\mu_k + O(u^2),$$

with

$$\theta_k = \frac{\|r^k\|^2}{\|r^{k-1}\|^2} \frac{\|A^{\frac{1}{2}} p^{k-1}\|}{\|A^{\frac{1}{2}} r^k\|},$$

and

$$|\mu_k| \leq 2\frac{|C_{Ap}^{k-1}|}{\lambda_1} + 2\kappa(A)(1 + C_p^{k-1}).$$

Proof. From the definition of p^k we have

$$(Ap^k, p^k) = (Ar^k, r^k) + 2\beta_k(Ar^k, p^{k-1}) + 2(Ar^k, \delta_p^{k-1}) \\ + 2\beta_k(Ap^{k-1}, \delta_p^{k-1}) + \beta_k^2(Ap^{k-1}, p^{k-1}) + (A\delta_p^{k-1}, \delta_p^{k-1}).$$

Using

$$(Ar^k, p^{k-1}) = (Ap^{k-1}, p^k) - \beta_k(Ap^{k-1}, p^{k-1}) - (\delta_p^{k-1}, Ap^{k-1}),$$

we obtain

$$(Ap^k, p^k) = (Ar^k, r^k) - \beta_k^2(Ap^{k-1}, p^{k-1}) + 2\beta_k(p^k, Ap^{k-1}) \\ + 2(Ar^k, \delta_p^{k-1}) + (A\delta_p^{k-1}, \delta_p^{k-1}).$$

Hence,

$$\frac{(Ap^k, p^k)}{(Ar^k, r^k)} = 1 - \frac{\|r^k\|^4}{\|r^{k-1}\|^4}\frac{(Ap^{k-1}, p^{k-1})}{(Ar^k, r^k)} + u\mu_k + O(u^2),$$

with

$$u\mu_k = 2u\frac{\|r^k\|^2 C_{Ap}^{k-1}}{(Ar^k, r^k)} + 2\frac{(Ar^k, \delta_p^{k-1})}{(Ar^k, r^k)}.$$

Bounding μ_k gives the result. □

Results similar to Proposition 5.20 are given in Bollen [11]. It shows that

$$\frac{(Ap^k, p^k)}{(Ar^k, r^k)} \le 1 + u\mu_k + O(u^2).$$

If $\mu_k < 0$, we have

$$\frac{(Ap^k, p^k)}{(Ar^k, r^k)} \le 1 + O(u^2).$$

Proposition 5.21.

$$\frac{\|p^k\|}{\|r^k\|} \le \sqrt{\kappa(A)}\left(1 + u\frac{\mu_k}{2} + O(u^2)\right).$$

Proof.

$$\lambda_1\|p^k\|^2 \le \frac{(Ap^k, p^k)}{\|p^k\|^2}\|p^k\|^2 \le (Ar^k, r^k)(1 + u\mu_k + O(u^2)).$$

Therefore,

$$\|p^k\|^2 \le \kappa(A)\|r^k\|^2(1 + u\mu_k + O(u^2)). \quad \square$$

This is to be compared with the result of Proposition 5.15.

5.7 CG convergence in finite precision

So far we do not have a direct proof of the convergence of the CG algorithm in finite precision arithmetic in the sense of the limit of the norm of the iterative residual being 0. We cannot use the optimality results of Chapter 2 since their proofs require global orthogonality.

There are not many results about CG convergence in finite precision arithmetic in the literature. One of the earliest work on the usual forms of CG is the Ph.D. thesis of Bollen [11]. However, he got results for the so-called natural form of the coefficients where

$$\gamma_k = \frac{(r^k, p^k)}{(p^k, Ap^k)}, \quad \beta_k = -\frac{(r^k, Ap^{k-1})}{(p^{k-1}, Ap^{k-1})}.$$

Moreover, there are some restrictions on the condition number to obtain convergence which are not clearly stated or, at least, are difficult to understand. Some results were also obtained by Wozniakowski [201], but he did not analyze the usual form of CG. There are results in the book by Cullum and Willoughby [27], but it is not clear if the hypothesis they were doing led to a constraint on the condition number of A, and, moreover, they chose $(A^{-1}r^k, r^k)$ as a measure of the error, where $r^k = b - Ax^k$. We know that this cannot converge to zero in finite precision arithmetic. Hence, as they are formulated in [27], their results are doubtful. Some interesting results have been obtained by Anne Greenbaum. They are summarized in Chapter 3. Her results show that at a given iteration k, the l_2 norm of the error is strictly decreasing from iterations 1 to k. However, one cannot consider the $\bar{A}$-norm since the larger matrix $\bar{A}$ does not define the same norm as A. See also results by Notay [125] about CG in finite precision arithmetic. Notice that Theorem 2.32 proved convergence under the hypothesis of exact local orthogonality (and no local rounding errors), that is,

$$(p^k, r^{k+1}) = 0,$$

$$(p^k, r^k) = (r^k, r^k),$$

$$(r^{k+1}, r^k) = 0.$$

Here we are interested in convergence when local orthogonality holds only approximately. We consider what we have chosen as a measure of the error $\|\varepsilon^k\|_A^2 = (A^{-1}r^k, r^k) = \|A^{-\frac{1}{2}}r^k\|^2$, where r^k is the iterative residual.

Proposition 5.22.

$$\|\varepsilon^{k+1}\|_A^2 = \|\varepsilon^k\|_A^2 - \frac{\|r^k\|^4}{(Ap^k, p^k)} + u\nu_k\|r^k\|^2 + u^2\|r^k\|^2 C_{(2)}^k,$$

with

$$|\nu_k| \le 2|C_{pr}^k|\frac{\|r^k\|^2}{(Ap^k, p^k)} + 2\|A^{-1}\|(1 + C_r^k) + 2\frac{\|r^k\|\,\|p^k\|}{(Ap^k, p^k)}(1 + C_r^k),$$

$$|C_{(2)}^k| \le \|A^{-1}\|(1 + C_r^k)^2.$$

Proof. Using the definition of r^{k+1} we have

$$
\begin{aligned}
(A^{-1}r^{k+1}, r^{k+1}) = (A^{-1}r^k, r^k) + \frac{\|r^k\|^4}{(Ap^k, p^k)} - 2\frac{\|r^k\|^2}{(Ap^k, p^k)}(p^k, r^k) + 2(A^{-1}r^k, \delta_r^k) \\
- 2\frac{\|r^k\|^2}{(Ap^k, p^k)}(p^k, \delta_r^k) + (A^{-1}\delta_r^k, \delta_r^k).
\end{aligned}
$$

We use the fact that $(p^k, r^k) = \|r^k\|^2(1 + uC_{pr}^k)$ and bound the perturbation terms. □

We note that ν_k can be bounded by

$$
|\nu_k| \le 2\frac{|C_{pr}^k|}{\lambda_1}\frac{\|r^k\|^2}{\|p^k\|^2} + 2\frac{1 + C_r^k}{\lambda_1}\left(1 + \frac{\|r^k\|}{\|p^k\|}\right).
$$

By using upper bounds for γ_k we shall prove later on, we can get rid of the factor $\|r^k\|^2/\|p^k\|^2$ by modifying the $O(u^2)$ term.

Notice that, fortunately, the previous result does not prevent $\|\varepsilon^k\|$ and $\|r^k\|$ from going to zero since the first and second order perturbation terms have the norm of the residual as a factor. However, we cannot directly obtain that $\|r^k\| \to 0$ from Proposition 5.22 since we do not know the signs of the factors of u.

Theorem 5.23. *Let $C_\nu^k = \lambda_1 \nu_k$. If*

$$
\kappa(A) < \frac{1}{u|C_\nu^k|} + O(u) \quad \forall k,
$$

then

$$
\|\varepsilon^{k+1}\|_A < \|\varepsilon^k\|_A.
$$

Proof. The proof is by contradiction. Suppose $\|\varepsilon^{k+1}\|_A \ge \|\varepsilon^k\|_A$; then

$$
0 \le -\frac{\|r^k\|^4}{(Ap^k, p^k)} + u\nu_k\|r^k\|^2 + u^2\|r^k\|^2 C_{(2)}^k.
$$

This gives

$$
\frac{\|r^k\|^2}{(Ap^k, p^k)} \le u\frac{|C_\nu^k|}{\lambda_1} + u^2 C_{(2)}^k.
$$

Then,

$$
\|r^k\|^2 \le (Ar^k, r^k)(1 + u|\mu_k| + O(u^2))\left(u\frac{|C_\nu^k|}{\lambda_1} + u^2 C_{(2)}^k\right),
$$

which implies

$$
\kappa(A) \ge \frac{1}{u|C_\nu^k|} + O(u). \quad \square
$$

If the condition number is not too large, $\|\varepsilon^k\|_A$ is strictly decreasing as it is doing in exact arithmetic and therefore the error (and the residual) converges. But, this does not prove that it converges to zero. To obtain this result, we need a uniform bound with a constant smaller than 1 independent of k. However, we can prove that the limit is the zero vector by carefully looking at the equations.

Theorem 5.24. *Let $\varepsilon^k \to \varepsilon$ and $r^k \to r$. If $\|A^{-1}\| \ll 1/u$, the limit is $r = 0$; that is, the computed iterative residual converges to zero.*

Proof. The proof is by contradiction. We suppose that $r \neq 0$. The idea is to pass to the limit in the equations. Unfortunately, Theorem 5.23 does not prove that p^k is converging. But, by what we have seen and some results below, $|p^k|$ is certainly uniformly bounded. Then we have to look carefully at the details of the roundoff error terms in the CG equations. Let us denote $fl(Ap^k) = (A + \Delta A)p^k$, where ΔA is of order u times a given matrix. We have

$$p^k = r^k + \beta_k p^{k-1} + r^k \delta_k + O_1^k p^{k-1},$$

where

$$O_1^k = \beta_k(\delta_k + \delta_k^1) + \beta_k \delta_k \delta_k^1$$

is an (at least) order 1 term since $|\delta_k|$, $|\delta_k^1| < u$. For the other equation we have

$$\begin{aligned} r^{k+1} &= r^k - \gamma_k A p^k + r^k \delta_k^2 - \gamma_k A p^k(\delta_k^2 + \delta_k^3) - \gamma_k \Delta A p^k + \bar{O}_2^k p^k, \\ &= r^k - \gamma_k A p^k + r^k \delta_k^2 + \bar{O}_1^k p^{k-1}, \end{aligned}$$

with $|\delta_k^2|$, $|\delta_k^3| < u$. Writing all the details and all the terms is cumbersome. So, here and in what follows O_1^k and $\bar{O}_1^k$ (resp., O_2^k) denote different terms (or matrices for $\bar{O}_1^k$) of order 1 (resp., 2) in u that could change during the proof. Their precise values are not important; the only relevant thing is that they do not depend on p^k and p^{k-1} and they are at least order 1 in u.

As we said before, the vector p^k is certainly bounded as well as γ_k. Since r^k converges to r, there exists a K such that for $k > K$ we have $r^k = r + \delta_r^k$ with $\delta_r^k = up^{k-1}$ and $\beta_k = 1 - \delta_4^k$, $|\delta_4^k| = O(u)$. Therefore,

$$r = p^k - p^{k-1} - r\delta_k + O_1^k p^{k-1}$$

and

$$r = (p^k - p^{k-1} + O_1^k p^{k-1})/(1 + \delta_k) = p^k - p^{k-1} + O_1^k p^{k-1} + \bar{O}_1^k p^k.$$

The other equation gives

$$\delta_r^{k+1} = \delta_r^k - \gamma_k A p^k + r\delta_k^2 + \bar{O}_1^k p^{k-1} + O_1^k p^k.$$

Inserting the value of r in the last equation gives

$$\gamma_k A p^k = O_1^k p^k + \bar{O}_1^k p^{k-1}$$

Multiplying by A^{-1}/γ_k,

$$p^k = O_1^k A^{-1} p^k + A^{-1} \bar{O}_1^k p^{k-1}.$$

As long as $\|A^{-1}\| \ll 1/u$, $\|A^{-1}\bar{O}_1^k\|$ is small. Moreover, the norm of the matrix $O_1^k A^{-1}$ is of order u and therefore much less than 1. Clearly $\|O_1^k A^{-1}\| < 1$ and $1 - \|O_1^k A^{-1}\| \approx 1$, and hence

$$\|p^k\| \leq \frac{\|A^{-1}\bar{O}_1^k\|}{1 - \|O_1^k A^{-1}\|} \|p^{k-1}\|.$$

This shows that when $k > K$, the multiplying factor is strictly smaller than 1 and p^k converges to zero, which implies that r^k converges to zero as well as ε^k. The important point in this sketch of the proof is that the rounding error terms are proportional to r^k or p^k. □

Having a limitation on $\kappa(A)$ for the decrease of the norm and the proof that the iterative residual r^k converges to zero is not very satisfactory since in numerical computations we never observe a strict increase of $\|\varepsilon^k\|_A$. The main problem we have is that we do not know the signs of the perturbation terms. However, we can derive other bounds for $\|\varepsilon^k\|_A$. We have

$$\|\varepsilon^{k+1}\|_A^2 = \|\varepsilon^k\|_A^2 - \frac{\|r^k\|^4}{(Ar^k, r^k)} \frac{(Ar^k, r^k)}{(Ap^k, p^k)} + u\nu_k \|r^k\|^2 + u^2 \|r^k\|^2 C_{(2)}^k.$$

Since $u\mu_k \ll 1$,

$$\frac{(Ar^k, r^k)}{(Ap^k, p^k)} \geq \frac{1}{1 + u\mu_k + O(u^2)} = 1 - u\mu_k + O(u^2),$$

it follows that

$$\|\varepsilon^{k+1}\|_A^2 \leq \|\varepsilon^k\|_A^2 - \frac{\|r^k\|^4}{(Ar^k, r^k)} (1 - u\mu_k + O(u^2)) + u\nu_k \|r^k\|^2 + u^2 \|r^k\|^2 C_{(2)}^k.$$

Let

$$\omega_k = [\mu_k + O(u^2)] \frac{\|r^k\|^2}{(Ar^k, r^k)} + \nu_k + uC_{(2)}^k.$$

Then,

$$\|\varepsilon^{k+1}\|_A^2 \leq \|\varepsilon^k\|_A^2 - \frac{\|r^k\|^4}{(Ar^k, r^k)} + u\omega_k \|r^k\|^2.$$

This leads to the following result for the residual norms.

Proposition 5.25.

$$\frac{\|r^{k+1}\|^2}{\|r^k\|^2} \leq \kappa(A) - 1 + u\lambda_n \omega_k.$$

Proof.

$$\|\varepsilon^{k+1}\|_A^2 \leq \|r^k\|^2 \left[\frac{1}{\lambda_1} - \frac{\|r^k\|^2}{(Ar^k, r^k)} + u\omega_k \right],$$

$$\leq \|r^k\|^2 \left[\frac{1}{\lambda_1} - \frac{1}{\lambda_n} + u\omega_k \right].$$

Then we use

$$\frac{1}{\lambda_n}\|r^{k+1}\|^2 \le \|\varepsilon^{k+1}\|_A^2. \quad \square$$

Proposition 5.26. *If* $\|\varepsilon^k\|_A < \|\varepsilon^j\|_A$, $j < k$,

$$\|r^k\| < \sqrt{\kappa(A)}\|r^j\|.$$

Proof. We have

$$\frac{1}{\lambda_n}\|r^k\|^2 \le \|\varepsilon^k\|_A^2 < \|\varepsilon^j\|_A^2.$$

Then $\varepsilon^j = A^{-1}r^j$ and $\|\varepsilon^j\|_A^2 = (A^{-1}r^j, r^j)$. This gives

$$\|r^k\|^2 < \kappa(A)\|r^j\|^2. \quad \square$$

In particular, if $\|\varepsilon^k\|_A < \|\varepsilon^0\|_A$ for all k, then the norm of the residual is bounded. We can write

$$\frac{\|r^k\|^4}{(r^k, Ar^k)} = \frac{\|r^k\|^4\|\varepsilon^k\|_A^2}{(r^k, Ar^k)(r^k, A^{-1}r^k)}.$$

Therefore,

$$\|\varepsilon^{k+1}\|_A^2 \le \left(1 - \frac{\|r^k\|^4}{(r^k, Ar^k)(r^k, A^{-1}r^k)}\right)\|\varepsilon^k\|_A^2 + u\omega_k\|r^k\|^2.$$

We can now use the Kantorovitch inequality

$$\frac{\|r^k\|^4}{(r^k, Ar^k)(r^k, A^{-1}r^k)} > \frac{4\lambda_1\lambda_n}{(\lambda_1+\lambda_n)^2} = \frac{4\kappa(A)}{(\kappa(A)+1)^2}.$$

From this we get the following result.

Theorem 5.27.

$$\|\varepsilon^{k+1}\|_A^2 < \left(\frac{\kappa(A)-1}{\kappa(A)+1}\right)^2\|\varepsilon^k\|_A^2 + u\omega_k\|r^k\|^2.$$

Proof. Using the lower bound given by the Kantorovitch inequality and

$$1 - \frac{4\kappa(A)}{(\kappa(A)+1)^2} = \left(\frac{\kappa(A)-1}{\kappa(A)+1}\right)^2$$

we obtain the result. $\square$

In this result, we have a bound with a uniform constant smaller than 1, but unfortunately there is an additional term proportional to the roundoff unit u. We also have

$$\frac{\|\varepsilon^k\|_A^2 - \|\varepsilon^{k+1}\|_A^2}{\|\varepsilon^k\|_A^2} = \frac{\|r^k\|^4}{(Ap^k, p^k)(A^{-1}r^k, r^k)} - \frac{u\nu_k\|r^k\|^2 + u^2C_{(2)}^k\|r^k\|^2}{(A^{-1}r^k, r^k)}.$$

We can factor this as

$$\frac{\|\varepsilon^k\|_A^2 - \|\varepsilon^{k+1}\|_A^2}{\|\varepsilon^k\|_A^2} = \frac{\|r^k\|^4}{(Ar^k, r^k)(A^{-1}r^k, r^k)} \left[\frac{(Ar^k, r^k)}{(Ap^k, p^k)} - \frac{(Ar^k, r^k)}{\|r^k\|^2}[u\nu_k + u^2 C^k_{(2)}] \right]$$
$$\geq \frac{4\kappa(A)}{(\kappa(A)+1)^2}[1 - u\mu_k - u\lambda_n \nu_k + O(u^2)]$$

and this proves the following result.

Proposition 5.28. *If* $1 > 1 - u\mu_k - u\lambda_n \nu_k + O(u^2) > 0$, *we have*

$$\frac{\|\varepsilon^k\|_A^2 - \|\varepsilon^{k+1}\|_A^2}{\|\varepsilon^k\|_A^2} \geq \theta,$$

with $0 < \theta < 1$. *Then,*

$$\|\varepsilon^{k+1}\|_A^2 \leq (1-\theta)\|\varepsilon^k\|_A^2.$$

This proves the convergence to zero of the norm, provided the condition $1 > 1 - u\mu_k - u\lambda_n \nu_k + O(u^2) > 0$ is fulfilled. Unfortunately, it is not really straightforward to interpret this condition. To try to obtain a more direct proof of convergence, let us write

$$\|\varepsilon^{k+1}\|_A^2 < C_\kappa \|\varepsilon^k\|_A^2 + u\omega_k \|r^k\|^2,$$

with

$$C_\kappa = \left(\frac{\kappa(A) - 1}{\kappa(A) + 1} \right)^2 < 1.$$

To obtain further insight we write the inequality for $\|\varepsilon^{k+1}\|_A^2$ as

$$C_\kappa \frac{\|\varepsilon^{k+1}\|_A^2}{C_\kappa^{k+1}} \leq C_\kappa \|\varepsilon^0\|_A^2 + u \left[\omega_k \frac{\|r^k\|^2}{C_\kappa^k} + \omega_{k-1} \frac{\|r^{k-1}\|^2}{C_\kappa^{k-1}} + \cdots + \omega_0 \|r^0\|^2 \right].$$

Bounding $\|\varepsilon^{k+1}\|_A^2$ from below and denoting

$$y_i = \frac{\|r^i\|^2}{C_\kappa^i},$$

we can write

$$y_{k+1} \leq \lambda_n \|\varepsilon^0\|_A^2 + u \frac{\lambda_n}{C_\kappa} \sum_{i=0}^{k} |\omega_i| y_i.$$

To bound y_{k+1} we apply the discrete Gronwall lemma.

Lemma 5.29. *Let* $\tilde{\omega}_i$, $i = 0, \ldots, k$, *be a sequence of positive numbers,* g_0 *given and*

$$z_0 \leq g_0, \quad z_k \leq g_0 + \sum_{i=0}^{k-1} \tilde{\omega}_i z_i.$$

Then for all k

$$z_k \le g_0 \exp\left(\sum_{i=0}^{k-1} \tilde{\omega}_i\right).$$

Applying this lemma to our inequality for y_{k+1} gives

$$y_{k+1} \le \lambda_n \|\varepsilon^0\|_A^2 \exp\left(u\frac{\lambda_n}{C_\kappa}\sum_{i=0}^{k}|\omega_i|\right).$$

Translating this to the norm of the residual we have the following result.

Proposition 5.30.

$$\|r^{k+1}\|^2 \le C_\kappa^{k+1}\lambda_n \|\varepsilon^0\|_A^2 \exp\left(u\frac{\lambda_n}{C_\kappa}\sum_{i=0}^{k}|\omega_i|\right).$$

If we suppose that

$$u\frac{\lambda_n}{C_\kappa}\sum_{i=0}^{k}|\omega_i|$$

is uniformly bounded by a constant independent of k, then we are done. However, it is not really clear that the sum over i can be uniformly bounded. To allow for some possible growth of this term we notice that $C_\kappa^{k+1} = \exp((k+1)\log(C_\kappa))$. Then,

$$\|r^{k+1}\|^2 \le \lambda_n \|\varepsilon^0\|_A^2 \exp\left((k+1)\log(C_\kappa) + u\frac{\lambda_n}{C_\kappa}\sum_{i=0}^{k}|\omega_i|\right).$$

To obtain something converging to zero on the right-hand side, we must have the argument of the exponential negative and going to $-\infty$.

Theorem 5.31. *If for all k,*

$$\log(C_\kappa) + u\frac{\lambda_n}{(k+1)C_\kappa}\sum_{i=0}^{k}|\omega_i| < C < 0,$$

where C is a constant independent of k, then $\|r^k\| \to 0$ *when* $k \to \infty$.

Since $C_\kappa < 1$, $\log(C_\kappa) < 0$. Therefore, we have convergence if

$$u\frac{\lambda_n}{(k+1)}\sum_{i=0}^{k}|\omega_i| < C_\kappa(C - \log C_\kappa).$$

A necessary condition is $C - \log C_\kappa > 0$. This gives a restriction on the condition number

$$\kappa(A) < \frac{1+\exp(C)}{1-\exp(C)}.$$

If the absolute value of C is small, the restriction on the condition number is not too severe. Supposing the necessary condition is satisfied, we still have to fulfill the previous inequality. Remember that

$$\omega_k = [\mu_k + O(u^2)]\frac{\|r^k\|^2}{(Ar^k, r^k)} + \nu_k + uC^k_{(2)}.$$

So, we are essentially interested in

$$\mu_k \frac{\|r^k\|^2}{(Ar^k, r^k)} + \nu_k,$$

with

$$|\mu_k| \le 2\frac{|C^{k-1}_{Ap}|}{\lambda_1} + 2\kappa(A)(1 + C^{k-1}_p),$$

and

$$|\nu_k| \le 2|C^k_{pr}|\frac{\|r^k\|^2}{(Ap^k, p^k)} + 2\|A^{-1}\|(1 + C^k_r) + 2\frac{\|r^k\|\,\|p^k\|}{(Ap^k, p^k)}(1 + C^k_r).$$

Unfortunately, once again, the condition for convergence is not easy to interpret. For further reference we write

$$\|\varepsilon^{k+1}\|^2_A = \|\varepsilon^k\|^2_A - \gamma_k\|r^k\|^2 + u\zeta_k\|r^k\|^2,$$

with $\zeta_k = \nu_k + uC^k_{(2)}$.

Proposition 5.32.

$$\|\varepsilon^{k+1}\|^2_A = \|\varepsilon^0\|^2_A - \sum_{i=0}^{k}(\gamma_i - u\zeta_i)\|r^i\|^2.$$

These results lead to a bound for γ_k.

Proposition 5.33.

$$\gamma_k \le \frac{1}{\lambda_1} + u\zeta_k.$$

Proof. Writing that $\|\varepsilon^{k+1}\|^2_A \ge 0$, we have

$$\gamma_k\|r^k\|^2 \le \|\varepsilon^k\|^2_A + u\zeta_k\|r^k\|^2 \le \frac{1}{\lambda_1}\|r^k\|^2 + u\zeta_k\|r^k\|^2. \quad \square$$

5.8 Numerical examples of convergence in variable precision

In this section we report numerical results for some examples of small dimension in variable precision. The right-hand sides were random vectors and the initial vectors were zero. Results are shown in Figures 5.28 to 5.37. All the dotted curves are the $\log_{10}$ of the norms of the x-residuals. As predicted they stagnate at different levels depending on the norm of A and the precision. The other curves are the $\log_{10}$ of the norms of the iterative residuals. The numbers on the figures give the value of *Digits* for variable precision; *stand* stands

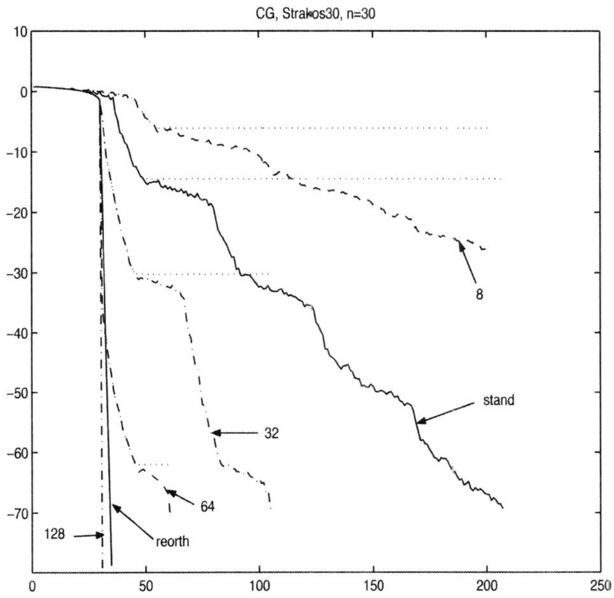

Figure 5.28. *Strakos*30

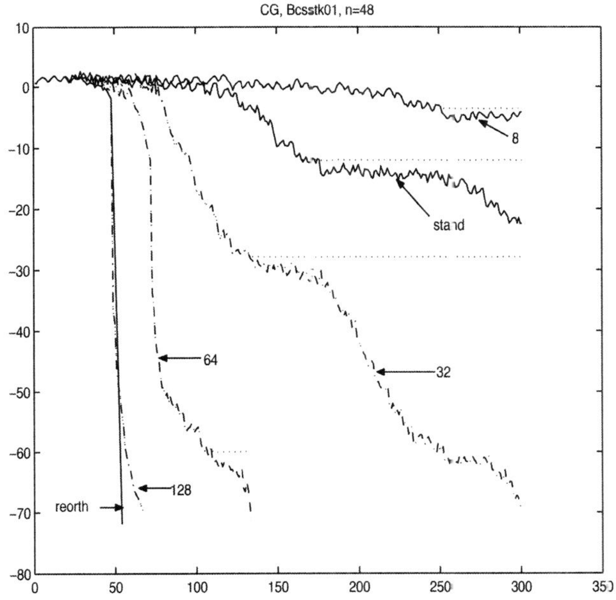

Figure 5.29. *Bcsstk*01

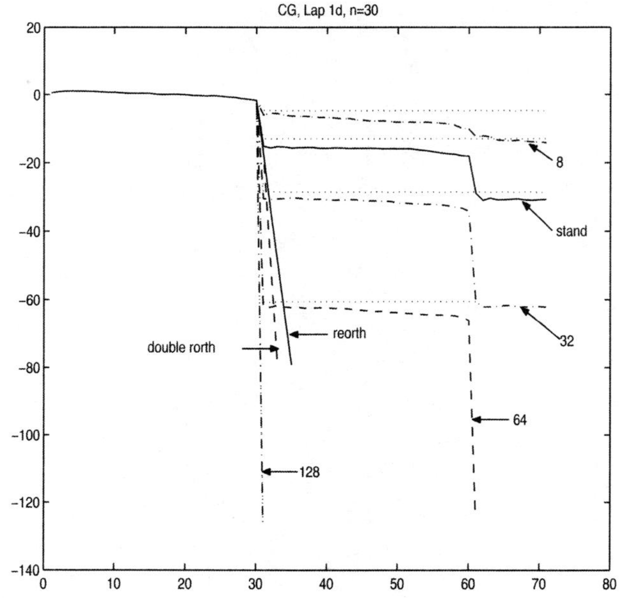

Figure 5.30. $Lap1d30$

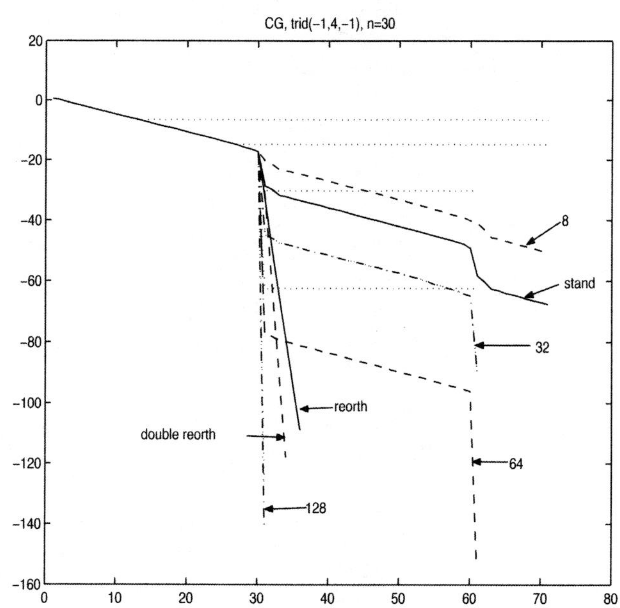

Figure 5.31. $T4$-30

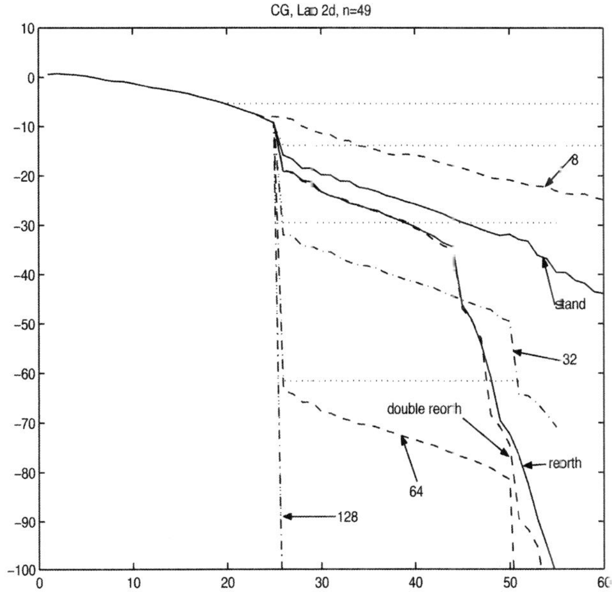

Figure 5.32. *Lap2d49*

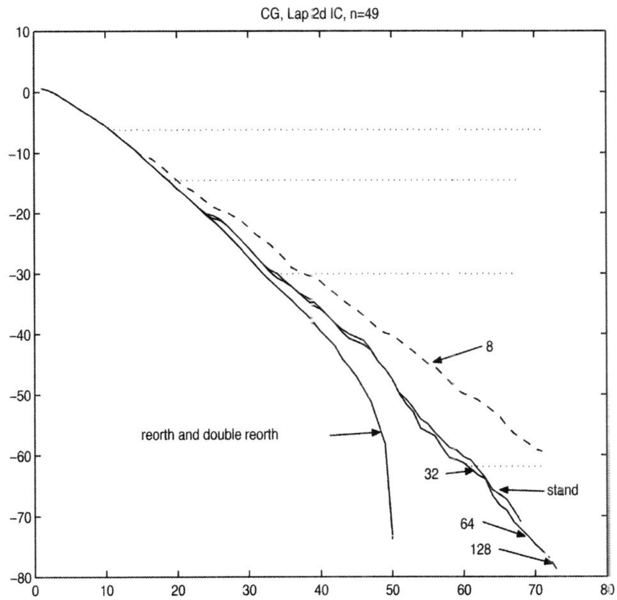

Figure 5.33. *Lap2dic49*

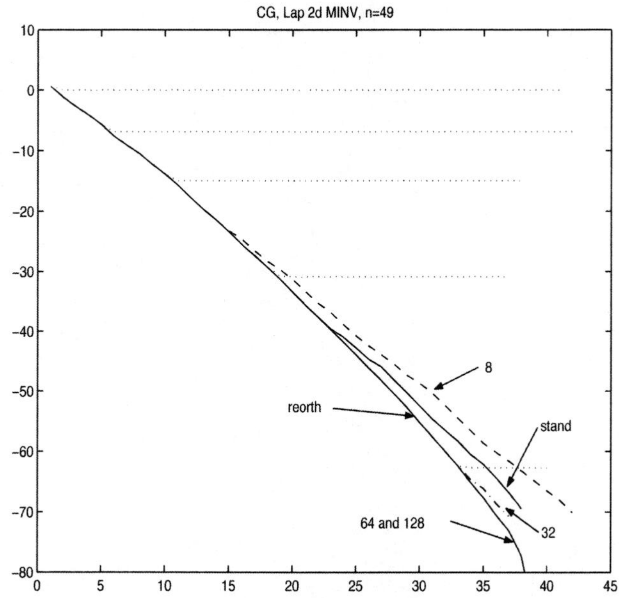

Figure 5.34. *Lap2dminv49*

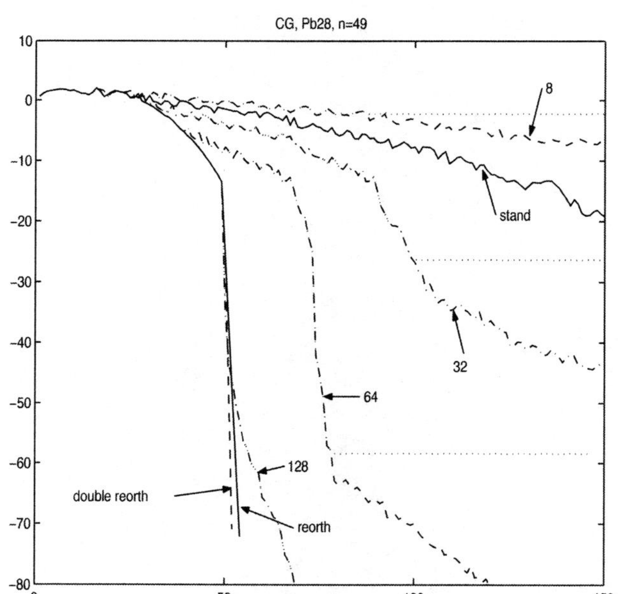

Figure 5.35. *Pb28-49*

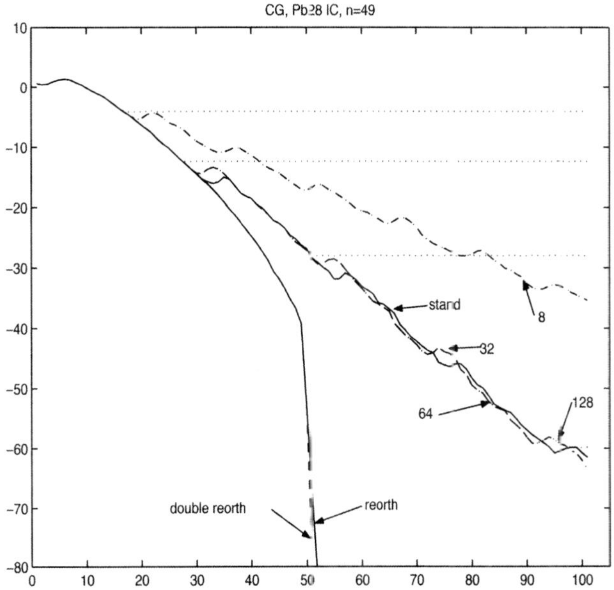

Figure 5.36. *Pb*28*ic*49

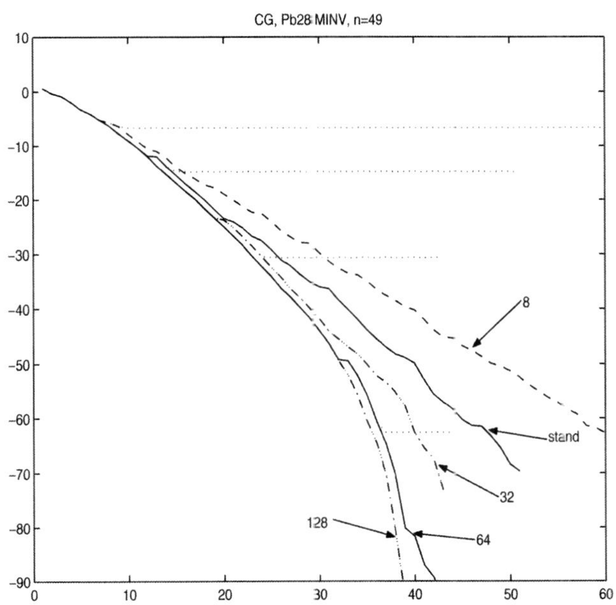

Figure 5.37. *Pb*28*minv*49

for the classical double precision floating point CG whose results are close to $Digits = 16$. *Reorth* or *doublereorth* curves show the results in floating point with reorthogonalization or double reorthogonalization every iteration. Of course, it may seem useless to reduce the norms of the residual to 10^{-70} or less, but this shows that the iterative residual converges to zero and it allows us to observe some interesting behaviors as well as stagnation with $Digits = 64$.

For the Strakos30 problem, the curves "converges." They all give the same residuals at the beginning, but only when using $Digits = 64$ or 128 do we get a very rapid convergence at iteration 30 (the dimension of the problem). Reorthogonalization gives almost the same behavior as $Digits = 128$. This is why we used double reorthogonalization results as the "exact" results for comparison purposes. The general trend is more or less similar for the matrix Bcsstk01.

Lap1d30 is the tridiagonal matrix of the one-dimensional Poisson equation of order 30, that is, with coefficients $(-1, 2, -1)$. There is a large decrease at iteration 30, then almost stagnation, and again a decrease at iteration 60. All results are almost the same; that is, there is no convergence until iteration 30. The reorthogonalization curves are different from $Digits = 128$.

T4-30 is the tridiagonal matrix $(-1, 4, -1)$. We have almost the same behavior as for the Poisson equation matrix, but after the large decrease at iteration 30, the norms of the residuals continue to decrease. In log scale, the slopes of the decrease are the same whatever the precision.

Lap2d49 is the matrix of the two-dimensional Poisson equation in a square with a 7×7 mesh giving a matrix of order 49. The main difference with previous examples is the result for CG with reorthogonalization. It is much different from any variable precision result. Note that in this example we have only 26 distinct eigenvalues. This explains the rapid decrease after 26 iterations.

Lap2dic49 is the same example but using an incomplete Cholesky preconditioner. We computed the incomplete decomposition LL^T and apply CG to $L^{-T}AL^{-1}$. Except $Digits = 8$, all the curves in variable precision are almost the same. This is another example for which we can say that the iterative residual is the true residual since with a better precision we are able to compute the solution corresponding to an iterative residual which is almost the same as with a smaller precision. Surprisingly with reorthogonalization we obtain better results. Depending on the precision that is required, not using any preconditioning can be more efficient than using IC with a large value of $Digits$. Remember that A has many multiple eigenvalues while this is not true for the preconditioned matrix. Of course, this would not happen in examples of larger dimension for which convergence without a preconditioner is much worse.

Lap2dminv49 is the same example but with a block modified incomplete decomposition preconditioner; see [24], [25]. Convergence is faster than with the earlier point incomplete Cholesky decomposition.

Figures 5.35 to 5.37 are concerned with another problem. This is the finite difference approximation of a diffusion equation with discontinuous coefficients in the unit square with a 7×7 mesh. The diffusion coefficients in x and y are 1 everywhere except in the square $[0.1, 0.6] \times [0.2, 0.7]$, where the value is 1000. The conclusions are the same as for the Poisson model problem, but convergence is slower. For the incomplete Cholesky preconditioner, even with $Digits = 128$, we do not have convergence in 49 iterations, as it seems we have with reorthogonalization. It is likely that even with so many digits of precision there are still some eigenvalues which are not approximated to working precision. For this example, computations with reorthogonalization better represent exact arithmetic, but we have seen this is not always the case. It would be interesting to explain these differences. See Higham's remarks on extended precision computations [94]. For computations using Krylov methods in extended precision, see the Ph.D. thesis of Facius [49].

Chapter 6

The maximum attainable accuracy

In the numerical experiments of the previous chapters we have seen that we cannot obtain a norm of the error $x-x^k$ (where x is the unknown solution) which is as small as we may want. At some point in the iterations the norm of the error $x - x^k$ and the norm of the residual $b - Ax^k$ stagnate. In this chapter we shall study this problem and obtain computable estimates of the maximum attainable accuracy. This helps stopping the iterations when there is no more improvement of the solution. We shall also consider ways to improve the maximum attainable accuracy.

6.1 Difference of residuals

In the CG algorithm the residuals are computed iteratively rather than computing $b - Ax^k$ for saving a matrix-vector product at each iteration. In previous chapters we denoted the computed residual by r^k and called it the iterative residual. The residual given by the definition $\hat{r}^k = b - Ax^k$ is called the x-residual. In fact, as we have seen, what we could have access to is the computed x-residual $\tilde{r}^k = fl(b - Ax^k)$. We have already noticed that

$$\tilde{r}^k = fl(b - Ax^k) = b - Ax^k + \mathbf{u}C^k,$$

where C^k is a vector such that

$$|C^k| \leq (m+1)|A|\,|x^k| + |b|,$$

m being the maximum number of nonzero entries in any row of A. Even if x^k would be the exact solution x, there are rounding errors in the computation of the x-residual. So, even if x^k is very close to x, the computed x-residual cannot converge to zero unless $b = 0$. We also notice that we may have problems if the computed x-residual is used in the stopping criteria because if the given threshold is too small, the criteria can eventually never be satisfied. We have seen in our numerical experiments that there exists a discrepancy between the iterative and computed x-residuals during the iterations, at least after the beginning of stagnation. The norm of the computed x-residual goes down until a certain iteration after which it stagnates while the norm of the iterative residual continues to decrease. There are even some cases in which after stagnating for awhile the x-residual may increase before going back to the stagnation level. This behavior was analyzed by several researchers. We shall briefly review the results of Greenbaum [79], who obtained the most significant results about this problem. We have seen in the previous chapter that the computed iterates and residuals satisfy

$$x^k = x^{k-1} + \gamma_{k-1} p^{k-1} + \delta_x^{k-1},$$
$$r^k = r^{k-1} - \gamma_{k-1} A p^{k-1} + \delta_r^{k-1},$$

with

$$|\delta_x^{k-1}| \le u[|x^{k-1}| + 2\gamma_{k-1}|p^{k-1}|] + |\delta_\gamma^{k-1}|\,|p^{k-1}| + O(u^2),$$
$$|\delta_r^{k-1}| \le u[(m+2)\gamma_{k-1}|A|\,|p^{k-1}| + |r^{k-1}|] + |\delta_\gamma^{k-1}|\,|A|\,|p^{k-1}| + O(u^2).$$

If we use the computed value of γ_{k-1}, we should take $\delta_\gamma^{k-1} = 0$ in these inequalities.

Proposition 6.1. *The difference between the iterative residual r^k and the x-residual $\hat{r}^k = b - Ax^k$ is given by*

$$r^k - \hat{r}^k = (r^0 - \hat{r}^0) + \sum_{j=0}^{k-1} (A\delta_x^j + \delta_r^j).$$

Proof. We have

$$\hat{r}^k = \hat{r}^{k-1} - \gamma_{k-1} A p^{k-1} - A\delta_x^{k-1}.$$

By taking the difference of residuals, we obtain

$$r^k - \hat{r}^k = r^{k-1} - \hat{r}^{k-1} + \delta_r^{k-1} + A\delta_x^{k-1}.$$

Summing up for all k, we obtain the result. $\square$

Taking norms on both sides,

$$\|r^k - \hat{r}^k\| \le \|r^0 - \hat{r}^0\| + \sum_{j=0}^{k-1} (\|A\|\,\|\delta_x^j\| + \|\delta_r^j\|).$$

Since $\tilde{r}^0 = r^0 = fl(b - Ax^0) = \hat{r}^0 + uC^0$,

$$|C^0| \le (m+1)|A|\,|x^0| + |b|,$$

it turns out that

$$\|r^0 - \hat{r}^0\| \le u(1+c)\|A\|\,\|x^0\| + u\|b\|,$$

with $c = m\sqrt{n}$. Greenbaum proved the following result [79].

Proposition 6.2. *Let $\theta_k = \max_{j \le k} \|x^j\|$; then*

$$\sum_{j=0}^{k-1} \|\delta_x^j\| \le 5uk\theta_k + O(u^2)$$

and

$$\sum_{j=0}^{k-1} \|\delta_r^j\| \le uk(1 + (5 + 2c)\theta_k)\|A\| + O(u^2).$$

Because of the way it has been obtained, this last bound is an overestimate since we have seen that the norm of δ_r^k converges to zero with r^k. Putting all these results together, Greenbaum obtained a bound for the norm of the difference between the iterative residual and the x-residual. As we said before, what we are able to observe is some norm of the difference between the computed iterative residual r^k and the computed x-residual $\tilde{r}^k$.

Proposition 6.3.

$$r^k - \tilde{r}^k = u(C^0 - C^k) + \sum_{j=0}^{k-1} \delta_r^j + A\delta_x^j.$$

Proof. The term C^0 arises from $\tilde{r}^0 - \hat{r}^0$ and the term C^k from $\tilde{r}^k - \hat{r}^k$. $\square$

The upper bound suggests that the difference between both residuals arises from summation of local roundoff errors at every iteration and is proportional to $u\|A\|$ and/or $u\|b\|$. Following this analysis after k iterations we must have something of the order of

$$u\left(\|b\| + Ck\|A\| \max_k \|x^k\|\right),$$

where C is a constant independent of k. According to these results the norm of the difference could increase during the iterations. We shall see in the numerical experiments that this analysis is usually too pessimistic. In particular, if we are in the situation where $\|r^k\|$ is rapidly decreasing and $\|\hat{r}^k\|$ stagnates, at some point the computed difference must be numerically constant even though the difference could be theoretically increasing. Numerically, there is only a small increase of the difference of the residuals for most examples. Sometimes there is even a decrease or oscillations.

6.2 Numerical experiments for $Ax = 0$

We start by looking at solving $Ax = 0$, whose exact solution is, of course, $x = 0$. This provides us with a problem for which we are able to compute the l_∞ (max) norm of the error given by an approximation x^k without any additional floating point operations by looking at $\max_i |x_i^k|$. Moreover, ideally we must be able to drive the residuals and the approximate solutions to be as small as we would like.

The first problem we solve is arising from the Poisson equation in the unit square with homogeneous Dirichlet boundary conditions discretized using standard finite differences. We use a random initial vector with components between 0 and 1. We stop the iterations as soon as $\|r^k\| \le 10^{-20}\|r^0\|$. This should not cause any problem since the norm of the computed iterative residual decreases to 0. Here, the stopping criteria does not matter as long as it allows us to reach stagnation for the computed x-residual. We computed as a function of the iteration number the l_2 norms of the computed residual, the computed x-residual, the error (which is x^k), and the A-norm of the error. A typical situation is exemplified in Figure 6.1, which shows the $\log_{10}$ of the norms for a 20×20 mesh which gives a matrix of order $n = 400$. We see that the l_2 norm of the computed x-residual stagnates at a value $7.1\ 10^{-15}$ after approximately 93 iterations. This is also true (with slightly different values) of the norms of the errors. The norm of the computed residual continues to decrease to 0. The norms of both residuals start to differ significantly at iteration 90. Before this point (at least on a log scale) they are indistinguishable. To be more precise, the max norm of the difference of the residuals over the first 90 iterations is $1.14\ 10^{-15}$ at iteration 13. The l_2 norm of the initial residual is 24.56 and after stagnation $\|\tilde{r}^k\|/\|r^0\| = 2.9\ 10^{-16}$.

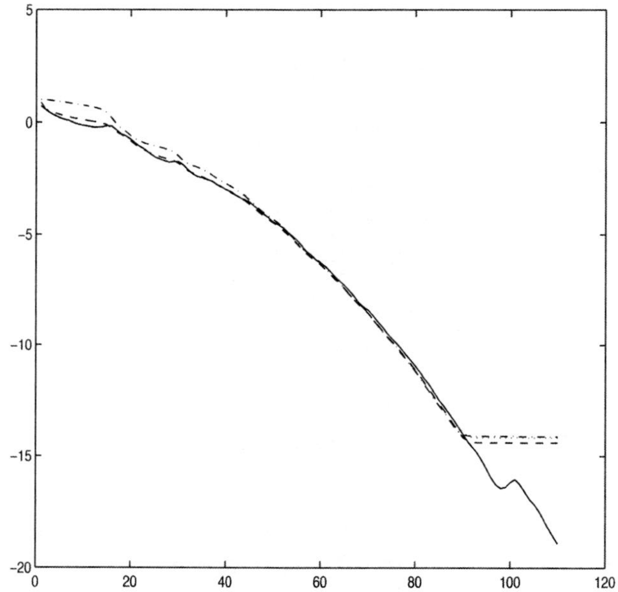

Figure 6.1. *Norms for the Poisson problem with $b = 0$, $n = 400$. Solid: residual; dotted: computed x-residual; dashed: A-norm of the error; dot-dashed: l_2 norm of the error*

Table 6.1 shows the same information for different problem sizes for the Poisson equation. It gives the norms at convergence. We solve problems of order ranging from $n = 400$ to $n = 250000$. We see that the maximum attainable accuracy measured in the suitably scaled l_2 norm (because the size of the problem is increasing) is more or less constant from the first to the last experiment. The scaled A-norm and l_2 norm of the error are only very slowly growing. The l_∞ norm grows only by a factor of 18. Notice that the number of iterations is multiplied by 20 and the order of the problem grows by a factor 625, but the norm of A is bounded by 8 whatever the order of the matrix. The number of floating point operations per iteration in this implementation of CG is $19n$. The number of floating point operations has been multiplied by a factor 13900 (which is $1.16\ 10^{10}/836000$).

Table 6.1. *CG for the Poisson problem,* $Ax = 0$

n	nb. it.	$\|r^k\|/\sqrt{n}$	$\|\tilde{r}^k\|/\sqrt{n}$	$\|e^k\|_A/\sqrt{n}$	$\|e^k\|/\sqrt{n}$	$\|e^k\|_\infty$
400	110	$6.4\ 10^{-21}$	$3.55\ 10^{-16}$	$2.07\ 10^{-16}$	$3.94\ 10^{-15}$	$1.04\ 10^{-15}$
2500	267	$1.08\ 10^{-20}$	$4.96\ 10^{-16}$	$2.5\ 10^{-16}$	$5.6\ 10^{-16}$	$1.68\ 10^{-15}$
4900	370	$1.1\ 10^{-20}$	$6.21\ 10^{-16}$	$3.08\ 10^{-16}$	$7.86\ 10^{-15}$	$2.65\ 10^{-15}$
10000	520	$1.17\ 10^{-20}$	$7.18\ 10^{-16}$	$3.45\ 10^{-16}$	$7.60\ 10^{-15}$	$2.42\ 10^{-15}$
22500	771	$1.29\ 10^{-20}$	$8.6\ 10^{-16}$	$4.1\ 10^{-16}$	$1.15\ 10^{-15}$	$3.69\ 10^{-15}$
40000	1020	$1.29\ 10^{-20}$	$9.7\ 10^{-16}$	$4.58\ 10^{-16}$	$2.03\ 10^{-15}$	$6.81\ 10^{-15}$
90000	1507	$1.27\ 10^{-20}$	$1.16\ 10^{-15}$	$5.37\ 10^{-16}$	$1.85\ 10^{-15}$	$6.04\ 10^{-15}$
160000	1989	$1.27\ 10^{-20}$	$1.3\ 10^{-15}$	$6.02\ 10^{-16}$	$3.35\ 10^{-15}$	$1.32\ 10^{-14}$
250000	2447	$1.26\ 10^{-20}$	$1.5\ 10^{-15}$	$6.92\ 10^{-16}$	$6.26\ 10^{-15}$	$1.86\ 10^{-14}$

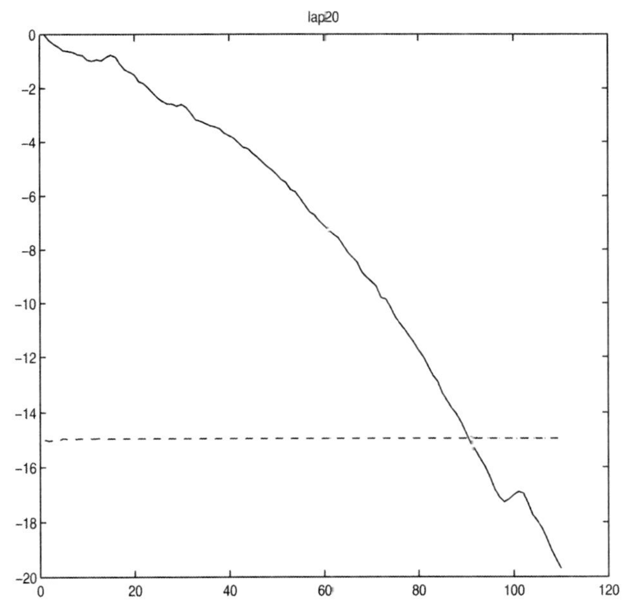

Figure 6.2. $\log_{10}$ *of max norms for Poisson* $n = 400$. *Solid: residual; dotted: computed* x*-residual; dashed: difference*

Figure 6.2 gives the $\log_{10}$ of the max norms of the two residuals and their difference for $n = 400$. The dotted curve is behind the solid one until the point of stagnation. We can see that the norm of the difference is almost constant during the iterations. So, there is no accumulation of roundoff errors in this example. This experimentally shows that the problem of the differences between the computed iterative and computed x-residuals and the stagnation does not come from a pure accumulation of rounding errors even when we do a large number of iterations on a large problem. Otherwise the differences must have been increasing all the time.

We now look at a set of symmetric positive definite matrices arising from the Matrix Market [115]. The characteristics of these matrices are given in Table 6.2. We give the order of the matrix n, the number of nonzero entries nnz, the minimum and maximum eigenvalues and their ratio, and the condition number κ. Some matrices are symmetrically scaled to have ones on the diagonal. This corresponds to using a diagonal preconditioner. They are denoted by an s at the end of the matrix name. The initial vector is random with components in [0, 1]. The stopping criterion is $\|r^k\| \leq 10^{-20}\|r^0\|$. Let us consider some of the examples whose results are given in Table 6.3, starting with the smaller ones.

Table 6.2. *Characteristics of the examples (in alphabetical order)*

Matrix	n	nnz	λ_{min}	λ_{max}	κ
1138-bus	1138	4054	$3.52\ 10^{-3}$	$3.01\ 10^{4}$	$8.6\ 10^{6}$
1138-bus s	1138	4054	$4.08\ 10^{-6}$	2	$4.9\ 10^{5}$
bcspwr01	39	131	1.2	4.55	3.8
bcsstk01	48	400	$3.42\ 10^{3}$	$3.02\ 10^{9}$	$8.8\ 10^{5}$
bcsstk01 s	48	400	$1.54\ 10^{-3}$	2.1	$1.36\ 10^{3}$
bcsstk09	1083	18437	$7.1\ 10^{3}$	$6.8\ 10^{7}$	$9.52\ 10^{3}$
bcsstk18 s	11948	149090	$< 8.6\ 10^{-5}$	4.63	$> 5.37\ 10^{4}$
msc04515 s	4515	97707	$< 5.6\ 10^{-6}$	4.77	$> 8.5\ 10^{5}$
nos1	237	1017	$1.23\ 10^{2}$	$2.46\ 10^{9}$	$1.99\ 10^{7}$
nos1 s	237	1017	$5.08\ 10^{-7}$	2	$3.93\ 10^{6}$
nos2 s	957	4137	$1.99\ 10^{-9}$	2	$1.00\ 10^{9}$
nos3	960	15844	$1.83\ 10^{-2}$	$6.90\ 10^{2}$	$3.77\ 10^{4}$
nos4	100	594	$5.38\ 10^{-4}$	$8.49\ 10^{-1}$	$1.58\ 10^{3}$
nos5	468	5172	$5.29\ 10^{1}$	$5.82\ 10^{5}$	$1.10\ 10^{4}$
nos6	675	3255	1.00	$7.65\ 10^{6}$	$7.65\ 10^{6}$
nos6 s	675	3255	$5.74\ 10^{-7}$	2	$3.48\ 10^{6}$
nos7 s	729	4617	$1.55\ 10^{-8}$	2	$1.29\ 10^{8}$

-*bcspwr*01. This is an example where everything works fine. The order is small ($n = 39$), and the condition number and the norms of the matrix are small, for instance, $\|A\|_\infty = 5.4$. The norm of the initial residual is 8.9; therefore using the stopping criteria gives a computed residual of the same order as $\epsilon = 10^{-20}$. The l_∞ norm of the difference of residuals is $4.6\ 10^{-16}$. The norms of the errors are of the order of 10^{-15}.

-*bcsstk*01. This is a small example ($n = 48$) where things are not so nice. The condition number is of the order 10^6. The size of the largest element of the matrix is of the order 10^9 as well as the norms of the matrix. The norm of the initial residual is $5.9\ 10^9$ and we stop when the norm of the computed residual is 10^{-11}. This gives a norm of the computed x-residual of the order of 10^{-6}. However, the norms of the errors are much smaller. For instance, the max norm of the error is 10^{-13}. The behavior of the l_2 and the max norms is almost the same during the iterations while the A-norm is larger and the residual norm is even larger and oscillating. This is an example where using the residual norm to stop the iterations is misleading. Moreover, there is almost no decrease in the norm of the error for

Table 6.3. *Norms at convergence solving $Ax = 0$ (in alphabetical order)*

Matrix	nb. it.	$\|r^k\|$	$\|\tilde{r}^k\|$	$\|e^k\|_A$	$\|e^k\|$	$\|e^k\|_\infty$
1138-bus	5301	$2.01\ 10^{-16}$	$5.57\ 10^{-11}$	$8.51\ 10^{-11}$	$1.36\ 10^{-10}$	$4.68\ 10^{-12}$
1138-bus s	1457	$1.28\ 10^{-19}$	$1.27\ 10^{-16}$	$4.24\ 10^{-14}$	$1.88\ 10^{-11}$	$2.7\ 10^{-12}$
bcspwr01	29	$2.08\ 10^{-20}$	$1.01\ 10^{-15}$	$6.57\ 10^{-16}$	$4.74\ 10^{-16}$	$2.04\ 10^{-16}$
bcsstk01	200	$1.04\ 10^{-11}$	$8.67\ 10^{-7}$	$3.78\ 10^{-11}$	$3.7\ 10^{-13}$	$2.26\ 10^{-13}$
bcsstk01 s	88	$1.73\ 10^{-20}$	$9.49\ 10^{-16}$	$2.61\ 10^{-15}$	$4.72\ 10^{-14}$	$2.34\ 10^{-14}$
bcsstk09	475	$2.71\ 10^{-12}$	$1.64\ 10^{-7}$	$6.93\ 10^{-11}$	$4.18\ 10^{-13}$	$4.39\ 10^{-14}$
bcsstk18 s	2941	$5.88\ 10^{-19}$	$2.23\ 10^{-14}$	$6.75\ 10^{-14}$	$3.86\ 10^{-12}$	$1.62\ 10^{-12}$
msc04515 s	6014	$2.61\ 10^{-19}$	$3.22\ 10^{-14}$	$3.27\ 10^{-13}$	$2.01\ 10^{-10}$	$2.46\ 10^{-11}$
nos1	4459	$2.37\ 10^{-11}$	$7.65\ 10^{-6}$	$7.27\ 10^{-10}$	$1.56\ 10^{-11}$	$3.01\ 10^{-12}$
nos1 s	959	$6.2\ 10^{-20}$	$6.47\ 10^{-15}$	$5.61\ 10^{-14}$	$7.31\ 10^{-11}$	$1.27\ 10^{-11}$
nos2 s	11202	$1.59\ 10^{-19}$	$3.92\ 10^{-14}$	$2.73\ 10^{-12}$	$5.94\ 10^{-8}$	$5.24\ 10^{-9}$
nos3	451	$1.92\ 10^{-17}$	$1.66\ 10^{-12}$	$1.99\ 10^{-13}$	$1.24\ 10^{-12}$	$1.02\ 10^{-13}$
nos4	126	$7.70\ 10^{-21}$	$4.75\ 10^{-16}$	$1.13\ 10^{-15}$	$1.36\ 10^{-14}$	$3.41\ 10^{-15}$
nos5	558	$2.69\ 10^{-14}$	$6.22\ 10^{-10}$	$2.06\ 10^{-12}$	$1.68\ 10^{-13}$	$5.06\ 10^{-14}$
nos6	2638	$8.28\ 10^{-14}$	$1.36\ 10^{-8}$	$9.26\ 10^{-11}$	$8.46\ 10^{-11}$	$5.79\ 10^{-12}$
nos6 s	189	$6.77\ 10^{-20}$	$3.72\ 10^{-15}$	$1.53\ 10^{-13}$	$2.02\ 10^{-10}$	$1.92\ 10^{-11}$
nos7 s	174	$7.26\ 10^{-20}$	$2.69\ 10^{-15}$	$1.24\ 10^{-13}$	$9.97\ 10^{-10}$	$1.92\ 10^{-10}$

the first 100 iterations. When we normalize the matrix we reduce the number of iterations although it is still larger than the order of the matrix, but there is a large decrease in the norms of the errors and the residual after 50 iterations. The norms of the matrix are $O(1)$, but the condition number has been reduced by only two orders of magnitude. The norm of the initial residual is 4.8. Therefore, we stop when the computed residual is of the order 10^{-20}, the computed x-residual being 10^{-15}. The ratio between both residuals is almost the same as in the nonnormalized case. In this case, the norms of the errors and the residual are much closer. When normalizing the matrix the l_∞ norm of the error does not improve too much. The norms of both residuals are about the same until iteration 50.

-*nos*1. This is also an interesting small example ($n = 237$) where the norms of the matrix are large (10^9) and we are doing a very large number of iterations (19 times the order!). The initial residual norm is 10^9. Therefore, when stopping the computed residual norm is only 10^{-11} and the computed x-residual norm is 10^{-5} when the norms of the errors are around 10^{-11}, much like the computed residual (this is different from the previous examples). The norms of the residuals are very oscillating and much larger than the norms of the errors up to stagnation (around 3400 iterations). It is interesting to note that the computed residual continues to decrease up to 3500 iterations but then stagnates for awhile before decreasing again. When the matrix is normalized the number of iterations is much smaller but is still four times larger than the matrix order. The initial residual norm is 8.4. When we stop the iterations the norms of the residuals as well as the A-norm of the error are much smaller than before. This is not true of the l_2 and max norms, which are of the same order of magnitude as in the nonnormalized case. Here the residual norms are oscillating and they are smaller than the errors norms.

-nos7 s. This matrix has to be normalized. Otherwise, the convergence is very slow. We can see an interesting feature since after the stagnation point, the computed residual norm after decreasing for 10 iterations starts increasing and, for awhile, "synchronizes" again with the computed x-residual before decreasing. Moreover, the l_2 and max norms of the error are much larger in the end than the A-norm and the residual norm.

Figures 6.3 to 6.10 show the differences between the iterative and computed x-residuals for some of our examples corresponding to solving $Ax = 0$. For each problem the first figure shows the max norms of the residuals and of their difference on a log scale.

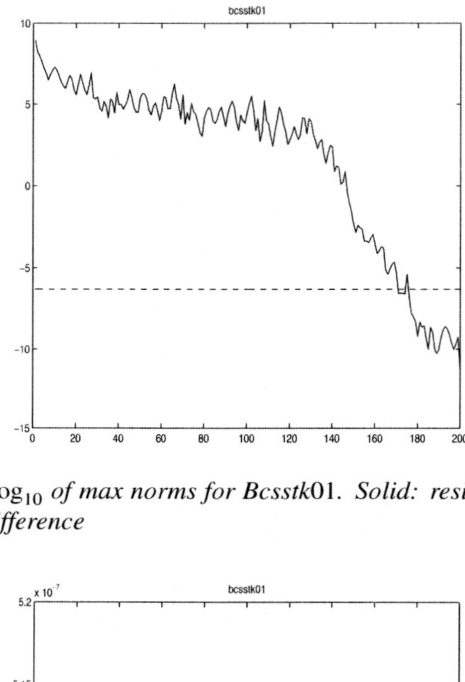

Figure 6.3. $\log_{10}$ *of max norms for Bcsstk*01*. Solid: residual; dotted: computed x-residual; dashed: difference*

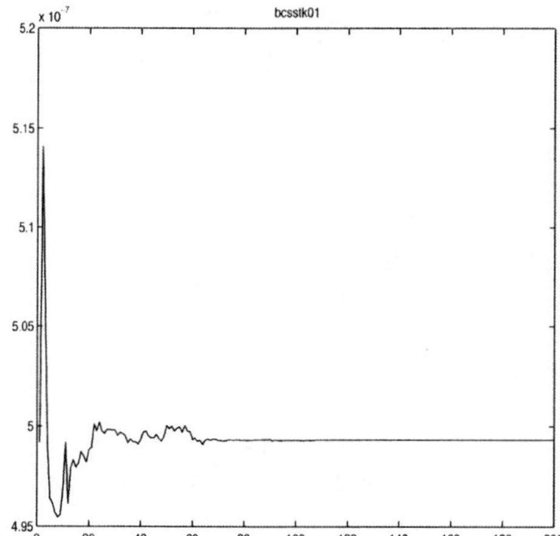

Figure 6.4. *Max norm of the difference of the residuals for Bcsstk*01

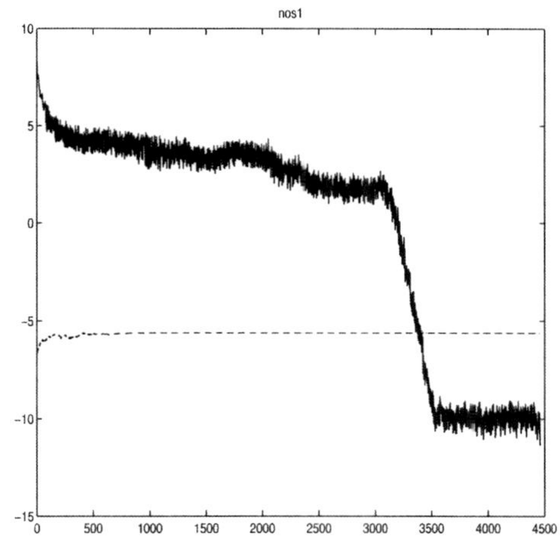

Figure 6.5. $\log_{10}$ *of max norms for* nos1. *Solid: residual; dotted: computed* x*-residual; dashed: difference*

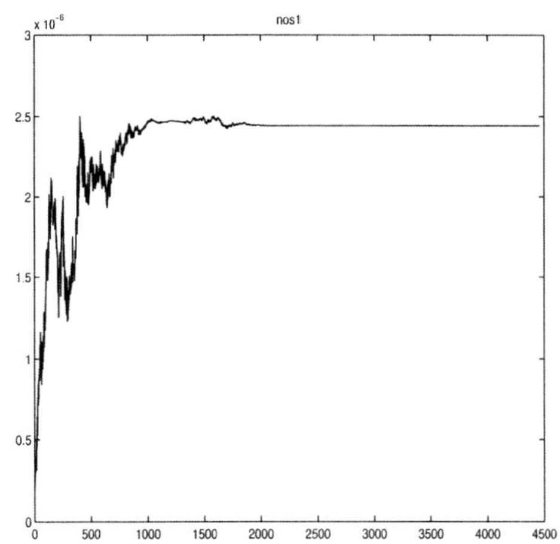

Figure 6.6. *Max norm of the difference of the residuals for* nos1

The second figure is again the max norm of the difference to be able to look at the growth during the first iterations.

As long as the residual norms are larger than the initial differences, they are quite close (since usually the initial difference is small, but this depends on the norm of the matrix and the initial vector), and at some point the iterative residual continues to decrease when the computed x-residual norm stagnates. We see that there is not much increase for the norm of the difference of the residuals during the iterations. The problem of stagnation arises with the vector $\gamma_k p^k$ becoming too small, the vector p^k being almost equal to the residual r^k.

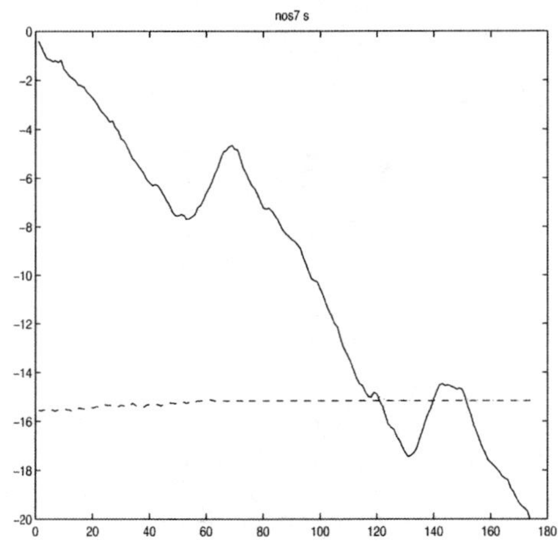

Figure 6.7. $\log_{10}$ *of max norms for nos7s. Solid: residual; dotted: computed x-residual; dashed: difference*

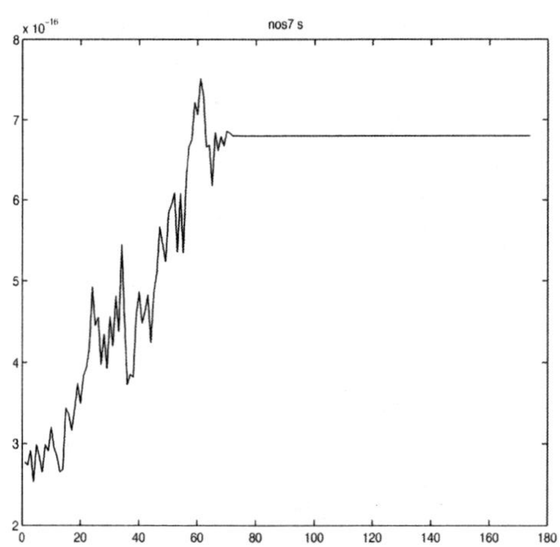

Figure 6.8. *Max norm of the difference of the residuals for nos7s*

Then everything is like having

$$r^{k+1} \simeq (I - \gamma_k A) r^k,$$

plus the perturbation terms (proportional to r^k). Since this corresponds mainly to the steepest descent algorithm in finite precision and A is positive definite, this is a convergent iteration; see Bollen [12].

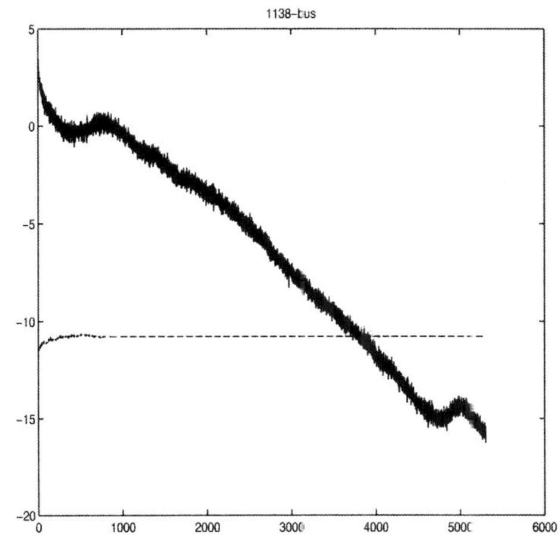

Figure 6.9. $\log_{10}$ *of max norms for* 1138-*bus. Solid: residual; dotted: computed* x-*residual; dashed: difference*

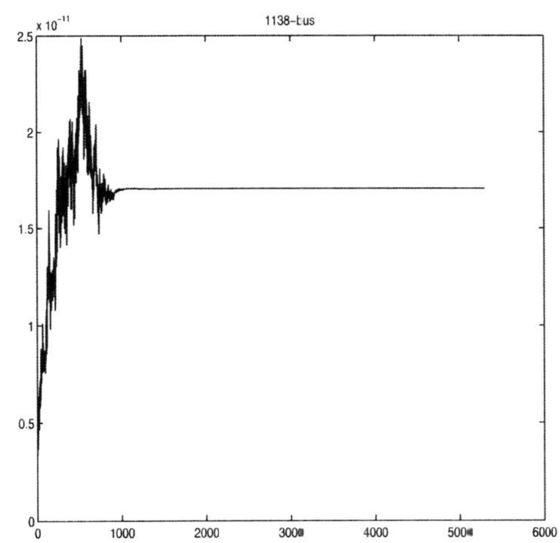

Figure 6.10. *Max norm of the difference of the residuals for* 1138-*bus*

6.3 Estimate of the maximum attainable accuracy

We have seen experimentally that the residual differences and the stagnation of the computed x-residual do not come from an accumulation of roundoff errors during the iterations. The differences increase but not by a multiplicative factor as large as the number of iterations and moreover they finally stagnate. What we are really interested in is estimating the maximum attainable accuracy (the stagnation level) of the residual $\tilde{r}^k$ in the maximum norm without having to compute the x-residual at each iteration. We have seen that

$$fl((Ax)_i) = (Ax)_i + a_{i,:}\Delta x, \quad |\Delta x| \le Cu|x|,$$

where C is $m + O(u)$, m being the maximum number of nonzero entries per row. This implies

$$\|\tilde{r}^k - \hat{r}^k\|_\infty \leq u\|b\|_\infty + (1+m)u\|A\|_\infty\|x^k\|_\infty + O(u^2).$$

Starting at iteration k_l when the computed x-residual starts to stagnate and noting that at this stage we still have $r^k \simeq \tilde{r}^k$, we have

$$r^k - \tilde{r}^k \simeq \sum_{j=k_l}^{k-1}(\delta_r^j + A\delta_x^j) + u(C^{k_l} - C^k).$$

We are interested in the situation where $r^k \to 0$. Then we neglect r^k and δ_r^j. Moreover, since p^k and r^k are close and $p^k \to 0$, we must have $\delta_x \simeq ux^k$. It is likely that $\|\tilde{r}^k\|_\infty$ is of the order of

$$u(k - k_l + 1)\|A\|_\infty\|x^k\|_\infty + 2u\|b\|_\infty + u(m+1)\|A\|_\infty(\|x^{k_l}\|_\infty + \|x^k\|_\infty).$$

Since $\|r^k\|$ is globally decreasing, it is not necessary to take $k - k_l$ large to estimate the accuracy. We may suppose $k - k_l$ to be a small constant, probably smaller than $m + 1$. To get rid of the dependence on k we can bound $\|x^k\|_\infty$. Finally, our guess is that the computed x-residual max norm at stagnation is of the order of

$$2u\|b\|_\infty + 2u(m+1)\|A\|_\infty \max_k \|x^k\|_\infty.$$

In most of our previous examples, the term coming from the right-hand side b is zero or much smaller than the other term. So, let us look at the l_∞ norm of the difference of residuals at stagnation and its ratio to $u\|A\|_\infty \max_k \|x^k\|_\infty$. Let us start with the Poisson model problem $Ax = 0$. Results are given in Table 6.4. We see that the ratio grows a little bit with the dimension of the problem but the difference of residual at stagnation is clearly of order $u\|A\|_\infty$ in this case since $\max_k \|x^k\|_\infty$ is 1. Note that $m + 1 = 6$.

Let us now consider our set of examples with $Ax = 0$. We see in Table 6.5 that the same conclusions apply as for the Poisson model problem. The ratios are $O(1)$, the smallest being 0.8 and the largest being 15. We remark that in most cases using the factor $m + 1$ would be an overestimate.

For our set of examples, let us also consider solving $Ax = e$, where e is a vector with all components equal to 1. Results are given in Table 6.6. The ratios are also $O(1)$, although some are larger than for the problems where $b = 0$. Taking into account

Table 6.4. *Poisson problem, $Ax = 0$*

n	$\|\tilde{r}^k - r^k\|_\infty$	$\max \|\tilde{r}^k - r^k\|_\infty$	$\max \|x^k\|_\infty$	Ratio
400	$1.17\ 10^{-15}$	$1.29\ 10^{-15}$	0.9995	1.31
2500	$1.96\ 10^{-15}$	$2.57\ 10^{-15}$	0.9997	2.21
4900	$2.90\ 10^{-15}$	$3.05\ 10^{-15}$	0.9998	3.27
10000	$3.94\ 10^{-15}$	$3.98\ 10^{-15}$	0.9999	4.44
22500	$4.78\ 10^{-15}$	$4.99\ 10^{-15}$	0.9999	5.38
40000	$5.19\ 10^{-15}$	$5.45\ 10^{-15}$	1	5.85
90000	$6.68\ 10^{-15}$	$6.78\ 10^{-15}$	1	7.52
160000	$7.79\ 10^{-15}$	$8.60\ 10^{-15}$	1	8.77
250000	$9.25\ 10^{-15}$	$1.01\ 10^{-14}$	1	10.41

Table 6.5. $Ax = 0$

Matrix	$\|\tilde{r}^k - r^k\|_\infty$	$\max \|\tilde{r}^k - r^k\|_\infty$	$\max \|x^k\|_\infty$	$m+1$	Ratio
1138-bus	$1.71\ 10^{-11}$	$2.49\ 10^{-11}$	0.9995	19	3.81
1138-bus s	$3.68\ 10^{-15}$	$3.89\ 10^{-15}$	1.97	19	4.65
bcspwr01	$4.62\ 10^{-16}$	$4.62\ 10^{-16}$	0.9501	7	0.81
bcsstk01	$4.99\ 10^{-7}$	$5.14\ 10^{-7}$	0.9501	13	1.33
bcsstk01 s	$3.81\ 10^{-16}$	$3.90\ 10^{-16}$	0.9501	13	1.36
bcsstk09	$3.40\ 10^{-8}$	$3.67\ 10^{-8}$	0.9995	24	3.02
bcsstk18 s	$2.19\ 10^{-15}$	$2.66\ 10^{-15}$	1.63	50	2.22
msc04515 s	$3.70\ 10^{-15}$	$3.81\ 10^{-15}$	1.20	28	3.94
nos1	$2.44\ 10^{-6}$	$2.50\ 10^{-6}$	0.9943	6	8.73
nos1 s	$1.68\ 10^{-15}$	$1.76\ 10^{-15}$	1.13	6	4.68
nos2 s	$5.28\ 10^{-15}$	$5.96\ 10^{-15}$	1.12	6	14.77
nos3	$2.97\ 10^{-13}$	$2.98\ 10^{-13}$	0.9995	19	3.48
nos4	$1.17\ 10^{-16}$	$1.51\ 10^{-16}$	0.9883	8	1.13
nos5	$1.54\ 10^{-10}$	$1.60\ 10^{-10}$	0.9994	24	2.04
nos6	$3.90\ 10^{-9}$	$4.61\ 10^{-9}$	0.9995	6	4.39
nos6 s	$9.41\ 10^{-16}$	$9.49\ 10^{-16}$	1.0116	6	3.35
nos7 s	$6.79\ 10^{-16}$	$7.51\ 10^{-16}$	1.25	8	1.80

Table 6.6. $Ax = e$

Matrix	$\|\tilde{r}^k - r^k\|_\infty$	$\max \|\tilde{r}^k - r^k\|_\infty$	$\max \|x^k\|_\infty$	Ratio
1138-bus	$2.37\ 10^{-8}$	$3.39\ 10^{-8}$	304.31	17.41
1138-bus s	$1.63\ 10^{-9}$	$1.74\ 10^{-9}$	$6.78\ 10^{5}$	5.97
bcspwr01	$2.22\ 10^{-15}$	$2.22\ 10^{-15}$	1.0306	3.59
bcsstk01	$4.15\ 10^{-7}$	$4.17\ 10^{-7}$	0.9501	1.10
bcsstk01 s	$2.91\ 10^{-13}$	$3.23\ 10^{-13}$	630.52	1.56
bcsstk09	$3.06\ 10^{-8}$	$3.34\ 10^{-8}$	0.9995	2.72
bcsstk18 s	$1.34\ 10^{-10}$	$1.39\ 10^{-10}$	$2.19\ 10^{4}$	10.06
msc04515 s	$7.60\ 10^{-9}$	$8.24\ 10^{-9}$	$8.66\ 10^{5}$	11.20
nos1	$2.63\ 10^{-6}$	$3.35\ 10^{-6}$	0.9943	9.41
nos1 s	$7.95\ 10^{-9}$	$7.95\ 10^{-9}$	$2.57\ 10^{6}$	9.77
nos2 s	$7.83\ 10^{-6}$	$8.31\ 10^{-6}$	$6.55\ 10^{8}$	37.54
nos3	$6.96\ 10^{-11}$	$7.07\ 10^{-11}$	75.86	10.78
nos4	$6.25\ 10^{-13}$	$7.67\ 10^{-13}$	$2.37\ 10^{3}$	2.52
nos5	$1.61\ 10^{-10}$	$1.64\ 10^{-10}$	0.9994	2.13
nos6	$1.61\ 10^{-10}$	$1.74\ 10^{-10}$	1.26	14.34
nos6 s	$2.62\ 10^{-9}$	$2.64\ 10^{-9}$	$1.94\ 10^{6}$	4.86
nos7 s	$9.13\ 10^{-8}$	$9.15\ 10^{-8}$	$9.32\ 10^{7}$	3.24

the factor $m + 1$ will improve some of these estimates. From this set of experiments we can conclude that computing $u(\|b\|_\infty + \|A\|_\infty \max_k \|x^k\|_\infty)$ gives a good idea of the order of the attainable accuracy for the maximum norm of the computed x-residual. Inserting some small multiplicative constants in front of both terms may eventually improve the estimate. When we reach this level of x-residual, CG by itself no longer can improve the solution.

6.4 Some ways to improve the maximum attainable accuracy

A way to improve the norm of x-residuals at stagnation has been considered by van der Vorst and Ye [196]. The idea is to monitor the growth of the difference of residuals without computing $b - Ax^k$ and to reset the iterative residual to the computed x-residual from time to time to better synchronize the residuals. At the same time the increments for the computation of x^k are also reset to 0. The algorithm is the following without the computations of the coefficients and the descent directions p^k, which are the same as in CG.

ALGORITHM 6.1.

$z = x^0, r^0 = b - Ax^0, \hat{x}^0 = 0$
$d_{init} = d_0 = u(\|r^0\| + \|A\| \, \|x^0\|)$
for $k = 1, 2, \ldots$ until convergence
$\hat{x}^k = \hat{x}^{k-1} + \gamma_{k-1} p^{k-1}$
$r^k = r^k - \gamma_{k-1} A p^{k-1}$
$d_k = d_{k-1} + u(\|r^k\| + \|A\| \, \|\hat{x}^k\|)$
if $d_{k-1} \leq \epsilon \|r^{k-1}\|$, $d_k > \epsilon \|r^k\|$ and $d_k > 1.1 d_{init}$
$z = z + \hat{x}^k$
$\hat{x}^k = 0$
$r^k = b - Az$
$d_{init} = d_k = u(\|r^k\| + \|A\| \, \|z\|)$
end if
end

In [196] it is suggested to use $\epsilon = \sqrt{u}$. Results using the van der Vorst and Ye algorithm are given in Table 6.7 for the Matrix Market set of examples when solving $Ax = 0$. We see that this algorithmic modification allows in these cases for both residual norms to be the same and to drive the residual norm to 0, although this happens at the expense of an increase in the number of iterations. Then we consider applying this correction to the problem $Ax = e$. We see that although there is still stagnation the level at which the residual norm $\|\tilde{r}^k\|$ stagnates is smaller than for the classical CG algorithm. Results are given in Table 6.8. We note that there is an increase in the number of iterations and sometimes the difference with the Gaussian elimination solution is larger.

Table 6.9 gives the results for solving $Ax = e$ with the van der Vorst and Ye correction using the max norm instead of the l_2 norm. We consider this because it is much cheaper to compute the max norm of the matrix than the l_2 norm. The results which are denoted with a (*) do not use any correction. This could be because $\epsilon = \sqrt{u}$ is not well adapted for the max norm. For the other problems the results are as good as with the l_2 norm, the number of operations being smaller.

Another way to control the stagnation problem is to monitor $\gamma_k p^k$ relative to x^k since we have seen that stagnation occurs when $\gamma_k p^k$ is relatively small. We use the same correction as in the van der Vorst and Ye algorithm but with a different test. We compute the minimum of the absolute value of $\gamma_k p_i^k / \hat{x}_i^k$. If it is smaller than ω, we apply the correction. Experimentally, we found that using $\omega = 10^4 u$ seems to give good results, but the choice of ω remains an open question. An advantage of this method is that we do not need to compute any norm of A.

Table 6.7. *Norms at convergence solving $Ax = 0$ with the van der Vorst and Ye correction*

Matrix	nb. it.	$\|r^k\|$	$\|\tilde{r}^k\|$	$\|e^k\|_A$	$\|e^k\|$	$\|e^k\|_\infty$
1138-bus	5437	$2.58\ 10^{-16}$	$2.58\ 10^{-16}$	$4.07\ 10^{-17}$	$4.26\ 10^{-17}$	$1.18\ 10^{-17}$
1138-bus s	1653	$1.06\ 10^{-19}$	$1.06\ 10^{-19}$	$1.30\ 10^{-18}$	$7.83\ 10^{-17}$	$4.04\ 10^{-17}$
bcspwr01	29	$5.78\ 10^{-20}$	$5.78\ 10^{-20}$	$4.63\ 10^{-20}$	$3.99\ 10^{-20}$	$2.18\ 10^{-20}$
bcsstk01	270	$3.70\ 10^{-11}$	$3.70\ 10^{-11}$	$1.56\ 10^{-13}$	$1.22\ 10^{-15}$	$6.10\ 10^{-16}$
bcsstk01 s	93	$3.10\ 10^{-20}$	$3.10\ 10^{-20}$	$2.48\ 10^{-20}$	$2.32\ 10^{-20}$	$9.30\ 10^{-21}$
bcsstk09	512	$2.34\ 10^{-12}$	$2.34\ 10^{-12}$	$1.28\ 10^{-15}$	$3.30\ 10^{-18}$	$4.22\ 10^{-19}$
bcsstk18 s	2950	$6.14\ 10^{-19}$	$6.14\ 10^{-19}$	$2.50\ 10^{-18}$	$5.51\ 10^{-17}$	$9.17\ 10^{-18}$
msc04515 s	6794	$2.76\ 10^{-19}$	$2.76\ 10^{-19}$	$2.43\ 10^{-18}$	$4.54\ 10^{-16}$	$4.0\ 10^{-17}$
nos1	4740	$2.30\ 10^{-9}$	$2.30\ 10^{-9}$	$4.52\ 10^{-12}$	$1.87\ 10^{-13}$	$2.95\ 10^{-14}$
nos1 s	1047	$6.32\ 10^{-20}$	$6.32\ 10^{-20}$	$9.67\ 10^{-20}$	$5.83\ 10^{-19}$	$8.31\ 10^{-20}$
nos2 s	13155	$1.33\ 10^{-19}$	$1.33\ 10^{-19}$	$7.05\ 10^{-18}$	$1.21\ 10^{-14}$	$1.37\ 10^{-15}$
nos3	482	$1.94\ 10^{-17}$	$1.94\ 10^{-17}$	$3.94\ 10^{-18}$	$2.12\ 10^{-18}$	$2.88\ 10^{-19}$
nos4	146	$9.63\ 10^{-21}$	$9.63\ 10^{-21}$	$3.36\ 10^{-20}$	$1.85\ 10^{-19}$	$4.73\ 10^{-20}$
nos5	619	$3.06\ 10^{-14}$	$3.06\ 10^{-14}$	$8.84\ 10^{-16}$	$9.64\ 10^{-17}$	$2.08\ 10^{-17}$
nos6	2753	$7.97\ 10^{-14}$	$7.97\ 10^{-14}$	$2.49\ 10^{-14}$	$2.32\ 10^{-14}$	$4.49\ 10^{-15}$
nos6 s	190	$7.73\ 10^{-20}$	$7.73\ 10^{-20}$	$1.46\ 10^{-19}$	$4.06\ 10^{-19}$	$8.83\ 10^{-20}$
nos7 s	174	$7.24\ 10^{-20}$	$7.24\ 10^{-20}$	$1.06\ 10^{-19}$	$1.08\ 10^{-17}$	$2.09\ 10^{-18}$

Table 6.8. *Norms at convergence solving $Ax = e$ with the van der Vorst and Ye correction*

Matrix	nb. it.	$\|r^k\|$	$\|\tilde{r}^k\|$	$\|e^k\|_A$	$\|e^k\|$	$\|e^k\|_\infty$
1138-bus	5669	$5.47\ 10^{-16}$	$3.53\ 10^{-9}$	$2.33\ 10^{-9}$	$3.68\ 10^{-8}$	$1.28\ 10^{-9}$
1138-bus s	1806	$4.23\ 10^{-16}$	$4.27\ 10^{-10}$	$1.64\ 10^{-8}$	$6.98\ 10^{-6}$	$1.00\ 10^{-6}$
bcspwr01	25	$1.01\ 10^{-16}$	$1.18\ 10^{-15}$	$1.39\ 10^{-15}$	$8.19\ 10^{-16}$	$5.55\ 10^{-16}$
bcsstk01	318	$3.82\ 10^{-11}$	$3.82\ 10^{-11}$	$1.28\ 10^{-13}$	$7.78\ 10^{-16}$	$3.53\ 10^{-16}$
bcsstk01 s	89	$2.32\ 10^{-16}$	$2.37\ 10^{-13}$	$4.37\ 10^{-12}$	$1.10\ 10^{-10}$	$7.38\ 10^{-11}$
bcsstk09	548	$2.41\ 10^{-12}$	$8.56\ 10^{-12}$	$1.47\ 10^{-13}$	$1.73\ 10^{-15}$	$2.11\ 10^{-16}$
bcsstk18 s	2938	$5.52\ 10^{-16}$	$3.94\ 10^{-11}$	$5.72\ 10^{-9}$	$4.92\ 10^{-7}$	$2.80\ 10^{-8}$
msc04515 s	7974	$5.47\ 10^{-16}$	$2.39\ 10^{-9}$	$2.18\ 10^{-7}$	$1.2\ 10^{-4}$	$1.08\ 10^{-5}$
nos1	4740	$2.01\ 10^{-8}$	$1.13\ 10^{-8}$	$4.30\ 10^{-11}$	$1.86\ 10^{-12}$	$3.11\ 10^{-13}$
nos1 s	1021	$4.36\ 10^{-16}$	$1.48\ 10^{-9}$	$4.12\ 10^{-7}$	$5.7\ 10^{-4}$	$1.00\ 10^{-4}$
nos2 s	16671	$5.40\ 10^{-16}$	$7.93\ 10^{-7}$	$3.00\ 10^{-3}$	68.05	6.06
nos3	482	$4.10\ 10^{-16}$	$2.56\ 10^{-11}$	$2.47\ 10^{-11}$	$1.43\ 10^{-10}$	$1.02\ 10^{-11}$
nos4	127	$4.09\ 10^{-16}$	$3.85\ 10^{-13}$	$3.22\ 10^{-12}$	$1.23\ 10^{-10}$	$2.68\ 10^{-11}$
nos5	754	$2.59\ 10^{-14}$	$1.23\ 10^{-12}$	$7.91\ 10^{-14}$	$7.31\ 10^{-15}$	$1.79\ 10^{-15}$
nos6	2840	$1.14\ 10^{-13}$	$3.32\ 10^{-9}$	$1.91\ 10^{-9}$	$1.75\ 10^{-9}$	$2.10\ 10^{-10}$
nos6 s	181	$5.56\ 10^{-16}$	$1.33\ 10^{-9}$	$2.89\ 10^{-7}$	$3.8\ 10^{-4}$	$3.62\ 10^{-5}$
nos7 s	166	$3.64\ 10^{-16}$	$5.08\ 10^{-8}$	$3.6\ 10^{-4}$	2.93	0.56

Results are given on the set of Matrix Market examples in Table 6.10. Our algorithm tends to better synchronize the computed residual and computed x-residual. Doing this we may reach stagnation of the iterative residual and error norms. Sometimes it is difficult to

Table 6.9. *Norms at convergence solving $Ax = e$ with the van der Vorst and Ye correction using the max norm*

Matrix	nb. it.	$\|r^k\|$	$\|\tilde{r}^k\|$	$\|e^k\|_A$	$\|e^k\|$	$\|e^k\|_\infty$
1138-bus	5589	$5.34\ 10^{-16}$	$3.33\ 10^{-9}$	$9.69\ 10^{-10}$	$1.26\ 10^{-8}$	$4.80\ 10^{-10}$
1138-bus s	1692	$4.97\ 10^{-16}$	$4.29\ 10^{-10}$	$1.33\ 10^{-8}$	$3.57\ 10^{-6}$	$6.57\ 10^{-7}$
bcspwr01	25	$1.01\ 10^{-16}$	$1.35\ 10^{-15}$	$1.34\ 10^{-15}$	$7.71\ 10^{-16}$	$3.33\ 10^{-16}$
bcsstk01(*)	189	$3.79\ 10^{-11}$	$2.18\ 10^{-6}$	$5.58\ 10^{-11}$	$1.31\ 10^{-13}$	$4.69\ 10^{-14}$
bcsstk01 s	84	$5.40\ 10^{-16}$	$1.65\ 10^{-13}$	$4.77\ 10^{-12}$	$1.18\ 10^{-10}$	$7.74\ 10^{-11}$
bcsstk09(*)	477	$1.59\ 10^{-12}$	$4.14\ 10^{-7}$	$9.57\ 10^{-11}$	$2.98\ 10^{-13}$	$4.76\ 10^{-14}$
bcsstk18 s	2933	$5.50\ 10^{-16}$	$4.07\ 10^{-11}$	$5.73\ 10^{-9}$	$4.90\ 10^{-7}$	$2.80\ 10^{-8}$
msc04515 s	7520	$5.12\ 10^{-16}$	$2.34\ 10^{-9}$	$1.50\ 10^{-7}$	$5.89\ 10^{-5}$	$5.46\ 10^{-6}$
nos1(*)	3961	$3.21\ 10^{-11}$	$1.57\ 10^{-5}$	$1.71\ 10^{-9}$	$6.47\ 10^{-11}$	$1.13\ 10^{-11}$
nos1 s	973	$5.27\ 10^{-16}$	$1.38\ 10^{-9}$	$4.50\ 10^{-7}$	$6.30\ 10^{-4}$	$1.22\ 10^{-4}$
nos2 s	13460	$4.32\ 10^{-16}$	$7.80\ 10^{-7}$	$3.09\ 10^{-3}$	68.89	6.14
nos3	462	$4.60\ 10^{-16}$	$2.54\ 10^{-11}$	$2.34\ 10^{-11}$	$1.17\ 10^{-10}$	$7.62\ 10^{-12}$
nos4	132	$5.22\ 10^{-16}$	$3.74\ 10^{-13}$	$4.57\ 10^{-12}$	$1.85\ 10^{-10}$	$4.18\ 10^{-11}$
nos5(*)	558	$2.27\ 10^{-14}$	$2.58\ 10^{-9}$	$5.34\ 10^{-12}$	$1.83\ 10^{-13}$	$4.47\ 10^{-14}$
nos6(*)	2587	$1.20\ 10^{-13}$	$4.56\ 10^{-8}$	$2.30\ 10^{-9}$	$2.10\ 10^{-9}$	$1.44\ 10^{-10}$
nos6 s	178	$5.40\ 10^{-16}$	$1.50\ 10^{-9}$	$2.94\ 10^{-7}$	$3.88\ 10^{-4}$	$3.68\ 10^{-5}$
nos7 s	162	$7.95\ 10^{-16}$	$5.42\ 10^{-8}$	$3.76\ 10^{-4}$	3.03	0.58

Table 6.10. *Norms at convergence solving $Ax = e$ with our correction*

Matrix	nb. it.	$\|r^k\|$	$\|\tilde{r}^k\|$	$\|e^k\|_A$	$\|e^k\|$	$\|e^k\|_\infty$
1138-bus	3867	$5.32\ 10^{-10}$	$3.88\ 10^{-9}$	$7.15\ 10^{-10}$	$2.61\ 10^{-9}$	$2.16\ 10^{-10}$
1138-bus s	1300	$2.55\ 10^{-10}$	$4.62\ 10^{-10}$	$9.75\ 10^{-9}$	$9.18\ 10^{-7}$	$2.21\ 10^{-7}$
bcspwr01	25	$1.02\ 10^{-16}$	$1.03\ 10^{-15}$	$1.44\ 10^{-15}$	$8.78\ 10^{-16}$	$4.44\ 10^{-16}$
bcsstk01	185	$4.60\ 10^{-9}$	$4.60\ 10^{-9}$	$1.44\ 10^{-11}$	$1.25\ 10^{-13}$	$4.52\ 10^{-14}$
bcsstk01 s	68	$5.20\ 10^{-14}$	$1.85\ 10^{-13}$	$4.28\ 10^{-12}$	$1.07\ 10^{-10}$	$7.22\ 10^{-11}$
bcsstk09	393	$1.01\ 10^{-8}$	$1.01\ 10^{-8}$	$2.10\ 10^{-11}$	$1.46\ 10^{-13}$	$2.78\ 10^{-14}$
bcsstk18 s	2882	$2.02\ 10^{-11}$	$4.28\ 10^{-11}$	$5.76\ 10^{-9}$	$4.96\ 10^{-7}$	$2.85\ 10^{-8}$
msc04515 s	4649	$7.05\ 10^{-10}$	$2.40\ 10^{-9}$	$2.25\ 10^{-7}$	$1.33\ 10^{-4}$	$9.27\ 10^{-6}$
nos1	4362	$8.66\ 10^{-8}$	$8.66\ 10^{-8}$	$1.09\ 10^{-10}$	$2.61\ 10^{-12}$	$4.70\ 10^{-13}$
nos1 s	847	$3.53\ 10^{-10}$	$1.79\ 10^{-9}$	$4.70\ 10^{-4}$	$8.37\ 10^{-5}$	-
nos2 s(+)	7773	$1.80\ 10^{-6}$	$1.92\ 10^{-6}$	$2.64\ 10^{-3}$	59.13	5.27
nos3	608	$4.94\ 10^{-12}$	$2.51\ 10^{-11}$	$2.80\ 10^{-11}$	$1.73\ 10^{-10}$	$1.27\ 10^{-11}$
nos4	102	$5.84\ 10^{-14}$	$3.58\ 10^{-13}$	$3.37\ 10^{-12}$	$1.33\ 10^{-10}$	$2.91\ 10^{-11}$
nos5	535	$3.50\ 10^{-11}$	$3.51\ 10^{-11}$	$9.23\ 10^{-13}$	$1.92\ 10^{-13}$	$3.33\ 10^{-14}$
nos6	3216	$5.48\ 10^{-10}$	$3.15\ 10^{-9}$	$1.84\ 10^{-9}$	$1.68\ 10^{-9}$	$1.15\ 10^{-10}$
nos6 s	125	$2.14\ 10^{-10}$	$1.46\ 10^{-9}$	$1.48\ 10^{-7}$	$1.95\ 10^{-4}$	$1.85\ 10^{-5}$
nos7 s	127	$1.51\ 10^{-8}$	$4.45\ 10^{-8}$	$3.78\ 10^{-4}$	3.04	0.59

reach a given stopping criterion if we choose too small a threshold ϵ. Hence we add another stopping criterion

$$\|r^k\| \leq 0.1u\|A\|_\infty \max \|\hat{x}^k\|_\infty.$$

The result denoted by a (+) was the only one in this set of examples that did not converge with this criterion. We had to use a larger value than 0.1. The numbers of iterations are smaller than with the van der Vorst and Ye correction and the errors are of the same order of magnitude.

Chapter 7
Estimates of norms of the error in finite precision

In Chapter 2 we have obtained expressions for the A-norm and the l_2 norm of the error in the CG algorithm. In this chapter, we first would like to show how to use these expressions to obtain computational bounds or estimates of the norms of the error. We shall also study if we can obtain similar formulas (up to rounding errors) in finite precision arithmetic.

7.1 Computation of estimates of the norms of the error in exact arithmetic

In this section we study how we can use the formulas of Chapter 2 to estimate the norms of the error. How can we approximately compute $\|\epsilon^k\|_A^2 = r^{k^T} A^{-1} r^k$? We can use the formula that relates the A-norm of the error at step k and the inverse of the matrix T_k,

$$\begin{aligned}\|\epsilon^k\|_A^2 &= r^{0^T} A^{-1} r^0 - \|r^0\|^2 (T_k^{-1})_{(1,1)} \\ &= \|r^0\|^2((T_n^{-1})_{(1,1)} - (T_k^{-1})_{(1,1)}).\end{aligned}$$

This formula was used computationally in Fischer and Golub [51], but the computations of $\|\epsilon^k\|_A$ were not calculated below 10^{-5}. A partial analysis in finite precision was done by Golub and Strakoš [67]. A more complete analysis was given by Strakoš and Tichý [186]. We shall show below that reliable estimates of $\|\epsilon^k\|_A$ can be computed during CG iterations. What can be done in finite precision arithmetic will be studied in the next sections. We shall see that up to $O(u)$ perturbation terms, we can still use the same formulas. For variants of these estimates, see [20]. Use of our CG error estimates for obtaining reliable stopping criterion in finite element problems has been studied by Arioli and his coworkers in [1], [2]. Other techniques for obtaining estimates of error norms are presented in Brezinski [17].

For the sake of simplicity, let us just consider the lower bound computed by the Gauss rule. Of course, the previous formula cannot be used directly since, at CG iteration k, we do not know $(T_n^{-1})_{1,1}$. But we have seen that the absolute values of $(T_k^{-1})_{1,1}$ are an increasing sequence bounded above by $|(T_n^{-1})_{1,1}|$. So, we shall use the current value of $(T_k^{-1})_{1,1}$ to approximate the final value. Let d be a given integer (to be named the delay); this can also be understood as writing

$$\|\epsilon^{k-d}\|_A^2 - \|\epsilon^k\|_A^2 = \|r^0\|^2((T_k^{-1})_{(1,1)} - (T_{k-d}^{-1})_{(1,1)})$$

and supposing that $\|\epsilon^k\|_A$ is negligible against $\|\epsilon^{k-d}\|_A$. Therefore, we shall use

$$\|\epsilon^{k-d}\|_A^2 \approx \|r^0\|^2((T_k^{-1})_{(1,1)} - (T_{k-d}^{-1})_{(1,1)}).$$

Let b_k be the computed value of $(T_k^{-1})_{1,1}$. It can be obtained in an additive way from the previous iteration by using the Sherman–Morrison formula [68]. This was suggested in [118]. Let $t^k = T_k^{-1}e^k$ be the last column of the inverse of T_k. Then

$$(T_{k+1}^{-1})_{1,1} = (T_k^{-1})_{1,1} + \frac{\eta_{k+1}^2 (t^k(t^k)^T)_{1,1}}{\alpha_{k+1} - \eta_{k+1}^2 (t^k)_k},$$

but $(t^k(t^k)^T)_{1,1} = [(t^k)_1]^2$. The first and last elements of the last column of the inverse of T_k that we need can be computed using the Cholesky decomposition of T_k whose diagonal elements are the last pivot function at 0: $\delta_1 = \alpha_1$ and

$$\delta_i = \alpha_i - \frac{\eta_i^2}{\delta_{i-1}}, \quad i = 2, \dots, k.$$

Then,

$$(t^k)_1 = (-1)^{k-1}\frac{\eta_2 \cdots \eta_k}{\delta_1 \cdots \delta_k}, \quad (t^k)_k = \frac{1}{\delta_k}.$$

Using these results for iteration $k-1$, we have

$$b_k = b_{k-1} + f_k, \quad f_k = \frac{\eta_k^2 c_{k-1}^2}{\delta_{k-1}(\alpha_k \delta_{k-1} - \eta_k^2)} = \frac{c_k^2}{\delta_k},$$

where

$$c_k = \frac{\eta_2 \cdots \eta_{k-1}}{\delta_1 \cdots \delta_{k-2}} \frac{\eta_k}{\delta_{k-1}} = c_{k-1}\frac{\eta_k}{\delta_{k-1}}.$$

Since T_k is positive definite, this shows that $f_k > 0$. Let s_k be the estimate of $\|\epsilon^k\|_A^2$ we are looking for at CG iteration number k; we set

$$s_{k-d} = \|r^0\|^2(b_k - b_{k-d}).$$

This will give us an estimate of the error d iterations before the current one k. It was shown in [65] that if we compute b_k in finite precision arithmetic and use the formula for $\|\epsilon^{k-d}\|_A^2$ straightforwardly, there exists a k_{max} such that if $k > k_{max}$, then $s_k = 0$. This happens because, when k is large enough, $\eta_{k+1}/\delta_k < 1$ and $c_k \to 0$ and consequently $f_k \to 0$. Therefore, when $k > k_{max}$, $b_k = b_{kmax}$. But, as it was noticed in [65], we can compute s_{k-d} in another way since we just need to sum up the last d values of f_j.

From what we have been doing when looking at the CG residual norms we have

$$c_k = \frac{\eta_2 \cdots \eta_k}{\delta_1 \cdots \delta_{k-1}} = \frac{\|r^{k-1}\|}{\|r^0\|},$$

and $\gamma_{k-1} = 1/\delta_k$. Therefore, $f_k = \gamma_{k-1}\|r^{k-1}\|^2/\|r^0\|^2$ which gives back the Hestenes and Stiefel formula [93] we have seen in Chapter 2 and a simpler way of computing the Gauss lower bound.

If we set λ_m and λ_M to be approximations of the smallest and largest eigenvalues of A, the CGQL (CG with quadrature using Lanczos) algorithm computing the iterates of CG and estimates from the Gauss (s_{k-d}), Gauss–Radau ($\underline{s}_{k-d}$ and $\bar{s}_{k-d}$), and Gauss–Lobatto ($\breve{s}_{k-d}$) rules (see [64]) is the following (with slight simplifications improving upon [65]).

ALGORITHM 7.1.

Let x^0 be given, $r^0 = b - Ax^0$, $p^0 = r^0$, $c_1 = 1$
for $k = 1, \dots$ until convergence
$\gamma_{k-1} = (r^{k-1}, r^{k-1})/(p^{k-1}, Ap^{k-1})$
$\alpha_k = 1/\gamma_{k-1} + \beta_{k-1}/\gamma_{k-2}$
if $k = 1$———————————————
$f_1 = 1/\alpha_1$
$\delta_1 = \alpha_1$
$\bar{\delta}_1 = \alpha_1 - \lambda_m$
$\underline{\delta}_1 = \alpha_1 - \lambda_M$,
else———————————————
$c_k = c_{k-1}\eta_k/\delta_{k-1} = \|r^{k-1}\|/\|r^0\|$
$\delta_k = \alpha_k - \eta_k^2/\delta_{k-1} = 1/\gamma_{k-1}$
$f_k = \gamma_{k-1}c_k^2$
$\bar{\delta}_k = \alpha_k - \lambda_m - \eta_k^2/\bar{\delta}_{k-1} = \alpha_k - \bar{\alpha}_{k-1}$
$\underline{\delta}_k = \alpha_k - \lambda_M - \eta_k^2/\underline{\delta}_{k-1} = \alpha_k - \underline{\alpha}_{k-1}$
endif———————————————
$x^k = x^{k-1} + \gamma_{k-1}p^{k-1}$
$r^k = r^{k-1} - \gamma_{k-1}Ap^{k-1}$
$\beta_k = (r^k, r^k)/(r^{k-1}, r^{k-1})$
$\eta_{k+1} = \sqrt{\beta_k}/\gamma_{k-1}$
$p^k = r^k + \beta_k p^{k-1}$
$\bar{\alpha}_k = \lambda_m + \eta_{k+1}^2/\bar{\delta}_k$
$\underline{\alpha}_k = \lambda_M + \eta_{k+1}^2/\underline{\delta}_k$
$\breve{\alpha}_k = (\bar{\delta}_k\underline{\delta}_k/(\underline{\delta}_k - \bar{\delta}_k))\left(\lambda_M/\bar{\delta}_k - \lambda_m/\underline{\delta}_k\right)$
$\breve{\eta}_{k+1}^2 = (\bar{\delta}_k\underline{\delta}_k/(\underline{\delta}_k - \bar{\delta}_k))(\lambda_M - \lambda_m)$
$\bar{f}_k = \eta_{k+1}^2 c_k^2/(\delta_k(\bar{\alpha}_k\delta_k - \eta_{k+1}^2))$
$\underline{f}_k = \eta_{k+1}^2 c_k^2/(\delta_k(\underline{\alpha}_k\delta_k - \eta_{k+1}^2))$
$\breve{f}_k = \breve{\eta}_{k+1}^2 c_k^2/(\delta_k(\breve{\alpha}_k\delta_k - \breve{\eta}_{k+1}^2))$
if $k > d$———————————————
$g_k = \sum_{j=k-d+1}^{k} f_j$
$s_{k-d} = \|r^0\|^2 g_k$
$\bar{s}_{k-d} = \|r^0\|^2(g_k + \bar{f}_k)$
$\underline{s}_{k-d} = \|r^0\|^2(g_k + \underline{f}_k)$
$\breve{s}_{k-d} = \|r^0\|^2(g_k + \breve{f}_k)$
endif———————————————
end

This algorithm gives lower bounds s_{k-d}, $\underline{s}_{k-d}$ and upper bounds $\bar{s}_{k-d}$, $\breve{s}_{k-d}$ of $\|\epsilon^{k-d}\|_A^2$. Notice that in the practical implementation we do not need to store all the f_k's but only the last d. We can also compute only some of the estimates, particularly if we do not have any estimates of the extreme eigenvalues. The additional number of operations for CG is approximately $50 + d$ if we compute the four estimates, which is almost nothing compared to the $10n$ operations plus the matrix-vector product of CG.

An interesting question is how large d has to be to get a reliable estimate of the error. Unfortunately the choice of d depends on the example. The faster the CG convergence the smaller d has to be. Moreover, d can generally be small if there are no oscillations of $\|r^k\|$. In many cases a value of $d = 1$ already gives good estimates of the norm of the error. Nevertheless, if we accept storing some more vectors whose lengths are the number of CG iterations, we can improve the bounds we compute. For instance, for the Gauss lower bound at iteration k we can compute f_k and sum it to what we got at all the previous iterations. This will improve our previous bounds and as a result we shall have a vector with bounds using $d = 1$ for iteration $k - 1$, $d = 2$ for iteration $k - 2$, and so on. This is interesting if we want to have an a posteriori look at the rate of convergence. Of course, it is not so useful if we just want to use the bound as a stopping criterion. A similar idea was proposed by Strakoš and Tichý [187].

In the CGQL algorithm λ_m and λ_M are lower and upper bounds of the smallest and largest eigenvalues of A. Notice that the value of s_k is independent of λ_m and λ_M, and $\bar{s}_k$ depends only on λ_m and $\underline{s}_k$ only on λ_M. Let us now prove that this algorithm does give lower and upper bounds for the A-norm of the error.

Theorem 7.1. *At iteration number k of CGQL, s_{k-d} and $\underline{s}_{k-d}$ are lower bounds of $\|\epsilon^{k-d}\|_A^2$, and $\bar{s}_{k-d}$ and $\breve{s}_{k-d}$ are upper bounds of $\|\epsilon^{k-d}\|_A^2$.*

Proof. We have

$$\|\epsilon^{k-d}\|_A^2 = \|r^0\|^2((T_n^{-1})_{1,1} - (T_{k-d}^{-1})_{1,1})$$

and

$$s_{k-d} = \|r^0\|^2((T_k^{-1})_{1,1} - (T_{k-d}^{-1})_{1,1}).$$

Therefore,

$$\|\epsilon^{k-d}\|_A^2 - s_{k-d} = \|r^0\|^2((T_n^{-1})_{1,1} - (T_k^{-1})_{1,1}) > 0,$$

showing that s_{k-d} is a lower bound of $\|\epsilon^{k-d}\|_A^2$. The same kind of proof applies for the other cases; see [65]. □

The quantities that we are computing in CGQL are indeed upper and lower bounds of the A-norm of the error. It turns out that the best bounds are generally the ones computed by the Gauss–Radau rule. It is unfortunate that estimates of the smallest eigenvalue are required to obtain upper bounds of the A-norm of the error. However, we have seen that the extreme eigenvalues of T_k are approximations of the extreme eigenvalues of A that are usually getting better and better as k increases. Therefore, we propose the following adaptive algorithm. We start the CGQL iterations with $\lambda_m = a_0$ an underestimate of $\lambda_{min}(A)$. An estimate of the smallest eigenvalue can be obtained by inverse iteration on T_k (see [68]) since, for computing the bounds of the norm, we already compute the Cholesky factorization of T_k. The smallest eigenvalue of T_k is obtained by repeatedly solving tridiagonal systems.

We use a fixed number n_a of (inner) iterations of inverse iteration at every CG iteration, obtaining a value λ_m^k. When λ_m^k is such that

$$\frac{|\lambda_m^k - \lambda_m^{k-1}|}{\lambda_m^k} \le \varepsilon_a,$$

with a prescribed threshold ε_a, we switch by setting $\lambda_m = \lambda_m^k$, we stop computing the eigenvalue estimate, and we go on with CGQL. Of course, this is cheating a little bit since the smallest Ritz value approximates the smallest eigenvalue from above and not from below, as is required by the theorem.

Our next goal is to be able to compute estimates of the l_2 norm of the error using the formula we have obtained previously:

$$\|\epsilon^k\|^2 = \|r^0\|^2[(e^1, T_n^{-2}e^1) - (e^1, T_k^{-2}e^1)] + (-1)^k 2\eta_{k+1}\frac{\|r^0\|}{\|r^k\|}(e^k, T_k^{-2}e^1)\|\epsilon^k\|_A^2.$$

We have already seen in Chapter 2 some formulas for doing this, starting from the Hestenes and Stiefel formula. We are now considering another way to obtain estimates of the norm of the error. Let us start by computing $(e^1, T_k^{-2}e^1)$. One way is to write this as $(T_k^{-1}e^1, T_k^{-1}e^1)$ and solve $T_k t^1 = e^1$. Another solution is to use a QR factorization of the tridiagonal matrix T_k

$$Q_k T_k = R_k,$$

where Q_k is an orthogonal matrix and R_k an upper triangular matrix. We have $T_k^2 = R_k^T R_k$; therefore

$$(e^1, T_k^{-2}e^1) = (R_k^{-T}e^1, R_k^{-T}e^1).$$

We just have to solve a linear system with matrix R_k^T and right-hand side e^1. To compute the decomposition of T_k we use the results of Fischer [50]. Let us look at the first steps of the reduction. To put a zero in the (2, 1) position of

$$\begin{pmatrix} \alpha_1 \\ \eta_2 \end{pmatrix}$$

we define $\hat{r}_{1,1} = \alpha_1$, $r_{1,1} = \sqrt{\hat{r}_{1,1}^2 + \eta_2^2}$ with

$$c_1 = \frac{\hat{r}_{1,1}}{r_{1,1}}, \quad s_1 = \frac{\eta_2}{r_{1,1}}.$$

When we apply this rotation to

$$\begin{pmatrix} \alpha_1 & \eta_2 \\ \eta_2 & \alpha_2 \end{pmatrix}$$

we obtain

$$\begin{pmatrix} r_{1,1} & r_{2,2} \\ 0 & \hat{r}_{1,2} \end{pmatrix} = \begin{pmatrix} r_{1,1} & c_1\eta_2 + s_1\alpha_2 \\ 0 & -s_1\eta_2 + c_1\alpha_2 \end{pmatrix}.$$

Then we reduce the column

$$\begin{pmatrix} r_{2,2} \\ \hat{r}_{1,2} \\ \eta_3 \end{pmatrix}$$

by a (s_2, c_2) rotation. We obtain

$$\begin{pmatrix} r_{1,1} & r_{2,2} & r_{3,3} \\ 0 & r_{1,2} & r_{2,3} \\ 0 & 0 & \hat{r}_{1,3} \end{pmatrix}.$$

The matrix R_k has only three nonzero diagonals whose entries are denoted as $r_{1,i}, r_{2,i}, r_{3,i}$. The general formulas are (see Fischer [50])

$$\hat{r}_{1,1} = \alpha_1, \quad \hat{r}_{1,2} = c_1\alpha_2 - s_1\eta_2, \quad \hat{r}_{1,i} = c_{i-1}\alpha_i - s_{i-1}c_{i-2}\eta_i, \quad i \geq 3,$$

$$r_{1,i} = \sqrt{\hat{r}_{1,i}^2 + \eta_{i+1}^2},$$

$$r_{3,i} = s_{i-2}\eta_i, \quad i \geq 3,$$

$$r_{2,2} = c_1\eta_2, \quad r_{2,i} = c_{i-2}c_{i-1}\eta_i + s_{i-1}\alpha_i, \quad i \geq 3,$$

$$c_i = \frac{\hat{r}_{1,i}}{r_{1,i}}, \quad s_i = \frac{\eta_{i+1}}{r_{1,i}}.$$

Now, we would like to incrementally compute the solution of the linear systems $R_k^T w^k = e^1$ for $k = 1, 2, \ldots$ R_k^T is a lower triangular matrix, but we have to be careful that even though the other elements stay the same during the iterations, the (k, k) element changes when we go from k to $k+1$. Hence changing notations $w = w^k$ and $\hat{w}$ being an auxiliary vector, we define

$$\hat{w}_1 = \frac{1}{\hat{r}_{1,1}}, \quad w_1 = \frac{1}{r_{1,1}},$$

$$\hat{w}_2 = -\frac{r_{2,2}w_1^2}{\hat{r}_{1,2}}, \quad w_2 = -\frac{r_{2,2}w_1^2}{r_{1,2}},$$

and more generally for $i \geq 3$

$$\hat{w}_i = -\frac{(r_{3,i}w_{i-2} + r_{2,i}w_{i-1})}{\hat{r}_{1,i}}, \quad w_i = -\frac{(r_{3,i}w_{i-2} + r_{2,i}w_{i-1})}{r_{1,i}}.$$

Therefore, $\hat{w}_k$ is the last component of the solution at iteration k and w_k will be used in the subsequent steps. Then,

$$\|R_k^{-T}e^1\|^2 = \sum_{j=1}^{k-1} w_j^2 + \hat{w}_k^2.$$

Now we proceed as we did before for the A-norm of the error. We introduce an integer delay d and we approximate $(r^0, A^{-2}r^0) - \|r^0\|^2(e^1, T_{k-d}^{-2}e^1)$ at iteration k by the difference of the k and $k - d$ terms computed from the solutions, that is,

$$\hat{w}_k^2 - \hat{w}_{k-d}^2 + \sum_{j=k-d}^{k-1} w_j^2, \quad k > d.$$

To approximate the last term

$$(-1)^{k-d} 2\eta_{k+1-d} \frac{\|r^0\|}{\|r^{k-d}\|}(e^{k-d}, T_{k-d}^{-2}e^1)\|\epsilon^{k-d}\|_A^2$$

we use the lower bound of $\|\epsilon^{k-d}\|_A$ from Gauss quadrature and the value $(e^{k-d}, T_{k-d}^{-2}e^1)$, which is $\hat{w}_{k-d}/\hat{r}_{1,k-d}$. We see that computing an estimate of the l_2 norm of the error adds only a few operations to each CG iteration. Ideally we have to subtract the same last term at iteration k, but this involves $\|\epsilon^k\|_A^2$ that we do not have. To approximate this term we would have to wait for d more iterations. We chose to neglect this term, which is positive, making our lower bound a little smaller. We shall look at some examples when we study computations in finite precision arithmetic.

7.2 The A-norm of the error in finite precision

In exact arithmetic we have seen that

$$\|\epsilon^k\|_A^2 = (A^{-1}r^k, r^k) = \|r^0\|^2[(T_n^{-1}e^1, e^1) - (T_k^{-1}e^1, e^1)],$$

where T_k is the matrix of the Lanczos coefficients at iteration k. This formula is in fact a quadrature rule. Another equivalent formula for the norm of the error is

$$\|\epsilon^k\|_A^2 = \sum_{j=k}^{n-1} \gamma_j \|r^j\|^2.$$

In finite precision arithmetic these formulas are no longer valid. The proof of the first formula uses global orthogonality through $A^{-1}V_n = V_n T_n^{-1}$ and $V_k^T A V_k = T_k$. These relations are not true in finite precision. The other formula also no longer makes sense. In finite precision we are interested in $\varepsilon^k = A^{-1}r^k$, where r^k is the iterative residual. We have shown in the previous chapter that

$$\|\varepsilon^{k+1}\|_A^2 = \|\varepsilon^0\|_A^2 - \sum_{i=0}^{k}(\gamma_i - u\zeta_i)\|r^i\|^2,$$

with $\zeta_k = \nu_k + uC_{(2)}^k$. The scalars γ_i involved in this formula are the ones given by the exact formula. If we want to refer instead to the computed γ's, we have to replace them by $\gamma_i - \delta_\gamma^i$. Let us consider

$$\sum_{i=0}^{k} \gamma_i \|r^i\|^2 = \|r^0\|^2 \left(\gamma_0 + \gamma_1 \frac{\|r^1\|^2}{\|r^0\|^2} + \cdots + \gamma_k \frac{\|r^k\|^2}{\|r^0\|^2}\right).$$

This can be expressed with the coefficients β_i. Again, if we refer to the computed β's, we have

$$\frac{\|r^k\|^2}{\|r^{k-1}\|^2} = \beta_k - \delta_\beta^k.$$

Therefore,

$$\frac{\|r^k\|^2}{\|r^0\|^2} = \frac{\|r^k\|^2}{\|r^{k-1}\|^2} \cdots \frac{\|r^1\|^2}{\|r^0\|^2} = \prod_{j=1}^{k}(\beta_j - \delta_\beta^j).$$

Hence,

$$\sum_{i=0}^{k} \gamma_i \|r^i\|^2 = \|r^0\|^2[\gamma_0 + \gamma_1\beta_1 + \cdots + \gamma_k\beta_1 \cdots \beta_k] + O(u) + O(u^2).$$

The $O(u)$ term is

$$\delta^k_{\beta\gamma} = \|r^0\|^2 \left[\delta^0_\gamma + \delta^1_\gamma \beta_1 + \delta^1_\beta \gamma_1 + \cdots + \delta^k_\gamma \beta_1 \cdots \beta_k + \gamma_k \sum_{j=1}^{k} \delta^j_\beta \prod_{i \neq j} \beta_i \right],$$

all the δ_γ's and δ_β's being of order u. This leads to the following result.

Theorem 7.2.

$$\|\varepsilon^{k+1}\|^2_A = \|\varepsilon^0\|^2_A - \|r^0\|^2 (T^{-1}_{k+1} e^1, e^1) - \delta^{k+1}_\varepsilon + O(u^2),$$

with

$$\delta^{k+1}_\varepsilon = \delta^k_{\beta\gamma} + \sum_{i=0}^{k} u \zeta_i \|r^i\|^2.$$

Proof. By denoting $x^{(k)}_1 = (T^{-1}_k e^1, e^1)$ and by using results in the proof of Theorem 2.14, we have

$$x^{(k+1)}_1 - x^{(j)}_1 = \beta_1 \cdots \beta_j [\gamma_j + \gamma_{j+1} \beta_{j+1} + \cdots + \gamma_k \beta_{j+1} \cdots \beta_k], \quad j \leq k.$$

We take $j = 1$ and we notice that $x^{(1)}_1 = \gamma_0$ to obtain

$$x^{(k+1)}_1 = \gamma_0 + \gamma_1 \beta_1 + \gamma_2 \beta_1 \beta_2 + \cdots + \gamma_k \beta_1 \cdots \beta_k.$$

This proves our result. □

We have two modifications from the exact arithmetic formula. The first term of the right-hand side is the same but written as $(A^{-1} r^0, r^0)$ and we have an additional $O(u)$ term. Of course we can still write the norm of the error with expressions involving the eigenvalues of A and the Ritz values,

$$\|\varepsilon^k\|^2_A = \|r^0\|^2 \left[\sum_{j=1}^{n} \frac{(\bar{r}^0_j)^2}{\lambda_j} - \sum_{j=1}^{k} \frac{[(z^j_{(k)})_1]^2}{\theta^{(k)}_j} \right] - \delta^k_\varepsilon + O(u^2),$$

where $\bar{r}^0_j$ are the components of the initial residual r^0 on the eigenvectors of A. According to this result, the convergence of CG in finite precision is still governed by the convergence of the Ritz values towards eigenvalues of A as in exact arithmetic. One may ask what happens when we get multiple copies of an eigenvalue λ_l. This question has been considered by Wülling [202], [203], who proved results on the stabilization of the sum of weights. Numerical experiments show that when we have multiple copies the sum of the corresponding weights $[(z^l_{(k)})_1]^2$ converges to $(\bar{r}^0_j)^2$. A study of the computation of estimates of quadratic forms in finite precision arithmetic using other techniques has also been done by Knizhnerman [105].

Of course the estimates of the norms can be used to stop the iterations rather than using the relative norm of the residual, which can sometimes be misleading. One has to be aware of the fact that since the norm estimate is based on the iterative residual we are approximating $\|A^{-\frac{1}{2}}r^k\|$ and not $\|x - x^k\|_A$. They are quite close until stagnation, but the estimate converges to zero with the iterative residual whence the A-norm of $x - x^k$ will stagnate. This is why if one wants to have a reliable algorithm, an estimate of the maximum attainable accuracy has to be computed (or at least we have to know that we have reached stagnation), since it does not make sense to continue computing when the solution can no longer be improved. However, in most practical cases, iterations are stopped before reaching stagnation.

7.3 The l_2 norm of the error in finite precision

Are the formulas we derived before for the l_2 norm of the error approximately valid in finite precision arithmetic? We are going to see that the situation is not as nice as for the A-norm. Since $\varepsilon^k = A^{-1}r^k$, we have

$$\varepsilon^{k+1} = \varepsilon^k - \gamma_k p^k + A^{-1}\delta_r^k.$$

Hence,

$$\|\varepsilon^{k+1}\|^2 = \|\varepsilon^k\|^2 - 2\gamma_k(\varepsilon^k, p^k) + \gamma_k^2\|p^k\|^2$$
$$-2(\varepsilon^k, A^{-1}\delta_r^k) - 2\gamma_k(p^k, A^{-1}\delta_r^k) + (A^{-1}\delta_r^k, A^{-1}\delta_r^k).$$

Since we have

$$p^k = r^k + \|r^k\|^2\left(\frac{r^{k-1}}{\|r^{k-1}\|^2} + \cdots + \frac{r^0}{\|r^0\|^2}\right) + uC^k\|r^k\|^2,$$

the term involving ε^k and p^k gives

$$(\varepsilon^k, p^k) = \|\varepsilon^k\|_A^2 + \|r^k\|^2\left(\frac{(\varepsilon^k, r^{k-1})}{\|r^{k-1}\|^2} + \cdots + \frac{(\varepsilon^k, r^0)}{\|r^0\|^2}\right) + u\|r^k\|^2(\varepsilon^k, C^k).$$

In exact arithmetic we have $(\varepsilon^k, r^j) = (\varepsilon^k, r^k)$. This not true in finite precision, but we obtain the following result.

Lemma 7.3.

$$(\varepsilon^k, r^j) = \|\varepsilon^k\|_A^2 + uC_\varepsilon^j,$$

with

$$C_\varepsilon^j = -u\sum_{i=j}^{k-1}\gamma_i C_{rp}^i\|r^i\|^2 + \sum_{i=j+1}^{k}(A^{-1}r^i, \delta_r^{i-1}).$$

Proof. This can be proved by induction. We have

$$(\varepsilon^k, r^k) = (\varepsilon^k, r^{k-1}) - \gamma_{k-1}(r^k, p^{k-1}) + (A^{-1}r^k, \delta_r^{k-1}).$$

But using our results about local orthogonality,

$$\gamma_{k-1}(r^k, p^{k-1}) = u\gamma_{k-1}C_{rp}^{k-1}\|r^{k-1}\|^2.$$

Iterating these results we obtain the proof. □

Proposition 7.4.

$$\begin{aligned}\|\varepsilon^{k+1}\|^2 = \|\varepsilon^k\|^2 - 2\gamma_k\|r^k\|^2\|\varepsilon^k\|_A^2\left(\frac{1}{\|r^k\|^2}+\cdots+\frac{1}{\|r^0\|^2}\right) + \gamma_k^2\|p^k\|^2 \\ -2u\gamma_k\|r^k\|^2\left(\frac{C_\varepsilon^{k-1}}{\|r^{k-1}\|^2}+\cdots+\frac{C_\varepsilon^0}{\|r^0\|^2}\right) - 2u\gamma_k\|r^k\|^2(\varepsilon^k, C^k) \\ -2(\varepsilon^k, A^{-1}\delta_r^k) - 2\gamma_k(p^k, A^{-1}\delta_r^k) + O(u^2).\end{aligned}$$

Proof. We simply collect the previous results. □

The trouble comes with the term $\gamma_k^2\|p^k\|^2$. The formula we used in exact arithmetic is not valid anymore in finite precision. Instead, we have the following.

Proposition 7.5.

$$\|p^k\|^2 = \|r^k\|^4\left(\frac{1}{\|r^k\|^2}+\cdots+\frac{1}{\|r^0\|^2}\right) + \|r^k\|^4\Delta_k^r + \|r^k\|^4 O(u),$$

with

$$\Delta_k^r = 2\sum_{i,j,i\neq j}\frac{(r^i, r^j)}{\|r^i\|^2\|r^j\|^2}.$$

Proof. In exact arithmetic because of the orthogonality of the residuals the last two terms in the expression for $\|p^k\|^2$ are zero. Here, we just compute the inner product (p^k, p^k). □

The extra term involving Δ_k^r is not small, but we have

$$\|r^k\|^4|\Delta_k^r| \le 2\|r^k\|^2\sum_{i\neq j}\frac{\|r^k\|}{\|r^i\|}\frac{\|r^k\|}{\|r^j\|} \le 2(k+1)^2\|r^k\|^2\left(\max_i\frac{\|r^k\|}{\|r^i\|}\right)^2.$$

This term goes to zero with $\|r^k\|$. Let

$$\pi_k = \gamma_k\|r^k\|^2\left(\frac{1}{\|r^k\|^2}+\cdots+\frac{1}{\|r^0\|^2}\right).$$

Then,

$$\gamma_k^2\|p^k\|^2 = \gamma_k\|r^k\|^2\pi_k + \gamma_k^2\|r^k\|^4\Delta_k^r + \gamma_k^2\|r^k\|^4 O(u).$$

This leads to the following result.

Theorem 7.6. *In finite precision arithmetic,*

$$\|\varepsilon^k\|^2 - \|\varepsilon^{k+1}\|^2 = \pi_k(\|\varepsilon^k\|_A^2 + \|\varepsilon^{k+1}\|_A^2) - \gamma_k^2\|r^k\|^4\Delta_k^r + O(u).$$

Proof.

$$-2\pi_k\|\varepsilon^k\|_A^2 + \gamma_k^2\|p^k\|^2 = -\pi_k(2\|\varepsilon^k\|_A^2 - \gamma_k\|r^k\|^2) + \gamma_k^2\|r^k\|^4\Delta_k^r + O(u).$$

Using

$$\|\varepsilon^{k+1}\|_A^2 = \|\varepsilon^k\|_A^2 - \gamma_k \|r^k\|^2 + O(u),$$

we prove the result. □

In finite precision, we have the same result as Hestenes and Stiefel [93] but with two modifications. First, the multiplying factor is π_k and not $1/\mu(p^k)$; second, there is an additional term $\gamma_k^2 \|r^k\|^4 \Delta_k^r$. The equivalence with the formula with inverses of tridiagonal matrices is still true modulo the additional term.

Proposition 7.7. *Let*

$$\omega_k = \gamma_0 + \gamma_1 \|r^1\|^2 \left(\frac{1}{\|r^0\|^2} + \frac{1}{\|r^1\|^2} \right) + \cdots + \gamma_{k-1} \|r^{k-1}\|^2 \left(\frac{1}{\|r^0\|^2} + \cdots + \frac{1}{\|r^{k-1}\|^2} \right).$$

Then,

$$\|r^0\|^2 [(T_{k+1}^{-2})_{1,1} - (T_k^{-2})_{1,1}] = 2\gamma_k \omega_k \|r^k\|^2 + \gamma_k \pi_k \|r^k\|^2.$$

Proof. We write the formulas for the entries of the inverses as in exact arithmetic but without using p^k. □

Proposition 7.8.

$$2\frac{(T_{k+1}^{-2})_{1,k+1}}{(T_{k+1}^{-1})_{1,k+1}} \|\varepsilon^{k+1}\|_A^2 - 2\frac{(T_k^{-2})_{1,k}}{(T_k^{-1})_{1,k}} \|\varepsilon^k\|_A^2 = 2\pi_k \|\varepsilon^k\|_A^2 - 2\omega_{k+1}\gamma_k \|r^k\|^2 + O(u).$$

Proof. As in exact arithmetic, we have

$$\frac{(T_k^{-2})_{1,k}}{(T_k^{-1})_{1,k}} = \omega_k.$$

The left-hand side of the formula is equal to

$$-2\omega_k \|\varepsilon^k\|_A^2 + 2\omega_{k+1}(\|\varepsilon^k\|_A^2 - \gamma_k \|r^k\|^2 + O(u)),$$

but $\omega_{k+1} = \omega_k + \pi_k$. □

Now, we collect these results.

Theorem 7.9. *In finite precision arithmetic*

$$\begin{aligned} \|\varepsilon^k\|^2 - \|\varepsilon^{k+1}\|^2 = {} & \|r^0\|^2 [(T_{k+1}^{-2})_{1,1} - (T_k^{-2})_{1,1}] \\ & +2\frac{(T_{k+1}^{-2})_{1,k+1}}{(T_{k+1}^{-1})_{1,k+1}} \|\varepsilon^{k+1}\|_A^2 - 2\frac{(T_k^{-2})_{1,k}}{(T_k^{-1})_{1,k}} \|\varepsilon^k\|_A^2 \\ & -\gamma_k^2 \|r^k\|^4 \Delta_k^r + O(u). \end{aligned}$$

Proof. We use the two previous results and adding we have

$$2\gamma_k \omega_k \|r^k\|^2 + \gamma_k \pi_k \|r^k\|^2 + 2\pi_k \|\varepsilon^k\|_A^2 - 2\omega_{k+1}\gamma_k \|r^k\|^2 = \pi_k (2\|\varepsilon^k\|_A^2 - \gamma_k \|r^k\|^2). \quad □$$

To obtain an approximation we write

$$\begin{aligned}\|\varepsilon^{k-d}\|^2 - \|\varepsilon^k\|^2 = &\|r^0\|^2[(T_k^{-2})_{1,1} - (T_{k-d}^{-2})_{1,1}] \\ &+2\frac{(T_k^{-2})_{1,k}}{(T_k^{-1})_{1,k}}\|\varepsilon^k\|_A^2 - 2\frac{(T_{k-d}^{-2})_{1,k-d}}{(T_{k-d}^{-1})_{1,k-d}}\|\varepsilon^{k-d}\|_A^2 \\ &-\sum_{j=k-d}^{k-1}\gamma_j^2\|r^j\|^4\Delta_j^r + O(u).\end{aligned}$$

As what we have done for the A-norm we suppose that $\|\varepsilon^k\|$ is negligible against $\|\varepsilon^{k-d}\|$. However, in the right-hand side we have a term $\|\varepsilon^k\|_A^2$. To have an approximation of this term we have to wait for d more iterations. So, it was proposed in [122] to drop this term. It is also too expensive to compute the term with Δ_k^r, so we neglect it. Hence as an approximation we take

$$\|\varepsilon^{k-d}\|^2 \simeq \|r^0\|^2[(T_k^{-2})_{1,1} - (T_{k-d}^{-2})_{1,1}] - 2\frac{(T_{k-d}^{-2})_{1,k-d}}{(T_{k-d}^{-1})_{1,k-d}}\|\varepsilon^{k-d}\|_A^2.$$

We have shown previously how to approximate the terms in the right-hand side using the QR factorization of T_k. We shall call this the lower estimate since generally it gives a lower bound. We can also just consider the terms arising from the difference of the elements of the inverses. We shall call this the upper estimate since when CG converges fast it generally gives an upper bound.

If we consider the Hestenes and Stiefel formula, we can use

$$\|\varepsilon^{k-d}\|^2 \simeq \sum_{j=k-d}^{k-1}\frac{1}{\mu(p^k)}(\|\varepsilon^j\|_A^2 + \|\varepsilon^{j+1}\|_A^2).$$

Here again, we need an approximation of $\|\varepsilon^k\|_A^2$.

7.4 Numerical examples

The first example is the Strakos30 matrix. We see in Figure 7.1 that a delay $d = 1$ already gives a very good approximation of the A-norm of the error. The Gauss–Radau upper bound was obtained using the exact value of the smallest eigenvalue of A. Figure 7.2 gives the approximations of the l_2 norm of the error. They are worse than the approximations of the A-norm when the norm almost stagnates. This is probably because we neglected some terms.

Figure 7.3 shows the estimates of the A-norm for the matrix Bcsstk01 with a delay $d = 1$. Since the Gauss lower bound is closely linked to the norm of the iterative residual which oscillates, the same phenomenon occurs for the estimate. This can be eliminated by choosing a delay d larger than the period of the oscillations of the norm of the residual. The results in Figure 7.4 use $d = 10$. The curves are smoother and closer to the norm of the error. Results for the l_2 norm are given in Figures 7.5 and 7.6.

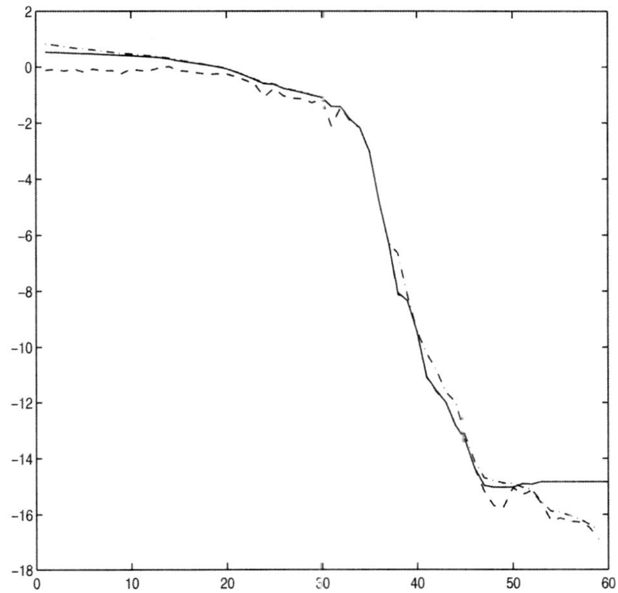

Figure 7.1. *Strakos30*: $\log_{10}$ *of A-norm of the error (solid), Gauss estimate (dashed), Gauss–Radau estimate (dot-dashed),* $d = 1$

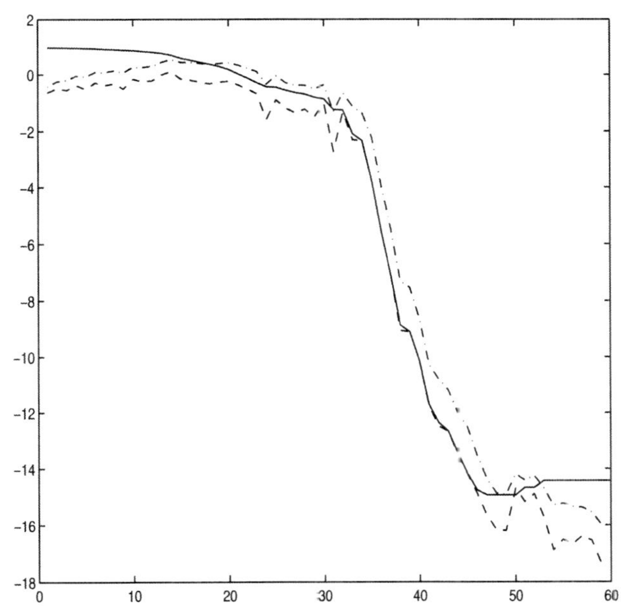

Figure 7.2. *Strakos30*: $\log_{10}$ *of* l_2 *norm of the error (solid), lower estimate (dashed), upper estimate (dot-dashed),* $d = 1$

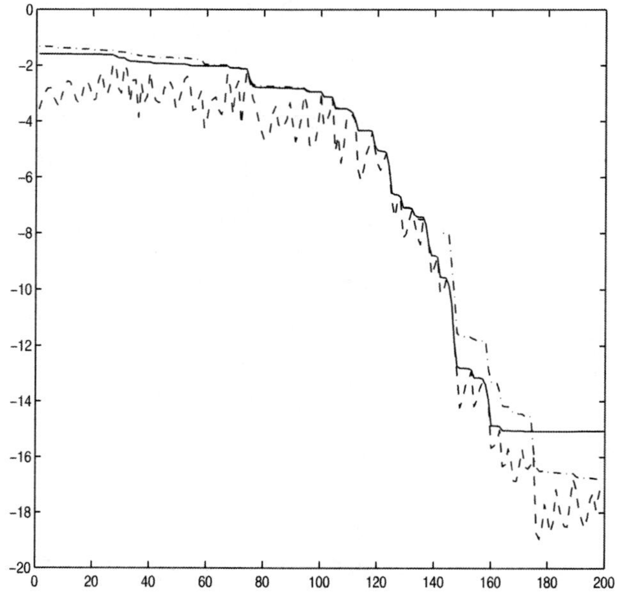

Figure 7.3. *Bcsstk*01: $\log_{10}$ *of A-norm of the error (solid), Gauss estimate (dashed), Gauss–Radau estimate (dot-dashed),* $d = 1$

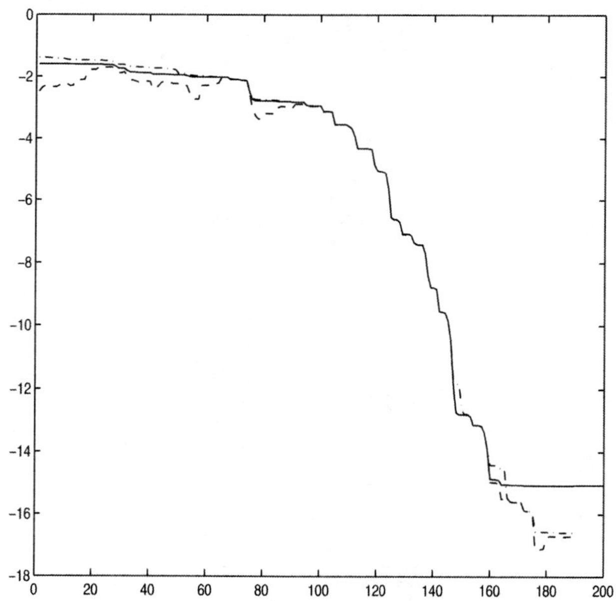

Figure 7.4. *Bcsstk*01: $\log_{10}$ *of A-norm of the error (solid), Gauss estimate (dashed), Gauss–Radau estimate (dot-dashed),* $d = 10$

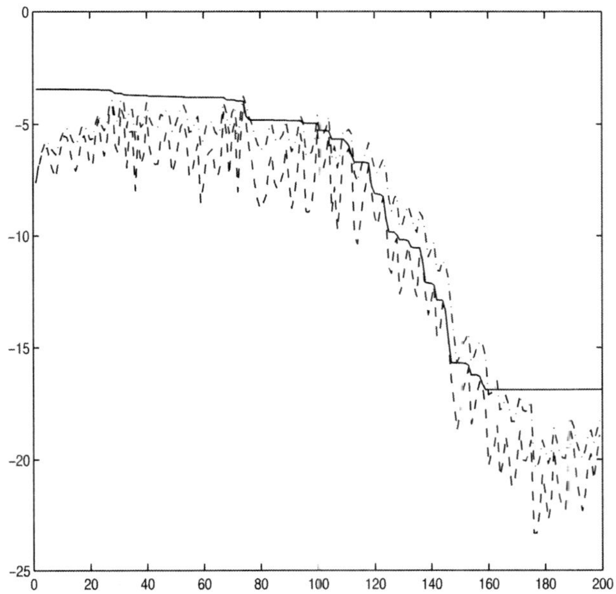

Figure 7.5. *Bcsstk01*: $\log_{10}$ *of* l_2 *norm of the error (solid), lower estimate (dashed), upper estimate (dot-dashed),* $d = 1$

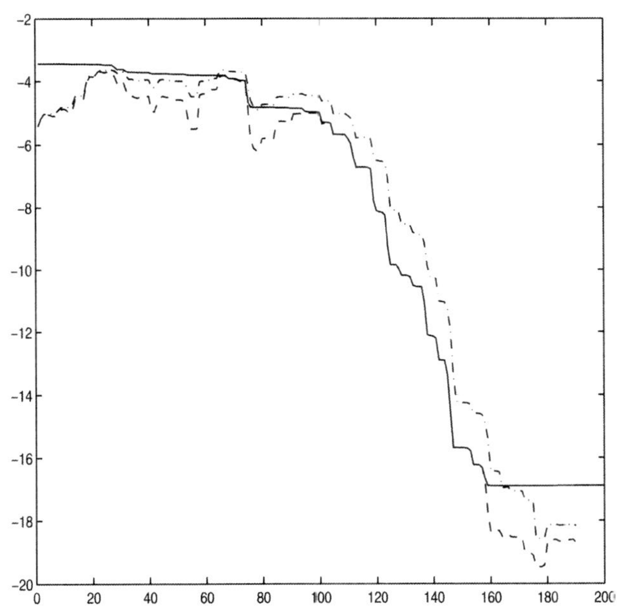

Figure 7.6. *Bcsstk01*: $\log_{10}$ *of* l_2 *norm of the error (solid), lower estimate (dashed), upper estimate (dot-dashed),* $d = 10$

Then we solve a larger problem with a matrix arising from the five-point discretization of the Poisson equation on the unit square with a 60×60 regular mesh which gives a matrix of order 3600. Figure 7.7 shows that the Gauss estimate with $d = 1$ is very close to the A-norm of the error. The l_2 norm of the error in Figure 7.8 is farther away, but this can be improved by choosing $d = 5$, as we can see in Figure 7.9. The upper estimate does not change too much (actually it is a little bit worse with $d = 5$), which means that the improvement comes from the other term. In fact, with a small value of d the term that we have neglected is probably still important. Increasing the value of d, this term is less and less important.

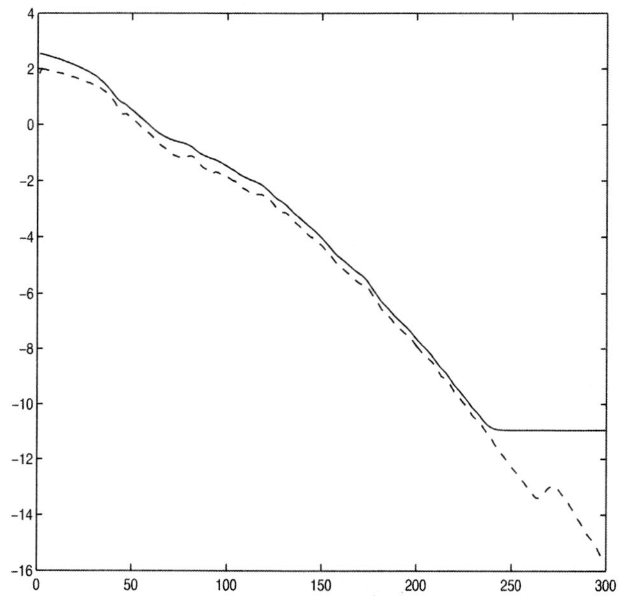

Figure 7.7. *Lap60*: $\log_{10}$ *of A-norm of the error (solid), Gauss estimate (dashed),* $d = 1$

The last example is Msc04515s of order 4515. The matrix has been diagonally normalized with 1's on the diagonal. Figure 7.10 shows the Gauss estimates for $d = 1$ and $d = 20$. We can see the improvement with a larger d. The estimates of the l_2 norm with $d = 1$ in Figure 7.11 are not very good. Using $d = 20$ in Figure 7.12 gives much better results.

In all the examples, we see that the estimates (with d large enough) are quite good until the norms of the error reach the level of stagnation. Beyond this point, the estimates are no longer valid. As we have said before, this is because the estimates are linked to the norm of the iterative residual which converges to zero in finite precision arithmetic. To have something really reliable we must use an estimate of the maximum attainable accuracy.

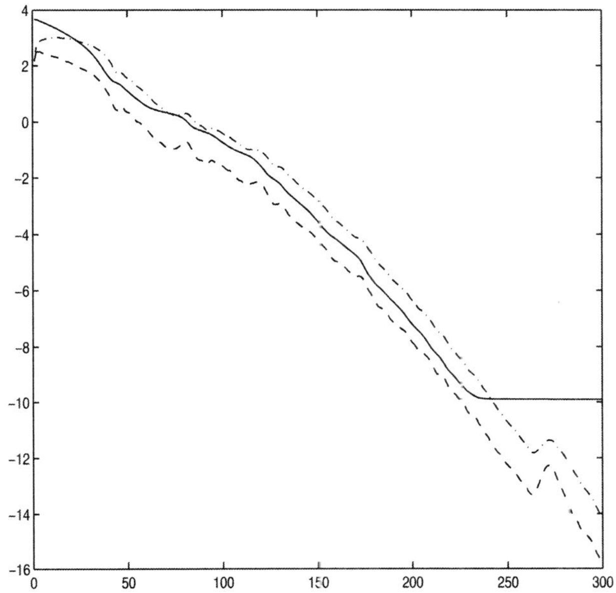

Figure 7.8. *Lap60*: $\log_{10}$ *of* l_2 *norm of the error (solid), lower estimate (dashed), upper estimate (dot-dashed),* $d = 1$

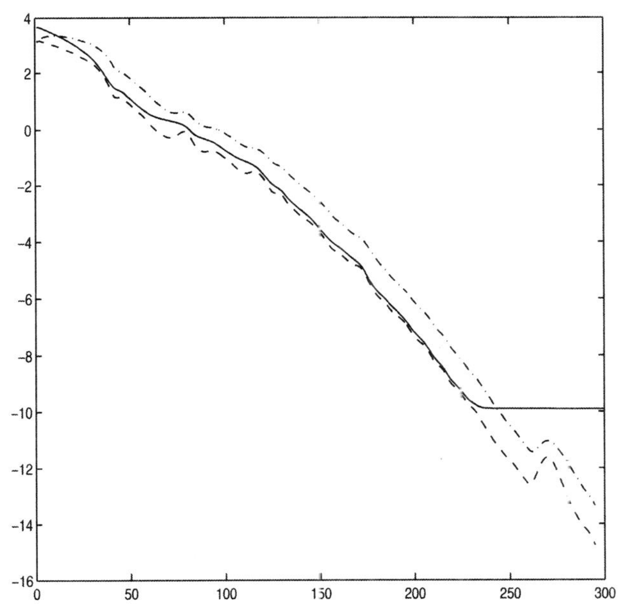

Figure 7.9. *Lap60*: $\log_{10}$ *of* l_2 *norm of the error (solid), lower estimate (dashed), upper estimate (dot-dashed),* $d = 5$

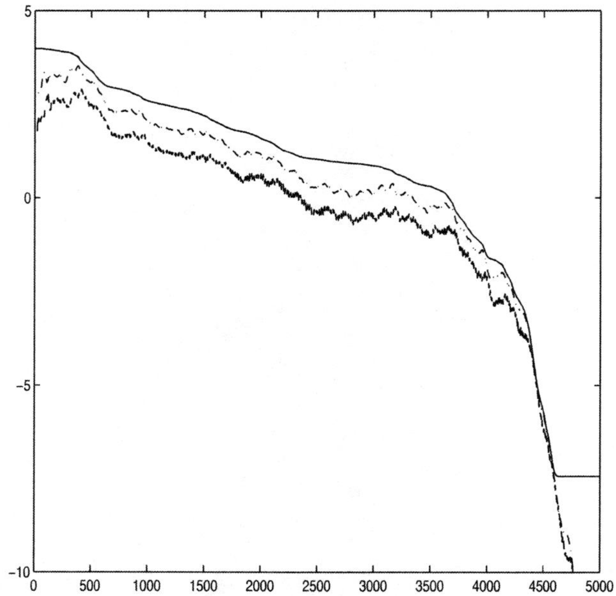

Figure 7.10. *Msc*04515*s*: $\log_{10}$ *of A-norm of the error (solid), Gauss estimate (dashed)* $d = 1$, $d = 20$ *(dot-dashed)*

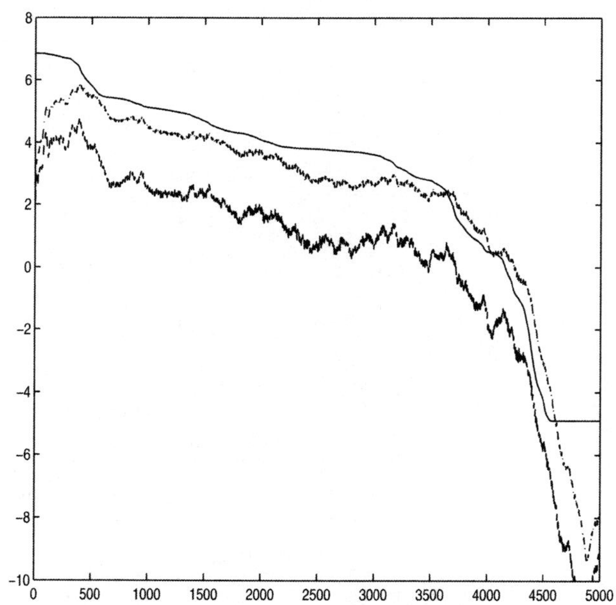

Figure 7.11. *Msc*04515*s*: $\log_{10}$ *of* l_2 *norm of the error (solid), lower estimate (dashed), upper estimate (dot-dashed),* $d = 1$

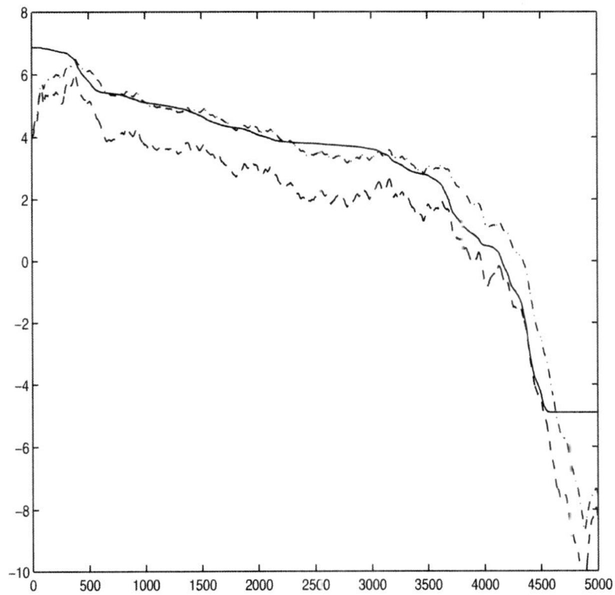

Figure 7.12. *Msc04515s*: $\log_{10}$ *of* l_2 *norm of the error (solid), lower estimate (dashed), upper estimate (dot-dashed),* $d = 20$

7.5 Comparison of two-term and three-term CG

In exact arithmetic two-term and three-term versions of CG are equivalent. The two-term version is generally preferred since the number of operations is smaller. As we have seen, this equivalence is no longer true in finite precision arithmetic since the rounding errors are different in the equivalent three-term recurrence of the two-term version and the genuine three-term version. This question has been theoretically investigated by Gutknecht and Strakoš [86]. They were interested in comparing the maximum attainable accuracies for the residual norm. Their conclusion was that the three-term version may stagnate at a worse level than the two-term version.

Figure 7.13 shows the residuals for the Strakos30 example. The level of stagnation of the x-residual norm for the three-term version is worse than for the two-term version. The norm of the difference of the solutions is 3.0708 10^{-14}. In Figure 7.14 the $\log_{10}$ of local residual errors, two-term (solid), three-term (dashed), are displayed. They are computed using 16 and 32 digits. Figure 7.15 shows the equivalent three-term local error of the two-term recurrence and the local error of the genuine three-term recurrence. We see that they are of the same order. So, this is not the explanation of the difference of the stagnation levels. As explained in [86] it arises from the way these local errors are differently amplified in the equations for the differences of the iterative and x-residuals in the two versions of CG. Figure 7.16 shows the local errors in the equations for x^k. The local error for the two-term version is larger than for the three-term version. Figures 7.17 to 7.20 give the same information for the matrix Bcsstk01. The difference of the stagnation levels is large. Therefore, unless we are using parallel computers for which the three-term version is more suited, it is better to use the two-term recurrences.

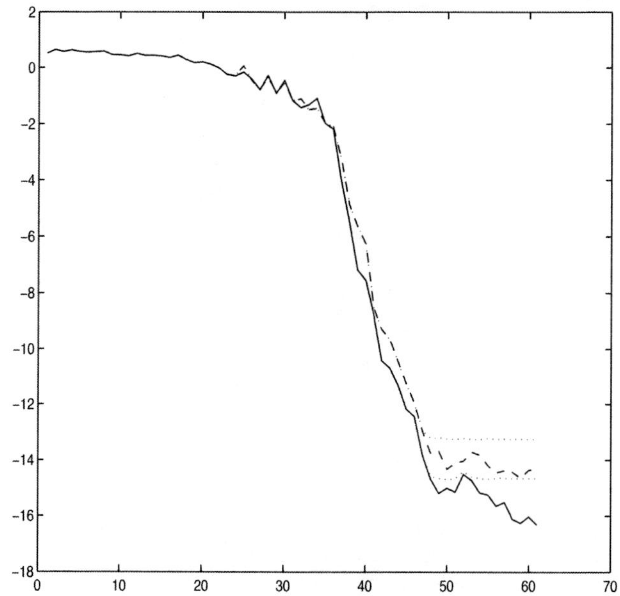

Figure 7.13. *Strakos30*: $\log_{10}$ *of x-residual norms (dotted) and it. residual norms, two-term (solid), three-term (dashed)*

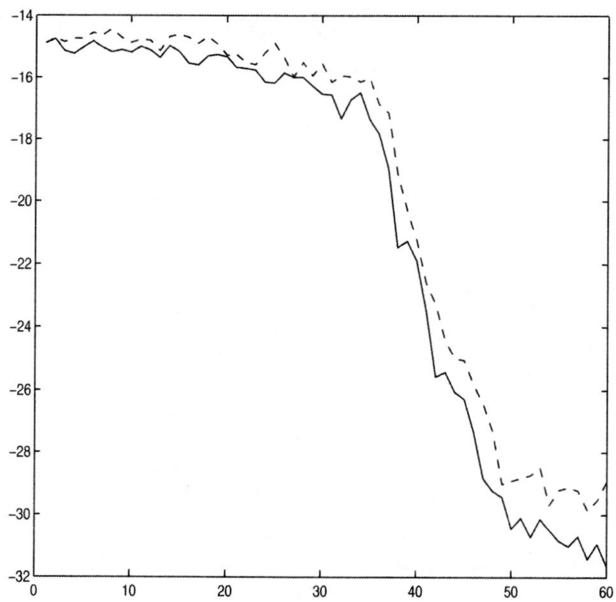

Figure 7.14. *Strakos30*: $\log_{10}$ *of local residual errors, two-term (solid), three-term (dashed)*

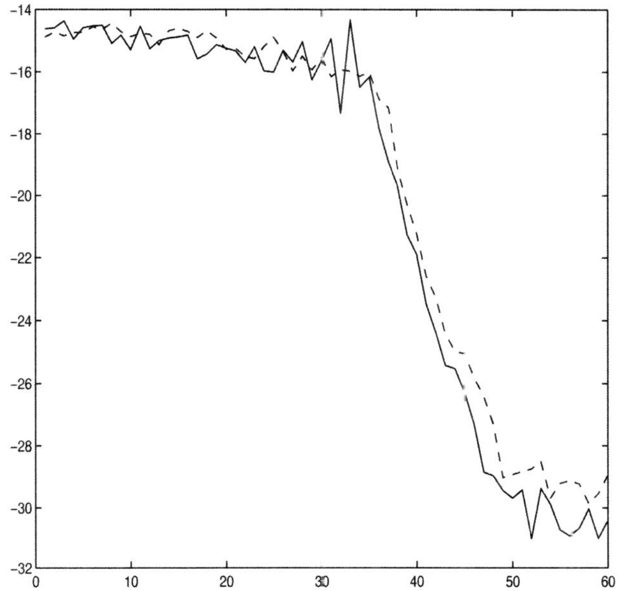

Figure 7.15. *Strakos30*: $\log_{10}$ *of local residual errors, two-term equivalent (solid), three-term (dashed)*

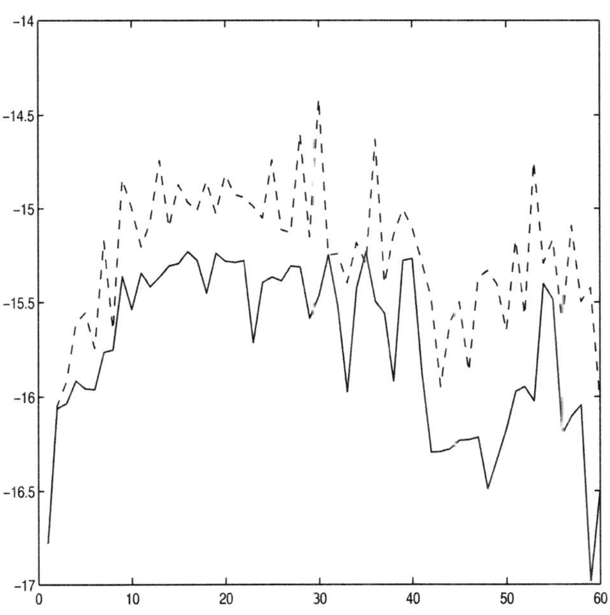

Figure 7.16. *Strakos30*: $\log_{10}$ *of local x errors, two-term (solid), three-term (dashed)*

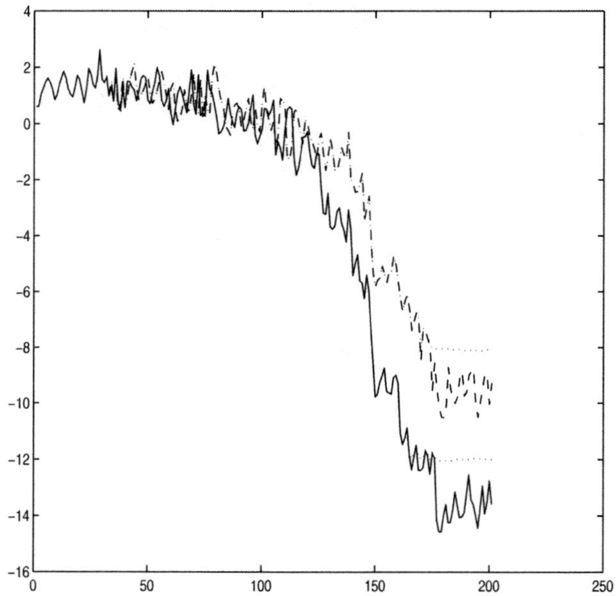

Figure 7.17. *Bcsstk*01: $\log_{10}$ *of* x*-residual norms (dotted) and it. residual norms, two-term (solid), three-term (dashed)*

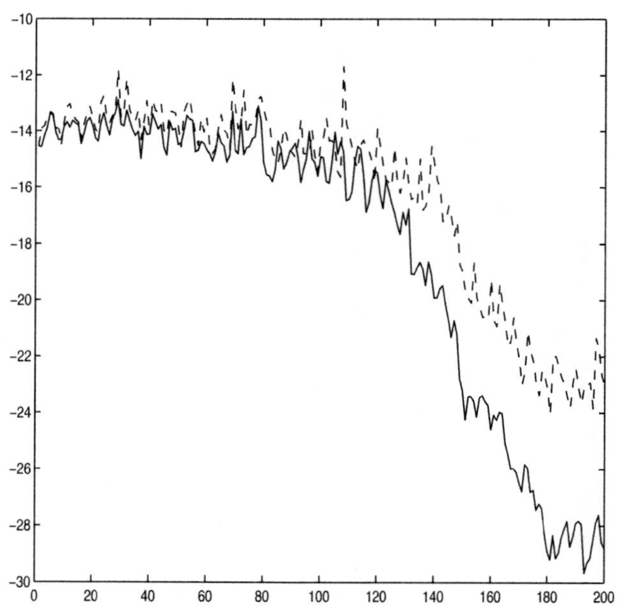

Figure 7.18. *Bcsstk*01: $\log_{10}$ *of local residual errors, two-term (solid), three-term (dashed)*

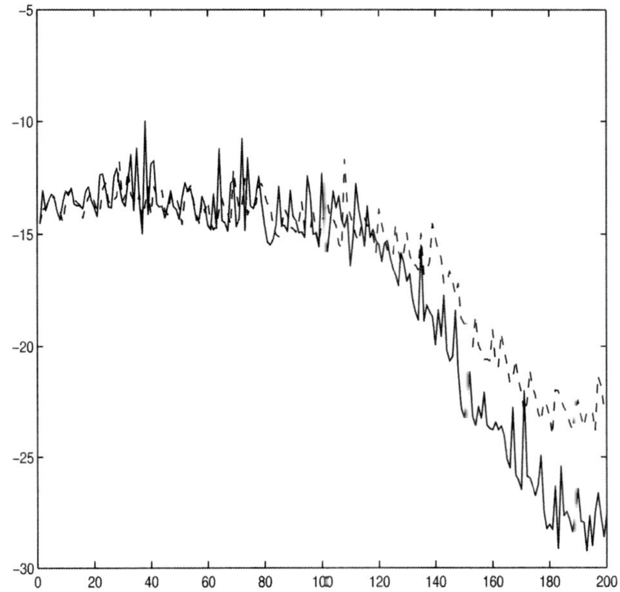

Figure 7.19. *Bcsstk*01: $\log_{10}$ *of local residual errors, two-term equivalent (solid), three-term (dashed)*

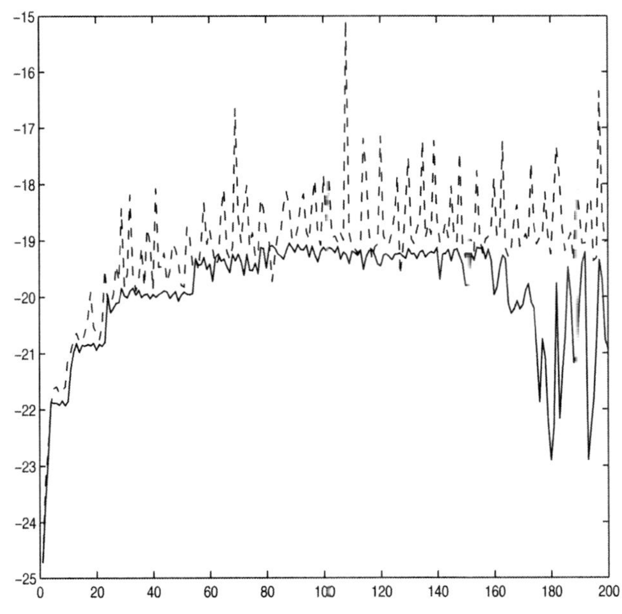

Figure 7.20. *Bcsstk*01: $\log_{10}$ *of local x errors, two-term (solid), three-term (dashed)*

Chapter 8

The preconditioned CG algorithm

To improve the rate of convergence, CG is almost always used with a preconditioner. In this chapter, we consider the modifications we have to make to the results of the previous chapters when a preconditioner M is used. We shall first derive the preconditioned CG (PCG) and give formulas for the computation of the norms of the error. Then we shall study PCG in finite precision arithmetic.

8.1 PCG in exact arithmetic

Let M be a symmetric positive definite matrix which is called the preconditioner. It is well known that the PCG algorithm for solving the linear system $Ax = b$ is obtained by applying CG to the transformed system

$$M^{-1/2}AM^{-1/2}(M^{1/2}x) = M^{-1/2}b,$$

for which the matrix $M^{-1/2}AM^{-1/2}$ is still symmetric positive definite. Notice that $M^{-1/2}AM^{-1/2}$ is similar to $M^{-1}A$, which is not symmetric but has real and positive eigenvalues. Then we obtain recurrences for the approximations to x by going back to the original variables. Let $r^k = b - Ax^k$ and $y^k = M^{1/2}x^k$ be the iterates for the preconditioned system. For the preconditioned equation the residual is

$$\hat{r}^k = M^{-1/2}b - M^{-1/2}AM^{-1/2}y^k = M^{-1/2}(b - Ax^k) = M^{-1/2}r^k.$$

Let z^k be given by solving $Mz^k = r^k$. Then the inner product we need in PCG is

$$(\hat{r}^k, \hat{r}^k) = (M^{-1}r^k, r^k) = (z^k, r^k).$$

Moreover, let $\hat{p}^k = M^{1/2}p^k$. Then

$$(\hat{p}^k, M^{-1/2}AM^{-1/2}\hat{p}^k) = (p^k, Ap^k).$$

By using this change of variable, the PCG algorithm is the following.

ALGORITHM 8.1.

Let x^0 be given, $r^0 = b - Ax^0$, $Mz^0 = r^0$, $p^0 = z^0$
for $k = 1, \dots$ until convergence
$\gamma_{k-1} = (z^{k-1}, r^{k-1})/(p^{k-1}, Ap^{k-1})$
$x^k = x^{k-1} + \gamma_{k-1} p^{k-1}$
$r^k = r^{k-1} - \gamma_{k-1} A p^{k-1}$
$Mz^k = r^k$
$\beta_k = (z^k, r^k)/(z^{k-1}, r^{k-1})$
$p^k = z^k + \beta_k p^{k-1}$
end

We can apply the same theory as before for CG. What is now involved are the eigenvalues and eigenvectors of $M^{-1/2}AM^{-1/2}$. Ideally, the preconditioner must be chosen to have a good distribution of the eigenvalues of $M^{-1}A$. In exact arithmetic we have orthogonality of the vectors $\hat{r}^k$ which translates into

$$(z^k, r^j) = (z^k, Mz^j) = 0, \quad j \neq k;$$

that is, the vectors z^k are orthogonal in the inner product defined by the matrix M. The conjugacy of the $\hat{p}^k$ relative to $M^{-1/2}AM^{-1/2}$ gives the A-conjugacy of the vectors p^k. Let $M^{-1/2}AM^{-1/2} = Q\Lambda Q^T$ be the spectral decomposition of the preconditioned matrix with Q orthogonal and Λ the diagonal matrix of the eigenvalues. Then we are interested in the components of the vectors $\hat{r}^k$ on the eigenvectors, that is,

$$\bar{\hat{r}}^k = Q^T \hat{r}^k = Q^T M^{-1/2} r^k = Q^T M^{1/2} z^k.$$

Let $S = Q^T M^{1/2}$; we have

$$M^{-1}A = M^{-1/2} Q \Lambda Q^T M^{1/2} = S^{-1} \Lambda S.$$

The columns of the matrix S^{-1} are the unnormalized right eigenvectors of $M^{-1}A$; moreover, $S^T S = M$ and $S^T \Lambda S = A$. We have

$$\bar{\hat{r}}^{k+1} = \bar{\hat{r}}^k - \gamma_k \Lambda \bar{\hat{p}}^k,$$

which translates into

$$Sz^{k+1} = Sz^k - \gamma_k \Lambda S p^k.$$

So the vectors we are interested in are the Sz^k. We also have

$$\|\varepsilon^k\|_A^2 = (A^{-1} r^k, r^k) = (\Lambda^{-1} S z^k, S z^k).$$

8.2 Formulas for norms of the error

Let $\hat{\epsilon}^k = y^k - y$, where $y = M^{1/2}x$ and $\epsilon^k = \varepsilon^k = A^{-1}r^k = x - x^k$. Then,

$$\|\hat{\epsilon}^k\|_{M^{-1/2}AM^{-1/2}}^2 = (M^{-1/2}AM^{-1/2}(y^k - y), y^k - y) = (A(x^k - x), x^k - x) = \|\epsilon^k\|_A^2.$$

This shows that for the A-norm we can use the formula

$$\|\epsilon^k\|_A^2 = (z^0, r^0)((T_r^{-1})_{1,1} - (T_k^{-1})_{1,1}),$$

where the Lanczos matrix T_k is constructed from the PCG coefficients. Since we have

$$(M^{1/2}A^{-1}M^{1/2}\hat{r}^0, \hat{r}^0) = (A^{-1}r^0, r^0),$$

the formula also reads

$$\|\epsilon^k\|_A^2 = (\Lambda^{-1}Sz^0, Sz^0) - (z^0, r^0)(T_k^{-1})_{1,1}.$$

Let $s^0 = Sz^0$. Then

$$\|\epsilon^k\|_A^2 = \sum_{j=1}^{n} \frac{(s_j^0)^2}{\lambda_j} - (z^0, r^0)\sum_{j=1}^{k} \frac{[(z_{(k)}^j)_1]^2}{\theta_j^{(k)}},$$

the vectors $z_{(k)}^j$ being the eigenvectors of T_k and the $\theta_j^{(k)}$ the Ritz values. Notice that $(z^k, r^k) = \|Sz^k\|^2$.

Theorem 8.1. *Since* $(z^0, r^0) = \|s^0\|^2$

$$\|\epsilon^k\|_A^2 = \|s^0\|^2 \left(\frac{1}{\|s^0\|^2} \sum_{j=1}^{n} \frac{(s_j^0)^2}{\lambda_j} - \sum_{j=1}^{k} \frac{[(z_{(k)}^j)_1]^2}{\theta_j^{(k)}} \right),$$

where $s^0 = Q^T M^{\frac{1}{2}} z^0 = Q^T M^{-\frac{1}{2}} r^0$.

The Hestenes and Stiefel-like formula becomes

$$\|\epsilon^k\|_A^2 = \sum_{j=k}^{n-1} \gamma_j (z^j, r^j).$$

From these formulas estimates of the A-norm of the error can be derived in the same spirit as without preconditioning. Unfortunately, things are not so nice for the l_2 norm since

$$\|\hat{\epsilon}^k\|^2 = (y^k - y, y^k - y) = (M(x^k - x), x^k - x) = \|\epsilon^k\|_M^2.$$

Directly translating the formula for the l_2 norm will provide us with only the M-norm of the error. However, let us suppose that $M = LL^T$, where L is a triangular matrix. We have $\epsilon^k = M^{-1/2}\hat{\epsilon}^k$; therefore

$$\|\epsilon^k\|^2 = (M^{-1}\hat{\epsilon}^k, \hat{\epsilon}^k) = \|L^{-1}\hat{\epsilon}^k\|^2.$$

Then,

$$\|\epsilon^k\| \leq \|L^{-1}\| \, \|\hat{\epsilon}^k\| = \|L^{-1}\| \, \|\epsilon^k\|_M.$$

It is usually difficult to compute or estimate the l_2 norm of L^{-1}. We will replace the l_2 norm by the l_∞ norm of this matrix. Notice that we have $\|L^{-1}\| \leq \sqrt{n}\|L^{-1}\|_\infty$. If we suppose that M is an M-matrix, the matrix L^{-1} has positive elements. If w is the solution of $Lw = e$, where e is the vector of all ones, then $l = \|L^{-1}\|_\infty = \max_i w_i$. Hence, we take

$$\|\epsilon^k\| \simeq l\|\epsilon^k\|_M.$$

When the matrix A is symmetrically scaled with 1's on the diagonal, it turns out that it is generally not too bad to use $l = 1$. For error estimation in PCG, see also Strakoš and Tichý [187].

8.3 PCG in finite precision

In this section we consider the modifications to be done when using a preconditioner in finite precision arithmetic. As in CG without preconditioning we have

$$r^{k+1} = r^k - \gamma_k A p^k + \delta_r^k.$$

Things are different for p^k which is computed with z^k and not r^k. The vector z^k is computed by solving $Mz^k = r^k$ and, of course, there are rounding errors in solving this system even though this is done with a "direct" method. These errors depend on the choice of the preconditioner M. It is outside the scope of this book to describe preconditioners; see, for instance, [120]. One of the most popular preconditioners is the incomplete Cholesky decomposition. Then $M = LL^T$, where L is a lower sparse triangular matrix. The solution z^k is obtained through two triangular solves,

$$Ly^k = r^k, \quad L^T z^k = y^k.$$

Results for triangular solves in finite precision arithmetic with dense matrices are given in Higham [94]. Here L is a sparse matrix. There are several algorithms for triangular solves. Let us consider the one using inner products. In exact arithmetic the components of y^k are given by

$$y_j^k = \frac{1}{l_{j,j}} \left(r_j^k - \sum_{i=1}^{j-1} l_{j,i} y_i^k \right).$$

The sum (in which most of the terms are zero) is the inner product of the nonzero part of the jth row of L with a part of the vector y^k. Let us denote the sum over i by s_j and suppose that there are at most m_L nonzero entries in each row of L. We can use the results for inner products in finite precision. Removing the index k for simplicity, we have

$$fl(y_j) = \frac{1}{l_{j,j}} \left[r_j - fl\left(\sum_{i=1}^{j-1} l_{j,i} fl(y_i) \right) \right] (1+\delta_1)(1+\delta_2), \quad |\delta_i| \le u, \quad i = 1, 2.$$

For the inner product we know that

$$fl\left(\sum_{i=1}^{j-1} l_{j,i} fl(y_i) \right) = \sum_{i=1}^{j-1} l_{j,i} fl(y_i) + uC_j(m_j - 1) + O(u^2),$$

with $|C_j| \le \sum_{i=1}^{j-1} |l_{j,i}|\,|fl(y_i)|$ and m_j is the number of nonzero elements in row j of L. Let us suppose that $fl(y_i) = y_i + \delta_i^y$, $i = 1, \dots, j-1$. Then

$$fl(y_j) = y_j - \frac{1}{l_{j,j}} \sum_{i=1}^{j-1} l_{j,i}\delta_i^y - \frac{uC_j(m_j - 1)}{l_{j,j}} + (\delta_1 + \delta_2)y_j + O(u^2) = y_j + \delta_j^y.$$

Therefore, we have a recursion for the roundoff terms,

$$\delta_j^y = -\frac{1}{l_{j,j}} \sum_{i=1}^{j-1} l_{j,i}\delta_i^y - \frac{uC_j(m_j - 1)}{l_{j,j}} + (\delta_1 - \delta_2)y_j + O(u^2).$$

Multiplying by $l_{j,j}$ and putting the first term on the right-hand side to the left-hand side, δ_y being a vector with components δ_i^y, we have

$$L\delta_y = -uC + (\delta_1 + \delta_2)D_L y + O(u^2),$$

where C is a vector with components $(m_j - 1)\sum_{i=1}^{j-1} |l_{j,i}|\,|y_i|$ and D_L is a diagonal matrix with the diagonal elements of L supposed to be positive. Thus, $fl(y^k) = y^k + \delta_y^k$ with

$$\begin{aligned} |\delta_y^k| &\le u|L^{-1}|[(m_L - 1)(|L| - D_L)|y^k| + 2D_L|y^k|] + O(u^2) \\ &\le u(m_L - 1)|L^{-1}|\,|L|\,|y^k| + O(u^2), \end{aligned}$$

provided $m_L > 2$. This is consistent with the results in Higham [94]. We apply the same results for the backward solve and we obtain $fl(z^k) = z^k + \delta_z^k$ with

$$|\delta_z^k| \le |\delta_y^k| + u(m_L - 1)|L^{-T}|\,|L^T|\,|z^k| + O(u^2).$$

We notice that we can replace $|y^k|$ and $|z^k|$ by their computed values and the bounds on the right-hand side implicitly contain a factor $|r^k|$.

Another popular preconditioner is using sparse approximate inverses; see [8]. These algorithms directly construct the inverse of M as an approximation of A^{-1} and the solve for z^k is reduced to a matrix-vector multiply $z^k = M^{-1}r^k$; see [8]. We can use the results for matrix multiply. Again we can write

$$fl(z^k) = z^k + \delta_z^k, \quad |\delta_z^k| \le \frac{m_M u}{1 - m_M u}|M^{-1}|\,|r^k|,$$

where m_M is the maximum number of nonzero entries in any row of M^{-1}. We obtain

$$p^k = z^k + \beta_k p^{k-1} + \delta_p^{k-1},$$

where z^k is the computed solution. But now,

$$|\delta_p^{k-1}| \le u[|z^k| + 2\beta_k|p^{k-1}|] + |\delta_\beta^k|\,|p^{k-1}|,$$

$$|\delta_\beta^k| \le u\frac{(z^k, r^k)}{(z^{k-1}, r^{k-1})}(1 + 2n) + O(u^2).$$

We have $Mz^k = r^k + M\delta_z^k$. From this we obtain the equation of z^k,

$$Mz^{k+1} = Mz^k - \gamma_k Ap^k + \delta_r^k + M\delta_z^{k+1} - M\delta_z^k.$$

Eliminating p^k we have the three-term equation

$$\begin{aligned} z^{k+1} &= -\gamma_k M^{-1}Az^k + \left(1 + \frac{\gamma_k\beta_k}{\gamma_{k-1}}\right) z^k - \frac{\gamma_k\beta_k}{\gamma_{k-1}} z^{k-1} \\ &+ M^{-1}\delta_r^k - \frac{\gamma_k\beta_k}{\gamma_{k-1}} M^{-1}\delta_r^{k-1} - \gamma_k M^{-1}A\delta_p^{k-1} \\ &- \left(1 + \frac{\gamma_k\beta_k}{\gamma_{k-1}}\right)\delta_z^k + \delta_z^{k+1} + \frac{\gamma_k\beta_k}{\gamma_{k-1}}\delta_z^{k-1}. \end{aligned}$$

Proposition 8.2. *Let*

$$\begin{aligned} \bar{\delta}_r^k &= M^{-1}\delta_r^k - \frac{\gamma_k\beta_k}{\gamma_{k-1}} M^{-1}\delta_r^{k-1} - \gamma_k M^{-1}A\delta_p^{k-1} \\ &- \left(1 + \frac{\gamma_k\beta_k}{\gamma_{k-1}}\right)\delta_z^k + \delta_z^{k+1} + \frac{\gamma_k\beta_k}{\gamma_{k-1}}\delta_z^{k-1}, \\ \bar{\delta}_r^0 &= M^{-1}\delta_r^0 + \delta_z^1 - \delta_z^0. \end{aligned}$$

The equivalent three-term recurrence of PCG in finite precision is

$$\begin{aligned} z^{k+1} &= -\gamma_k M^{-1}Az^k + \left(1 + \frac{\gamma_k\beta_k}{\gamma_{k-1}}\right) z^k - \frac{\gamma_k\beta_k}{\gamma_{k-1}} z^{k-1} + \bar{\delta}_r^k, \\ z^1 &= z^0 - \gamma_0 M^{-1}Az^0 + \bar{\delta}_r^0. \end{aligned}$$

From this proposition we can obtain the same results as for CG. We define the PCG-Lanczos vectors as

$$w^{k+1} = (-1)^k \frac{z^k}{\|z^k\|_M}.$$

However, the quantities of interest are now $\bar{z}^k = Sz^k$ or $\bar{w}^k = Sw^k$, where $M^{-1}A = S^{-1}\Lambda S$ and $S = Q^TM^{1/2}$, Q being the matrix of the eigenvectors of $M^{-1/2}AM^{-1/2}$. These are expansions of z^k and w^k in a nonorthogonal basis whose vectors are the columns of S. The relevant norms for PCG are $\|\epsilon^k\|_A$ and $\|z^k\|_M$. Notice that the norm of the residual $\|r^k\|$ is not even computed by the algorithm (except when $M = I$). If it has to be used in the stopping criteria, this is an inner product which has to be computed with $2n - 1$ additional operations.

Theorem 8.3. *The equivalent three-term recurrence for the PCG-Lanczos vectors is*

$$\eta_{k+1}w^{k+1} = Aw^k - \alpha_k w^k - \eta_k w^{k-1} + (-1)^k \frac{\bar{\delta}_r^{k-1}}{\gamma_{k-1}\|z^{k-1}\|_M},$$

$$\eta_{k+1} = \frac{\sqrt{\beta_k}}{\gamma_{k-1}}, \quad \alpha_k = \frac{1}{\gamma_{k-1}} + \frac{\beta_{k-1}}{\gamma_{k-2}}, \quad \alpha_1 = \frac{1}{\gamma_0},$$

with

$$\gamma_k = \frac{(r^k, z^k)}{(Ap^k, p^k)}, \quad \beta_k = \frac{(r^k, z^k)}{(r^{k-1}, z^{k-1})}.$$

This leads to the following results, which are simple translations of what we proved for the nonpreconditioned CG.

Theorem 8.4. *Let*

$$\tilde{\delta}^k = S\left((-1)^k \frac{\bar{\delta}_r^{k-1}}{\gamma_{k-1}\|z^{k-1}\|_M}\right),$$

j given, and $p_{j,k}$ be the polynomial determined by

$$p_{j,j-1} = 0, \quad p_{j,j} = 1,$$

$$\eta_{k+1} p_{j,k+1}(\lambda) = (\lambda - \alpha_k) p_{j,k}(\lambda) - \eta_k p_{j,k-1}(\lambda), \quad k = j, \dots.$$

The solution of the perturbed recurrence

$$\eta_{k+1}\bar{w}_i^{k+1} = (\lambda_i - \alpha_k)\bar{w}_i^k - \eta_k \bar{w}_i^{k-1} + \tilde{\delta}_i^k$$

starting from $\bar{w}_i^0 = 0$ and $\bar{w}_i^1$ is given by

$$\bar{w}_i^{k+1} = p_{1,k+1}(\lambda_i)\bar{w}_i^1 + \sum_{l=1}^{k} p_{l+1,k+1}(\lambda_i)\frac{\tilde{\delta}_i^l}{\eta_{l+1}}.$$

Theorem 8.5. *Let*

$$\delta^k = (-1)^k \frac{\bar{\delta}_r^{k-1}}{\gamma_{k-1}\|z^{k-1}\|_M},$$

$$w^{k+1} = p_{1,k+1}(M^{-1}A)w^1 + \sum_{l=1}^{k} p_{l+1,k+1}(M^{-1}A)\frac{\bar{\delta}^l}{\eta_{l+1}}.$$

Theorem 8.6.

$$z^k = (-1)^k \frac{\|z^k\|_M}{\|z^0\|_M} p_{1,k+1}(M^{-1}A)z^0 + (-1)^k\|z^k\|_M \sum_{l=1}^{k} p_{l+1,k+1}(M^{-1}A)\frac{\bar{\delta}^l}{\eta_{l+1}},$$

where the polynomials are defined in Theorem 8.4.

8.4 Numerical examples of convergence

We are interested in looking at the components of Sz^k.

Proposition 8.7.

$$(Sz^k)_i = \bar{z}_i^k = (-1)^k \frac{\|z^k\|_M}{\|z^0\|_M} p_{1,k+1}(\lambda_i)\bar{z}_i^0 + (-1)^k\|z^k\|_M \sum_{l=1}^{k} p_{l+1,k+1}(\lambda_i)\frac{\tilde{\delta}_i^l}{\eta_{l+1}},$$

where λ_i is an eigenvalue of $M^{-1}A$.

The polynomial $p_{1,k}$ at λ_i has the behavior that was described when looking at CG in finite precision. Remember that $\|Sz^k\| = \|z^k\|_M$.

The first example is the Poisson equation in a square discretized by finite differences on a 20×20 mesh giving $n = 400$. The preconditioner is IC(0), the incomplete Cholesky factorization without any fill-in. The spectrum of $M^{-1}A$ is shown in Figure 8.1. The smallest eigenvalues are well separated and they are the ones for which Ritz values converge first. Figure 8.2 displays the M-norm of z^k as well as two components of Sz^k, the first and the last ones. We see in Figure 8.3 that, when normalized, the first component (which corresponds to the first Ritz value to converge) goes down approximately to $\sqrt{u}$ following $p_{1,k}(\lambda_1)$ and then oscillates as predicted by theory when the perturbation terms become large. The perturbation terms stay small for the eigenvalues for which no Ritz value has converged yet. Figure 8.4 shows the norms of the iterative residual, the x-residual and norms of the error $x - x^k$. Convergence is quite fast and all the norms have almost the same behavior.

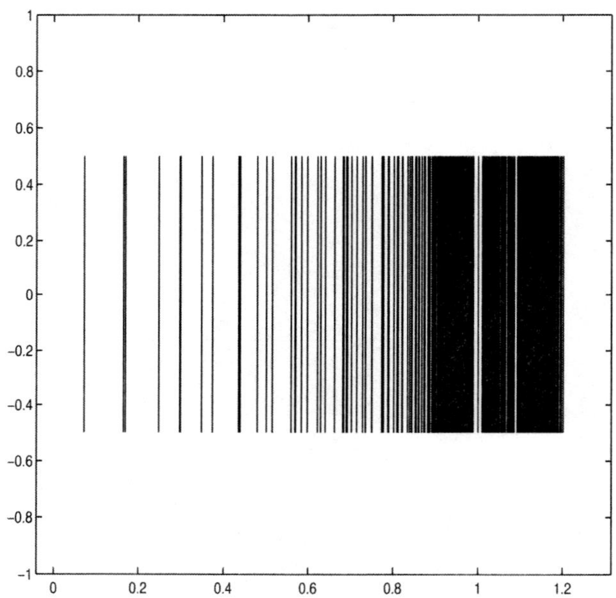

Figure 8.1. *Lap20-IC: eigenvalues of $M^{-1}A$ for IC*

We then look at an example of convergence, following the result of Theorem 8.1. Figure 8.5 shows the first 10 components,

$$\frac{1}{\|s^0\|^2}\frac{(s_j^0)^2}{\lambda_j}, \qquad j = 1, \ldots, 10.$$

The stars are the values of

$$\frac{[(z_{(k)}^j)_1]^2}{\theta_j^{(k)}}, \qquad j = 1, \ldots, 10,$$

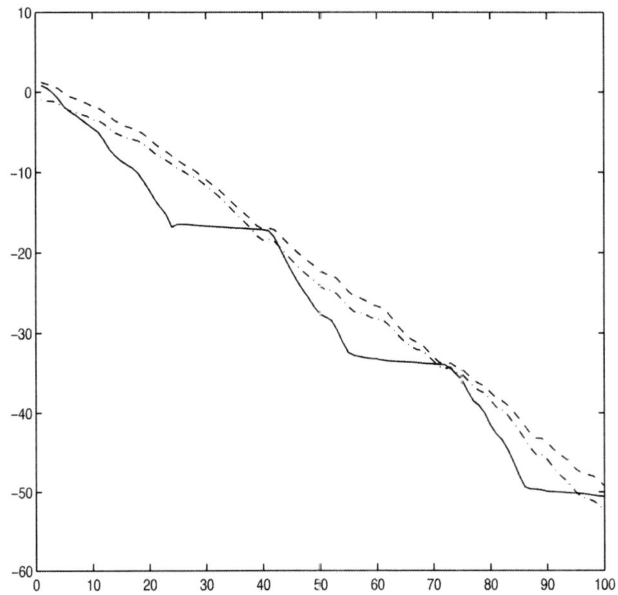

Figure 8.2. *Lap20-IC:* $\log_{10}$ *of* $\|z^k\|_M$ *(dashed),* $(Sz^k)_1$ *(solid),* $(Sz^k)_{400}$ *(dot-dashed)*

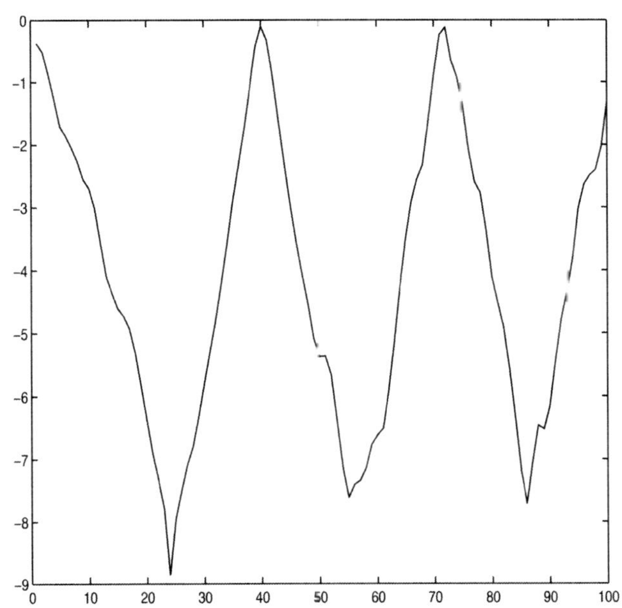

Figure 8.3. *Lap20-IC:* $\log_{10}$ *of* $(Sz^k)_1$ *normalized by* $\|z^k\|_M$

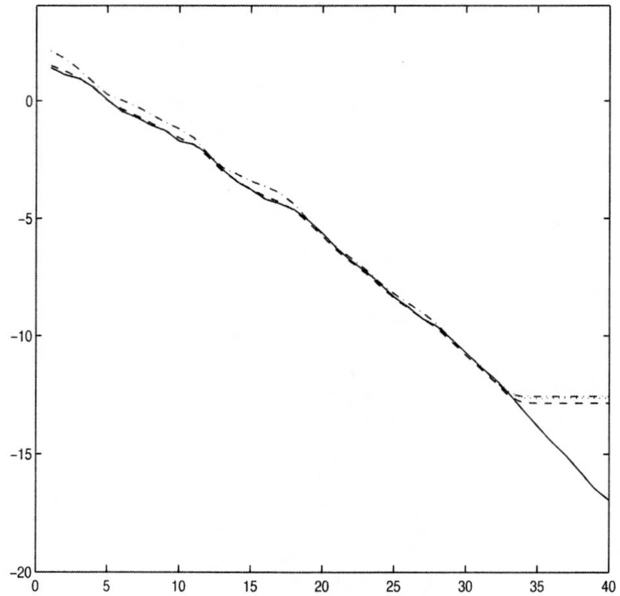

Figure 8.4. *Lap20-IC:* $\log_{10}$ *of* $\|r^k\|$ *(solid), norm of x-residual (dotted),* $\|\epsilon^k\|_A$ *(dashed),* $\|\epsilon^k\|$ *(dot-dashed)*

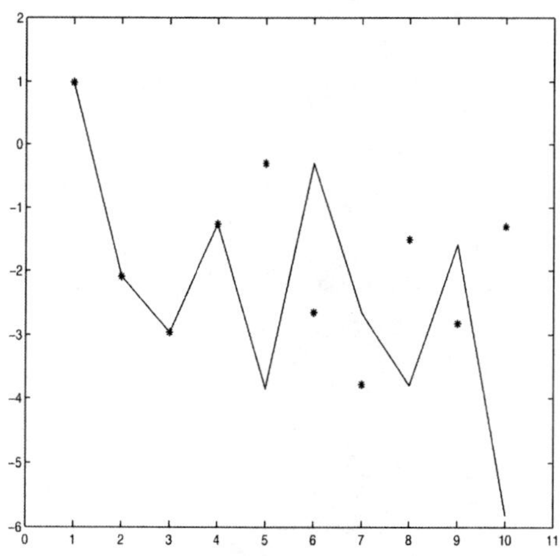

Figure 8.5. *Lap20-IC: convergence* 1

at iteration $k = 25$. We see that the first four values corresponding to the first four eigenvalues are well approximated. It seems that this is not the case for the other ones. However, if we shift the stars starting from the fifth one by one position to the right (see Figure 8.6), we see that the entries 6 to 9 are also well approximated. The fifth component is not approximated yet. It corresponds to an eigenvalue $\lambda_5 = 0.2971$ which is close to $\lambda_6 = 0.2994$ and to which no Ritz value has converged so far. The y-axis has a logarithmic scale. Most of

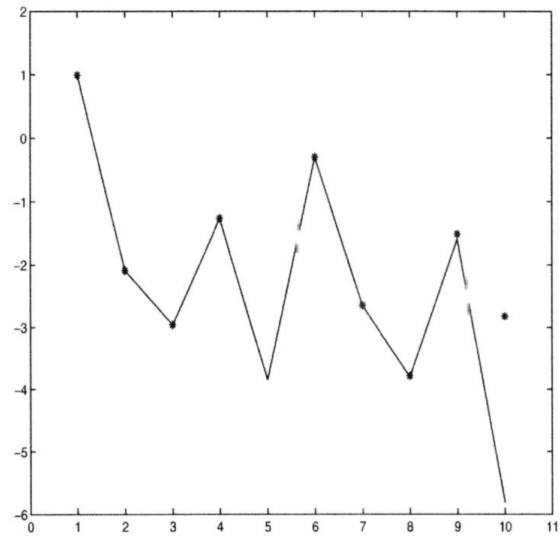

Figure 8.6. *Lap*20-*IC: convergence* 2

the other components are small. In this example the largest part of the norm of the error corresponds to the smallest eigenvalue. Hence the A-norm of the error is already small at iteration 25.

The second example is Bcsstk01. Convergence was very slow without preconditioning and the norm of the residual was oscillating. Convergence is much faster with the incomplete Cholesky preconditioner IC(0). The spectrum of $M^{-1}A$ in Figure 8.7 shows

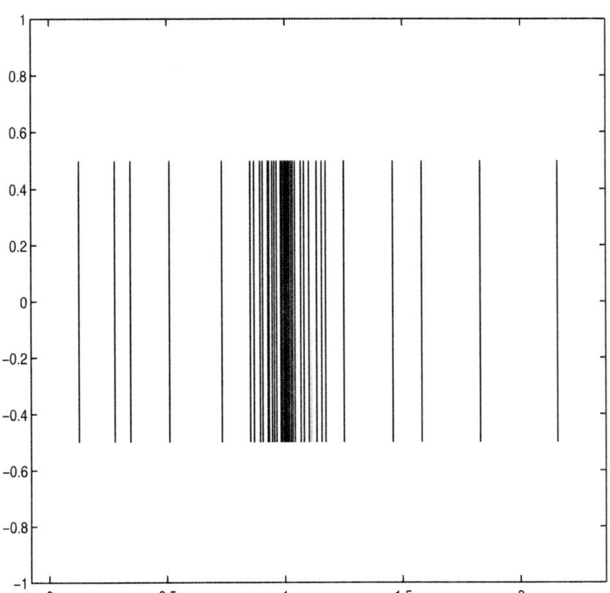

Figure 8.7. *Bcsstk*01-*IC: eigenvalues of* $M^{-1}A$ *for IC*

that both the smallest and the largest eigenvalues are well separated. The smallest and largest Ritz values converge first (see Figure 8.8); correspondingly, the components of the PCG-Lanczos vectors oscillate; as we see in Figure 8.9. Figure 8.10 gives the norms of residuals and errors. In this example, their values are quite different. Using the residual norm to stop the iterations is misleading, giving too many iterations.

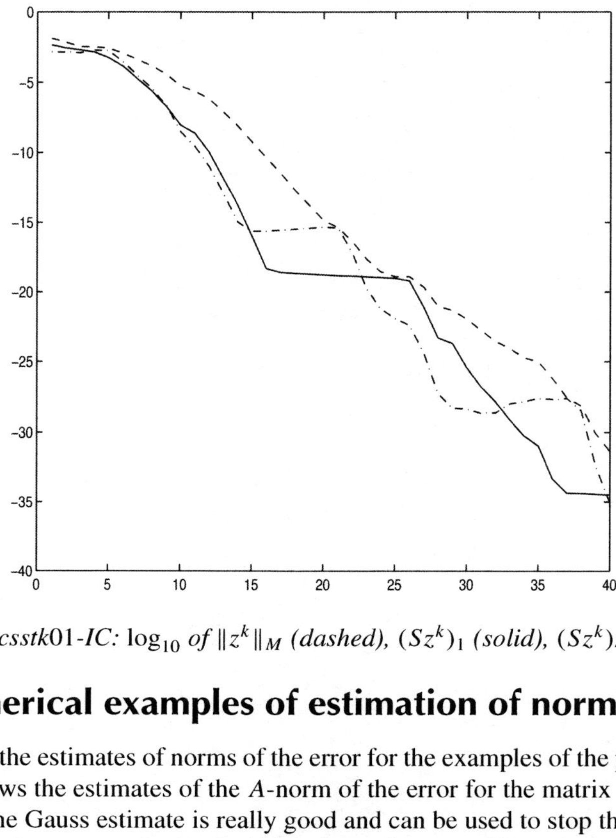

Figure 8.8. *Bcsstk*01*-IC:* $\log_{10}$ *of* $\|z^k\|_M$ *(dashed),* $(Sz^k)_1$ *(solid),* $(Sz^k)_{48}$ *(dot-dashed)*

8.5 Numerical examples of estimation of norms

Let us consider the estimates of norms of the error for the examples of the previous section. Figure 8.11 shows the estimates of the A-norm of the error for the matrix Lap20-IC with a delay $d = 1$. The Gauss estimate is really good and can be used to stop the iterations. The Gauss–Radau estimate was obtained with an approximate smallest eigenvalue of 0.01, the exact value being 0.0724. Estimates of the l_2 norm are given in Figure 8.12.

Figure 8.13 displays the estimates of the A-norm of the error for Bcsstk01 with an incomplete Cholesky preconditioner. Results are satisfying even with $d = 1$. The results for the l_2 norm in Figure 8.14 are not as good as for the other example. This is a case where the estimate is not a lower bound. Figure 8.15 shows the norm of the iterative residual (solid) and the x-residual (dotted); the horizontal dot-dashed lines are lower and upper estimates of the maximum attainable accuracy for the residual norm, the dashed curve is the A-norm of the error and the dotted one its estimate, and the dot-dashed curve is the l_2 norm of the error and the dotted one its estimate. Therefore, we have good estimates of the norms and of the maximum attainable accuracy. We can use this to build a reliable stopping criterion. We can stop when our estimate of the A-norm has reached the level we need or when the residual norm is smaller than our lower estimate of the maximum attainable accuracy since then the residual norm and the solution stagnate.

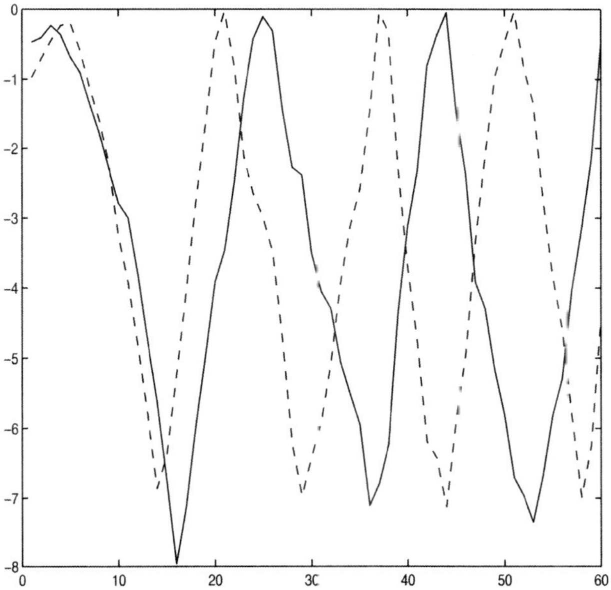

Figure 8.9. *Bcsttk01-IC:* $\log_{10}$ *of* $(Sz^k)_1$ *(solid) and* $(Sz^k)_{48}$ *(dashed) normalized by* $\|z^k\|_M$

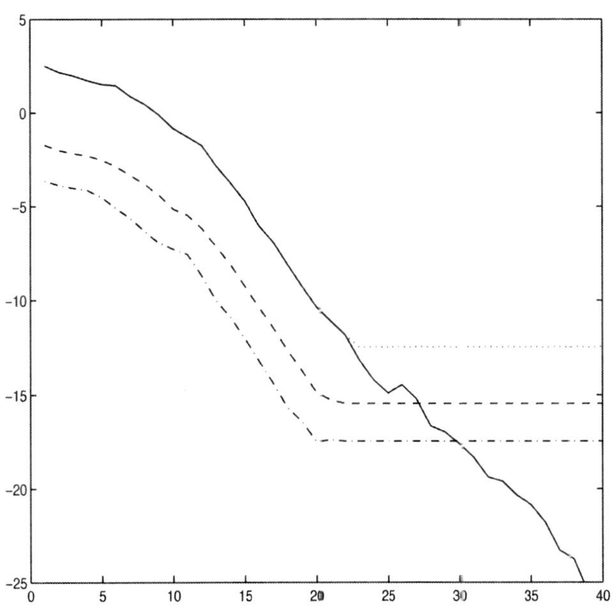

Figure 8.10. *Bcsttk01-IC:* $\log_{10}$ *of* $\|r^k\|$ *(solid), norm of x-residual (dotted),* $\|\epsilon^k\|_A$ *(dashed),* $\|\epsilon^k\|$ *(dot-dashed)*

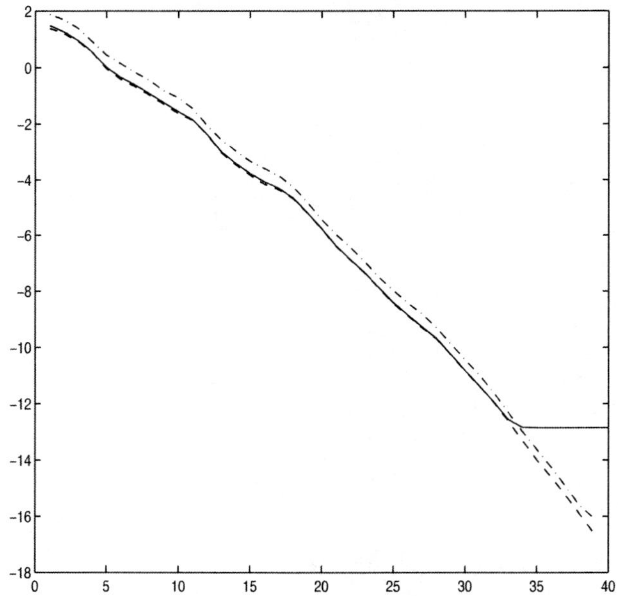

Figure 8.11. *Lap20-IC:* $\log_{10}$ *of* $\|\epsilon^k\|_A$ *(solid), Gauss (dashed), Gauss–Radau (dot-dashed),* $d = 1$

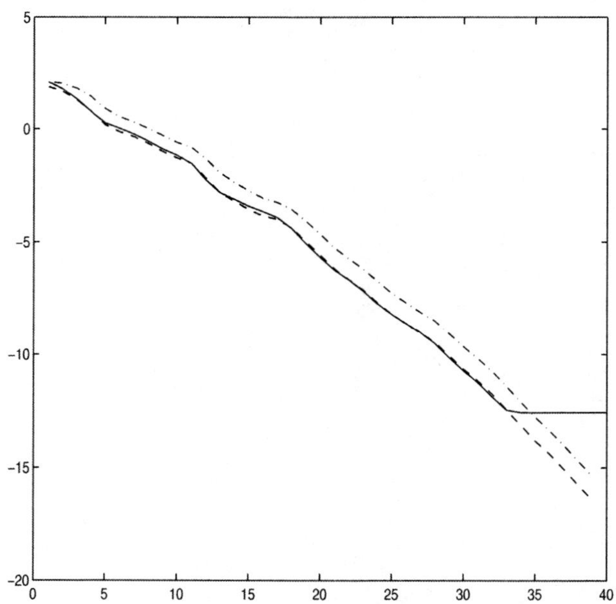

Figure 8.12. *Lap20-IC:* $\log_{10}$ *of* $\|\epsilon^k\|$ *(solid), lower (dashed), upper (dot-dashed),* $d = 1$

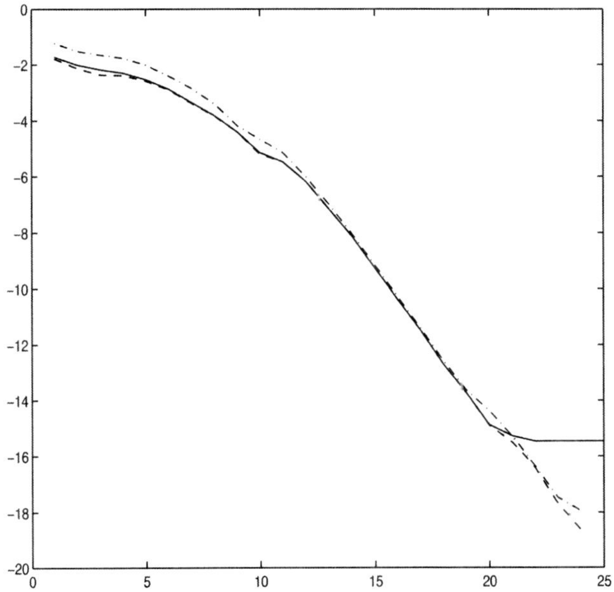

Figure 8.13. *Bcsstk01-IC:* $\log_{10}$ *of* $\|\varepsilon^k\|_A$ *(solid), Gauss (dashed), Gauss–Radau (dot-dashed),* $d = 1$

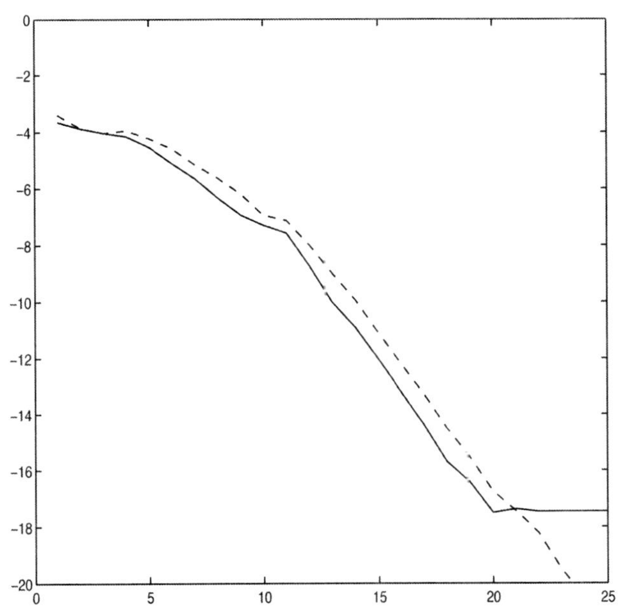

Figure 8.14. *Bcsstk01-IC:* $\log_{10}$ *of* $\|\epsilon^k\|$ *(solid), lower (dashed),* $d = 1$

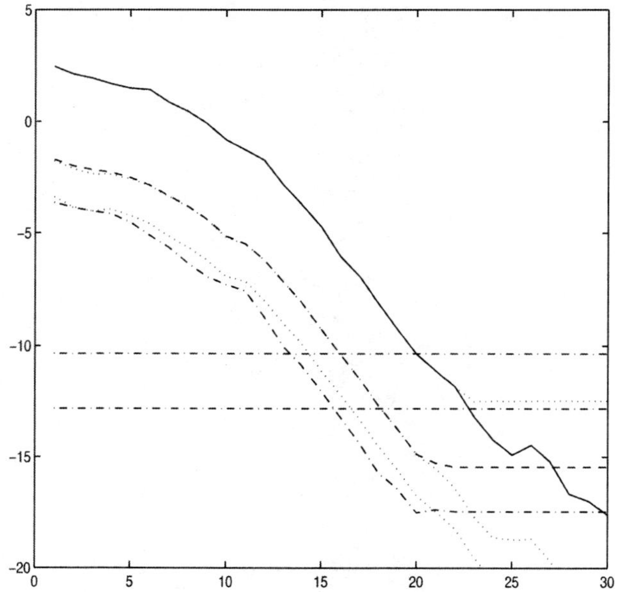

Figure 8.15. *Bcsstk*01-*IC:* $\log_{10}$ *of norms and estimates,* $d = 1$

Chapter 9

Miscellaneous

In this chapter we shall consider different topics related to CG, like the choice of the starting vector, solving a set of linear systems with the same matrix but different right-hand sides, residual smoothing, inner-outer iterations, indefinite systems, and so forth.

9.1 Choice of the starting vector

CG converges for any initial vector x^0. Therefore, one can raise the question, What is a good starting vector (if not the best one)?

We note that what is really important for CG is the initial residual $r^0 = b - Ax^0$ since this is what determines (together with the matrix A) the behavior of the algorithm.

Proposition 9.1. *Let q be an eigenvector of A corresponding to an eigenvalue μ. If x^0 is such that $r^0 = \gamma q$, where γ is a real number, then CG converges in one iteration.*

Proof. If we have $p^0 = r^0 = \gamma q$, we obtain

$$\alpha_0 = \frac{1}{\mu}.$$

Therefore,

$$r^1 = r^0 - \alpha_0 A r^0 = \gamma q - \frac{1}{\mu}(\gamma \mu q) = 0. \qquad \square$$

This seems quite nice. However, even if by chance we know an eigenvector q of A, having $r^0 = \gamma q$ translates into

$$b - Ax^0 = \gamma q \Longrightarrow x^0 = A^{-1}b - \frac{\gamma}{\mu} q.$$

This choice is not feasible since we have to know the solution. Another possible choice if we know an eigenvector q or an approximation of it is to require $q^T r^0 = 0$. This can be obtained by choosing

$$x^0 = x^{-1} + q(q^T Aq)^{-1} q^T (b - Ax^{-1}),$$

where x^{-1} is any given vector. If we know the eigenvalue μ associated with q we have

$$x^0 = x^{-1} + \frac{1}{\mu}(b - Ax^{-1}).$$

We shall see a generalization of this in the next sections. Commonly used starting vectors are $x^0 = 0$ and x^0 random. It would be interesting to have starting vectors giving large residual components on the eigenvectors of A corresponding to small eigenvalues. This would give a better convergence of the Ritz values to the smallest eigenvalues. However, this is difficult to construct, as we shall see. A possibility could be to use, in a preprocessing phase, other iterative methods leading to residuals which are rich in eigenvectors corresponding to the smallest eigenvalues; see, for instance, Touhami [193].

We shall report on numerical experiments using different starting vectors after we have studied some other variants of CG in the next sections giving other choices for the starting vector.

9.2 Variants of CG and multiple right-hand sides

In this section we shall describe some variants of CG which are mainly used to solve a series of linear systems with the same matrix and different right-hand sides. There are two categories of such methods depending on whether we know all the right-hand sides before starting the solves or if the right-hand sides are computed incrementally, like in methods where one right-hand side depends on the solutions of the previous systems.

For the first category the main method is the block CG algorithm, which has been introduced by O'Leary [126]. The second category uses information from one system (denoted as the seed system) to solve the next ones; see, for instance, Chan and Wan [22] and Joly [100].

But first we are going to describe a method which has been derived to impose some constraints on the residual. Generalizations of this method will be used to solve systems with multiple right-hand sides.

9.2.1 Constrained CG

Nicolaides [124] proposed a variant of CG in which the residuals verify linear constraints. This method was called "deflation" by Nicolaides; however, deflation has a slightly different meaning for eigenvalue computations and we prefer to call it constrained CG. The method was formulated in [124] for the three-term recurrence form of CG. Let C be a given $n \times m$ matrix ($m < n$) of rank m. We would like to modify CG to have

$$C^T r^k = 0 \quad \forall k.$$

Let

$$x^{k+1} = \nu_{k+1} x^k + (1 - \nu_{k+1}) x^{k-1} + \mu_k \nu_{k+1} (r^k - Cu^k),$$

where u^k is to be defined. It follows that the residuals are defined by

$$r^{k+1} = \nu_{k+1} r^k + (1 - \nu_{k+1}) r^{k-1} - \mu_k \nu_{k+1} A(r^k - Cu^k).$$

The vector u^k is defined by minimizing

$$(A(r^k - Cu^k), r^k - Cu^k).$$

The solution of this problem is given by solving a linear system of order m

$$C^T ACu^k = C^T Ar^k.$$

Denoting $\mathcal{A}_C = C^T AC$, the equation for the residuals becomes

$$r^{k+1} = \nu_{k+1} r^k + (1 - \nu_{k+1}) r^{k-1} - \mu_k \nu_{k+1} A(I - C\mathcal{A}_C{}^{-1} C^T A) r^k.$$

Theorem 9.2. *If $C^T r^0 = 0$, then $C^T r^k = 0$ for all k.*

Proof. This is easily proved by induction. □

The matrix $P_C = I - C\mathcal{A}_C{}^{-1} C^T A$ is the projector onto span($(AC)^\perp$) along span(C). Let $A_C = A(I - P_C)$, this matrix is symmetric and singular. The coefficients of CG are determined by requiring that r^{k+1} is orthogonal to r^k and r^{k-1}. This gives

$$\mu_k = \frac{(r^k, r^k)}{(r^k, A_C r^k)},$$

and the formula for ν_{k+1} is the same as for CG. Note that at each iteration, we have to solve an $m \times m$ linear system to compute u^k. Constrained CG can also be combined with preconditioning. The initial constraint $C^T r^0 = 0$ can be satisfied in the following way: let u be an arbitrary vector, $s = b - Au$, and

$$C^T ACt = C^T s.$$

We set $x^0 = u + Ct$. This gives $r^0 = b - A(u + Ct)$ and $C^T r^0 = C^T b - C^T Au - C^T ACt = C^T b - C^T Au - C^T (b - Au) = 0$.

The matrix of constraints C can be chosen in many different ways. Nicolaides' suggestion is related to problems arising from discretization of PDEs. It uses a coarse mesh as if the set of unknowns is partitioned into m disjoint subsets Ω_k, $k = 1, \dots, m$; then

$$c_{i,j} = \begin{cases} 1, & i \in \Omega_j, \\ 0, & i \notin \Omega_j. \end{cases}$$

This method can also be used to enforce column sums constraints.

9.2.2 Block CG

O'Leary [126] proposed several block methods, including a block CG algorithm. The Hestenes and Stiefel (two-term recurrence) version of the method for solving $AX = B$, where X and B are n by s, is the following.

ALGORITHM 9.1.

Let X^0 be an $n \times s$ matrix, $R^0 = B - AX^0$, $P^0 = R^0\gamma_0$ where γ_0 is $s \times s$
for $k = 0, 1, \dots$ until convergence
$\alpha_k = ((P^k)^T A P^k)^{-1}\gamma_k^T (R^k)^T R^k$,
$X^{k+1} = X^k + P^k\alpha_k$
$R^{k+1} = R^k - AP^k\alpha_k$
$\beta_k = \gamma_k{}^{-1}((R^k)^T R^k)^{-1}(R^{k+1})^T R^{k+1}$
$P^{k+1} = (R^{k+1} + P^k\beta_k)\gamma_k$
end

The block matrices satisfy

$$(R^k)^T R^j = 0, \quad j \neq k,$$

$$(P^k)^T A P^j = 0, \quad j \neq k.$$

The matrices γ_k can be chosen to reduce roundoff errors; see [126]. One can also derive a three-term recurrence version of block CG and a block minimum residual algorithm. O'Leary [126] has shown that the rate of convergence is governed by the ratio λ_n/λ_s. The block CG method can be derived from the block Lanczos algorithm by using a block LU decomposition of the block tridiagonal matrix that is generated. Block CG can be used when solving systems with different right-hand sides which are not related to each other.

9.2.3 Init, Augmented, and Deflated CG

As we have said before, a problem that arises frequently is to solve a series of linear systems with the same matrix A and different right-hand sides. Many algorithms have been proposed to speed up CG (or more generally Krylov methods) convergence for one system by using information obtained during solving the previous systems; see [22], [23], [100], [154].

Let us now describe two propositions by Erhel and Guyomarc'h (see [87], [46]), which are very close to the work of Nicolaides [124]. Here the constraint matrix is obtained from the first (seed) system to help solving the subsequent ones. Moreover, the methods are derived for the two-term version of CG. Consider solving two linear systems $Ay = c$ and $Ax = b$. The first system is solved giving residuals $s^0, \dots, s^j$ and descent directions $w^0, \dots, w^j$ which give matrices S_j and W_j. We have

$$S_j^T S_j = \Delta_j, \quad W_j^T A W_j = D_j,$$

where Δ_j and D_j are diagonal matrices. The matrix

$$H_j = I - W_j D_j^{-1}(AW_j)^T$$

is the matrix of the A-orthogonal projection on the A-orthogonal set of $\mathcal{K}(s^0, A)$. The residuals are linked by $s^{j+1} = H_j^T s^0$. The idea to speed up the solution of the second system is to devise an initial vector x^0 such that the residual $r^0 = b - Ax^0$ is orthogonal to the Krylov subspace $\mathcal{K}_m(s^0, A)$, enforcing the condition $W_m^T r^0$. This is obtained by choosing

$$x^0 = x^{-1} + W_m D_m^{-1} W_m^T r^{-1}, \quad r^0 = H_m^T r^{-1},$$

where x^{-1} is given. Then we apply CG obtaining algorithm InitCG.

Augmented CG starts with the same vector, but the method uses two subspaces $\mathcal{K}_k(s^0, A)$ and $span(r^0, \dots, r^k)$. The residual r^{k+1} is constructed to be orthogonal to both subspaces and the descent direction p^{k+1} to be A-orthogonal to these subspaces. The initial descent direction is $p^0 = H_m r^0$. Let $\mathcal{K}_m(A, s^0, r^0) = \mathcal{K}_k(s^0, A) + span(r^0, \dots, r^k)$. The solution is sought to verify

$$x^{k+1} - x^k \in \mathcal{K}_m(A, s^0, r^0), \quad (r^{k+1}, z) = 0 \quad \forall z \in \mathcal{K}_m(A, s^0, r^0).$$

This algorithm denoted as AugCG in [46] is the following.

ALGORITHM 9.2.
$r^{-1} = b - Ax^{-1}$, $x^0 = x^{-1} + W_m D_m^{-1} W_m^T r^{-1}$
$r^0 = b - Ax^0$, $p^0 = (I - W_m D_m^{-1} (AW_m)^T) r^0$
for $k = 1, 2, \dots$ until convergence
$\gamma_k = (r^k, r^k)/(p^k, Ap^k)$
$x^{k+1} = x^k + \gamma_k p^k$
$r^{k+1} = r^k - \gamma_k A p^k$
$\beta_{k+1} = (r^{k+1}, r^{k+1})/(r^k, r^k)$
$\mu_{k+1} = (r^{k+1}, Aw^m)/(w^m, Aw^m)$
$p^{k+1} = r^{k+1} + \beta_{k+1} p^k - \mu_{k+1} w^m$
end

Note that only w^m the last column of W_m is involved in the iterations. In [46] it is proved that

$$span(r^0, \dots, r^k) = \mathcal{K}_k(r^0, H_m^T A H_m).$$

In exact arithmetic the A-norm of the error satisfies

$$\|\varepsilon^k\|_A \leq (1 - \epsilon) \|\varepsilon^{k-1}\|_A$$

for some $\epsilon > 0$. It is also proved that

$$\|\varepsilon^k\|_A \leq 2\|\varepsilon^0\|_A \left(\frac{\sqrt{\kappa_1} - 1}{\sqrt{\kappa_1} + 1} \right)^k,$$

where κ_1 is the condition number of $H_m^T A H_m$. The residuals verify $W^T r^k = 0$ like in Nicolaides' method.

Deflated CG [156] uses the same ideas. The constraint matrix being W, the descent directions are obtained as

$$W^T A W \mu_k = W^T A r^k,$$
$$p^k = r^k + \beta_k p^{k-1} - W \mu_k.$$

W is chosen as the matrix of the approximate eigenvectors corresponding to the smallest eigenvalues of A. This makes sense since we have seen that when we have small eigenvalues which are not well separated CG convergence can be very slow. These approximate eigenvectors are obtained when solving the first system and can be improved during the solves with the other systems. Numerical experiments are given in [156]. For other techniques, see Carpentieri, Duff, and Giraud [21] and the Ph.D. thesis of Langou [110].

Some of these techniques were also considered in the Ph. D. thesis of Touhami [193], where some numerical comparisons are given. In this work another method is proposed combining CG and Chebyshev polynomial filters for solving systems with several right-hand sides. This allows construction of an initial residual which is rich in components corresponding to the smallest eigenvalues. This is used in subsequent systems.

9.2.4 Global CG

Global Krylov methods for solving systems with multiple right-hand sides have been introduced by Sadok and his coauthors [98], [99], [44], [43]. If we have s right-hand sides, we introduce an inner product on the vector space of $n \times s$ rectangular matrices: let U and V be $n \times s$ matrices; then we define

$$\langle U, V\rangle_F = tr(U^T V),$$

where tr is the trace and the associated Frobenius norm $\|U\|_F$. Using this inner product we can define the global Lanczos algorithm. Starting from a matrix $\tilde{V}_1$ we have the following algorithm.

ALGORITHM 9.3.

for $k = 1, 2, \ldots$ until convergence
$\eta_k = \|\tilde{V}_k\|_F$
$V_k = \tilde{V}_k / \eta_k$
$\alpha_k = \langle V_k, AV_k\rangle_F = tr(V_k^T A V_k)$
$\tilde{V}_{k+1} = AV_k - \alpha_k V_k - \eta_k V_{k-1}$
end

This gives a tridiagonal matrix T_k. Using this basis we can solve linear systems with a matrix A and several right-hand sides B, an $n \times s$ matrix. Let

$$\Gamma_k = (\, V_1 \quad \cdots \quad V_k \,)$$

and the product

$$\Gamma_k * y = \sum_{i=1}^{k} y_i V_i,$$

where $y^k = (\, y_1 \quad \cdots \quad y_k \,)$. With the Lanczos algorithm the solution X_k is sought starting from X_0 as $X_k = X_0 + \Gamma_k * y^k$. The vector y^k is given as the solution of

$$T_k y^k = \|R_0\|_F e^1,$$

with $R_0 = B - AX_0$. From this global Lanczos algorithm, a global CG method can also be derived.

9.3 Residual smoothing

In our numerical experiments we have seen that there are problems for which the residuals are highly oscillating even though they could decrease in average. It will be nice if we can simply remove these oscillations. This is the goal of residual smoothing. This technique was introduced by Schönauer (see [157], [158]); see also Weiss [198]. It turns out that it can be viewed as a special case of hybrid methods that we briefly describe below.

Hybrid methods were introduced by Brezinski and Redivo-Zaglia [18]. The basic principle of the methods is quite simple. Suppose we are given two approximations x^1 and x^2 of the solution of $Ax = b$. Then we combine these two approximate solutions as

$$y = \omega x^1 + (1 - \omega)x^2$$

to obtain a better approximation. This is done by computing the parameter ω to minimize the norm of the residual $r = b - Ay$. If r^1 and r^2 are the residuals corresponding to x^1 and x^2, the value of ω is given by

$$\omega = -\frac{(r^1 - r^2, r^2)}{(r^1 - r^2, r^1 - r^2)}.$$

We have

$$\|r\|^2 \leq \min\{\|r^1\|^2, \|r^2\|^2\},$$

since

$$(r, r) = (r^1, r^1) - \frac{(r^1 - r^2, r^1)^2}{(r^1 - r^2, r^1 - r^2)},$$

$$(r, r) = (r^2, r^2) - \frac{(r^1 - r^2, r^2)^2}{(r^1 - r^2, r^1 - r^2)}.$$

Different choices for the approximations were proposed in [18]. If we have two given sequences x^k and z^k, $k = 1, \ldots$, the combination being denoted by y^k, then we can do the following:

- Compute x^k and z^k by two different iterative methods and combine them. The total cost is the sum of the costs of both methods or sometimes lower if both methods are related.
- Compute x^k and take $z^k = x^{k-1}$. This leads to semi-iterative algorithms. If CG is the primary method, then the norm of the residual is reduced.
- Compute x^k and take $z^k = y^{k-1}$. Then the norm of the residual is monotonely decreasing and the algorithm is called a smoothing procedure. This was introduced by Schönauer and his coworkers.
- Compute x^k from y^{k-1} and take $z^k = y^{k-1}$.
- Compute x^k by some method and z^k from x^k.

Numerical experiments provided by Brezinski show that smoothing methods are efficient in removing the wriggles in the convergence histories of some methods at the expense of having to compute additional inner products.

A somewhat related but different smoothing procedure was proposed by Zhou and Walker [208]. Residual smoothing techniques were also investigated by Gutknecht and Rozložník [84], [85]. They did a rounding error analysis of the smoothing algorithms and showed that the smoothed residuals do not converge faster than the primary residuals and their maximum attainable accuracy is of the same order. Numerical experiments are provided in [84].

9.4 Inner-outer iterations and relaxation strategies

When we introduced PCG the preconditioner M was supposed to be constructed in such a way that we can "exactly" solve the linear systems $Mz^k = r^k$ at each CG iteration up to rounding errors. This is the case when $M = LL^T$ where L is a triangular matrix or when M^{-1} is directly constructed. However, it can be that the preconditioner is not explicitly computed or is not in a form such that $Mz = r$ can be easily solved. Then, at every iteration, this system can be approximately solved by another iterative method. These iterations are denoted as inner iterations as opposed to the CG iterations, which are the outer iterations. The problem is to know under which conditions the algorithm is converging and when the inner iterations have to be stopped to minimize the total cost for a given accuracy.

This problem has been considered for CG by Golub and Ye [70]. They also modified PCG in the following way.

Algorithm 9.4.

Let x^0 be given, $r^0 = b - Ax^0$
for $k = 0, 1, \dots$ until convergence
$Mz^k = r^k + y^k, \; \|y^k\|_{M^{-1}} \le \epsilon \|r^k\|_{M^{-1}}$
$\beta_k = (z^k, r^k - r^{k-1})/(z^{k-1}, r^{k-1}), \quad \beta_0 = 0$
$p^k = z^k + \beta_k p^{k-1}$
$\gamma_k = (z^k, r^k)/(p^k, Ap^k)$
$x^{k+1} = x^k + \gamma_k p^k$
$r^{k+1} = r^k - \gamma_k Ap^k$
end

This form of PCG allows one to better maintain local orthogonality properties, that is, $(p^{k-1}, r^{k+1}) = 0$, $(p^k, r^{k+1}) = 0$, $(z^k, r^{k+1}) = 0$, $(p^k, Ap^{k+1}) = 0$. For both theoretical and experimental reasons, Golub and Ye [70] suggested using

$$\epsilon = \frac{1}{2\sqrt{\kappa(M^{-1}A)}}.$$

This is related to what has been done in recent years on inexact Krylov methods. These methods have been initiated by Bouras and Frayssé [15]; see also Bouras, Frayssé, and Giraud [16]. The idea is to be able to use approximate matrix-vector products since there are applications where these products are expensive but computed by algorithms whose precision (and cost) can be controlled. In the generalized minimum residual algorithm (GMRES) Bouras and Frayssé replaced the product Av by $(A + \Delta A_k)v$. They show that it is possible to let the perturbations ΔA_k grow throughout the iterations, still obtaining a residual norm smaller than a given criterion. The perturbation is allowed to be $\|\Delta A_k\| = \nu_k \|A\|$ with

$$\nu_k = \min\left(\frac{\eta}{\min(\|r^{k-1}\|, 1)}, 1\right),$$

where η is a given threshold. We see that if $\|r^k\|$ decreases, the allowed perturbations can be large. Of course, there is a slowdown of the decrease of residual norms when compared to the unperturbed case. The same kind of strategy has been applied to CG for domain decomposition problems using a Schur complement approach and another CG iteration to compute the matrix-vector products in [16].

This kind of technique has been analyzed theoretically by Sleijpen and his coauthors [48], [47]. Several methods were studied in [48], in particular, the CG algorithm, which is considered to be the following.

Algorithm 9.5.

for $k = 0, 1, \dots$ until convergence
$c = Ap^k + g^k$
$\gamma_k = \|r^k\|/(p^k, c)$
$r^{k+1} = r^k - \gamma_k c$
$x^{k+1} = x^k + \gamma_k p^k$
$\beta_k = (r^{k+1}, r^{k+1})/(r^k, r^k)$
$p^{k+1} = r^{k+1} + \beta_k p^k$
end

The perturbation g^k is controlled through

$$\|g^k\| \leq \eta_k \, |A\| \, \|p^k\|.$$

The strategy analyzed in [48] is slightly different from that of Bouras and Frayssé. The threshold is given by $\eta_k = \varepsilon/\|r^k\|$. Sleijpen analyzed the difference between the iterative residual r^k and the x-residual $b - Ax^k$ for which he obtained the bound

$$\|r^k - (b - Ax^k)\| \leq \varepsilon k \kappa(A) \max_j (1 + \sqrt{\beta_j}).$$

The inexact Krylov methods (in particular CG) have also been analyzed using different mathematical techniques by Simoncini and Szyld [166], [167], [168], [169]. The behavior of these inexact CG algorithms can also be explained by our analysis of CG in finite precision done in Chapter 5. When we exhibited the solution of the perturbed three-term recurrences for the components of the residual on the eigenvectors of A, nowhere were we using the fact that the perturbations were small. If we have perturbations arising either from the matrix-vector product or eventually from the preconditioner, we can formally write the same solution. The problem is to keep the perturbations from growing too much. Since in CG we have a factor $\|r^k\|$ in front of the perturbation terms, if we want to drive the residual norm to ε it makes sense to require the perturbations not to be larger than $\varepsilon/\|r^k\|$ with eventually some constant multiplicative factor. This shows that these relaxation techniques can eventually be applied in a more general setting than just for the matrix-vector product.

9.5 Numerical experiments with starting vectors

Figure 9.1 shows the norms of the iterative residuals $\|r^k\|$ for $k = 1, \dots$ for the Strakos30 matrix, computed with CG and different starting vectors x^0, the right-hand side b being the same random vector. The solid curve uses $x^0 = 0$, and the dashed curve corresponds to x^0 random with components in $[0, 1]$. The results are not much different. Remember

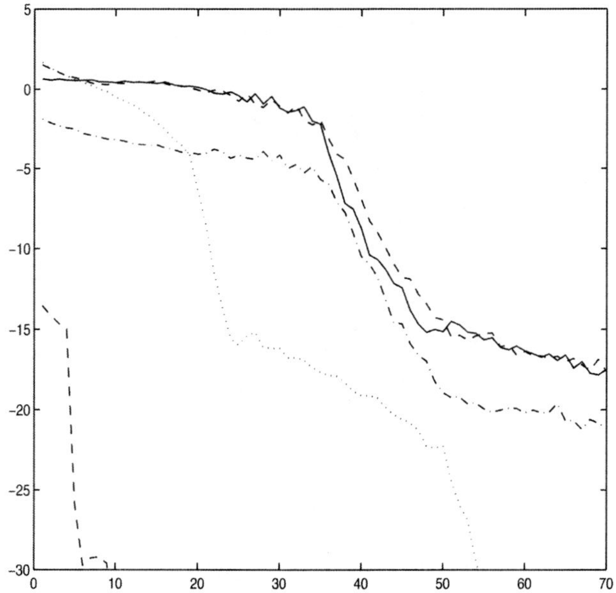

Figure 9.1. *Strakos*30: $\log_{10}$ *of residual norms for different* x^0

that with the Lanczos algorithm on this example we get a good approximation of all the eigenvalues by iteration 40 and two copies of all eigenvalues by iteration 80. The choice for the dot-dashed curve is $x^0 = A^{-1}b + \xi w$, w random in $[-0.5, 0.5]$ and $\xi = 0.001$ to explore what happens when the starting vector is close to the exact solution. The initial residual is $r^0 = -\xi Aw$. Although the norm of the residual is smaller in the 30 first iterations, the shape of the curve is the same as before. Using a starting vector $x^0 = A^{-1}b - \frac{1}{\lambda_{min}} q_{min}$, where λ_{min} and q_{min} are the smallest eigenvalue and the corresponding eigenvector, gives the dashed left curve. The initial residual is $r^0 = q_{min}$ and $\|r^0\| = 1$. The drop of $\|r^k\|$ is very fast and at iteration 1 the norm of the residual is of the order 10^{-14}. Of course, this choice is not feasible for practical problems since we need the exact solution (!) and an eigenvector. Let x^{-1} be random, choosing $x^0 = x^{-1} + W(W^TAW)^{-1}W^T(b - Ax^{-1})$, where W is a 30×10 matrix whose columns are the eigenvectors corresponding to the 10 smallest eigenvalues (in our case this is the 10 first columns of the identity matrix), giving the dotted curve. This is the choice made in InitCG. The initial residual has very small components on the first 10 eigenvectors. This gives a rapid decrease of the norm of the residual in the first iterations. However, because the perturbations caused by rounding errors are growing, components of the residual vectors on the first 10 eigenvectors come back into play after awhile, which explains the slowdown in the decrease of the residual norm after iteration 20. This is illustrated in Figure 9.2, where the solid curve is the $\log_{10}$ of the absolute value of the first component of the projection of the residual. It is of the order 10^{-16} at the beginning, but then it comes back into play, as can be seen in Figure 9.3, where we plot the first component of the normalized residual.

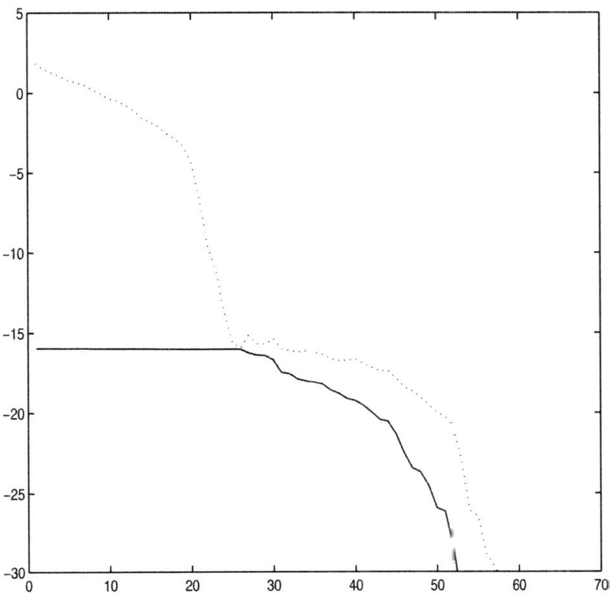

Figure 9.2. *Strakos30 with InitCG:* $\log_{10}$ *of residual norm (dotted) and first component of the residual (solid)*

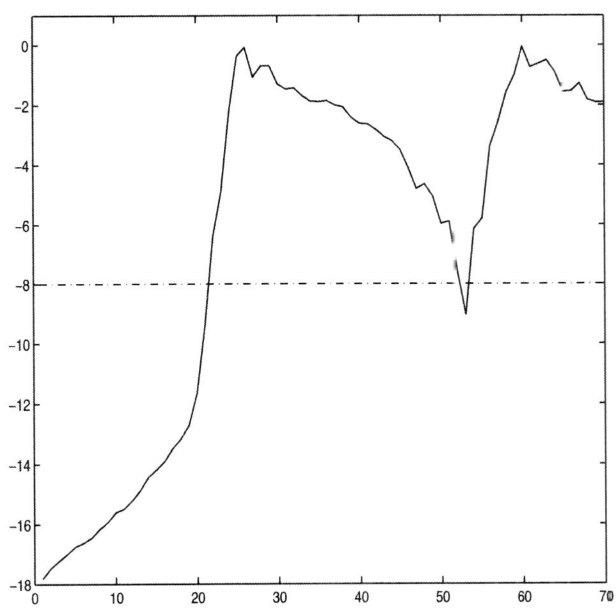

Figure 9.3. *Strakos30 with InitCG:* $\log_{10}(|r_1^k|/\|r^k\|)$

These computations also show that we may have very different norms of the initial residual. This means that we have to be very careful if we are using a stopping criteria like

$$\frac{\|r^k\|}{\|r^0\|} \leq \xi,$$

where ξ is a given threshold because the denominators can be very different when varying x^0.

The results for AugCG on matrix Strakos30 using the same W are given in Figure 9.4. The dotted curve is InitCg and the solid one AugCG.

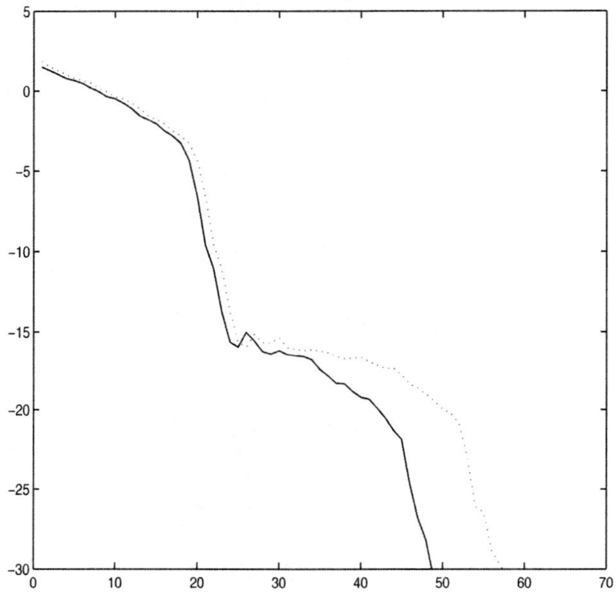

Figure 9.4. *Strakos30 with InitCG (dotted) and AugCG (solid)*

9.6 Shifted matrices

The Lanczos algorithm is, so to speak, "invariant" if we apply it to a shifted matrix $A + \sigma I$, where σ is a real number. The matrix V_k is the same as for A. If T_k is the Lanczos matrix related to A, $T_k + \sigma I$ is the Lanczos matrix related to $A + \sigma I$. The Lanczos matrix relation can be written as

$$(A + \sigma I)V_k = V_k(T_k + \sigma I) + \eta_{k+1} v^{k+1} (e^k)^T.$$

In exact arithmetic the coefficients η_k are the same in both cases and the coefficients α_k are shifted by σ. The Lanczos algorithm will deliver Ritz values shifted by σ. However, things are different for CG. Of course, since

$$r^k = (-1)^k \|r^k\| v^{k+1},$$

the directions of the residual are shift invariant, but the convergence of CG is different when the matrix is shifted (positively or negatively). Without shift the CG matrix relation is

$$AR_k = R_k \bar{T}_k + \frac{1}{\gamma_{k-1}} r^k (e^k)^T,$$

and the relation with the Lanczos matrix T_k is $T_k = D_k^{-1} \bar{T}_k D_k$, where D_k is the diagonal matrix of the inverses of the norms of the residual vectors. Let us denote with a tilde the variables when we have a shift σ. Then,

$$\begin{aligned} \tilde{T}_k &= \tilde{D}_k (T_k + \sigma I) \tilde{D}_k^{-1}, \\ &= \tilde{D}_k T_k \tilde{D}_k^{-1} + \sigma I, \\ &= \tilde{D}_k D_k^{-1} \bar{T}_k D_k \tilde{D}_k^{-1} + \sigma I. \end{aligned}$$

Proposition 9.3.

$$\frac{\|\tilde{r}^{k+1}\|}{\|\tilde{r}^k\|} = \frac{\tilde{\gamma}_k}{\gamma_k} \frac{\|r^{k+1}\|}{\|r^k\|}.$$

Proof. We have seen that the subdiagonal of T_k is shift invariant. The proof is obtained by identification of these terms in T_k and $\tilde{T}_k$. □

The proposition can also be proved by noticing that we have $\delta_{k+1} = 1/\gamma_k$, δ_k being the diagonal entries of the Cholesky decomposition of T_k and

$$\frac{\|r^{k-1}\|}{\|r^k\|} = \frac{\delta_k}{\eta_{k+1}}.$$

It is interesting to compare δ_k and $\tilde{\delta}_k$. This has already been done in previous chapters and we have the following relation:

$$\tilde{\delta}_k - \delta_k = \sigma \left(1 + \frac{\eta_k^2}{\delta_{k-1}\tilde{\delta}_{k-1}} + \cdots + \frac{\eta_k^2 \cdots \eta_2^2}{\delta_1 \cdots \delta_{k-1} \tilde{\delta}_1 \cdots \tilde{\delta}_{k-1}} \right).$$

This leads to the following result.

Theorem 9.4.

$$\begin{aligned} \frac{\|\tilde{r}^{k-1}\|}{\|\tilde{r}^k\|} - \frac{\|r^{k-1}\|}{\|r^k\|} = \sigma \frac{\|r^{k-1}\| \, \|\tilde{r}^{k-1}\|}{\eta_{k+1}} & \left(\frac{1}{\|r^{k-1}\| \, \|\tilde{r}^{k-1}\|} \right. \\ & \left. + \frac{1}{\|r^{k-2}\| \, \|\tilde{r}^{k-2}\|} + \cdots + \frac{1}{\|r^0\| \, \|\tilde{r}^0\|} \right). \end{aligned}$$

Hence, if $\sigma > 0$,

$$\frac{\|\tilde{r}^{k-1}\|}{\|\tilde{r}^k\|} > \frac{\|r^{k-1}\|}{\|r^k\|},$$

the larger the shift, the faster the reduction of the residual norm. The fact that convergence gets better when we increase the shift is also obvious from the bounds on the A-norm of the error involving the condition number since if $\sigma \to \infty$, the condition number tends to 1. We also know that generally the larger the smallest eigenvalue λ_1, the better we are. This is illustrated in Figure 9.5, where we see the norm of the residual for different shifts and the Strakos30 problem. When we have a very small smallest eigenvalue (dashed curve), convergence is very bad since there is even an increase of the norm of the residual in the beginning. When increasing the shift, convergence gets better and better. We remark that for moderate values of the shift we have peaks in the norm of the residual at the same iterations. This arises from oscillations of some components of the residual on the eigenvectors of A. Of course, when we shift the matrix we do not solve the same linear system since we keep the same right-hand side.

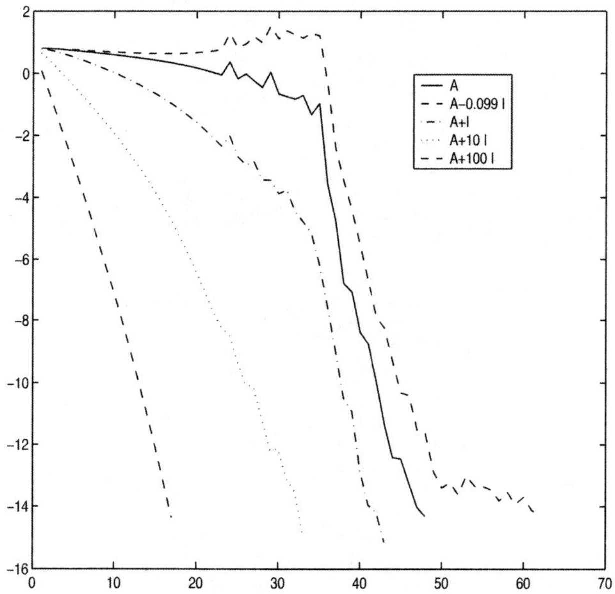

Figure 9.5. *Norms of residual for Strakos30 and* $A + \sigma I$

9.7 CG on indefinite systems

We have seen that theoretically CG is feasible for positive definite symmetric matrices. Of course, we can also apply CG to symmetric negative definite matrices. An interesting question is to know what happens if we use CG for an indefinite matrix with both positive and negative eigenvalues. When A is indefinite, $(Ax, x)^{\frac{1}{2}}$ is no longer a norm. We can just look at $\|r^k\|$ and $\|\epsilon^k\|$. Remembering the expressions we obtained for these norms, we might anticipate that we can expect troubles if Ritz values go from the positive real axis to the negative one when they converge to a negative eigenvalue of A. Of course, if we are unlucky we may even have a zero Ritz value at some iteration, but this is unlikely to happen in finite precision arithmetic. Let us look at some examples which are constructed by negatively shifting the Strakos30 matrix. Figure 9.6 shows the norms of the iterative residuals for $A + \sigma I$ with $\sigma = 0, -0.2, -0.9, -11$. Figure 9.7 displays the l_2 norm of the errors. These shifts correspond to 0, 1, 4, and 15 negative eigenvalues. In the beginning of the Lanczos iterations the Ritz values are positive, and then some of them move to the negative real axis. With $\sigma = -0.2$ the smallest Ritz value becomes negative at iteration 15. The second smallest Ritz value converges to a second copy of the smallest eigenvalue of A and become negative at iteration 54. For $\sigma = -0.9$ there are four Ritz values which become negative before iteration 40.

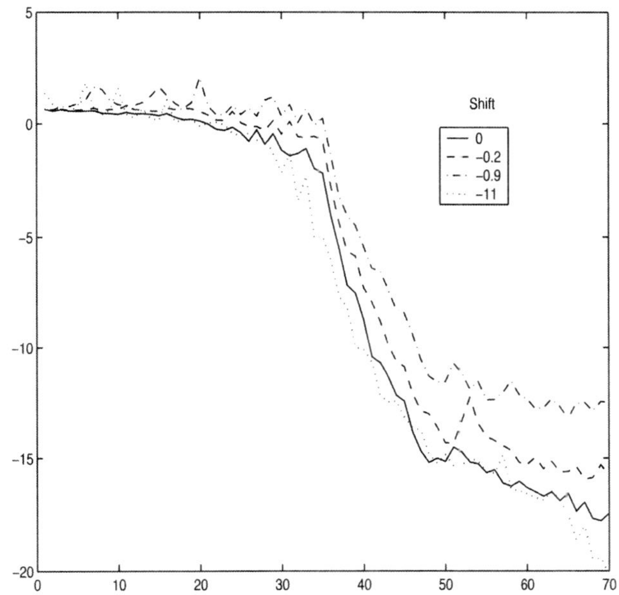

Figure 9.6. $\log_{10}$ *of norms of residuals for Strakos*30 *negatively shifted*

We see that there are some peaks in the norm of the residuals. Moreover, the l_2 norms of the errors are not monotonely decreasing. However, the convergence is not much worse when the matrix is indefinite. This is not always the case, as is seen by considering the Poisson equation on a 30×30 mesh that is $n = 900$. The solid curves in Figures 9.8 and 9.9 correspond to a zero shift. Then we shifted the matrix by -0.2, which leads to seven negative eigenvalues, some of them being multiple. The norms are the dashed curves. We have seven peaks in the norms of the residual and of the l_2 error before iteration 100. The convergence is substantially delayed even though the slope of the decrease is the same when convergence occurs.

Even if the CG algorithm can eventually be used to solve indefinite symmetric linear systems, especially when there are only a few negative eigenvalues and no eigenvalue too close to zero, it is usually better to use methods which were specially devised for these kinds of matrices. The most well-known method is the SYMMLQ algorithm, which was proposed by Paige and Saunders in 1973–1975 [136], [137]. Paige and Saunders used an LQ factorization of the Lanczos matrix T_k. Another possibility is to use a QR factorization;

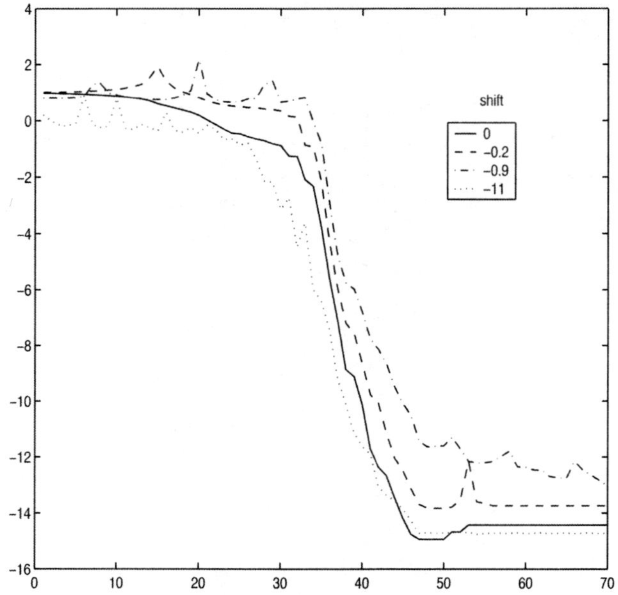

Figure 9.7. $\log_{10}$ *of norms of l_2 errors for Strakos30 negatively shifted*

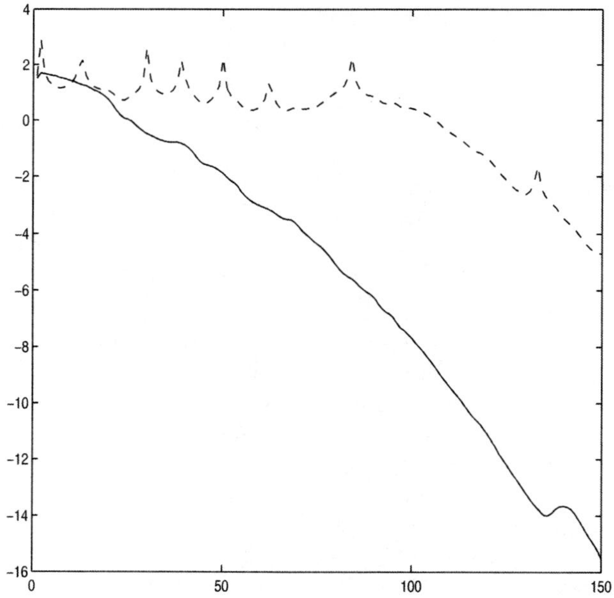

Figure 9.8. $\log_{10}$ *of norms of residuals for Lap30 negatively shifted*

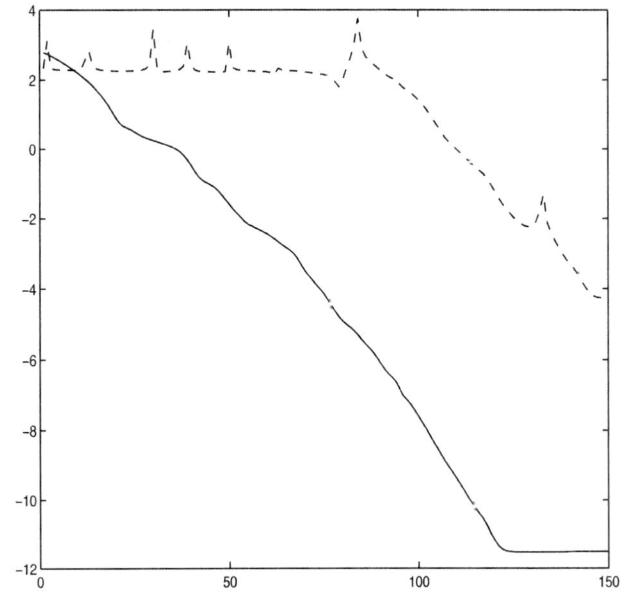

Figure 9.9. $\log_{10}$ *of norms of* l_2 *errors for Lap30 negatively shifted*

see Fischer [50]. The same principles as those used by Paige and Saunders can be used to derive a stable implementation of the conjugate residual method which minimizes the norm of the residual. This method is denoted as MINRES [136].

9.8 Examples with PCG

Although the main purpose of this work is not to compare preconditioners, we are going to give some results with CG using some classical and newer preconditioners for solving some two-dimensional problems to show the efficiency of PCG in finite precision arithmetic. Let us now describe the test problems we are using.

The first problems are diffusion equations,

$$-\text{div}(\lambda(x, y)\nabla u) = f,$$

in a domain $\Omega =]0, 1[^2$ with Dirichlet boundary conditions $u|_{\partial\Omega} = 0$. The PDE is discretized using standard finite differences with a five-point scheme on an $m \times m$ mesh. The first example is the Poisson equation, the diffusion coefficient being 1. Then we solve a discontinuous problem (Pb14): x- and y-diffusion coefficients are 1 except in the square $[1/4, 3/4]^2$, where the value is 1000. For $m = 50$, $\lambda_{min} = 8.83\ 10^{-3}$, $\lambda_{max} = 7.97\ 10^3$, $\kappa = 9.03\ 10^5$. Finally, we solve the sine-sine problem (Pb26): x- and y-coefficients are equal. The diffusion coefficient is

$$\frac{1}{(2 + p \sin \frac{x}{\eta})(2 + p \sin \frac{y}{\eta})},$$

where $p = 1.99$ and $\eta = 0.01$. For $m = 50$, $\lambda_{min} = 1.81\ 10^{-2}$, $\lambda_{max} = 8.86\ 10^3$, $\kappa = 4.89\ 10^5$. Then we use matrices from the Matrix Market collection (http://math.nist.gov). Their characteristics are given in Table 9.1.

Table 9.1. *Matrix Market matrices*

Matrix	n	λ_{min}	λ_{max}	$\kappa(A)$
Bcsstk01	48	$3.42\ 10^3$	$3.02\ 10^9$	$8.82\ 10^5$
Bcsstk14	1806	1	$1.19\ 10^{10}$	$1.19\ 10^{10}$
Nos1	237	$1.23\ 10^2$	$2.46\ 10^9$	$1.99\ 10^7$
Nos7	729	$4.15\ 10^{-3}$	$9.86\ 10^6$	$2.37\ 10^9$

In Tables 9.2 to 9.8 we give the results for different methods. Just to show what residual norm we may expect we give the result using Gaussian elimination. Jacobi, Gauss–Seidel, and SOR (with an optimal parameter ω) are the classical iterative methods. All the other

Table 9.2. *Poisson, $m = 50$, $epss = 10^{-6}$*

Method	nb it	$\|r^k\|$	$\|\epsilon^k\|$	Time (s)
Gauss	–	$1.9285\ 10^{-12}$		
Jacobi	7090	$2.9171\ 10^{-5}$	0.0038	13.8440
Gauss–Seidel	3546	$2.9162\ 10^{-5}$	0.0038	20.9060
SOR $\omega_{opt} = 1.89$	146	$3.1065\ 10^{-5}$	$8.1825\ 10^{-4}$	1.1720
CG M = I	127	$2.6037\ 10^{-5}$	$8.4147\ 10^{-5}$	0.297
CG M = diag	127	$2.6037\ 10^{-5}$	$8.4147\ 10^{-5}$	0.375
CG M = IC(0)	39	$1.2263\ 10^{-5}$	$5.2022\ 10^{-4}$	0.375
CG M = IC(l = 2)	26	$1.8554\ 10^{-5}$	$2.6769\ 10^{-4}$	0.25
CG M = IC(l = 3)	22	$1.9601\ 10^{-5}$	$1.6566\ 10^{-4}$	0.235
CG M = SSOR $\omega = 1$	46	$2.2925\ 10^{-5}$	$1.4882\ 10^{-4}$	0.687
CG M = POL $deg = 1$	70	$2.9105\ 10^{-5}$	$1.3250\ 10^{-4}$	0.438
CG M = POL $deg = 2$	49	$2.7637\ 10^{-5}$	$1.3848\ 10^{-4}$	0.484
CG M = POL $deg = 3$	38	$2.5000\ 10^{-5}$	$1.1778\ 10^{-4}$	0.515
CG M = AINV $\tau = 0.25$	65	$1.1334\ 10^{-5}$	$1.7105\ 10^{-4}$	0.328
CG M = AINV $\tau = 0.1$	41	$1.0740\ 10^{-5}$	$3.7991\ 10^{-4}$	0.359
CG M = AINV $\tau = 0.05$	30	$1.0079\ 10^{-5}$	$4.0654\ 10^{-4}$	0.422
CG M = SAINV $\tau = 0.25$	65	$1.2247\ 10^{-5}$	$2.0315\ 10^{-4}$	0.329
CG M = ml 'gs,b,st,st' $\tau = 0.05$	4	$6.0406\ 10^{-6}$	$3.2352\ 10^{-5}$	0.235
CG M = ml 'ic,b,st,st' $\tau = 0.05$	4	$2.3068\ 10^{-7}$	$1.5569\ 10^{-6}$	0.171

results are using CG with different preconditioners M. For their complete descriptions, see, for instance, [120].

The first result is the standard CG without preconditioner (M = I). The second is a diagonal preconditioner (M = diag(A)). Then we have incomplete Cholesky decompositions without any fill-in (M = IC(0)) and using a level of fill-in l. SSOR is the symmetric successive overrelaxation preconditioner. The next results are with least squares polynomial preconditioners of several polynomial degrees. AINV is an approximate inverse preconditioner with different values of the parameter τ controlling the fill-in; see [8]. The smaller the τ, the more fill-ins we keep. SAINV is a stabilized version of INV. These preconditioners are applied to a matrix that has been symmetrically scaled by putting 1's on the diagonal. The last two results are with multilevel preconditioners described in [121]. The parameters are standard ones, the first parameter being the smoother that is used, either Gauss–Seidel or incomplete Cholesky.

The right-hand side b is a random vector. The starting vector is $x^0 = 0$. The stopping criteria is $\|r^k\| \leq 10^{-6}\|b\|$, even though this can be misleading for some examples. We give the norm of the residual at convergence and the norm of the difference of the iterate x^k at convergence with the solution given by Gaussian elimination. The computing times are just given as an indication since when using MATLAB software they are not really meaningful, the efficiency depending very much on the way the algorithms are coded.

We do not want to comment on all the results. However, we see that the multilevel preconditioners are very efficient but using them is expensive and the initialization phase which is not reported here is also quite expensive. The incomplete Cholesky decomposition gives good results on most of the problems. The only exceptions are cases for which the matrix does not possess the properties needed to compute the factorization in a stable way. To illustrate the effect of the preconditioner on the spectrum of A, we show the spectra without preconditioning and using IC(0) in Figure 9.10. A zoom on the spectrum of $M^{-1}A$ is given in Figure 9.11. We see that the smallest eigenvalues are well separated and not too close to zero. According to our analysis, this is a good spectrum for CG explaining the good convergence when using IC(0) as a preconditioner. The smallest eigenvalues are well separated and are the first to be approximated by the Ritz values removing the corresponding terms in the error norm.

From what we have seen in previous chapters, we can draw the following conclusions on preconditioners:

- If the preconditioner reduces the condition number, there is an improvement in the convergence rate.
- For a given condition number, it is better to have well-separated eigenvalues, especially the smallest ones.
- Having small eigenvalues is not good unless they are well separated.

Of course, it is difficult to satisfy all these constraints, since constructing preconditioners is more an art than a science.

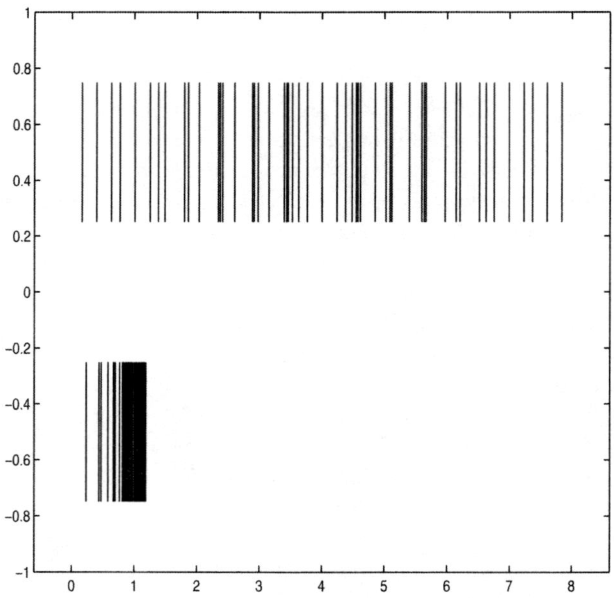

Figure 9.10. *Spectra of A (top) and $IC(0)$ $M^{-1}A$ (bottom) for the Poisson problem*

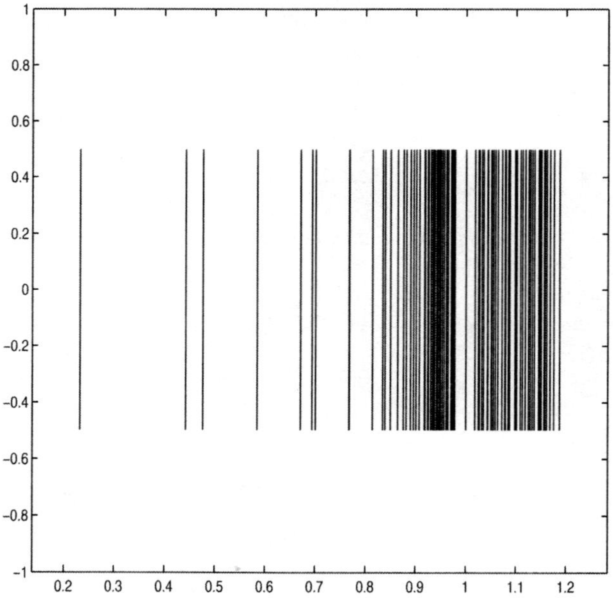

Figure 9.11. *Spectra of $M^{-1}A$ for the Poisson problem with $M = IC(0)$*

Table 9.3. *Pb*14, $m = 50$, $epss = 10^{-6}$

Method	nb it	$\|r^k\|$	$\|\epsilon^k\|$	Time (s)
Gauss	–	$1.4521\ 10^{-9}$		
Jacobi	v.s.c.			
Gauss–Seidel	v.s.c.			
SOR	v.s.c.			
CG M = I	1551	$2.6211\ 10^{-5}$	$2.7909\ 10^{-5}$	3.859
CG M = diag	166	$2.4297\ 10^{-5}$	$2.3925\ 10^{-5}$	0.469
CG M = IC(0)	53	$1.0033\ 10^{-5}$	$1.4868\ 10^{-4}$	0.5
CG M = IC(l = 2)	36	$1.5683\ 10^{-5}$	$5.1312\ 10^{-6}$	0.344
CG M = IC(l = 3)	29	$1.8448\ 10^{-5}$	$7.1641\ 10^{-6}$	0.297
CG M = SSOR $\omega = 1$	65	$2.0698\ 10^{-5}$	$3.0055\ 10^{-5}$	0.953
CG M = POL $deg = 1$	901	$2.1013\ 10^{-5}$	$2.7195\ 10^{-5}$	6.015
CG M = POL $deg = 2$	635	$2.7254\ 10^{-5}$	$3.3552\ 10^{-5}$	6.578
CG M = POL $deg = 3$	499	$2.9084\ 10^{-5}$	$4.3737\ 10^{-5}$	7.047
CG M = AINV $\tau = 0.25$	87	$1.0230\ 10^{-5}$	$1.4061\ 10^{-4}$	0.469
CG M = AINV $\tau = 0.1$	54	$9.7391\ 10^{-6}$	$1.5163\ 10^{-4}$	0.469
CG M = AINV $\tau = 0.05$	37	$7.6959\ 10^{-6}$	$1.4915\ 10^{-4}$	0.532
CG M = SAINV $\tau = 0.25$	80	$9.6212\ 10^{-6}$	$2.3974\ 10^{-4}$	0.422
CG M = SAINV $\tau = 0.1$	51	$1.0737\ 10^{-5}$	$1.9716\ 10^{-4}$	0.453
CG M = SAINV $\tau = 0.05$	34	$7.7236\ 10^{-6}$	$8.6639\ 10^{-5}$	0.547
CG M = ml 'gs,b,st,st' $\tau = 0.05$	8	$3.1880\ 10^{-7}$	$3.3869\ 10^{-6}$	0.469
CG M = ml 'ic,b,st,st' $\tau = 0.05$	6	$7.5153\ 10^{-6}$	$3.8023\ 10^{-5}$	0.265

v.s.c. = very slow convergence

Table 9.4. *Pb26*, $m = 50$, $epss = 10^{-6}$

Method	nb it	$\|r^k\|$	$\|\epsilon^k\|$	Time (s)
Gauss	–	$5.9053\ 10^{-11}$		
Jacobi	> 50000	$9.4847\ 10^{-4}$	0.0105	169.2960
Gauss–Seidel	33353	$2.9172\ 10^{-5}$	$2.3137\ 10^{-4}$	206.2340
SOR $\omega_{opt} = 1.96$	481	$2.9623\ 10^{-5}$	$8.0038\ 10^{-7}$	3.7180
CG M = I	2013	$2.3420\ 10^{-5}$	$4.1623\ 10^{-5}$	4.688
CG M = diag	436	$2.5696\ 10^{-5}$	$3.1074\ 10^{-5}$	1.188
CG M = IC(0)	83	$2.0339\ 10^{-5}$	0.0013	0.766
CG M = IC(l = 2)	28	$2.3474\ 10^{-5}$	$3.1773\ 10^{-5}$	0.266
CG M = IC(l = 3)	24	$1.8063\ 10^{-5}$	$2.6646\ 10^{-5}$	0.25
CG M = SSOR $\omega = 1$	214	$2.8148\ 10^{-5}$	$5.3301\ 10^{-5}$	3.078
CG M = POL $deg = 1$	1577	$2.9007\ 10^{-5}$	$4.1807\ 10^{-5}$	9.719
CG M = POL $deg = 2$	1262	$1.9681\ 10^{-5}$	$3.9558\ 10^{-5}$	12.141
CG M = POL $deg = 3$	1038	$2.6075\ 10^{-5}$	$4.4625\ 10^{-5}$	13.5
CG M = AINV $\tau = 0.25$	103	$1.8260\ 10^{-5}$	0.0015	0.484
CG M = AINV $\tau = 0.1$	70	$1.7278\ 10^{-5}$	0.0011	0.5
CG M = MAINV $\tau = 0.05$	47	$1.6103\ 10^{-5}$	$7.8652\ 10^{-4}$	0.531
CG M = SAINV $\tau = 0.25$	78	$1.8207\ 10^{-5}$	0.0021	0.359
CG M = SAINV $\tau = 0.1$	58	$1.7891\ 10^{-5}$	$9.6876\ 10^{-4}$	0.406
CG M = SAINV $\tau = 0.05$	40	$1.4228\ 10^{-5}$	$5.8608\ 10^{-4}$	0.438
CG M = ml 'gs,b,st,st' $\tau = 0.05$	23	$1.3891\ 10^{-5}$	0.0010	1.5
CG M = ml 'ic,b,st,st' $\tau = 0.05$	13	$0.609\ 10^{-5}$	$6.3135\ 10^{-4}$	0.172

Table 9.5. *Bcsstk01*, $epss = 10^{-6}$

Method	nb it	$\|r^k\|$	$\|\epsilon^k\|$	Time (s)
Gauss	–	$2.4910\ 10^{-13}$		
Jacobi	div	†		
Gauss–Seidel	4305	$4.1343\ 10^{-6}$	$5.3731\ 10^{-10}$	1.1090
SOR $\omega_{opt} = 1.9$	168	$5.2143\ 10^{-6}$	$4.4650\ 10^{-12}$	0.0310
CG M = I	139	$3.8910\ 10^{-6}$	$7.4606\ 10^{-13}$	0.031
CG M = diag	48	$1.9818\ 10^{-6}$	$5.3852\ 10^{-15}$	0.015
CG M = IC(0)	16	$9.3495\ 10^{-7}$	$6.2030\ 10^{-14}$	0.015
CG M = SSOR $\omega = 1$	25	$2.2969\ 10^{-7}$	$9.7950\ 10^{-16}$	0.016
CG M = POL $deg = 1$	111	$9.3578\ 10^{-7}$	$1.6543\ 10^{-13}$	0.031
CG M = POL $deg = 2$	91	$3.8830\ 10^{-6}$	$2.9354\ 10^{-13}$	0.047
CG M = POL $deg = 3$	81	$1.1262\ 10^{-6}$	$2.0599\ 10^{-13}$	0.047
CG M = AINV $\tau = 0.25$	20	$2.1759\ 10^{-9}$	$7.9490\ 10^{-9}$	–
CG M = AINV $\tau = 0.1$	14	$2.6290\ 10^{-9}$	$1.4125\ 10^{-8}$	–
CG M = AINV $\tau = 0.05$	13	$1.3292\ 10^{-9}$	$8.7294\ 10^{-9}$	–
CG M = SAINV $\tau = 0.25$	16	$2.5931\ 10^{-9}$	$1.0420\ 10^{-8}$	–
CG M = SAINV $\tau = 0.1$	13	$2.4287\ 10^{-9}$	$1.0611\ 10^{-8}$	–
CG M = ml 'gs,b,st,st' $\tau = 0.05$	10	$1.0272\ 10^{-9}$	$1.0107\ 10^{-8}$	–
CG M = ml 'ic,b,st,st' $\tau = 0.05$	9	$1.9440\ 10^{-9}$	$5.6457\ 10^{-9}$	–

div = divergence
† $2D - A$ nonSPD, D diagonal of A

Table 9.6. *Bcsstk*14, $epss = 10^{-6}$

Method	nb it	$\|r^k\|$	$\|\epsilon^k\|$	Time (s)
Gauss	–	$1.5488\ 10^{-11}$		
Jacobi	div	†		
Gauss–Seidel	11696	$2.4581\ 10^{-5}$	$3.8638\ 10^{-10}$	148.89
SOR $\omega_{opt} = 1.86$	1059	$2.4814\ 10^{-5}$	$1.7961\ 10^{-10}$	19.9370
CG M = I	14924	$2.1901\ 10^{-5}$	$1.5159\ 10^{-10}$	108.469
CG M = diag	457	$2.1686\ 10^{-5}$	$1.8488\ 10^{-11}$	3.156
CG M = IC(0) *	258	$1.5376\ 10^{-5}$	$8.5034\ 10^{-12}$	4.718
CG M = IC(lev = 2) *	48	$1.7116\ 10^{-5}$	$9.1611\ 10^{-12}$	1.141
CG M = IC(lev = 3)	33	$1.8114\ 10^{-5}$	$1.2675\ 10^{-11}$	0.766
CG M = SSOR $\omega = 1$	186	$2.4450\ 10^{-5}$	$2.3062\ 10^{-11}$	4.61
CG M = POL	slow conv			
CG M = AINV $\tau = 0.25$	79	$3.0003\ 10^{-6}$	$1.7563\ 10^{-4}$	0.703
CG M = AINV $\tau = 0.1$	50	$3.4530\ 10^{-6}$	$4.8175\ 10^{-4}$	0.625
CG M = AINV $\tau = 0.05$	35	$2.5568\ 10^{-6}$	$4.2597\ 10^{-5}$	0.656
CG M = SAINV $\tau = 0.25$	77	$3.1263\ 10^{-6}$	$8.0122\ 10^{-5}$	0.703
CG M = SAINV $\tau = 0.1$	45	$3.4120\ 10^{-6}$	$7.6038\ 10^{-5}$	0.562
CG M = SAINV $\tau = 0.05$	32	$2.4114\ 10^{-6}$	$3.2010\ 10^{-5}$	0.578
CG M = ml 'gs,b,st,st' $\tau = 0.05$	69	$2.5627\ 10^{-6}$	$7.7924\ 10^{-4}$	8.375
CG M = ml 'ic,b,st,st' * $\tau = 0.05$	no conv			

div = divergence
† $2D - A$ nonSPD, D diagonal of A
* Pb with the incomplete decomposition

Table 9.7. *Nos*1, *epss* $= 10^{-6}$

Method	nb it	$\|r^k\|$	$\|\varepsilon^k\|$	Time (s)
Gauss	–	$1.2792\ 10^{-9}$		
Jacobi	v.s.c.			
Gauss–Seidel	v.s.c.			
SOR	v.s.c.			
CG M = I	3188	$1.7741\ 10^{-6}$	$9.5519\ 10^{-12}$	1.688
CG M = diag	491	$8.1794\ 10^{-6}$	$4.1244\ 10^{-11}$	0.171
CG M = IC(0) *	3077	$1.2460\ 10^{-8}$	$1.1284\ 10^{-5}$	3.109
CG M = IC(lev = 2)†	1	$6.9882\ 10^{-10}$	$1.0990\ 10^{-12}$	–
CG M = SSOR $\omega = 1$	242	$6.7752\ 10^{-6}$	$6.3632\ 10^{-11}$	0.297
CG M = POL $deg = 1$	2036	$7.3983\ 10^{-6}$	$1.8460\ 10^{-11}$	1.75
CG M = POL $deg = 2$	1566	$5.6527\ 10^{-6}$	$5.2065\ 10^{-12}$	1.938
CG M = POL $deg = 3$	1379	$5.0420\ 10^{-6}$	$4.7894\ 10^{-12}$	2.203
CG M = AINV $\tau = 0.25$	190	$1.1433\ 10^{-8}$	$3.0821\ 10^{-7}$	0.219
CG M = AINV $\tau = 0.1$	97	$1.2095\ 10^{-8}$	$7.8411\ 10^{-8}$	0.141
CG M = AINV $\tau = 0.05$	68	$6.4919\ 10^{-9}$	$3.9781\ 10^{-8}$	0.11
CG M = SAINV $\tau = 0.25$	165	$1.1886\ 10^{-8}$	$6.3023\ 10^{-7}$	0.125
CG M = SAINV $\tau = 0.1$	91	$4.8988\ 10^{-9}$	$3.9719\ 10^{-8}$	0.125
CG M = SAINV $\tau = 0.05$	59	$1.1979\ 10^{-8}$	$9.3062\ 10^{-8}$	0.11
CG M = ml 'gs,b,st,st' $\tau = 0.05$	5	$4.3710\ 10^{-9}$	$6.5420\ 10^{-7}$	0.047
CG M = ml 'ic,b,st,st' $\tau = 0.05$ *	282	$8.2578\ 10^{-9}$	$3.5641\ 10^{-8}$	1.891

v.s.c. = very slow convergence
* Pb with the incomplete factorization
† Decomposition is complete

Table 9.8. *Nos7, epss* $= 10^{-6}$

Method	nb it	$\|r^k\|$	$\|\epsilon^k\|$	Time (s)
Gauss	–	$8.6520\ 10^{-7}$		
Jacobi	v.v.s.c.			
Gauss–Seidel	v.v.s.c.			
SOR	v.v.s.c.			
CG M = I	3776	$1.0649\ 10^{-5}$	$1.9434\ 10^{-4}$	4.266
CG M = diag	89	$9.3199\ 10^{-6}$	$6.3311\ 10^{-6}$	0.078
CG M = IC(0)	23	$2.2085\ 10^{-5}$	0.0018	0.062
CG M = IC(lev = 2)	25	$3.2361\ 10^{-6}$	$5.8652\ 10^{-6}$	0.078
CG M = SSOR $\omega = 1$	38	$3.4097\ 10^{-6}$	$2.7747\ 10^{-6}$	0.156
CG M = POL $deg = 1$	2979	$1.3054\ 10^{-5}$	$1.9465\ 10^{-4}$	8.422
CG M = POL $deg = 2$	2543	$1.1969\ 10^{-5}$	$1.9432\ 10^{-4}$	10.156
CG M = POL $deg = 3$	2245	$9.9299\ 10^{-6}$	$9.1589\ 10^{-5}$	12.062
CG M = AINV $\tau = 0.25$	60	$2.1879\ 10^{-5}$	0.0021	0.094
CG M = AINV $\tau = 0.1$	33	$3.6340\ 10^{-5}$	0.0013	0.078
CG M = AINV $\tau = 0.05$	27	$3.4963\ 10^{-5}$	0.0011	0.094
CG M = SAINV $\tau = 0.25$	62	$3.7568\ 10^{-5}$	0.0016	0.094
CG M = SAINV $\tau = 0.1$	33	$1.5978\ 10^{-5}$	0.0029	0.078
CG M = SAINV $\tau = 0.05$	26	$2.1718\ 10^{-5}$	0.0036	0.094
CG M = ml 'gs,b,st,st' $\tau = 0.05$	8	$1.6675\ 10^{-5}$	0.0042	0.203
CG M = ml 'ic,b,st,st' $\tau = 0.05$	7	$5.1063\ 10^{-7}$	$7.8618\ 10^{-4}$	0.125

v.v.s.c. = very, very slow convergence

Appendix

A.1 A short biography of Cornelius Lanczos

For these biographical notes to be as complete as possible we used an account of Lanczos' life prepared by Dianne P. O'Leary and the historical Web pages at the University of St. Andrews, Scotland. We also used the nice biographical essay by Gellai [56] in [148].

Cornelius Lanczos was born as Kornél Löwy on February 2, 1893, in the small town of Székesfehérvár (Hungary). He was the eldest son of a Jewish lawyer, Károly Löwy. His father changed the name of his children in 1906 in the process of Hungarization of the surnames. The young Lanczos attended Jewish elementary school, Catholic secondary school, from which he graduated in 1910, and then the University of Budapest in the Faculty of Arts. His physics teacher was Baron Roland von Eötvös, who interested him in the theory of relativity, and his mathematics teacher was Lipót Fejér, who was already a well-known mathematician. Lanczos graduated in 1915 and was then an assistant of Károly Tangl at the Technical University in Budapest. His Ph.D. thesis, obtained in 1921 under Prof. Ortvay, was about the theory of relativity; its title was "The Function Theoretical Relation to the Maxwellian Aetherequation." Then the political situation in Hungary led him to move to Germany at the University of Freiburg, where he spent 3 years continuing his work in physics. He moved to Frankfurt in 1924. During this period he spent a year (1928–1929) as Einstein's assistant in Berlin and married a German woman, Maria Rupp. In the fall of 1929, he returned to Frankfurt. In 1931 he spent a year as a visiting professor at Purdue University. In 1932 because of the economic and political troubles in Germany, he returned to the United States, where he received a professorship at Purdue University. Unfortunately, he had to travel without his wife, who contracted tuberculosis soon after their wedding and could not get a visa. She died in 1939 and Lanczos brought his son Elmar to the United States. In 1938 Lanczos obtained U.S. citizenship. Most of the members of his family died in concentration camps during World War II. During this period he worked on Einstein's field equations and Dirac's equation but was increasingly interested in mathematical techniques, particularly for computations with mechanical calculators. In 1940 he published a method for quickly evaluating Fourier coefficients that is equivalent to the FFT algorithm, which had not yet been discovered. During 1943–1944 Lanczos became associated with the National Bureau of Standards (NBS). After some time spent at Boeing Aircraft Company in Seattle in 1946, in January 1949 he joined NBS, which had founded an Institute for Numerical Analysis at the University of California campus in Los Angeles. Lanczos turned his attention to the solution of linear systems and matrix eigenvalue problems with which he had been interested when working in physics and at Boeing. He investigated his method of "minimized iterations," a continuation of some work begun at Boeing. In a paper published in 1950 he proposed constructing an orthogonal basis of what is now called a Krylov subspace. He demonstrated the algorithm's effectiveness on practical problems: the lateral vibration of a bar, the vibrations of a membrane, and a string through numerical computations. In 1952 he discussed the solution of linear systems using the recurrence of minimized iterations and recognized that it was equivalent to the method of conjugate gradients of Hestenes and Stiefel also published in 1952. At this time Lanczos came under investigation for allegedly being a communist sympathizer and in 1954 he decided to move permanently to the Dublin Institute for Advanced Studies, where he had been invited by Erwin Schrödinger in 1952. He resumed his physics research, including the geometry of space time, and married another German woman, Ilse Hildebrand. He died in Budapest of a heart attack on his second visit back to Hungary, on June 25, 1974. A six-volume edition of his collected works was published by North Carolina State University in 1998. Lanczos published more than 120 papers during his career.

A.2 A short biography of M.R. Hestenes and E. Stiefel

These notes rely on information obtained during the historical lectures given by several people at the meeting "Iterative Solvers for Large Linear Systems" organized at ETH Zürich, February 18–21, 2002, to celebrate the fiftieth anniversary of the conjugate gradient and the seventieth birthday of Gene Golub. The paper by O'Leary and Golub [66] is also a very interesting source of information on CG history.

Magnus Rudolph Hestenes was born in Brycelyn, Minnesota, in 1906. He obtained a master's degree from the University of Wisconsin in 1928 and his Ph.D. thesis from the University of Chicago in 1932. The title of his thesis was "Sufficient Conditions for the General Problem of Mayer with Variable End-Points," and his advisor was G. Bliss. From 1932 to 1937 he was an instructor at various universities. After a year in Chicago, he left for Harvard as a National Research Fellow to work with Marston Morse. According to his own words [91], "in 1936 I developed an algorithm for constructing a set of mutually conjugate directions in Euclidean space for the purpose of studying quadric surfaces. I showed my results to Prof. Graustein, a geometer at Harvard University. His reaction was that it was too obvious to merit publication." In 1937 he took a position as an assistant professor at the University of Chicago, later becoming an associate professor. During the latter years of World War II, he was a member of the Applied Mathematics Group at Columbia University, concerned with aerial gunnery. He left Chicago in 1947 to join UCLA, where he stayed until his retirement in 1973. From 1948 to 1952 he was a consultant at the Rand Corporation and from 1949 to 1954 a consultant at the National Bureau of Standards. About the discovery of the conjugate gradient algorithm, Hestenes wrote, "In June or July 1951, after almost two years of studying algorithms for solving systems of linear equations, we finally hit upon a conjugate-gradient method. I had the privilege of first formulation of this new method. However, it was an outgrowth of my discussions with my colleagues at

INA. In particular, my conversations with George Forsythe had a great influence on me. During the month of July 1951, I wrote an INA report on this new development." About his meeting with Eduard Stiefel, Hestenes wrote, "When E. Stiefel arrived at INA to attend the conference on solutions of linear equations, he was given a copy of my paper. Shortly thereafter he came to my office and said about the paper, 'This is my talk.' He too has invented the conjugate-gradient algorithm and had carried out successful experiments using this algorithm. Accordingly, I invited Stiefel to remain at UCLA and INA for one semester so that we could write an extensive paper on this subject." From 1950 to 1958 Hestenes was chairman of the UCLA mathematics department and director of the university's computing facility from 1961 to 1963. During the sixties and seventies he was also a consultant at the Institute for Defense Analyses and at the IBM Watson Research Laboratory. In 1973 he became professor emeritus. During his career he had a total of 35 Ph.D. students. He died on May 31, 1999.

Eduard Ludwig Stiefel was born on April 21, 1909, in Zürich, the son of Eduard Stiefel, a well-known Swiss artist. He did his studies in mathematics and physics at the Swiss Federal Institute of Technology (ETH) in Zürich. In 1932–1933 he visited the universities of Hamburg and Göttingen in Germany. From 1933 to 1935 he was an assistant in mathematics at ETH. He got his Ph.D. thesis from ETH in 1935 in topology, the title of his thesis being "Richtungsfelder und Fernparallelismus in n-dimensionalen Mannigfaltigkeiten," under the supervision of Heinz Hopf. From 1936 to 1943 he was lecturer in mathematics at ETH. In 1943 he became full professor of mathematics at ETH. From 1946 to 1948 he was head of the department of mathematics and physics and director of the Institute of Applied Mathematics (IAM) from 1948 to 1978. In 1949 he rented and moved to Zürich the Z4 computing engine of the German designer Konrad Zuse. Therefore, ETH became the first European university with an electronic computer. In 1956–1957 he was president of the Swiss Mathematical Society. Stiefel began his career as a topologist studying geometry and topology of Lie groups and made notable contributions in this field. In 1948 he completely changed the direction of his research and became interested in computers and numerical analysis. He gave a permanent place to computer sciences at ETH by creating and directing IAM for 30 years. He died on November 25, 1978.

A.3 Examples in "exact" arithmetic

The following examples are designed to show the influence of the starting vector and the distribution of the eigenvalues on the Lanczos algorithm. They were computed using double reorthogonalization, which we consider for these problems as being close to the "exact" result. All the eigenvalues of A are obtained at step n with an absolute precision of about $10u$, and the orthogonality of Lanczos vectors was preserved up to machine precision.

First example

This example is the diagonal Strakoš matrix that we have used throughout this book. This example was defined in [184]. The matrix is diagonal with eigenvalues

$$\lambda_i = \lambda_1 + \left(\frac{i-1}{n-1}\right)(\lambda_n - \lambda_1)\rho^{n-i}.$$

Here we choose $n = 10$, $\lambda_1 = 0.1$, $\lambda_n = 100$, $\rho = 0.9$ to obtain a small example. The eigenvalues are shown in Figure A.1 and their values are

$$\begin{array}{ccccc} 0.1000 & 4.8782 & 10.7182 & 17.7970 & 26.3178 \\ 36.5136 & 48.6514 & 63.0370 & 80.0200 & 100.0000; \end{array}$$

the smallest eigenvalues are more clustered than the large ones. The starting vector is the vector of all ones, meaning that each eigenvector has the same weight. Figures A.2 to A.10 show how the solutions of the secular equations are obtained. The crosses on the x-axis are the eigenvalues of A, the diamonds are the ends of the interval for α_k, and the dot-dashed vertical lines are the poles. In these figures, we clearly see why good approximations are first obtained for the outer eigenvalues, especially when they are well separated like the largest ones in our example. We see also, for instance in Figure A.2, the influence of the eigenvalue distribution and the starting vector. If the starting vector would have larger weights for the eigenvectors corresponding to the largest eigenvalues, the oblique straight line would move to the right and give a better approximation of the largest eigenvalue. Looking at the diamonds on the x-axis we see that the value of α_k is more and more constrained. Figure A.11 shows the approximate eigenvalues as a function of the iteration number. The rightmost stars are the eigenvalues of A.

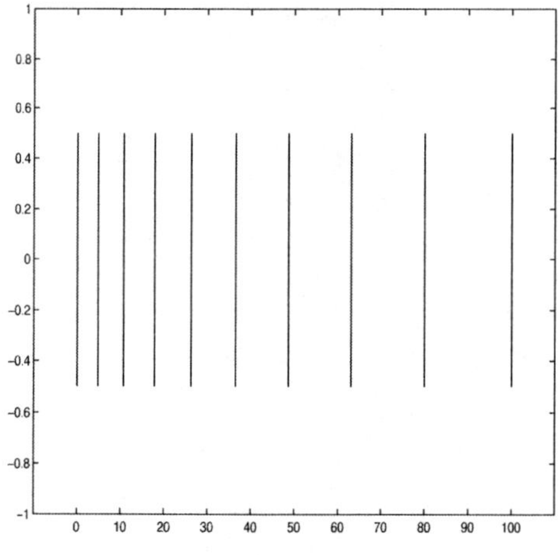

Figure A.1. *Eigenvalues of the Strakos*10 *matrix*

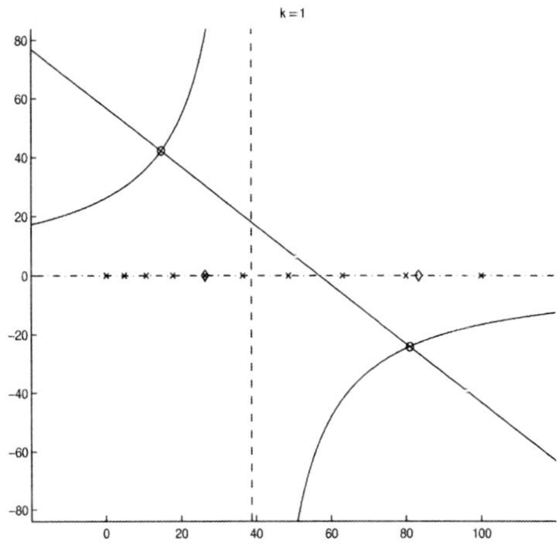

Figure A.2. *Strakos*10, $k = 1$, $v^1 = (1, \ldots, 1)^T$

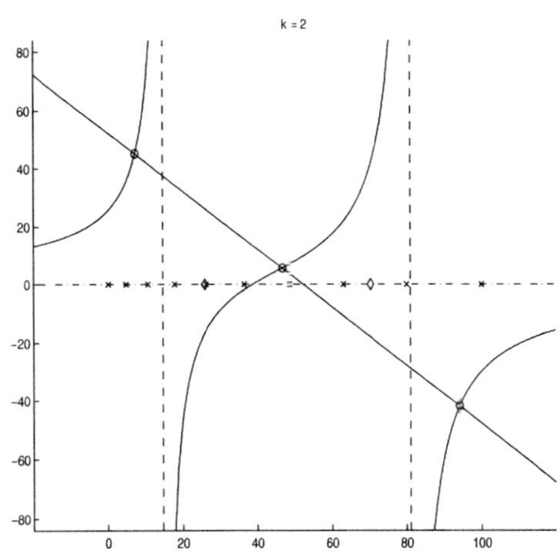

Figure A.3. *Strakos*10, $k = 2$, $v^1 = (1, \ldots, 1)^T$

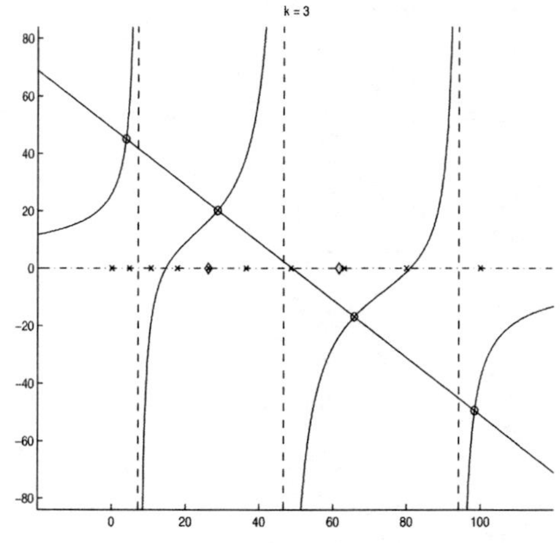

Figure A.4. *Strakos*10, $k = 3$, $v^1 = (1, \ldots, 1)^T$

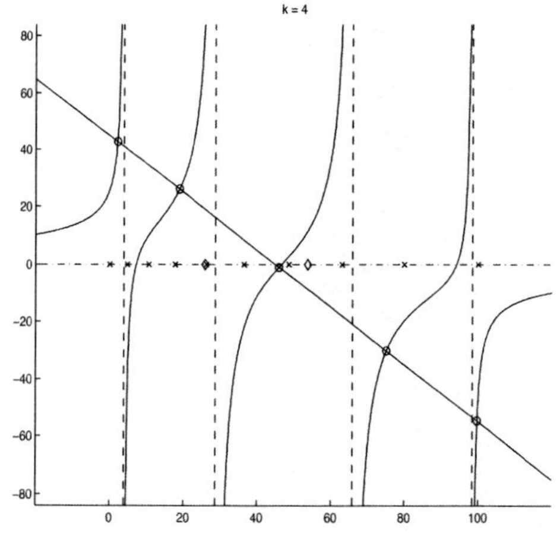

Figure A.5. *Strakos*10, $k = 4$, $v^1 = (1, \ldots, 1)^T$

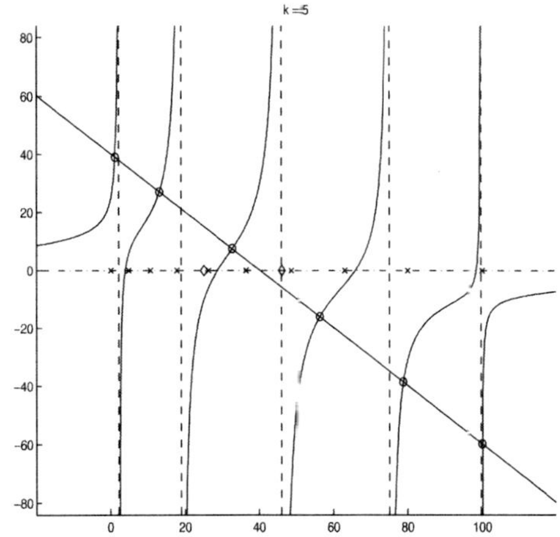

Figure A.6. *Strakos*10, $k = 5$, $v^1 = (1, \ldots, 1)^T$

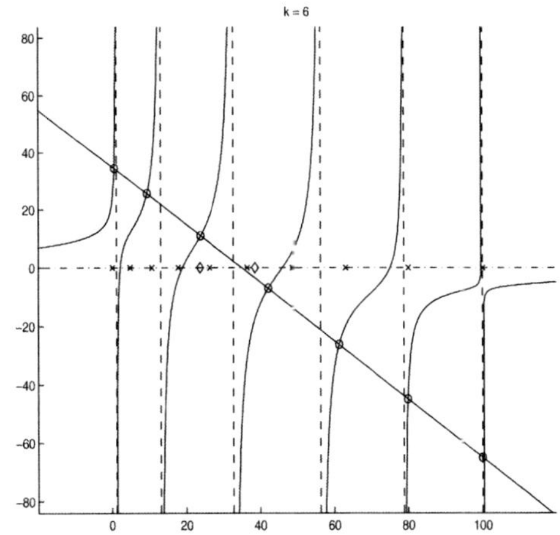

Figure A.7. *Strakos*10, $k = 6$, $v^1 = (1, \ldots, 1)^T$

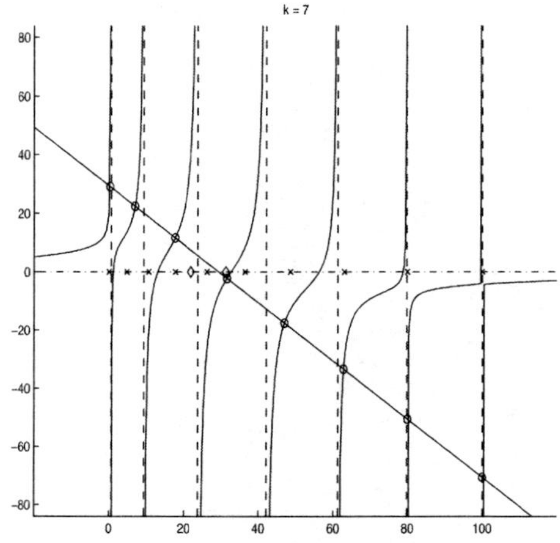

Figure A.8. *Strakos*10, $k = 7$, $v^1 = (1, \ldots, 1)^T$

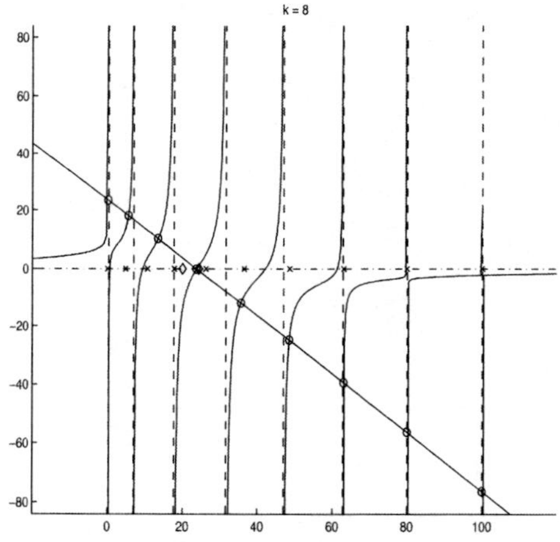

Figure A.9. *Strakos*10, $k = 8$, $v^1 = (1, \ldots, 1)^T$

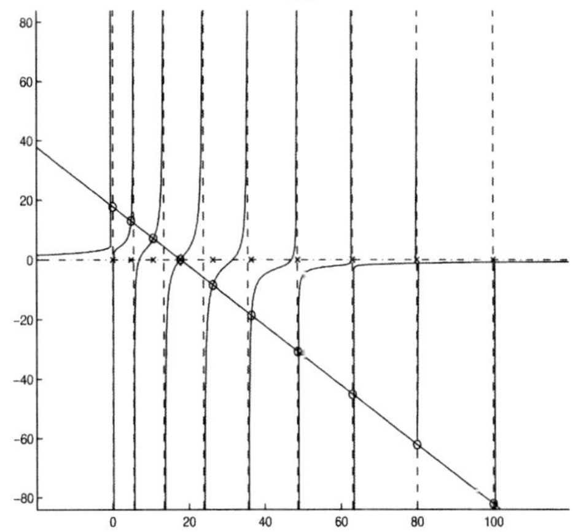

Figure A.10. *Strakos*10, $k = 9$, $v^1 = (1, \dots, 1)^T$

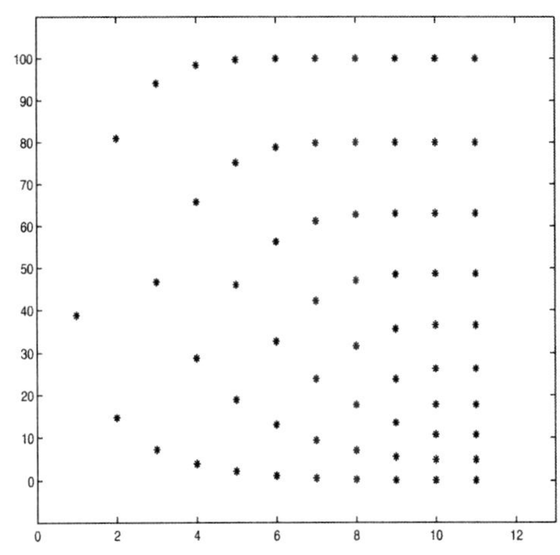

Figure A.11. *Strakos*10, $v^1 = (1, \dots, 1)^T$

Second example

The matrix Strakos10 is the same as in the first example. However, the starting vector is computed to impose Ritz values in the middle of the intervals in the next-to-last step to illustrate the results of Scott [160]. The starting vector with norm 1 is

$$\begin{array}{lllll} 3.8405\ 10^{-2} & 2.4091\ 10^{-1} & 5.2897\ 10^{-1} & 6.2952\ 10^{-1} & 4.6104\ 10^{-1} \\ 2.1748\ 10^{-1} & 6.6163\ 10^{-2} & 1.2439\ 10^{-2} & 1.2796\ 10^{-3} & 4.7796\ 10^{-5}. \end{array}$$

We see that the components corresponding to the largest eigenvalues are small. Comparing

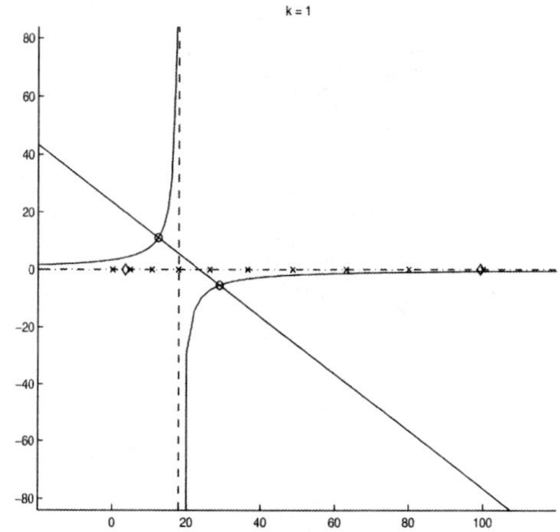

Figure A.12. *Strakos*10, $k = 1$, *slow conv.*

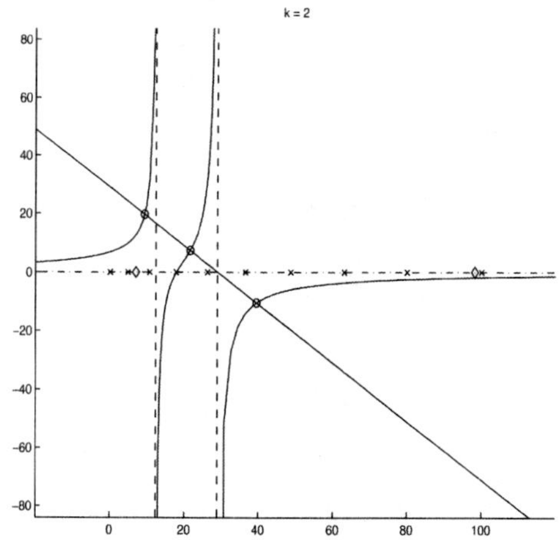

Figure A.13. *Strakos*10, $k = 2$, *slow conv.*

to the first example, this moves α_1 to the left of the spectrum, as seen in Figure A.12, so the largest Ritz value will move slowly to the largest eigenvalue of A.

Figures A.12 to A.17 are equivalent to the ones for the first example. We see that the largest eigenvalue is very slow to converge with this starting vector. The initial values of α_k are located in the left part of the spectrum, so there cannot be good approximations of the largest eigenvalues in the first stages.

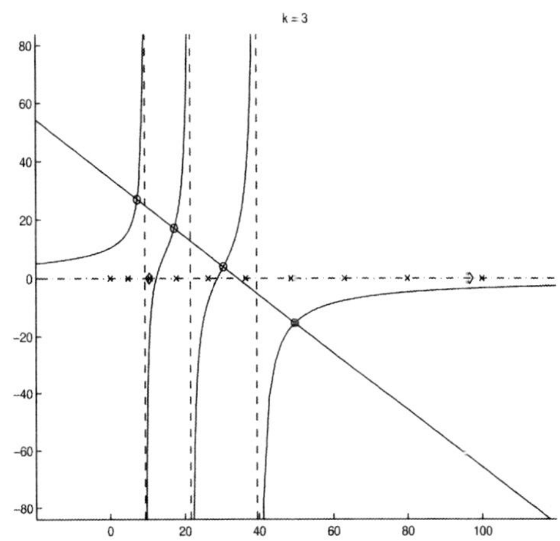

Figure A.14. *Strakos*10, $k = 3$, *slow conv.*

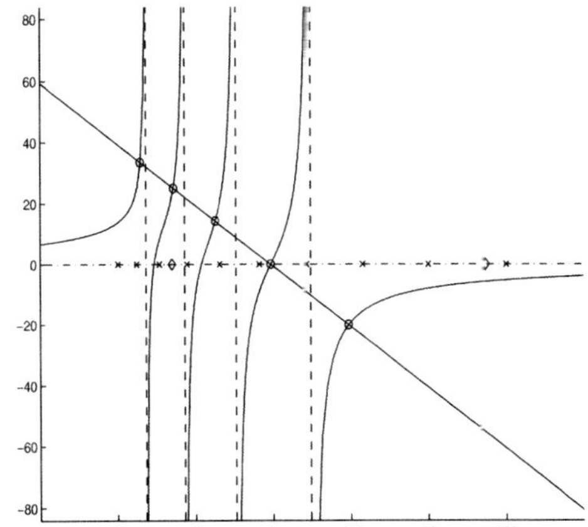

Figure A.15. *Strakos*10, $k = 4$, *slow conv.*

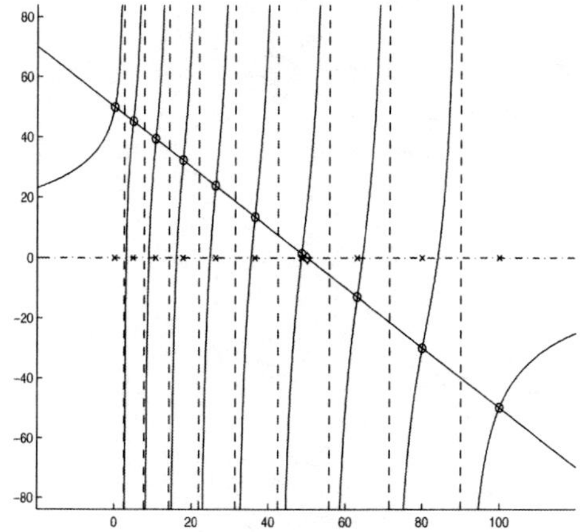

Figure A.16. *Strakos*10, $k = 9$, *slow conv.*

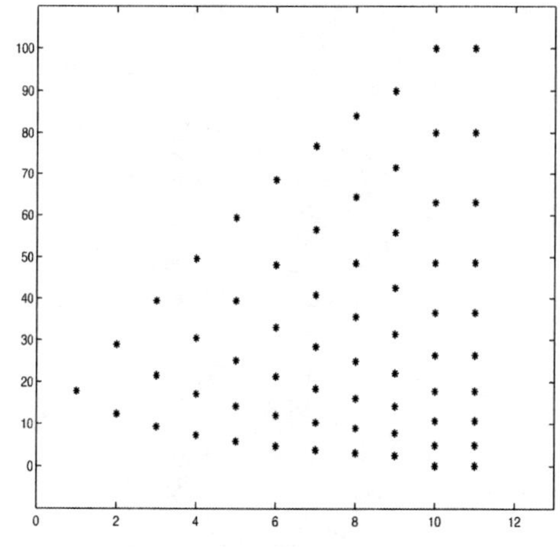

Figure A.17. *Strakos*10, *slow conv.*

Third example

The matrix is the same as before but the starting vector is linear between 0.1 and 100 and normalized

$$5.3086\ 10^{-3} \quad 6.3703\ 10^{-2} \quad 1.2210\ 10^{-1} \quad 1.8049\ 10^{-1} \quad 2.3889\ 10^{-1}$$

$$2.9728\ 10^{-1} \quad 3.5567\ 10^{-1} \quad 4.1407\ 10^{-1} \quad 4.7246\ 10^{-1} \quad 5.3086\ 10^{-1}.$$

Thus the largest elements of the initial vector are corresponding to the largest eigenvalues. We see (Figures A.18 to A.23) that this gives an initial value α_1 closer to the largest eigenvalues and a better initial approximation of the largest eigenvalues. Then α_k moves to the left. But, the smallest Ritz values are far from the smallest eigenvalues before the last iteration.

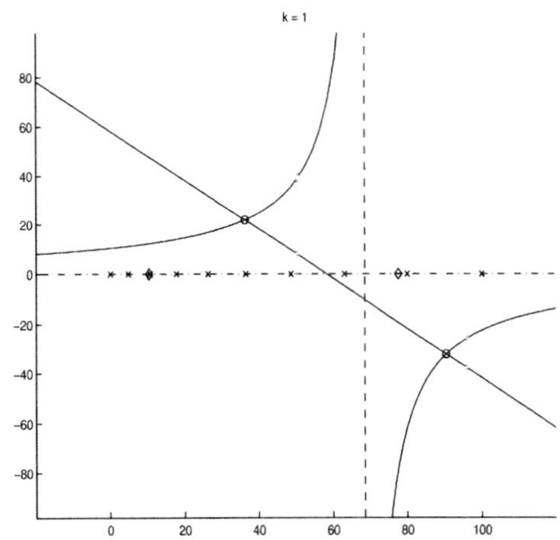

Figure A.18. *Strakos*10, $k = 1$

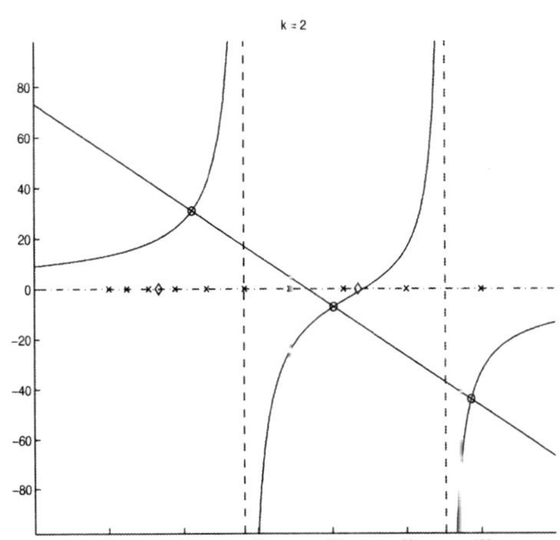

Figure A.19. *Strakos*10, $k = 2$

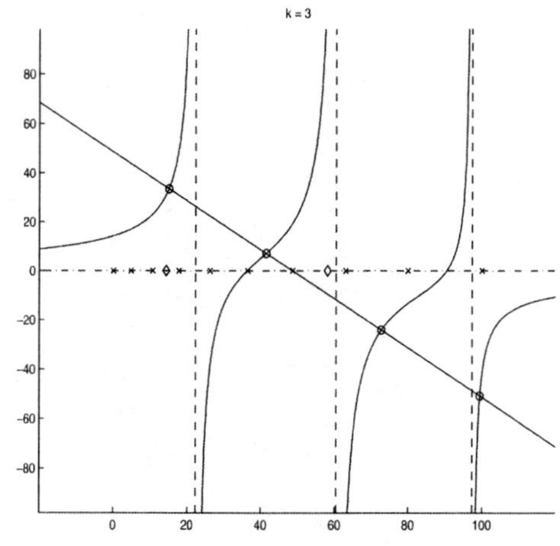

Figure A.20. *Strakos*10, $k = 3$

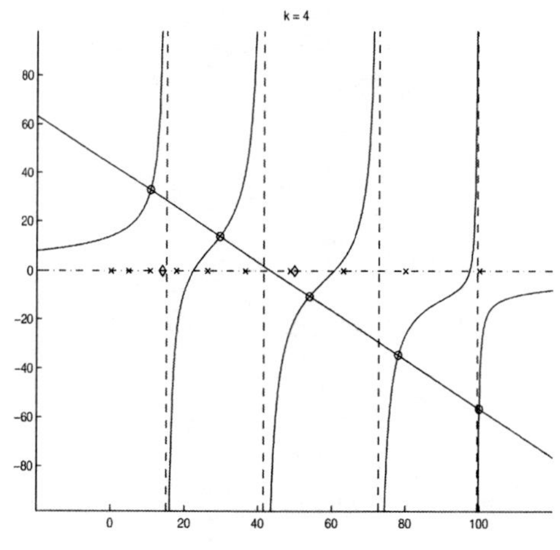

Figure A.21. *Strakos*10, $k = 4$

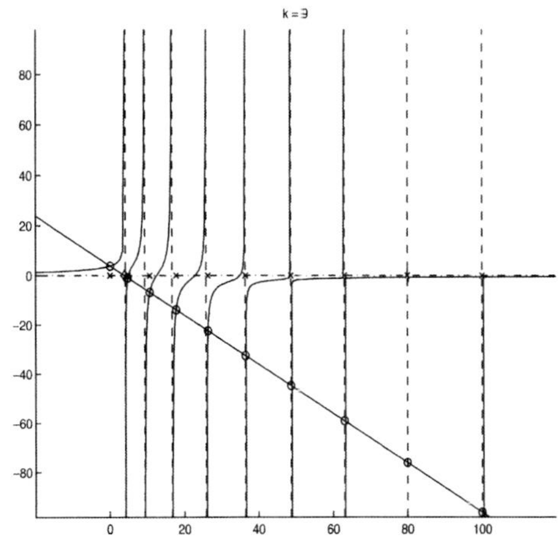

Figure A.22. *Strakos*10, $k = 9$

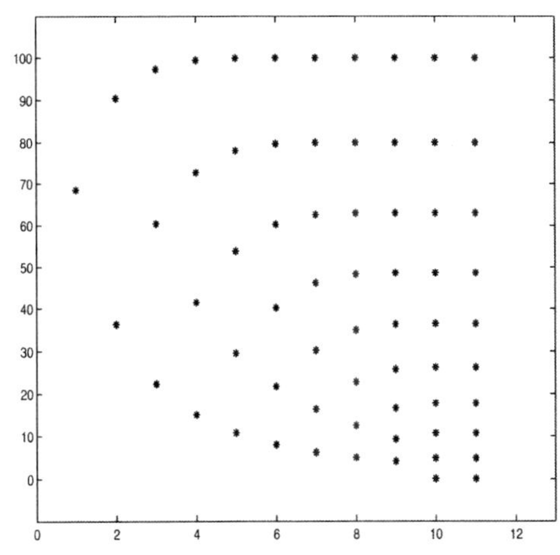

Figure A.23. *Strakos*10

Fourth example

The matrix is still the same, but the starting vector is the reverse of the previous one with more emphasis on the smallest eigenvalues

$$\begin{array}{ccccc} 5.3086\ 10^{-1} & 4.7246\ 10^{-1} & 4.1407\ 10^{-1} & 3.5567\ 10^{-1} & 2.9728\ 10^{-1} \\ 2.3889\ 10^{-1} & 1.8049\ 10^{-1} & 1.2210\ 10^{-1} & 6.3703\ 10^{-2} & 5.3086\ 10^{-3}. \end{array}$$

This starting vector gives a slow convergence even though it is a little faster than in the second example, which was especially devised for that purpose; see Figures A.24 to A.29.

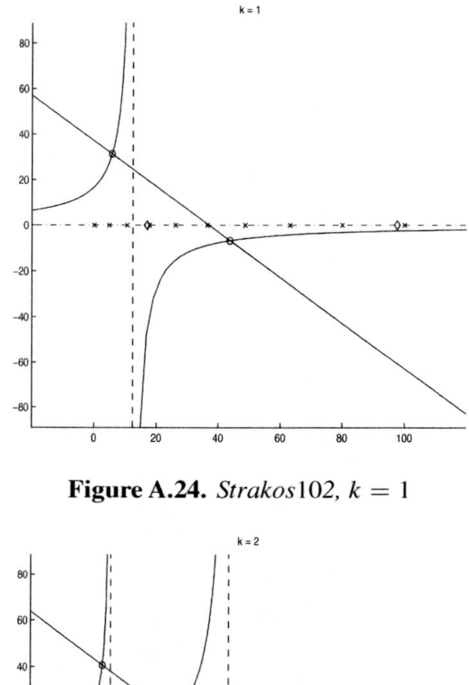

Figure A.24. *Strakos*102*, $k = 1$*

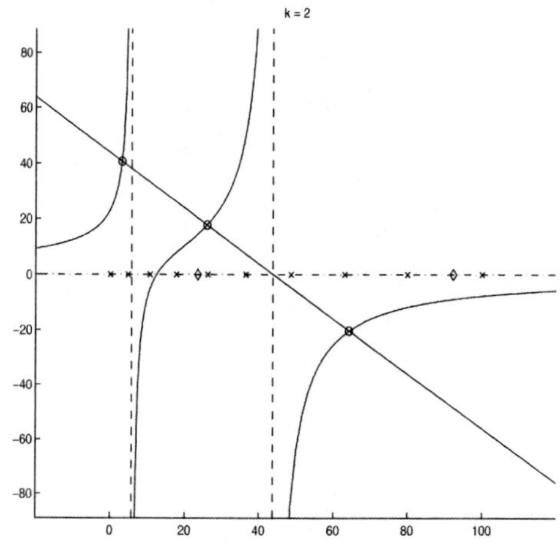

Figure A.25. *Strakos*10*, $k = 2$*

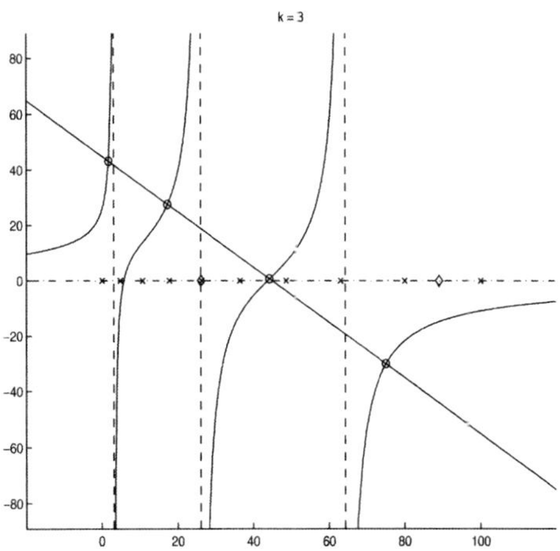

Figure A.26. *Strakos*10, $k = 3$

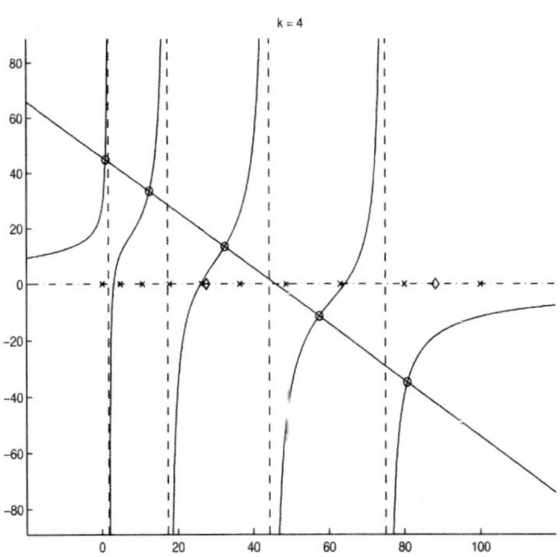

Figure A.27. *Strakos*10, $k = 4$

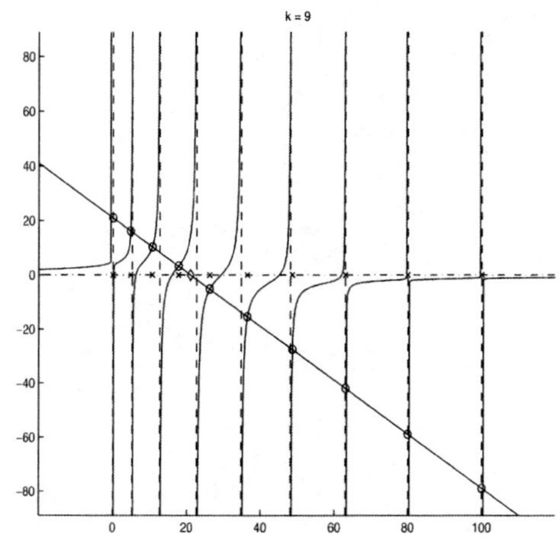

Figure A.28. *Strakos*10, $k = 9$

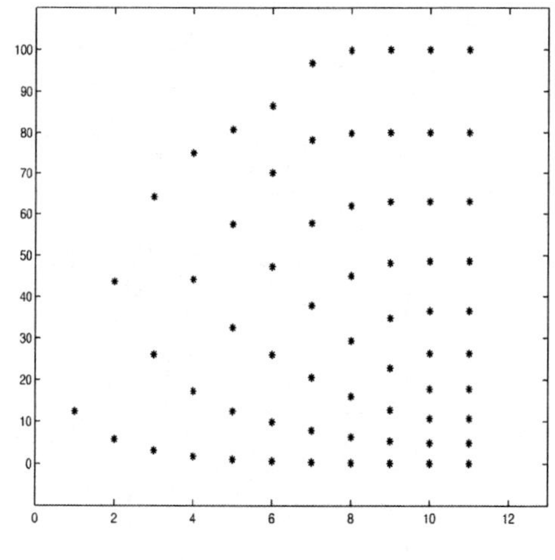

Figure A.29. *Strakos*10

Fifth example

The matrix is diagonal and the eigenvalues are computed from those of the Strakos10 matrix of example 1 (λ_i) as $\lambda_n + \lambda_1 - \lambda_i$. The eigenvalues are depicted in Figure A.30. The largest eigenvalues are more clustered than the smallest ones. Since the smallest eigenvalues are well separated they converge before the largest ones; see Figures A.31 to A.36. The starting vector is all ones.

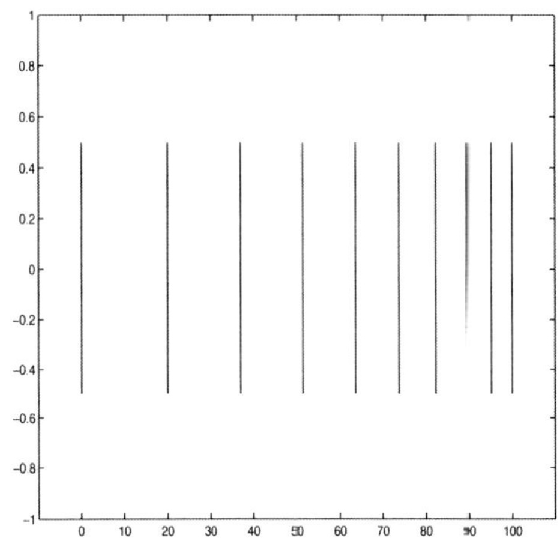

Figure A.30. *Eigenvalues of the fifth example matrix*

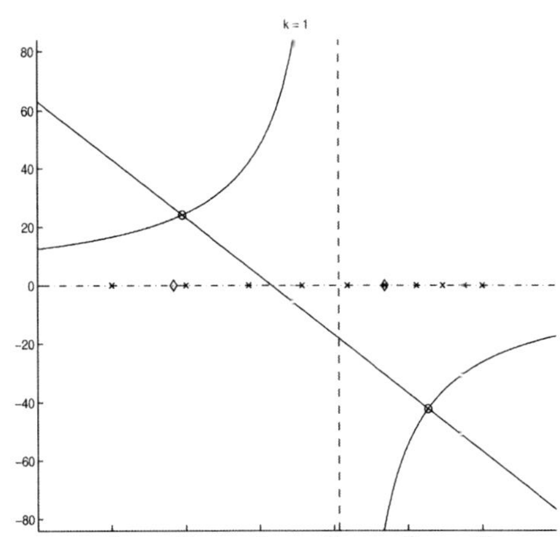

Figure A.31. *Fifth example, $k = 1$*

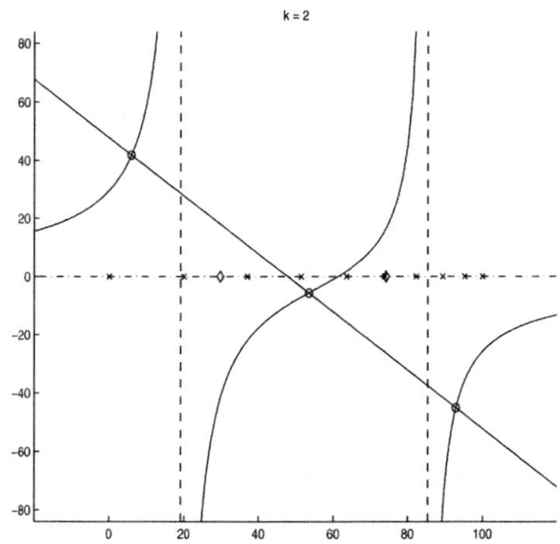

Figure A.32. *Fifth example, $k = 2$*

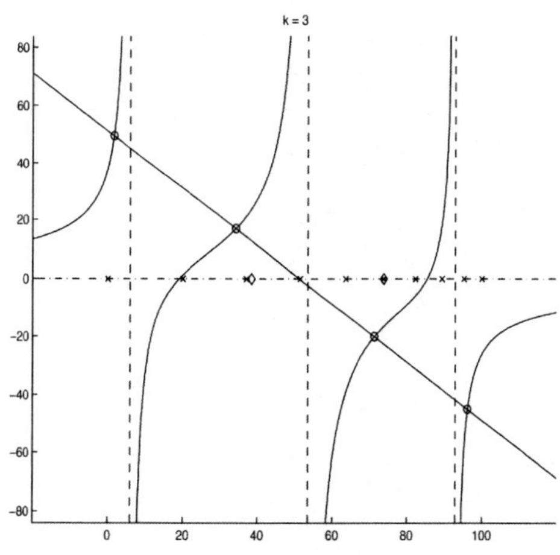

Figure A.33. *Fifth example, $k = 3$*

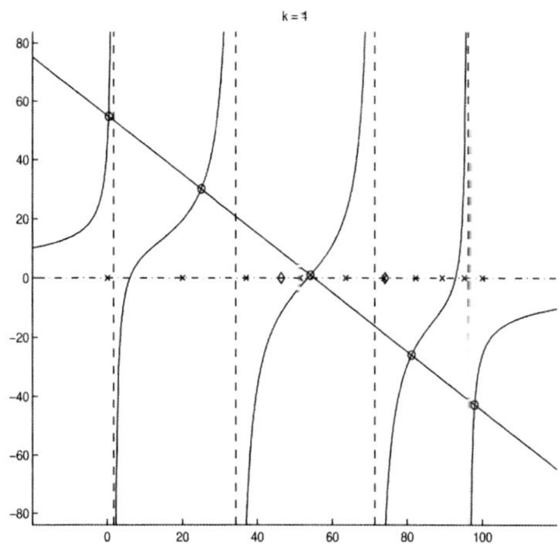

Figure A.34. *Fifth example,* $k = 4$

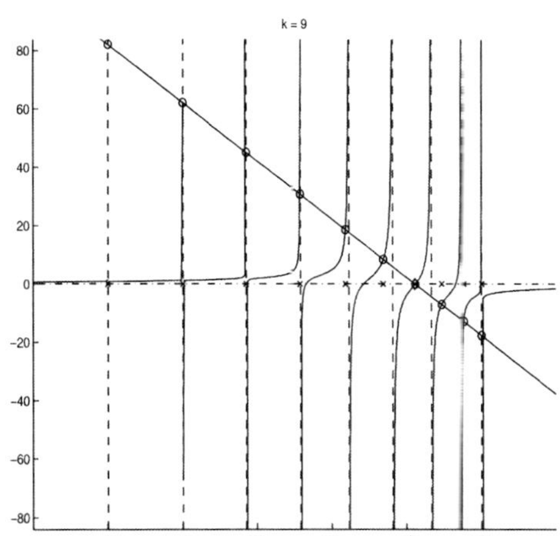

Figure A.35. *Fifth example,* $k = 9$

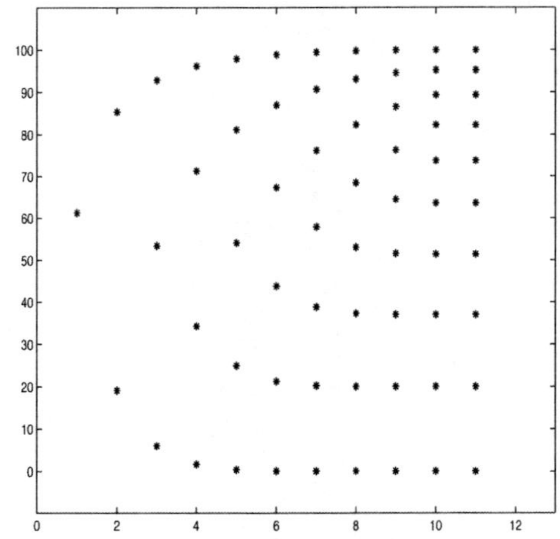

Figure A.36. *Fifth example*

Sixth example

In this example there are clustered eigenvalues in the center of the spectrum (see Figure A.37). The eigenvalues are

$$0.1 \quad 20 \quad 40 \quad 48 \quad 49 \quad 51 \quad 52 \quad 60 \quad 80 \quad 100.$$

The starting vector is all ones. We obtain good approximations of the outer eigenvalues in the first iterations; see Figures A.38 to A.43. We also notice that the value of α_k does not move too much during the iterations.

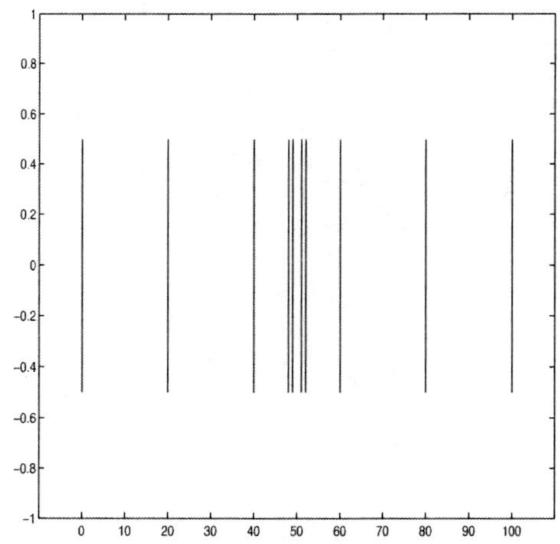

Figure A.37. *Eigenvalues of the sixth example matrix*

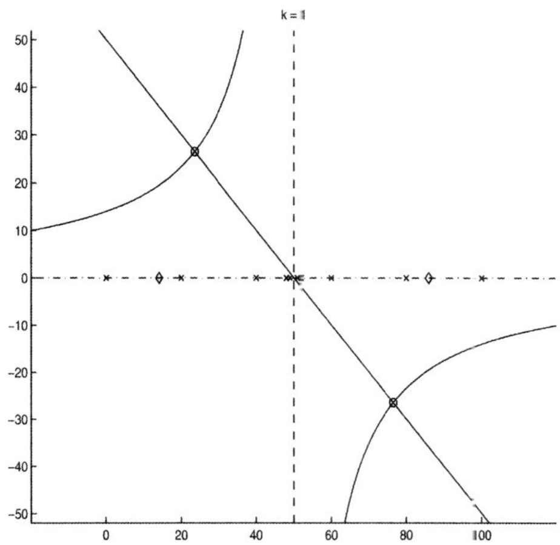

Figure A.38. *Sixth example, $k = 1$*

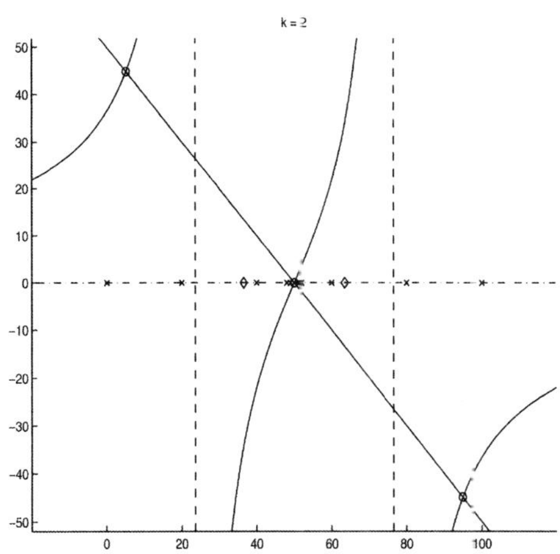

Figure A.39. *Sixth example, $k = 2$*

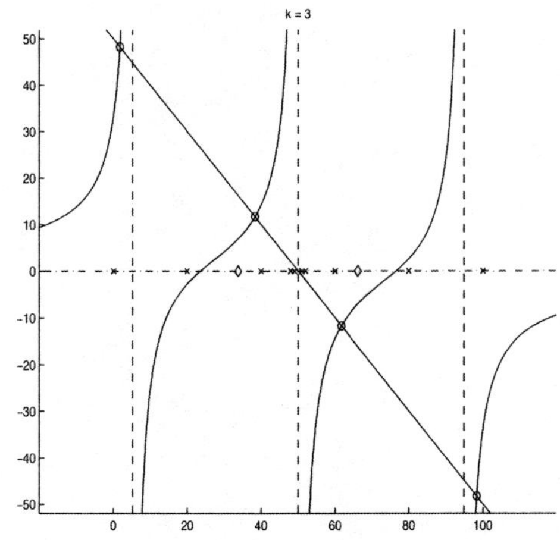

Figure A.40. *Sixth example,* $k = 3$

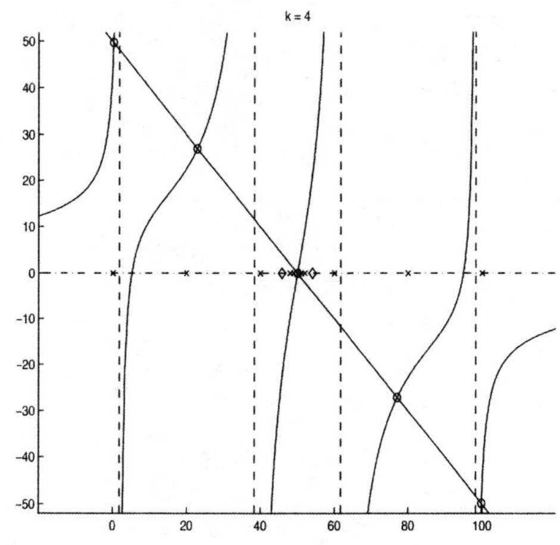

Figure A.41. *Sixth example,* $k = 4$

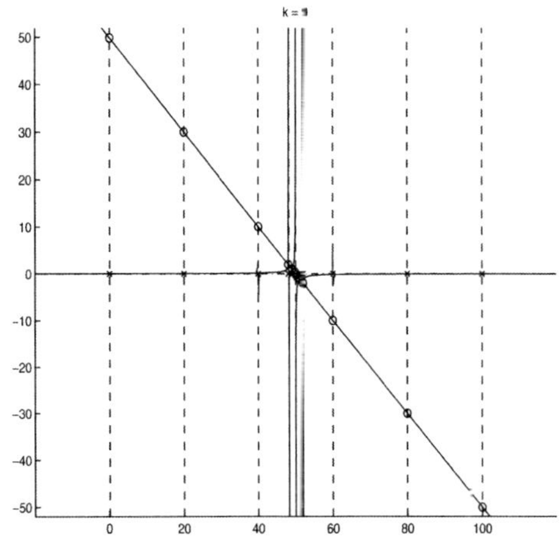

Figure A.42. *Sixth example, $k = 9$*

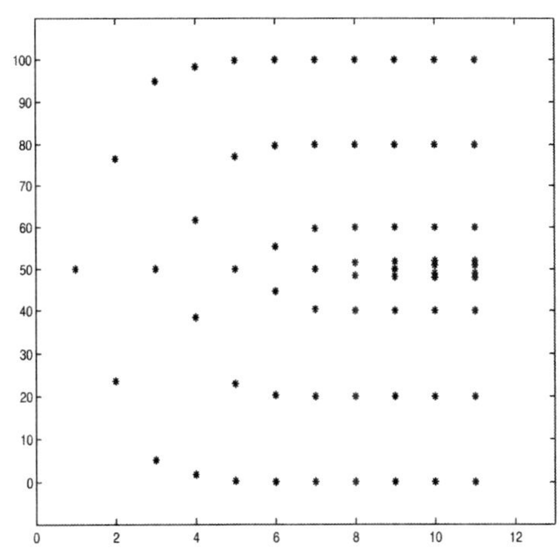

Figure A.43. *Sixth example*

Bibliography

[1] M. Arioli, *Stopping criterion for the conjugate gradient algorithm in a finite element method framework*, Numer. Math., v 97 (2004), pp 1–24.

[2] M. Arioli, D. Loghin, and A. Wathen, *Stopping criteria for iterations in finite element methods*, Numer. Math., v 99 (2005), pp 381–410.

[3] W.E. Arnoldi, *The principle of minimized iterations in the solution of the matrix eigenvalue problem*, Quart. Appl. Math., v 9 (1951), pp 17–29.

[4] O. Axelsson and G. Lindskog, *On the rate of convergence of the preconditioned conjugate gradient method*, Numer. Math., v 48 (1986), pp 499–523.

[5] J. Baglama, D. Calvetti, and L. Reichel, *IRBL: An implicitly restarted block-Lanczos method for large-scale Hermitian eigenproblems*, SIAM J. Sci. Comput., v 24 (2003), pp 1650–1677.

[6] I. Bar-On, *Interlacing properties of tridiagonal symmetric matrices with applications to parallel computing*, SIAM J. Matrix Anal. Appl., v 17 (1996), pp 548–562.

[7] S. Belmehdi, *On the associated orthogonal polynomials*, J. Comput. Appl. Math., v 32 (1990), pp 311–319.

[8] M. Benzi, C.D. Meyer, and M. Tůma, *A sparse approximate inverse preconditioner for the conjugate gradient method*, SIAM J. Sci. Comput., v 17 (1995), pp 1135–1149.

[9] P. Bientinesi, I.S. Dhillon, and R.A. van de Geijn, *A parallel eigensolver for dense symmetric matrices based on multiple relatively robust representations*, SIAM J. Sci. Comput., v 27 (2005), pp 43–66.

[10] D. Bini and V.Y. Pan, *Computing matrix eigenvalues and polynomial zeros where the output is real*, SIAM J. Comput., v 27 (1998), pp 1099–1115.

[11] J.A.M. Bollen, *Round-off error analysis of descent methods for solving linear equations*, Ph.D. thesis, Technische Hogeschool Eindhoven, The Netherlands (1980).

[12] J.A.M. Bollen, *Numerical stability of descent methods for solving linear equations*, Numer. Math, v 43 (1984), pp 361–377.

[13] M. Bonnet and G. Meurant, *Résolution de systèmes d'équations linéaires par la méthode du gradient conjugué avec préconditionnement*, Rapport CEA/DAM 80069 (1980).

[14] C. de Boor and G.H. Golub, *The numerically stable reconstruction of a Jacobi matrix from spectral data*, Linear Algebra Appl., v 21 (1978), pp 245–260.

[15] A. Bouras and V. Frayssé, *A relaxation strategy for inexact matrix-vector products for Krylov methods*, CERFACS Technical Report TR/PA/00/15 (2000). Published as *Inexact matrix-vector products in Krylov methods for solving linear equations: A relaxation strategy*, SIAM J. Matrix Anal. Appl., v 26 (2005), pp 660–678.

[16] A. Bouras, V. Frayssé, and L. Giraud, *A relaxation strategy for inner-outer linear solvers in domain decomposition methods*, CERFACS Technical Report TR/PA/00/17 (2000).

[17] C. Brezinski, *Error estimates for the solution of linear systems*, SIAM J. Sci. Comput., v 21 (1999), pp 764–781.

[18] C. Brezinski and M. Redivo-Zaglia, *Hybrid procedure for solving linear systems*, Numer. Math., v 67 (1994), pp 1–19.

[19] D. Calvetti, L. Reichel, and D.C. Sorensen, *An implicitly restarted Lanczos method for large symmetric eigenvalue problems*, Electron. Trans. Numer. Anal., v 2 (1994), pp 1–21.

[20] D. Calvetti, S. Morigi, L. Reichel, and F. Sgallari, *Computable error bounds and estimates for the conjugate gradient method*, Numer. Algorithms, v 25 (2000), pp 79–88.

[21] B. Carpentieri, I.S. Duff, and L. Giraud, *A class of spectral two-level preconditioners*, SIAM J. Sci. Comput., v 25 (2003), pp 749–765.

[22] T.F. Chan and W.L. Wan, *Analysis of projection methods for solving linear systems with multiple right-hand sides*, SIAM J. Sci. Comput., v 18 (1997), pp 1698–1721.

[23] A. Chapman and Y. Saad, *Deflated and augmented Krylov subspace techniques*, Numer. Linear Algebra Appl., v 4 (1997), pp 43–66.

[24] P. Concus, G.H. Golub, and G. Meurant, *Block preconditioning for the conjugate gradient method*, SIAM J. Sci. Comput., v 6 (1985), pp 220–252.

[25] P. Concus and G. Meurant, *On computing INV block preconditionings for the conjugate gradient method*, BIT, v 26 (1986), pp 493–504.

[26] P. Concus, G.H. Golub, and D.P. O'Leary,*A generalized conjugate gradient method for the numerical solution of elliptic partial differential equations*, in Sparse matrix computations, J.R. Bunch and D.J. Rose Eds., Academic Press (1976), pp 309–332.

[27] J.K. CULLUM AND R.A. WILLOUGHBY, *Lanczos algorithms for large symmetric eigenvalue computations*, vol. I Theory, vol. II Programs, Birkhäuser (1985). Vol. I reprinted by SIAM in the series Classics in Applied Mathematics, (2002).

[28] J.J.M. CUPPEN, *A divide and conquer method for the symmetric tridiagonal eigenproblem*, Numer. Math., v 36 (1981), pp 177–195.

[29] G. CYBENKO, *An explicit formula for Lanczos polynomials*, Linear Algebra Appl., v 88 (1987), pp 99–115.

[30] G. DAHLQUIST, S.C. EISENSTAT, AND G.H. GOLUB, *Bounds for the error of linear systems of equations using the theory of moments*, J. Math. Anal. Appl., v 37 (1972), pp 151–166.

[31] G. DAHLQUIST, G.H. GOLUB, AND S.G. NASH, *Bounds for the error in linear systems*, in Proc. of the Workshop on Semi-Infinite Programming, R. Hettich Ed., Springer (1978), pp 154–172.

[32] P. DAVIS AND P. RABINOWITZ, *Methods of numerical integration*, Second Edition, Academic Press (1984).

[33] I.S. DHILLON, *A new $O(N^2)$ algorithm for the symmetric tridiagonal eigenvalue/eigenvector problem*, Ph.D. thesis, University of California, Berkeley (1997).

[34] I.S. DHILLON AND B.N. PARLETT, *Orthogonal eigenvectors and relative gaps*, SIAM J. Matrix Anal. Appl., v 25 (2004), pp 858–899.

[35] I.S. DHILLON AND B.N. PARLETT, *Multiple representations to compute orthogonal eigenvectors of symmetric tridiagonal matrices*, Linear Algebra Appl., v 387 (2004), pp 1–28.

[36] J. J. DONGARRA AND D.C. SORENSEN, *A fully parallel algorithm for the symmetric eigenvalue problem*, SIAM J. Sci. Comput., v 8 (1987), pp 139–154.

[37] V. DRUSKIN AND L. KNIZHNERMAN, *Two polynomial methods of calculating functions of symmetric matrices*, U.S.S.R. Comput. Math. and Math. Phys., v 29 (1989), pp 112–121.

[38] V. DRUSKIN AND L. KNIZHNERMAN, *Error bounds in the simple Lanczos procedure for computing functions of symmetric matrices and eigenvalues*, Comput. Math. Math. Phys., v 31 (1991), pp 20–30.

[39] V. DRUSKIN AND L. KNIZHNERMAN, *An application of the Lanczos method to solution of some partial differential equations*, J. Comput. Appl. Math., v 50 (1994), pp 255–262.

[40] V. DRUSKIN AND L. KNIZHNERMAN, *Krylov subspace approximation of eigenpairs and matrix functions in exact and computer arithmetic*, Numer. Linear Algebra Appl., v 2 (1995), pp 205–217.

[41] V. Druskin and L. Knizhnerman, *Extended Krylov subspaces: Approximation of the matrix square root and related functions*, SIAM J. Matrix Anal. Appl., v 19 (1998), pp 755–771.

[42] V. Druskin, A. Greenbaum, and L. Knizhnerman, *Using nonorthogonal Lanczos vectors in the computation of matrix functions*, SIAM J. Sci. Comput., v 19 (1998), pp 38–54.

[43] A. El Guennouni, K. Jbilou, and H. Sadok, *A block version of BiCGStab for linear systems with multiple linear systems*, Electron. Trans. Numer. Anal., v 16 (2003), pp 129–142.

[44] A. El Guennouni, K. Jbilou, and H. Sadok, *The block-Lanczos method for linear systems with multiple right-hand sides*, Appl. Numer. Math., v 51 (2004), pp 243–256.

[45] S. Elhay, G.M.L. Gladwell, G.H. Golub, and Y.M. Ram, *On some eigenvector-eigenvalue relations*, SIAM J. Matrix Anal. Appl., v 20 (1999), pp 563–574.

[46] J. Erhel and F. Guyomarc'h, *An augmented conjugate gradient method for solving consecutive symmetric positive definite linear systems*, SIAM J. Matrix Anal. Appl., v 21 (2000), pp 1279–1299.

[47] J. van den Eshof and G.L.G. Sleijpen, *Inexact Krylov subspace methods for linear systems*, SIAM J. Matrix Anal. Appl., v 26 (2004), pp 125–153.

[48] J. van den Eshof, G.L.G. Sleijpen, and M.B. van Gijzen, *Relaxation strategies for nested Krylov methods*, J. Comput. Appl. Math., v 177 (2005), pp 347–365.

[49] A. Facius, *Iterative solution of linear systems with improved arithmetic and result verification*, Ph.D. thesis, Karlsruhe University (2000).

[50] B. Fischer, *Polynomial based iteration methods for symmetric linear systems*, Wiley–Tubner (1996).

[51] B. Fischer and G.H. Golub, *On the error computation for polynomial based iteration methods*, in Recent advances in iterative methods, A. Greenbaum and M. Luskin Eds., Springer (1993), pp 59–67.

[52] G.E. Forsythe, M.R. Hestenes, and J.B. Rosser, *Iterative methods for solving linear equations*, Bull. Amer. Math. Soc., v 57 (1951), p 480.

[53] W. Gautschi, *Construction of Gauss–Christoffel quadrature formulas*, Math. Comp., v 22 (1968), pp 251–270.

[54] W. Gautschi, *Orthogonal polynomials—constructive theory and applications*, J. Comput. Appl. Math., v 12 & 13 (1985), pp 61–76.

[55] W. Gautschi, *The interplay between classical analysis and (numerical) linear algebra—A tribute to Gene H. Golub*, Electron. Trans. Numer. Anal., v 13 (2002), pp 119–147.

[56] B. GELLAI, *Cornelius Lanczos, a biographical essay*, in Proceedings of the Cornelius Lanczos international centenary conference, J.D. Brown, M.T. Chu, D.C. Ellison, and R.J. Plemmons Eds., SIAM, (1994), pp xxi–xlviii.

[57] L. GIRAUD AND J. LANGOU, *Another proof for modified Gram-Schmidt with reorthogonalization*, CERFACS Working Notes WN/PA/02/53 (2002).

[58] L. GIRAUD AND J. LANGOU, *A robust criterion for the modified Gram–Schmidt algorithm with selective reorthogonalization*, SIAM J. Sci. Comput., v 25 (2003), pp 417–441.

[59] L. GIRAUD, J. LANGOU, AND M. ROZLOZNÍK, *On the round-off error analysis of the Gram-Schmidt algorithm with reorthogonalization*, CERFACS Technical Report TR/PA/02/33 (2002).

[60] L. GIRAUD, J. LANGOU, AND M. ROZLOZNÍK, *On the loss of orthogonality in the Gram-Schmidt orthogonalization process*, Comput. Math. Appl., v 50 (2005), pp 1069–1075.

[61] L. GIRAUD, J. LANGOU, M. ROZLOZNÍK, AND J. VAN DEN ESHOF, *Rounding error analysis of the classical Gram-Schmidt orthogonalization process*, Numer. Math., v 101 (2005), pp 87–100.

[62] G.H. GOLUB, *Some modified matrix eigenvalue problems*, SIAM Rev., v 15 (1973), pp 318–334.

[63] G.H. GOLUB, *Bounds for matrix moments*, Rocky Mountain J. Math., v 4 (1974), pp 207–211.

[64] G.H. GOLUB AND G. MEURANT, *Matrices, moments and quadrature*, in Numerical analysis 1993, D.F. Griffiths and G.A. Watson Eds., Pitman Research Notes in Mathematics, v 303 (1994), pp 105–156.

[65] G.H. GOLUB AND G. MEURANT, *Matrices, moments and quadrature* II *or how to compute the norm of the error in iterative methods*, BIT, v 37 (1997), pp 687–705.

[66] G.H. GOLUB AND D.P. O'LEARY, *Some history of the conjugate gradient and Lanczos algorithms:* 1948–1976, SIAM Rev., v 31 (1989), pp 50–102.

[67] G.H. GOLUB AND Z. STRAKOŠ, *Estimates in quadratic formulas*, Numer. Algorithms, v 8 (1994), pp. 241–268.

[68] G.H. GOLUB AND C. VAN LOAN, *Matrix computations*, The Johns Hopkins University Press (1989).

[69] G.H. GOLUB AND J.H. WELSCH, *Calculation of Gauss quadrature rule*, Math. Comp., v 23 (1969), pp 221–230.

[70] G.H. GOLUB AND Q. YE, *Inexact preconditioned conjugate gradient method with inner-outer iterations*, SIAM J. Sci. Comput., v 21 (1999), pp 1305–1320.

[71] W.B. GRAGG, *The Padé table and its relation to certain algorithms of numerical analysis*, SIAM Rev., v 14 (1972), pp 1–62.

[72] W.B. GRAGG AND W.J. HARROD, *The numerically stable reconstruction of Jacobi matrices from spectral data*, Numer. Math., v 44 (1984), pp 317–335.

[73] J. GRCAR, *Analysis of the Lanczos algorithm and of the approximation problem in Richardson's method*, Ph.D. thesis, University of Illinois at Urbana–Champaign (1981).

[74] A. GREENBAUM, *Comparison of splittings used with the conjugate gradient algorithm*, Numer. Math., v 33 (1979), pp 181–194.

[75] A. GREENBAUM, *Convergence properties of the conjugate gradient algorithm in exact and finite precision arithmetic*, Ph.D. thesis, University of California, Berkeley (1981).

[76] A. GREENBAUM, *Behavior of slightly perturbed Lanczos and conjugate gradient recurrences*, Linear Algebra Appl., v 113 (1989), pp 7–63.

[77] A. GREENBAUM, *The Lanczos and conjugate gradient algorithms in finite precision arithmetic*, in Proceedings of the Cornelius Lanczos international centenary conference, J.D. Brown, M.T. Chu, D.C. Ellison, and R.J. Plemmons Eds., SIAM, (1994), pp 49–60.

[78] A. GREENBAUM, *Iterative methods for solving linear systems*, SIAM (1997).

[79] A. GREENBAUM, *Estimating the attainable accuracy of recursively computed residual methods*, SIAM J. Matrix Anal. Appl., v 18 (1997), pp 535–551.

[80] A. GREENBAUM, *Private communication* (2004).

[81] A. GREENBAUM AND Z. STRAKOŠ, *Predicting the behavior of finite precision Lanczos and conjugate gradient computations*, SIAM J. Matrix Anal. Appl., v 13 (1992), pp 121–137.

[82] A. GREENBAUM, V.L. DRUSKIN, AND L.A. KNIZHNERMAN, *On solving indefinite symmetric linear systems by means of the Lanczos method*, Comput. Math. Math. Phys., v 39 (1999), pp 350–356.

[83] M. GU AND S.C. EISENSTAT, *A stable and efficient algorithm for the rank-one modification of the symmetric eigenproblem*, SIAM J. Matrix Anal. Appl., v 15 (1994), pp 1266–1276.

[84] M.H. GUTKNECHT AND M. ROZLOZNÍK, *Residual smoothing techniques: Do they improve the limiting accuracy of iterative solvers?*, BIT, v 41 (2001), pp 86–114.

[85] M.H. GUTKNECHT AND M. ROZLOZNÍK, *By how much can residual minimization accelerate the convergence of orthogonal residual methods?*, Numer. Algorithms, v 27 (2001), pp 189–213.

[86] M.H. Gutknecht and Z. Strakoš, *Accuracy of two three-term and three two-term recurrences for Krylov space solvers*, SIAM J. Matrix Anal. Appl., v 22 (2000), pp 213–229.

[87] F. Guyomarc'h, *Méthodes de Krylov: régularisation de la solution et accélération de la convergence*, Ph.D. thesis, Université de Rennes I (2000).

[88] M.R. Hestenes, *Iterative methods for solving linear equations*, Report 52–9, NAML (1951), National Bureau of Standards (1951). Reprinted in J. Optim. Theory Appl., v 11 (1973), pp 323–334.

[89] M.R. Hestenes, *The conjugate gradient method for solving linear systems*, Report INA 54–11, National Bureau of Standards (1954).

[90] M.R. Hestenes, *Conjugate directions methods in optimization*, Springer (1980).

[91] M.R. Hestenes, *Conjugacy and gradients*, in A history of scientific computing, S.G. Nash Ed., ACM Press (1990), pp 167–179.

[92] M.R. Hestenes and W. Karush, *A method of gradients for the calculation of the characteristic roots and vectors of a real symmetric matrix*, J. Res. Nat. Bur. Standards, v 47 (1961), pp 45–61.

[93] M.R. Hestenes and E. Stiefel, *Methods of conjugate gradients for solving linear systems*, J. Res. Nat. Bur. Standards, v 49 (1952), pp 409–436.

[94] N.J. Higham, *Accuracy and stability of numerical algorithms*, Second Edition, SIAM (2002).

[95] R.O. Hill Jr. and B.N. Parlett, *Refined interlacing properties*, SIAM J. Matrix Anal. Appl., v 13 (1992), pp 239–247.

[96] A.S. Householder, *The theory of matrices in numerical analysis*, Blaisdell (1964). Reprinted by Dover (1975).

[97] I.C.F. Ipsen, *Ritz value bounds that exploit quasi-sparsity*, North Carolina State Univ. Report CRSC-TR03-31 (2003).

[98] K. Jbilou, A. Messaoudi, and H. Sadok, *Global GMRES algorithm for solving nonsymmetric linear systems of equations with multiple right-hand sides*, Appl. Numer. Math., v 31 (1999), pp 49–63.

[99] K. Jbilou, H. Sadok, and A. Tinzefte, *Oblique projection methods for linear systems with multiple right-hand sides*, Electron. Trans. Numer. Anal., v 20 (2005), pp 119–138.

[100] P. Joly, *Résolution de systèmes linéaires avec plusieurs second membres par la méthode du gradient conjugué*, Technical Report R-91012, Université Paris VI, France (1991).

[101] W. Kahan and B.N. Parlett, *How far should you go with the Lanczos process?*, in Sparse matrix computations, J.R. Bunch and D.J. Rose Eds., Academic Press (1976), pp 131–144.

[102] S. Kaniel, *Estimates of some computational techniques in linear algebra*, Math. Comp. v 20 (1966), pp 369–378.

[103] L. Knizhnerman, *The quality of approximations to an isolated eigenvalue and the distribution of "Ritz numbers" in the simple Lanczos procedure*, Comput. Math. Math. Phys., v 35 (1995), pp 1175–1187.

[104] L. Knizhnerman, *On adaptation of the Lanczos method to the spectrum*, Report EMG-001-95-12, Schlumberger–Doll–Research (1995).

[105] L. Knizhnerman, *The simple Lanczos procedure: Estimates of the error of the Gauss quadrature formula and their applications*, Comput. Math. Math. Phys., v 36 (1996), pp 1481–1492.

[106] A.N. Krylov, *O Čislemnon rešenii uravnenija, kotorym v techničeskih voprasah opredeljajutsja častoy malyh kolebaniĭ material'nyh*, Izv. Adad. Nauk SSSR old. Mat. Estet., (1931), pp 491–539.

[107] A.B.J. Kuijlaars, *Which eigenvalues are found by the Lanczos method?*, SIAM J. Matrix. Anal. Appl., v 22 (2000), pp 306–321.

[108] C. Lanczos, *An iteration method for the solution of the eigenvalue problem of linear differential and integral operators*, J. Res. Nat. Bur. Standards, v 45 (1950), pp 255–282.

[109] C. Lanczos, *Solution of systems of linear equations by minimized iterations*, J. Res. Nat. Bur. Standards, v 49 (1952), pp 33–53.

[110] J. Langou, *Solving large linear systems with multiple right hand sides*, Ph.D. thesis, CERFACS (2003).

[111] R.B. Lehoucq, *Analysis and implementation of an implicitly restarted Arnoldi iteration*, Ph.D. thesis, Rice University (1995).

[112] R.B. Lehoucq, D.C. Sorensen, and C. Yang, *ARPACK User's guide: Solution of large-scale eigenvalue problems by implicitly restarted Arnoldi method*, SIAM (1998).

[113] R.-C. Li, *Solving secular equations stably and efficiently*, Report UCB/CSD-94-851, University of California, Berkeley (1994). See also LAPACK Working Notes 89.

[114] R.K. Mallik, *Solutions of linear difference equations with variable coefficients*, J. Math. Anal. Appl., v 222 (1998), pp 79–91.

[115] Matrix Market, http://math.nist.gov

[116] G. Meurant, *Multitasking the conjugate gradient method on the CRAY X-MP/48*, Parallel Comp., v 5 (1987), pp 267–280.

[117] G. Meurant, *A review of the inverse of symmetric tridiagonal and block tridiagonal matrices*, SIAM J. Matrix Anal. Appl., v 13 (1992), pp 707–728.

[118] G. Meurant, *The computation of bounds for the norm of the error in the conjugate gradient algorithm*, Numer. Algorithms, v 16 (1997), pp 77–87.

[119] G. Meurant, *Numerical experiments in computing bounds for the norm of the error in the preconditioned conjugate gradient algorithm*, Numer. Algorithms, v 22 (1999), pp 353–365.

[120] G. Meurant, *Computer solution of large linear systems*, North–Holland (1999).

[121] G. Meurant, *Numerical experiments on algebraic multilevel preconditioners*, Electron. Trans. Numer. Anal., v 12 (2001), pp 1–65.

[122] G. Meurant, *Estimates of the l_2 norm of the error in the conjugate gradient algorithm*, Numer. Algorithms, v 40 (2005), pp 157–169.

[123] G. Meurant and Z. Strakoš, *The Lanczos and conjugate gradient algorithms in finite precision arithmetic*, Acta Numerica, v 15 (2006), pp. 471–542.

[124] R.A. Nicolaides, *Deflation of conjugate gradients with applications to boundary value problems*, SIAM J. Numer. Anal., v 24 (1987), pp 355–365.

[125] Y. Notay, *On the convergence rate of the conjugate gradients in presence of rounding errors*, Numer. Math., v 65 (1993), pp 301–317.

[126] D.P. O'Leary, *The block conjugate gradient algorithm and related methods*, Linear Algebra Appl., v 29 (1980), pp 293–322.

[127] M.L. Overton, *Numerical computing with IEEE floating point arithmetic*, SIAM (2001).

[128] C.C. Paige, *Error analysis of the generalized Hessenberg processes*, London Univ. Inst. of Computer Science, Tech. Note ICSI 179 (1969).

[129] C.C. Paige, *Error analysis of the symmetric Lanczos process for the eigenproblem*, London Univ. Inst. of Computer Science, Tech. Note ICSI 248 (1970).

[130] C.C. Paige, *Eigenvalues of perturbed Hermitian matrices*, London Univ. Inst. of Computer Science, Tech. Note ICSI 179 (1969).

[131] C.C. Paige, *Practical use of the symmetric Lanczos process with reorthogonalization*, BIT, v 10 (1970), pp 183–195.

[132] C.C. Paige, *The computation of eigenvalues and eigenvectors of very large sparse matrices*, Ph.D. thesis, University of London (1971).

[133] C.C. Paige, *Computational variants of the Lanczos method for the eigenproblem*, J. Inst. Math. Appl., v 10 (1972), pp 373–381.

[134] C.C. Paige, *Error analysis of the Lanczos algorithm for tridiagonalizing a symmetric matrix*, J. Inst. Math. Appl., v 18 (1976), pp 341–349.

[135] C.C. Paige, *Accuracy and effectiveness of the Lanczos algorithm for the symmetric eigenproblem*, Linear Algebra Appl., v 34 (1980), pp 235–258.

[136] C.C. Paige and M.A. Saunders, *Solution of sparse indefinite systems of equations and least squares problems*, Report STAN-CS-73-399, Computer Science Department, Stanford University (1973).

[137] C.C. Paige and M.A. Saunders, *Solution of sparse indefinite systems of linear equations*, SIAM J. Numer. Anal., v 12 (1975), pp 617–629.

[138] C.C Paige, B.N. Parlett, and H. van der Vorst, *Approximate solutions and eigenvalue bounds from Krylov subspaces*, Numer. Linear Algebra Appl., v 2 (1995), pp 115–133.

[139] C.C. Paige and P. van Dooren, *Sensitivity analysis of the Lanczos reduction*, Numer. Linear Algebra Appl., v 6 (1999), pp 29–50.

[140] B.N. Parlett, *A new look at the Lanczos algorithm in solving symmetric systems of linear equations*, Linear Algebra Appl., v 29 (1980), pp 323–346.

[141] B.N. Parlett, *The symmetric eigenvalue problem*, Prentice–Hall (1980). Reprinted by SIAM in the series Classics in Applied Mathematics (1998).

[142] B.N. Parlett and D.S. Scott, *The Lanczos algorithm with selective orthogonalization*, Math. Comp., v 33 (1979), pp 217–238.

[143] B.N. Parlett and W.D. Wu, *Eigenvector matrices of symmetric tridiagonals*, Numer. Math. v 44 (1984), pp 103–110.

[144] B.N. Parlett and B. Nour Omid, *The use of refined error bound when updating eigenvalues of tridiagonals*, Linear Algebra Appl., v 68 (1985), pp 179–219.

[145] B.N. Parlett and J.K. Reid, *Tracking the progress of the Lanczos algorithm for large symmetric eigenproblem*, IMA J. Numer. Anal., v 1 (1981), pp 135–155.

[146] B.N. Parlett and I.S. Dhillon, *Fernando's solution to Wilkinson's problem: An application of double factorization*, Linear Algebra Appl., v 267 (1997), pp 247–279.

[147] M.J.D. Powell, *Some convergence properties of the conjugate gradient method*, Math. Prog., v 11 (1976), pp 42–49.

[148] *Proceedings of the Cornelius Lanczos international centenary conference* (1993), J.D. Brown, M.T. Chu, D.C. Ellison, and R.J. Plemmons Eds., SIAM (1994).

[149] J.K. REID, *On the method of conjugate gradients for the solution of large sparse systems of linear equations*, in Large sparse sets of linear equations, J.K. Reid Ed., Academic Press (1971), pp 231–254.

[150] A. RUHE, *Rational Krylov sequence methods for eigenvalue computation*, Linear Algebra Appl., v 58 (1984), pp 391–405.

[151] H. RUTISHAUSER, *Theory of gradient methods*, in Refined iterative methods for computation of the solution and the eigenvalues of self-adjoint boundary value problems, Mitt. Inst. Angew. Math. ETH Zürich, Birkhäuser (1959), pp 24–49.

[152] J. RUTTER, *A serial implementation of Cuppen's divide and conquer algorithm for the symmetric eigenvalue problem*, Report UCB/CSD 04/799, University of California, Berkeley (1994).

[153] Y. SAAD, *On the rates of convergence of the Lanczos and the block-Lanczos methods*, SIAM J. Numer. Anal., v 17 (1980), pp 687–706.

[154] Y. SAAD, *On the Lanczos method for solving symmetric linear systems with several right hand sides*, Math. Comp., v 178 (1987), pp 651–662.

[155] Y. SAAD, *Numerical methods for large eigenvalue problems*, Wiley (1992).

[156] Y. SAAD, M. YEUNG, J. ERHEL, AND F. GUYOMARC'H, *A deflated version of the conjugate gradient algorithm*, SIAM J. Sci. Comput., v 21 (2000), pp 1909–1926.

[157] W. SCHÖNAUER, *Scientific computing on vector computers*, North–Holland (1987).

[158] W. SCHÖNAUER, H. MÜLLER, AND E. SCHNEPF, *Numerical tests with biconjugate gradient type methods*, Z. Angew. Math. Mech., v 65 (1985), pp 400–402.

[159] D.S. SCOTT, *Analysis of the symmetric Lanczos algorithm*, Ph.D. thesis, University of California, Berkeley (1978).

[160] D.S. SCOTT, *How to make the Lanczos algorithm converge slowly*, Math. Comp., v 33 (1979), pp 239–247.

[161] D.S. SCOTT, *The Lanczos algorithm*, in Sparse matrices and their use, I.S. Duff Ed., Academic Press (1981), pp 139–159.

[162] H.D. SIMON, *The Lanczos algorithm for solving symmetric linear systems*, Ph.D. thesis, University of California, Berkeley (1982).

[163] H.D. SIMON, *The Lanczos algorithm with partial reorthogonalization*, Math. Comp., v 42 (1984), pp 115–142.

[164] H.D. SIMON, *Analysis of the symmetric Lanczos algorithm with reorthogonalization methods*, Linear Algebra Appl., v 61 (1984), pp 101–131.

[165] H.D. SIMON AND K. WU, *Thick-restart Lanczos method for large symmetric eigenvalue problems*, SIAM. J. Matrix Anal. Appl., v 22 (2000), pp 602–616.

[166] V. SIMONCINI AND D.B. SZYLD, *Flexible inner-outer Krylov subspace methods*, SIAM J. Numer. Anal., v 40 (2003), pp 2219–2239.

[167] V. SIMONCINI AND D.B. SZYLD, *Theory of inexact Krylov subspace methods and applications to scientific computing*, SIAM J. Sci. Comput., v 25 (2003), pp 454–477.

[168] V. SIMONCINI AND D.B. SZYLD, *On the occurrence of superlinear convergence of exact and inexact Krylov subspace methods*, SIAM Rev., v 47 (2005), pp 247–272.

[169] V. SIMONCINI AND D.B. SZYLD, *The effect of non-optimal bases on the convergence of Krylov subspace methods*, Numer. Math., v 100 (2005), pp 711–733.

[170] M.R. SKRZIPEK, *Polynomial evaluation and associated polynomials*, Numer. Math., v 79 (1998), pp 601–613.

[171] M.R. SKRZIPEK, *Generalized associated polynomials and their application in numerical differentiation and quadrature*, Calcolo, v 40 (2003), pp 131–147.

[172] G.L.G. SLEIJPEN AND J. VAN DEN ESHOF, *On the use of harmonic Ritz pairs in approximating internal eigenpairs*, Preprint 1184, University of Utrecht (2001).

[173] G.L.G. SLEIJPEN, J. VAN DEN ESHOF, AND P. SMIT, *Optimal a priori error bounds for the Rayleigh–Ritz method*, Math. Comp., v 72 (2003), pp 677–684.

[174] G.L.G. SLEIJPEN, H. VAN DER VORST, AND J. MODERSITSKI, *Differences in the effects of rounding errors in Krylov solvers for symmetric indefinite linear systems*, SIAM J. Matrix Anal. Appl., v 22 (2000), pp 726–751.

[175] G.W. STEWART, *Two simple residual bounds for the eigenvalues of a Hermitian matrix*, SIAM J. Matrix Anal. Appl., v 12 (1991), pp 205–208.

[176] G.W. STEWART, *Lanczos and linear systems*, in Proceedings of the Cornelius Lanczos international centenary conference, J.D. Brown, M.T. Chu, D.C. Ellison, and R.J. Plemmons Eds., SIAM, (1994), pp xxi–xlviii.

[177] G.W. STEWART, *A generalization of Saad's theorem on Rayleigh–Ritz approximation*, Linear Algebra Appl., v 327 (2001), pp 115–119.

[178] G.W. STEWART, *Matrix algorithms: Basic Decomposition*, SIAM (1998)

[179] G.W. STEWART, *Matrix algorithms, volume* II*: Eigensystems*, SIAM (2001)

[180] G.W. STEWART, *Adjusting the Rayleigh quotient in semiorthogonal Lanczos methods*, SIAM J. Sci. Comput., v 24 (2002), pp 201–207.

[181] G.W. STEWART, *Backward error bounds for approximate Krylov subspaces*, Linear Algebra Appl., v 340 (2002), pp 81–86.

[182] E. STIEFEL, *Über einige methoden der relaxationsrechnung*, Z.A.M.P., v 3 (1952), pp 1–33.

[183] J. STOER AND R. BULIRSCH, *Introduction to numerical analysis*, Second Edition, Springer Verlag (1983).

[184] Z. STRAKOŠ, *On the real convergence rate of the conjugate gradient method*, Linear Algebra Appl., v 154–156 (1991), pp 535–549.

[185] Z. STRAKOŠ AND A. GREENBAUM, *Open questions in the convergence analysis of the Lanczos process for the real symmetric eigenvalue problem*, IMA Preprint 934 (1992).

[186] Z. STRAKOŠ AND P. TICHÝ, *On error estimation in the conjugate gradient method and why it works in finite precision computation*, Electron. Trans. Numer. Anal., v 13 (2002), pp 56–80.

[187] Z. STRAKOŠ AND P. TICHÝ, *Error estimation in preconditioned conjugate gradient*, BIT, v 45 (2005), pp 789–817.

[188] G. SZEGÖ, *Orthogonal polynomials*, Third Edition, American Mathematical Society (1974).

[189] R.C. THOMPSON, *Principal submatrices of normal and hermitian matrices*, Illinois J. Math., v 10 (1966), pp 296–308.

[190] R.C. THOMPSON AND P. MCENTEGGERT, *Principal submatrices, II: The upper and lower quadratic inequalities*, Linear Algebra Appl., v 1 (1968), pp 211–243.

[191] F. TISSEUR AND J. DONGARRA, *A parallel divide and conquer algorithm for the symmetric eigenvalue problem on distributed memory architectures*, SIAM J. Sci. Comput., v 20 (1999), pp 2223–2236.

[192] J. TODD, *Basic numerical mathematics, volume 2, Numerical algebra*, Birkhauser (1977).

[193] A. TOUHAMI, *Utilisation des filtres de Tchebycheff et construction de préconditioneurs spectraux pour l'accélération des méthodes de Krylov*, Ph.D. thesis, Institut National Polytechnique de Toulouse (2005).

[194] H.A. VAN DER VORST, *Computational methods for large eigenvalue problems*, in Handbook of numerical analysis, P.G. Ciarlet and J.L. Lions Eds., v VIII (2002), pp 3–179.

[195] H.A. VAN DER VORST, *Iterative Krylov methods for large linear systems*, Cambridge University Press (2003).

[196] H.A. VAN DER VORST AND Q. YE, *Residual replacement strategies for Krylov subspace iterative methods for the convergence of true residuals*, SIAM J. Sci. Comput., v 22 (2000), pp 835–852.

[197] A. VAN DER SLUIS AND H.A. VAN DER VORST, *The convergence behavior of Ritz values in the presence of close eigenvalues*, Linear Algebra Appl., v 88 (1987), pp 651–694.

[198] R. WEISS, *Convergence behavior of generalized conjugate gradient methods*, Ph.D. thesis, University of Karlsruhe (1990).

[199] H.S. WILF, *Mathematics for the physical sciences*, Wiley (1962).

[200] J.H. WILKINSON, *The algebraic eigenvalue problem*, Oxford University Press (1965).

[201] H. WOZNIAKOWSKI, *Roundoff error analysis of a new class of conjugate gradient algorithms*, Linear Algebra Appl., v 29 (1980), pp 507–529.

[202] W. WÜLLING, *On stabilization and convergence of clustered Ritz values in the Lanczos method*, SIAM J. Matrix Anal., v 27 (2006), pp 891–908.

[203] W. WÜLLING, *The stabilization of weights in the Lanczos and conjugate gradient methods*, BIT, v 45 (2005), pp 395–414.

[204] Q. YE, *On close eigenvalues of tridiagonal matrices*, Numer. Math., v 70 (1995), pp 507–514.

[205] J.-P.M. ZEMKE, *Krylov subspace methods in finite precision: A unified approach*, Ph.D. thesis, Technical University of Hamburg (2003).

[206] J.-P.M. ZEMKE, *(Hessenberg) eigenvalue–eigenmatrix relations*, Report 78, Hamburg–Harburg Technical University, Linear Algebra Appl., v 414 (2006), pp. 589–606.

[207] J.-P.M. ZEMKE, *Abstract perturbed Krylov methods*, Report 89, Hamburg–Harburg Technical University (2005).

[208] L. ZHOU AND H.F. WALKER, *Residual smoothing techniques for iterative methods*, SIAM J. Sci. Comput., v 15 (1994), pp 297–312.

Index

A-norm, xiii, xiv, 23, 45, 58, 61, 62, 64–66, 68, 69, 72, 77, 78, 80, 130, 187, 242, 244, 257, 260, 265, 268, 272, 283, 291, 292, 301, 310
l_2 norm, xiv, 56, 65, 68, 69, 86, 88, 93, 187, 242, 252, 257, 261, 263, 265, 268, 272, 283, 292, 311

approximate inverse, 315

backward error analysis, 83, 121

Cauchy, 1, 18, 20
characteristic polynomial, 1, 15, 22
Chebyshev polynomials, 41, 78–80, 94, 131
Chebyshev recurrence, 132
Chebyshev series, 131
Cholesky, xiii, 5, 9–11, 46, 50, 52, 55, 114, 115, 158, 191, 236, 258, 260, 284, 288, 291, 292, 309, 315, 316
Christoffel–Darboux relation, 12, 29
condition number, xiv, 1, 45, 54, 57, 77, 78, 136, 187, 191, 223, 224, 244, 301, 310

determinant, 9–11, 15, 151, 158, 208
diagonal dominance, 56
diagonally dominant matrix, 86
double reorthogonalization, 109, 142–144, 177, 183, 236, 325

eigenvalues, xi, xii, 1, 2, 4, 6–13, 15, 16, 18–20, 25–29, 31–34, 36, 37, 39, 41–43, 45, 51, 57, 58, 60, 62, 78, 80, 81, 89, 93, 94, 96–99, 101, 102, 108, 113, 121, 122, 126, 127, 130, 135, 139, 140, 142, 151–153, 162, 174, 177, 180, 206–208, 236, 244, 260, 264, 281, 282, 288, 290, 292, 298, 301, 306, 310, 311, 316, 325, 326, 332, 335, 338, 341, 344
eigenvectors, xi, 2, 3, 9, 12, 16, 19–22, 26–28, 34, 39, 43, 57, 58, 60, 61, 63, 73, 96, 98, 105, 108, 113, 122–124, 126, 139, 140, 142, 149, 152, 153, 157, 174, 175, 182, 183, 198, 200, 203, 204, 210, 264, 282, 283, 286, 298, 301, 305, 306, 310, 326

finite precision arithmetic, xi, xiii, xiv, 30, 50, 81, 82, 94, 99, 100, 131, 143, 187, 201, 214, 223, 257, 258, 263–267, 272, 275, 281, 284, 310
forward analysis, 82, 105, 157
forward error, 84
Frobenius norm, 97, 302
full reorthogonalization, 108, 117

Gauss lower bound, 259, 260, 268
Gauss quadrature, xiii, 12, 45, 58, 60, 62, 263
Gauss rule, 59, 257
Gaussian elimination, 121, 187, 252, 314, 316
Gram–Schmidt algorithm, 1
Gram–Schmidt orthogonalization, 2, 81
Gram–Schmidt process, xii, 6, 81

IEEE arithmetic, xiii, 83
indefinite linear systems, xiv
indefinite matrix, 310
inner product, 62, 75, 82–86, 93, 95, 98, 111, 215, 217, 266, 281, 282, 284, 286, 302
irreducible, 171
iterative method, xiv, xv, 304
iterative methods, 77, 314

Lanczos polynomial, 11, 13, 15, 27, 56, 59, 60, 73, 107, 142, 143, 152
Laplace, 1
level of orthogonality, 82, 113, 114, 117, 118
local orthogonality, xiv, 54, 75, 79, 80, 90, 94, 102, 116, 187, 215, 265, 304
loss of orthogonality, xiii, 82, 90, 98, 101, 102, 105, 108, 113, 139, 187
LU decomposition, 11, 71, 300

machine epsilon, 83
matrix-vector product, xiv, 75, 86, 93, 239, 260, 304, 305
maximum attainable accuracy, xiv, 188, 239, 242, 249, 272, 292
maximum norm, 249, 251
MINRES, 313
multilevel preconditioners, 315, 316

Newton's method, 43
norm of the error, 62, 65, 126, 131, 188, 239, 241, 244, 260, 261, 263, 264, 268, 291
norm of the residual, 55, 57, 64, 66, 191, 202, 214, 224, 229, 239, 264, 268, 286, 291, 306, 310, 313, 315
norms of the error, 272

orthogonal basis, xi, 1, 45, 324
orthogonal matrix, xi, 2, 16, 201, 261
orthogonal polynomials, xii, 1, 139
orthogonality, xi–xiii, 3, 62, 74, 75, 81, 82, 97, 104, 108, 109, 124, 139, 223, 263, 266, 282, 325
orthogonalization, 1, 143

parallel computation, 75
partial differential equations, xiii
partial reorthogonalization, 117
periodic reorthogonalization, 109
Poisson equation, 140, 182, 191, 236, 241, 242, 272, 288, 311, 313
Poisson model problem, 236, 250
positive definite, xii, 9, 45, 46, 49, 50, 53–55, 58, 86, 112, 115, 136, 191, 244, 248, 258, 281

QR algorithm, 43
QR factorization, xii, 2–5, 50, 70, 114, 261, 268, 311

reorthogonalization, 1, 81, 90, 94, 108, 109, 118, 130, 142–144, 236
residual smoothing, 302
Ritz polynomial, 15, 32
Ritz value, xii, 8, 9, 13, 15, 25, 28–31, 34, 37, 41–43, 45, 60, 73, 81, 82, 94, 96, 100, 102, 103, 113, 123, 125, 129–133, 135, 136, 139, 142–144, 147, 150, 152, 159, 160, 162, 166, 176, 181, 182, 184, 186, 198–201, 204, 206–208, 210, 261, 264, 283, 288, 290, 292, 308, 310, 316, 332, 333, 335
Ritz vector, 8, 15, 28, 64, 82, 98, 101, 122–125
roundoff error, xiii, 139, 140, 197, 198, 214, 241, 243, 249

selective orthogonalization, 117
semiorthogonality, 82, 116
semiorthogonalization, 117
spectral decomposition, 16, 26, 33, 62, 64, 73, 97, 99, 123, 129, 210, 282
SYMMLQ algorithm, 50, 53, 311

three-term recurrence, xiii, 7, 9, 11, 12, 41, 74, 75, 82, 152, 155, 168, 172, 174, 195–198, 204, 207,

209, 214, 275, 286, 298, 300, 305
two-term recurrence, xiv, 75, 198, 210, 275

UL decomposition, 11, 51
unit roundoff, 82

variable precision, 142, 176, 177, 193, 204, 230, 236